T0213613

Lecture Notes in Computer Science 10263

Commenced Publication in 1973
Founding and Former Series Editors:
Gerhard Goos, Juris Hartmanis, and Jan van Leeuwen

More information about this series at http://www.springer.com/series/7412

Mihaela Pop · Graham A. Wright (Eds.)

Functional Imaging and Modelling of the Heart

9th International Conference, FIMH 2017
Toronto, ON, Canada, June 11–13, 2017
Proceedings

 Springer

Editors
Mihaela Pop
University of Toronto
Toronto, ON
Canada

Graham A. Wright
University of Toronto
Toronto, ON
Canada

ISSN 0302-9743   ISSN 1611-3349 (electronic)
Lecture Notes in Computer Science
ISBN 978-3-319-59447-7   ISBN 978-3-319-59448-4 (eBook)
DOI 10.1007/978-3-319-59448-4

Library of Congress Control Number: 2017941495

LNCS Sublibrary: SL6 – Image Processing, Computer Vision, Pattern Recognition, and Graphics

Printed on acid-free paper

This Springer imprint is published by Springer Nature
The registered company is Springer International Publishing AG
The registered company address is: Gewerbestrasse 11, 6330 Cham, Switzerland

Preface

FIMH 2017 was the 9th International Conference on Functional Imaging and Modelling of the Heart. It was held in Toronto, Canada, during June 11–13, 2017. This year's edition of FIMH followed the past eight editions held in Helsinki (2001), Lyon (2003) Barcelona (2005), Salt Lake City (2007), Nice (2009), New York (2011), London (2013), and Maastricht (2015). FIMH 2017 provided a unique forum for the discussion of the latest developments in the areas of functional cardiac imaging as well as computational modelling of the heart. The topics of the conference included (but were not limited to): advanced cardiac imaging and image processing techniques, myocardial tissue characterization and perfusion, robotics and image-guided therapy procedures, computational fluid dynamics, forward and inverse problems in electro-physiology, modelling of cardiac function across different patient populations, statistical atlases, computational physiology and biomechanics of the heart, parameterization of mathematical models from data, integrated functional and structural analyses, as well as the preclinical and clinical applicability of these methods.

FIMH 2017 drew many submissions from around the world. From the initial 63 registered papers, 49 contributions were accepted. Finally, 48 selected papers were invited to be published by Springer in this *Lecture Notes in Computer Science* proceedings volume. All submitted papers were peer-reviewed by two or three Program Committee members who are international experts in the field. The review process was double-blinded and all papers underwent a rebuttal phase, during which the authors addressed specific concerns and issues raised by reviewers and improved the scientific content and the quality of the manuscripts.

The conference was greatly enhanced by invited keynote lectures given by four world experts in various fields related to: parameterization of patient-specific cardiac biophysical models, novel magnetic resonance (MR) and computed tomography (CT) techniques for myocardial tissue characterization, advanced MR imaging and image reconstruction, as well as artificial agents and personalized medicine. We are extremely grateful to Dr. Maxime Sermesant (Inria, Sophia Antiplois, Asclepios project, France), Dr. Maria Drangova (Robarts Institute, London Ontario, Canada), Dr. Dorin Comaniciu (Siemens, Princeton, USA), and Dr. Debiao Li (Cedar-Sinai, UCLA, Los Angeles, USA) for their exceptional lectures.

We hope that all these papers, along with the keynote contributions and fruitful discussion during the conference, will act to accelerate progress in the important areas of functional imaging and modelling of the heart.

June 2017

Mihaela Pop
Graham A. Wright

Organization

We would like to thank all organizers, additional reviewers, contributing authors, and sponsors for their time, effort, and financial support in making FIMH 2017 a successful event.

Chairs

Mihaela Pop	University of Toronto, Sunnybrook Research Institute, Canada
Graham A. Wright	University of Toronto, Sunnybrook Research Institute, Canada

Program Committee

Elsa Angelini	Telecom ParisTech, France
Nicholas Ayache	Inria, France
Leon Axel	New York University, USA
Yves Bourgault	University of Ottawa, Canada
Peter Bovendeerd	Eindhoven University of Technology, The Netherlands
Yves Coudiere	Inria Bordeaux, France
Oscar Camara	University Pompeu Fabra, Barcelona, Spain
Dominique Chapelle	Inria, France
Patrick Clarysse	University of Lyon, France
Dorin Comaniciu	Siemens, USA
Tammo Delhaas	Maastricht University, The Netherlands
Herve Delingette	Inria, France
Jan D'hooge	KU Leuven, Belgium
Nicolas Duchateau	University of Lyon, France
James Duncan	Yale University, USA
Alejandro Frangi	Sheffield University, UK
Jean-Frederic Gerbeau	Inria, France
Arun Holden	Leeds University, UK
Pablo Lamata	King's College London, UK
Cristian Linte	RIT-Rochester, USA
Herve Lombaert	Inria, France
Cristian Lorenz	Philips Research, Germany
Chris Macgowan	SickKids Hospital Toronto, Canada
Rob MacLeod	University of Utah, USA
Isabelle Magnin	University of Lyon, France
Tommaso Mansi	Siemens, USA
Dimitris Metaxas	Rutgers University, USA

Martyn Nash	University of Auckland, New Zealand
Sebastien Ourselin	University College London, UK
Terry Peters	Robarts Institute, Canada
Caroline Petitjean	University of Rouen, France
Mihaela Pop	Sunnybrook, University of Toronto, Canada
Kawal Rhode	King's College London, UK
Idan Roifman	Sunnybrook Health Sciences Center, Canada
Daniel Rueckert	Imperial College London, UK
Frank Sachse	University of Utah, USA
Gunnar Seemann	University of Freiburg, Germany
Maxime Sermesant	Inria, France
Kaleem Siddiqi	McGill University, Canada
Larry Staib	Yale University, USA
Regis Vaillant	GE Healthcare, France
Hans van Assen	Carelabs, The Netherlands
Jurgen Weese	Phillips Research, Germany
Graham Wright	Sunnybrook, University of Toronto, Canada
Alistair Young	University of Auckland, New Zealand
Xiahai Zhuang	Fudan University, China

Sponsor Relations, OCS Submission System, Webmaster

Mihaela Pop	University of Toronto, Sunnybrook Research Institute, Canada

Local Organization

Jean Rookwood	Sunnybrook Research Institute, Canada
Kimberly Allen	Sunnybrook Research Institute, Canada
Labonny Biswas	Sunnybrook Research Institute, Canada
Sebastian Ferguson	Sunnybrook Research Institute, Canada

Sponsors

We are extremely grateful for the industrial and institutional funding support from the following Sponsors:

Siemens-Healthineers (http://siemens.com) **SIEMENS**

Fuji Film/VisualSonics (http://visualsonics.com)

Inria (https://www.inria.fr/en)

Imricor (http://imricor.com/)

SciMedia Ltd (http://www.scimedia.com/)

GE HealthCare (http://www3.gehealthcare.ca/en)  **GE Healthcare**

Shelly Medical Imaging Technologies (http://simutec.com)

HeartVista (http://www.heartvista.com/) **HEARTVISTA**

Contents

Contents   XV

Novel Imaging and Analysis Methods for Myocardial Tissue Characterization and Remodelling

Three-Dimensional Quantification of Myocardial Collagen Morphology from Confocal Images

Abdallah I. Hasaballa[1](✉), Gregory B. Sands[1,2], Alexander J. Wilson[1,2], Alistair A. Young[1,3], Vicky Y. Wang[1], Ian J. LeGrice[1,2], and Martyn P. Nash[1,4]

[1] Auckland Bioengineering Institute, University of Auckland, Auckland, New Zealand
ahas804@aucklanduni.ac.nz
[2] Department of Physiology, University of Auckland, Auckland, New Zealand
[3] Department of Anatomy with Radiology, University of Auckland, Auckland, New Zealand
[4] Department of Engineering Science, University of Auckland, Auckland, New Zealand

Abstract. The mechanical properties of myocardial tissue are primarily determined by the organisation of the collagen network. Quantitative measurements of collagen morphology can help to understand the structure-function relationship in cardiac tissue. In this study, we segmented collagen from high-resolution three-dimensional (3D) images of the left ventricle (LV) mid-wall myocardium obtained using extended-volume confocal microscopy. 3D shape analysis was used to compute the morphological parameters elongation (e), flatness (f), and anisotropy (a). We applied this analysis to both control and hypertensive rat hearts and showed distinct differences between the control and remodelled hearts, particularly in collagen elongation. The predominant form of collagen in the control rat is elongated with a value of e = 0.846 ± 0.041, whereas in the hypertensive rat collagen, is arranged mostly in a sheet-like form with e = 0.301 ± 0.023. Such quantitative information can be used to develop microstructural models of the myocardium that link the observed changes in cardiac microstructure to changes in mechanical function during the progression of heart diseases, which will help to elucidate the underlying pathological mechanisms.

Keywords: Collagen morphology · Seeded region-growing algorithm · Shape analysis · Moments of inertia · Confocal imaging

1 Introduction

The mechanical behaviour of the myocardium is mainly dependent on the organisation of myocytes and the composition of the cardiac extracellular matrix (ECM). Collagen is the predominant structural component of ECM and the major stress-bearing component of passive myocardium [1]. Consequently, any change in the collagen architecture is often associated with a pathological function such as is observed in hypertensive hearts [2]. Therefore, it is important to

© Springer International Publishing AG 2017
M. Pop and G.A. Wright (Eds.): FIMH 2017, LNCS 10263, pp. 3–12, 2017.
DOI: 10.1007/978-3-319-59448-4_1

have quantitative information on collagen morphology in order to develop a better understanding of the relationship between the microstructure and mechanical function of the myocardium.

The role of the collagen structure in myocardial function has been investigated by many research groups [3,4] using animal models, such as the spontaneously hypertensive rat (SHR) [5,6]. The SHR is a well-established model of genetic hypertension that develops with aging, and leads to myocardial changes that reflect those seen in hypertensive human hearts. In studies using high-resolution extended volume confocal microscopy, it has been shown that aged SHRs have significant changes in myocardial architecture when compared to age-matched control Wistar-Kyoto (WKY) rats [6]. These structural changes include myocyte hypertrophy, an increase in the amount of collagen, and a loss of myocardial laminar organisation because of the scarring together of collagen strands, which line the cleavage planes between laminae. These alterations in the myocardial structure are believed to lead to impaired myocardial mechanical function and, hence, reduced cardiac performance.

In this study, we have developed a novel method to quantify collagen morphology in the tissue blocks of the mid-wall of the left ventricle (LV) myocardium taken from a 12-month-old WKY rat and an age-matched SHR. Preliminary results on differences in structural and morphological parameters between WKY and SHR are presented and discussed.

2   Methods

In order to quantify collagen morphology, collagen was segmented from images of the rat LV free wall, and morphological parameters describing elongation, flatness, and anisotropy were computed at 50,000 randomly selected collagen locations within each image volume.

2.1   3D Tissue Images

Blocks of LV wall from a 12-month-old WKY rat and an age-matched SHR were labelled with picrosirius red to highlight collagen, embedded in resin, mounted on a high-precision three-axis stage and imaged using laser scanning confocal microscopy. The imaging protocol and technique have been described in detail elsewhere [7]. Although this imaging technique is time-consuming and generates large datasets, it provides precise information at high spatial resolution about the 3D organisation of the myocytes and collagen network [8]. Representative 3D image volumes of the WKY rat and SHR used in this study are shown in Fig. 1. The 3D tissue blocks were imaged at a resolution of 1 μm per voxel side and with a total volume of (293 μm x 256 μm x 237 μm).

2.2   Collagen Segmentation

In order to segment the collagen network in the 3D volume images, we developed an image processing framework, which includes four main steps:

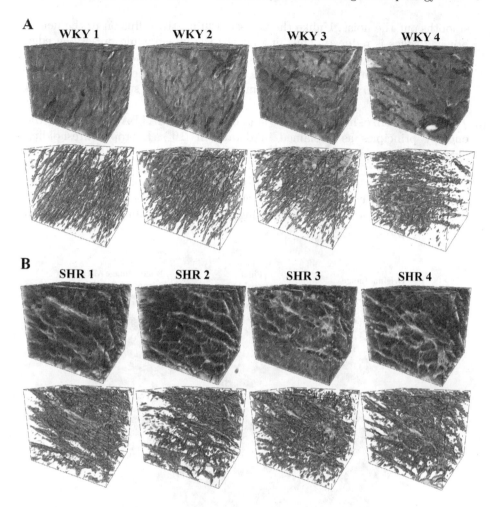

Fig. 1. 3D image volumes from (A) a 12-month-old WKY rat heart and (B) an age-matched SHR heart (*top*), together with the corresponding collagen structures (*bottom*). Collagen appears brighter, while myocytes have variable intensity.

(i) **Adjustment of image intensity** (using the MATLAB[1] *imadjust* function). The *imadjust* function improves the intensity distribution of the image by mapping the input image's intensity values to a new range such that 1% of the data is saturated at low and high intensities of the input image.

(ii) **Unsharp masking** (using the MATLAB *imsharpen* function). This function sharpens edges on the elements without increasing noise. The parameters used in this study were *radius* = 5, *amount* = 0.5 and *threshold* = 0.

(iii) **Edge-preserving smoothing** (customised code in MATLAB as described by Weickert et al. in [9]). This step is very similar to a Gaussian filter as it

---

[1] The MathWorks, Inc., Natick, Massachusetts, United States.

smooths out the noise. Unlike the Gaussian filter, the diffusion in the neighbouring edges is reduced. Thus, it smooths the image while preserving edge information. The behaviour of this filter is controlled by two parameters: *contrast*, which determines how much the smoothing is reduced in the vicinity of the edges, and *kernel size*. The parameters used in this study were *contrast* = 3.5 and *kernel size* = 3.

(iv) **Entropy thresholding** (customised code in MATLAB following the entropy principles as explained in details in [10]). Entropy thresholding extracts a binary segmentation of the collagen.

The combination of these steps helps to preserve as much of the collagen information that is contained in the original image as possible. An example of automated collagen segmentation using the proposed framework is given in Fig. 2.

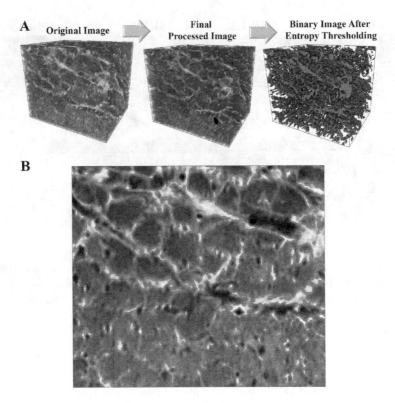

Fig. 2. Collagen segmentation in the 3D confocal images. (A) image segmentation steps. (B) cross section view of collagen segmentation boundaries overlaid on the original image. (Color figure online)

2.3   Region Extraction

Collagen morphology is described at a local level, therefore, we extracted a number of small sub-regions from the image block for quantification.

Distinct, random collagen locations were selected from the 3D segmented image block, and a cube (region of interest) was extracted with collagen voxel/point at the center of the cube (cubes were allowed to overlap). The collagen locations were randomly selected using the *randi* function in MATLAB, which creates uniformly distributed random locations within the 3D image bounds, and retained for analysis if the selected location had been classified as collagen. Finally, a 3D region-growing algorithm [11] was applied in each cube, which starts with the central collagen seed point identified above and grows with neighboring voxels of connected collagen.

The size of the region of interest was selected to be large enough to represent the microstructural organisation of the collagen accurately, yet small enough to only contain local information. In this study, the length of the cube was 25 μm. The number of cubes to be analysed was determined by progressively increasing the number of cubes until the results converged (i.e. no statistically significant differences with the addition of more cubes). Here, convergence was reached at 50,000 cubes.

2.4   Shape Analysis

Many imaging studies have used moments of inertia for shape analysis [12] and pattern recognition [13]. In a binary image, the first-order moments define the center of mass

$$M_{1x} = \sum_C x_i, \quad M_{1y} = \sum_C y_i, \quad M_{1z} = \sum_C z_i \tag{1}$$

where (x_i, y_i, z_i) is a point in the object C. Here, C is the segmented collagen object in the region of interest. The second-order moments are defined as

$$
\begin{aligned}
M_{2xx} &= \sum_C (x_i - M_{1x})^2, \\
M_{2yy} &= \sum_C (y_i - M_{1y})^2, \\
M_{2zz} &= \sum_C (z_i - M_{1z})^2, \\
M_{2xy} &= \sum_C (x_i - M_{1x})(y_i - M_{1y}), \\
M_{2yz} &= \sum_C (y_i - M_{1y})(z_i - M_{1z}), \\
M_{2xz} &= \sum_C (x_i - M_{1x})(z_i - M_{1z})
\end{aligned}
\tag{2}
$$

and the inertia matrix (covariance matrix) can be written as

$$M = \begin{bmatrix} M_{2xx} & M_{2xy} & M_{2xz} \\ M_{2xy} & M_{2yy} & M_{2yz} \\ M_{2xz} & M_{2yz} & M_{2zz} \end{bmatrix} \tag{3}$$

The eigenvalues of M provide a good indicator of shape. For example, when the three eigenvalues are similar, the shape tends to be spherical. The 3D morphological parameters in the range of [0,1] were derived from the eigenvalues: elongation (e), flatness (f), and anisotropy (a)

$$e = 1 - \frac{\lambda_2}{\lambda_1}, \quad f = 1 - \frac{\lambda_3}{\lambda_2}, \quad a = 1 - \frac{\lambda_3}{\lambda_1} \tag{4}$$

where λ_1 denotes the primary eigenvalue of M, λ_2 denotes the secondary eigenvalue, and λ_3 denotes the tertiary eigenvalue such that $\lambda_1 \geq \lambda_2 \geq \lambda_3 > 0$.

The proposed framework for quantifying the 3D collagen morphology from confocal images of healthy and diseased hearts is summarised in Fig. 3.

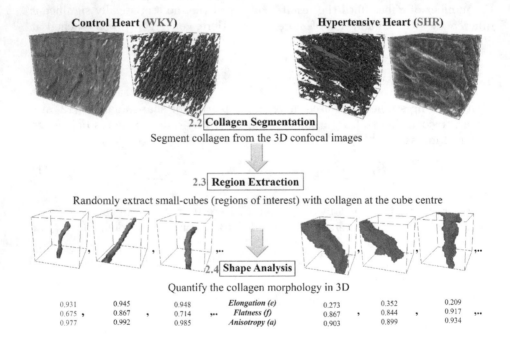

Fig. 3. A framework for quantification of collagen morphology in 3D.

3   Results and Discussion

The proposed framework for quantifying collagen morphology in 3D has been applied to four sets of confocal images from each of a control (WKY) rat heart and a hypertensive rat (SHR) heart (see Fig. 1). The distributions of the three morphological parameters for all eight tissue blocks are shown in Fig. 4 and are similar within each animal group. There are distinct differences in the distributions of elongation between the WKY and SHR groups, whereas distributions of flatness and anisotropy in both groups are similar.

Figure 5(A) shows the 3D distribution density of the morphological parameters for WKY and SHR using heatmaps. The core locations and the patterns of density distribution are substantially different between WKY and SHR. The core centres which indicate the most common collagen form in WKY and SHR are given in Table 1. To illustrate the structural differences between WKY and SHR, four small cubes from the core of each heatmap for each case are shown in Fig. 5 (B). This comparative analysis revealed that the dominant collagen shape in WKY was elongated, whereas in SHR it had a sheet-like form. Such differences in the structure may explain the higher stiffness observed in SHR functional studies [5,6]. From a mechanics perspective, the microstructural shape of composite materials has a considerable effect on the physical properties [14]. It seems likely that the increased stiffness observed in SHR hearts compared with those from normal animals is a consequence of collagen remodelling into sheet-like shapes in the SHR.

Table 1. Comparison of the mean and standard deviations (SDs) of the core centres for WKY and SHR.

	Elongation		Flatness		Anisotropy	
	Mean	SD	Mean	SD	Mean	SD
WKY	0.846	0.041	0.771	0.053	0.967	0.002
SHR	0.301	0.023	0.833	0.016	0.883	0.001

There is a dearth of quantitative information available in the literature regarding the 3D morphology of collagen in the heart. Nevertheless, microstructural studies [6,15] by LeGrice and colleagues have reported that the perimysial collagen organisation in WKY rat hearts is a network of collagen fibres and bundles, while the SHR, collagen forms dense sheets due to the fusion and thickening of perimysial collagen between adjacent myocardial layers. Our quantitative results are consistent with these observations.

A limitation of this study is that the automatic image analysis techniques are not able to distinguish between perimysial and endomysial collagen, and thus, the contributions of each type of collagen to the results could not be determined.

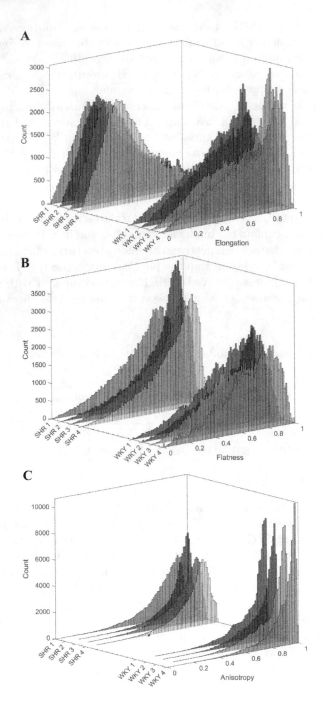

Fig. 4. 1D distributions of the (A) elongation, (B) flatness, (C) anisotropy parameters among tissue blocks from WKY (blues) and SHR (reds) hearts. (Color figure online)

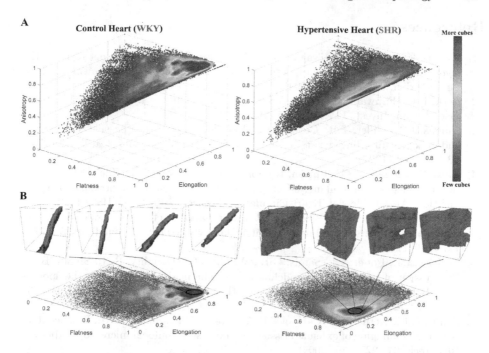

Fig. 5. Heatmap visualisation of the distributions of morphological parameters in (A) 3D and (B) in the elongation-flatness plane for heart tissue blocks from WKY (*left*) and SHR (*right*), with typical collagen shapes from the core of each heatmap

4 Conclusions

We have developed an automated method to quantify the 3D myocardial collagen morphology from confocal images using three morphological parameters. Using multiple confocal images from control and hypertensive hearts, we have shown that the proposed framework can effectively quantify the collagen shape in 3D, and can distinguish the structural remodelling of the collagen network during development and disease. In particular, our quantitative analysis revealed that the collagen structure in the diseased hearts was more sheet-like in comparison to the elongated collagen structure of the age-matched control hearts. The microstructural parameters measured in this study will be used in future computational models of the myocardium to link the observed changes in cardiac microstructure to changes in ventricular function. Such models could help to improve our understanding of the pathophysiological processes underpinning heart diseases and pave the way towards more effective treatments that target the underlying mechanisms.

References

1. Weber, K.T., Sun, Y., Tyagi, S.C., Cleutjens, J.P.: Collagen network of the myocardium: function, structural remodeling and regulatory mechanisms. J. Mol. Cell. Cardiol. **26**, 279–292 (1994)
2. Berk, B.C., Fujiwara, K., Lehoux, S.: ECM remodeling in hypertensive heart disease. J. Clin. Invest. **117**, 568–575 (2007)
3. Omens, J.H., Miller, T.R., Covell, J.W.: Relationship between passive tissue strain and collagen uncoiling during healing of infarcted myocardium. Cardiovasc. Res. **33**, 351–358 (1997)
4. Kato, S., Spinale, F.G., Tanaka, R., Johnson, W., Cooper, G., Zile, M.R.: Inhibition of collagen cross-linking: effects on fibrillar collagen and ventricular diastolic function. Am. J. Physiol. Heart Circulatory Physiol. **269**, H863–H868 (1995)
5. Brilla, C.G., Janicki, J., Weber, K.: Cardioreparative effects of lisinopril in rats with genetic hypertension and left ventricular hypertrophy. Circulation **83**, 1771–1779 (1991)
6. LeGrice, I.J., Pope, A.J., Sands, G.B., Whalley, G., Doughty, R.N., Smaill, B.H.: Progression of myocardial remodeling and mechanical dysfunction in the spontaneously hypertensive rat. Am. J. Physiol. Heart Circulatory Physiol. **303**, H1353–H1365 (2012)
7. Sands, G.B., Gerneke, D.A., Hooks, D.A., Green, C.R., Smaill, B.H., LeGrice, I.J.: Automated imaging of extended tissue volumes using confocal microscopy. Microsc. Res. Tech. **67**, 227–239 (2005)
8. Young, A., LeGrice, I., Young, M., Smaill, B.: Extended confocal microscopy of myocardial laminae and collagen network. J. Microsc. **192**, 139–150 (1998)
9. Weickert, J., Romeny, B.T.H., Viergever, M.A.: Efficient and reliable schemes for nonlinear diffusion filtering. IEEE Trans. Image Process. **7**, 398–410 (1998)
10. Kapur, J.N., Sahoo, P.K., Wong, A.K.: A new method for gray-level picture thresholding using the entropy of the histogram. Comput. Vision Graph. Image Process. **29**, 273–285 (1985)
11. Rangayyan, R.M.: Biomedical Image Analysis. CRC Press, Boca Raton (2004)
12. Loncaric, S.: A survey of shape analysis techniques. Pattern Recogn. **31**, 983–1001 (1998)
13. Flusser, J., Zitova, B., Suk, T.: Moments and Moment Invariants in Pattern Recognition. Wiley, Hoboken (2009)
14. El Moumen, A., Kanit, T., Imad, A., El Minor, H.: Effect of reinforcement shape on physical properties and representative volume element of particles-reinforced composites: Statistical and numerical approaches. Mech. Mater. **83**, 1–16 (2015)
15. Pope, A.J., Sands, G.B., Smaill, B.H., LeGrice, I.J.: Three-dimensional transmural organization of perimysial collagen in the heart. Am. J. Physiol. Heart Circulatory Physiol. **295**, H1243–H1252 (2008)

In Vivo Parametric T1 Maps Correlate with Structural and Molecular Characteristics of Focal Fibrosis

Mihaela Pop[1,2(✉)], Samuel Oduneye[1,2], Li Zhang[1,2], Susan Newbigging[3], and Graham Wright[1,2]

[1] Sunnybrook Research Institute, Toronto, Canada
mihaela.pop@utoronto.ca
[2] Medical Biophysics, University of Toronto, Toronto, Canada
[3] Centre for Modelling of Human Disease (CMHD) - Pathology Core,
Toronto Centre for Phenogenomics, Toronto, Canada

Abstract. The purpose of this work was to use noninvasive *in vivo* MRI to characterize the substrate (i.e., known as *gray zone*, GZ) of ventricular arrhythmia, a major cause of sudden death. Our aim was to use a preclinical model of chronic infarction to study the structural and molecular characteristics of infarcted areas. For this, we related parametric T_1 maps with the density of collagen and gap junctions (Cx43 proteins) in patchy fibrosis in n = 6 swine with chronic infarction. Specifically, *in vivo* T_1 relaxation maps were calculated from 2D multi-contrast late enhancement (MCLE) MR images obtained at $1 \times 1 \times 5$ mm resolution. Quantitative analysis and regression analysis demonstrated that the comparison between GZ and scar extent in MCLE to the corresponding areas identified in histology, yielded very good correlations in both cases (i.e., goodness of fit: $R^2 = 0.89$ for GZ, and 0.92 for dense scar, respectively). Furthermore, the gap junction Cx43 density was significantly reduced (i.e., by > 50%) in the ischemic GZ areas determined from MCLE. These novel results suggest that *in vivo* 2D parametric T_1 maps can be used to evaluate the biophysical properties of healing myocardium post-infarction, and to distinguish between the infarct categories (i.e., scar vs. GZ) with re-modelled structural and electrical characteristics.

Keywords: Myocardial infarct · Cardiac MRI · T1 maps · Gray zone · Fibrosis · Gap junctions

1 Introduction

The arrhythmogenic substrate for potentially lethal heart rhythms in patients with prior myocardial infarction is typically located in the peri-infarct, or so-*called gray zone* (GZ), which is found between the normal tissue and dense scar (i.e., mature collagen that replaced dead tissue during healing) [1]. GZ is comprised of a mixture of viable myocytes and collagen fibers, and has altered electrical properties (i.e., this tissue conducts the action potential wave slower than normal tissue) [2]. Currently, in the electrophysiology (EP) lab, the chronic infarct areas are identified using invasive electro-anatomical mapping. However, although the electrical signals help identify some of the

© Springer International Publishing AG 2017
M. Pop and G.A. Wright (Eds.): FIMH 2017, LNCS 10263, pp. 13–22, 2017.
DOI: 10.1007/978-3-319-59448-4_2

arrhythmogenic foci, the voltage maps are recorded only from the surface and do not provide transmural information [3]. Thus, an important task for the clinicians is to determine the location, extent and transmurality of infarction, and this is often done non-invasively by employing contrast-based MRI methods [4]. The conventional late gado-linium LGE MR imaging method is based on the elevation of signal intensity (SI) values within infarct regions, measured ~15 min after the contrast agent injection. The enhance-ment is due to different concentrations and kinetics of the Gd-based molecules in the extracellular space within the infarct (scar/GZ), where they shorten the T_1 relaxation time [5]. As a result, the infarct appears brighter than the healthy myocardium, other surrounding tissues and the blood.

To date, several studies fused LGE and EP maps, showing that critical sites of scar-related arrhythmias are confined to areas of elevated signal intensity (SI) in LGE images [6, 7]. However, they also suggested that discrepancies between GZ detected by LGE and EP maps could be due to: far-field influences from normal tissue, manual delineation of scar, and poor wall-catheter contact during EP mapping. In most of these LGE studies, SI threshold was set at 2–3 SD (standard deviation) higher than the mean SI selected from remote myocardium [8]. However, this algorithm is affected by noise, which is inherent in the MR signal. The noise in the inversion recovery gradient echo (IR-GRE) image plays a large role in determining the measured SI; thus, by changing the location of the remote region, the SD and peak SI can vary significantly due to a repeated sampling of the noise whose distribution has a large variance. This leads to different SI cut-off values for defining the scar and GZ, resulting in a high variability of the GZ size and less reproducible results. An alternative method involves our T_1-mapping method based on multi-contrast late enhancement MCLE, which employs a simultaneous nulling of MR signal from healthy tissue and blood for better identification of sub-endocardial lesions, has superior SNR compared to LGE and outperforms the SD and FWHM-based segmentation algorithms [9].

To date, there is no quantitative study demonstrating the relation between *in vivo* T1* parametric maps and quantitative histopathology. Thus, the purpose of this paper was to better characterize the heterogeneous fibrosis (i.e., dense scar and GZ) using a pre-clinical swine model of chronic infarction by employing an *in vivo* 2D MCLE imaging method that uses steady-state free precession SSFP readouts, along with quan-titative histopathology (i.e., density of collagen in the fibrotic areas, and density of gap junctions Cx43 responsible for electrical connections between myocardial cells). Such investigation might provide a better characterization of structural and molecular char-acteristics of the collagenous scar and ischemic GZ (i.e., the arrhythmogenic substrate).

2 Materials and Methods

A simplified diagram of the proposed framework is shown in Fig. 1. Briefly, we first acquired the *in vivo* MCLE data, followed by data post-processing (i.e., LV segmentation and tissue clustering into: healthy area, scar and GZ). Lastly, using histopathological and Cx43 fluorescence images we analyzed the density of collagen and gap junctions, and then related these measures with the segmented T1* parametric maps.

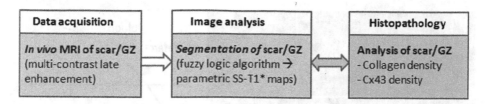

Fig. 1. Schematic diagram of the work-flow

2.1   Animal Model of Infarction and *in Vivo* MRI Study

All animal experiments included in this pre-clinical framework were approved by our research institute. The myocardial infarction was generated in n = 6 swine using an occlusion-reperfusion method as previously described in other studies [10]. The infarct lesions were allowed to heal for 5–6 weeks.

The MR imaging studies were performed on a dedicated 1.5T research scanner (GE SignaExcite) using a 5-inch surface coil. The MR scans had the following parameters: TE = 1.9 ms, TR = 5.5 ms, NEX = 1, FOV = 26 cm, 256×256 acquisition matrix. The spatial resolution was $1 \times 1 \times 5$ mm. Late enhancement imaging was started 10–15 min after the injection of 0.2 mmol/kg of Gd-DTPA. Several short-axis MCLE images were acquired through the infarct volume, with the inversion pulse placed such that the infarct-enhanced images were acquired with minimal motion during diastole, for which TI (inversion time) ranged from 175 to 250 ms.

The MCLE used SSFP readouts during the inversion recovery IR process, producing 20 images over the cardiac cycle. Notably, the MCLE images at early inversion times have varying contrast where the infarct can be visualized as an area of fast T_1^* recovery. Moreover, the observed T_1^* is usually shorter than the intrinsic T_1 relaxation time due to the nature of the b-SSFP measurement (readout), with the magnetization recovering to a steady-state (SS) value. Figures 2a–f show six (out of 20) MCLE images corresponding to one 2D short-axis MR slice in one of the animals studied. Approximately 6–8 images per heart cycle (in diastole) were selected and then used to extract the SI recovery curve for each pixel within the LV (including the blood pool pixels), given that SI obeys the Eq. (1):

$$\text{SI (time)} = \text{SS} \cdot [1 - 2\exp(-\text{time}/T_1^*)] \tag{1}$$

Specifically, T_1^* depends on the longitudinal relaxation T_1, the transverse relaxation T_2 and the flip angle α (i.e. the angle to which the net magnetization is tipped relative to the direction of main magnetic field after the application of an RF excitation pulse at Larmor frequency), as per the following equation [11]:

$$T_1^* = \left(\frac{1}{T_1} \cos^2 \frac{\alpha}{2} + \frac{1}{T_2} \sin^2 \frac{\alpha}{2} \right)^{-1} \tag{2}$$

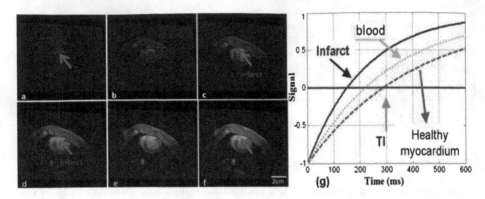

Fig. 2. Example of MCLE imaging in a pig with an LAD-infarct: (a–f) six consecutive short-axis MCLE images (the infarct is visualized as a hyper-enhanced region as indicated by arrow); (g) SI recovery plotted from three pixels (different tissue types).

The plot in Fig. 2g shows an example of the SI recovery based on Eq. (1), obtained in three different pixels from the infarct, blood and healthy myocardium, respectively (using a three-parameter fitting and a smooth function in Matlab).

2.2 Data Processing: Tissue Segmentation and Parametric T1*-SS Maps

For the 2D MCLE images, a corresponding spatial map of various tissue types in LV was generated using an in-house developed fuzzy clustering analysis (in Matlab). This algorithm determined the probability of each pixel belonging to each of the 3 clusters, based on a distance metric derived from the scatter plot of T_1^* versus steady-state (SS). A scatter plot of T_1^* versus SS value for each pixel was then used as input to a fuzzy C-means algorithm, and automatically classified each pixel as: either infarct, or healthy myocardium, or blood.

The FCM algorithm was based on the Gustav-Kessel (GK) clustering method, a robust tool used in system identification and in particular in some classification problems where it is desired to implement a local adjustment of the distance metric to the geometrical shape of the cluster(s). The main characteristic of the GK scheme is the estimation of the cluster covariance matrix and the distance-induce matrix. This was done here by optimizing an objective FCM-type of function using the *Fuzzy-logic Matlab toolbox*. For tissue segmentation we deter-mined the probability of each pixel belonging to a cluster using the distance metric derived from the input SS-T1* scatter plot. The pixels with probabilities > 75% of belonging to infarct or healthy tissue were classified as *dense scar* and *healthy* myocardium, respectively. Furthermore, the pixels classified as *GZ* (the mixture of dead and viable myocytes) had a significant probability for belonging to both the healthy and infarct clusters (less than a 75% probability of belonging to one of those clusters and greater than 25% probability for belonging to the other cluster).

2.3 Quantitative Histopathology Analysis

For histology, a total of N = 10 tissue slices (5-mm) were cut from each heart (1–2 slices/ heart), matching short-axis MCLE images. The samples were fixed in 10% formalin and embedded in paraffin for a few days, avoiding excessive shrinking.

First, thin slices (4–5 μm) underwent whole-mount staining on large glass slides, using collagen-sensitive stain Picrosirius Red, PR. The slides were scanned with a TISSUEscope™ 4000 confocal system (Huron Technologies International Inc.), saved as multi-resolution digital images and analyzed with the Aperio Image Scope software (Vista, CA). The GZ and scar areas were manually delineated by pathology expert (Fig. 3) based on score for fibrosis severity as in [12]. We defined three grades of fibrosis for our tissue types: F0 (< 20% fibrosis, *no or mild*) for healthy/normal myocardium (in remote areas); F1 (20–70% fibrosis, *moderate*) for GZ; and, F2 (≥ 70% fibrosis, *severe*) for scar.

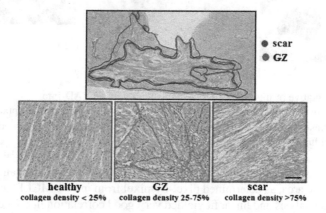

Fig. 3. Histopathology in an infarcted pig heart. Example of manual tissue segmentation (healthy zone, GZ and dense scar) based on grading fibrosis using collagen-sensitive P.R. stain (note: collagen stains red and the healthy myocytes stain yellow). (Color figure online)

Note that for the correlation between the scar and GZ areas (relative to LV area) and the corresponding areas obtained from the SS-T1* maps, we manually registered the histological and MR images via anatomical markers (e.g. RV/LV insertion points, papillary muscle, etc.), which were clearly visible in both type of images. Second, select samples were also prepared for fluorescence microscopic imaging of Cx43 as in [13], and scanned with a Hamamatsu scanner. The density of Cx43 was quantified using the *Visiopharm* software (www.visiopharm.com). ROIs were selected from zones defined as: healthy tissue, GZ and scar in the SS-T1* maps and collagen-based stains. We investigated the changes in cell-to-cell coupling (in the longitudinal direction of myocytes) and side-to-side coupling (in the transverse direction of myocytes). The software calculates the area stained for Cx43 in microscopy fluorescence images in a ROI (where myocytes stain in green and the gap junctions stain in red) relative to the area of that selected ROI. Here we used ROIs of 2 × 2 mm.

3   Results

Figure 4 shows a representative result from the analysis of MCLE images in one pig. SS image and T1* map are shown in Fig. 4a. Figure 4b shows a scatter plot of T1* vs. SS values for all pixels within the LV. The resulted 2D spatial map of various tissue types classified using the fuzzy clustering algorithm is shown in Fig. 4c: dense scar (green), BZ (yellow), healthy tissue (blue), and blood (red).

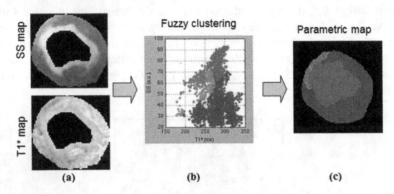

Fig. 4. Representative results obtained in a swine heart with an LAD-infarct: (a) SS image (*up*) and T_1^* map (*bottom*); (b) fuzzy clustering scatter plot of T_1^* vs. SS values for all pixels within the LV; and, (c) corresponding 2D parametric map of various tissue categories classified by the fuzzy clustering algorithm.

Figure 5 shows the quantitative comparison between % GZ area (a) and dense scar (b) relative to LV area, as determined pixel-by-pixel from *in vivo* MCLE and histology. The regression analysis and linear fit yielded very good correlation in both cases (goodness of fit: $R^2 = 0.89$ for GZ, and 0.92 for scar, respectively).

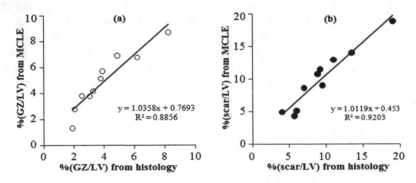

Fig. 5. Correlation between GZ (a) and scar areas (b) relative to LV area determined from 2D parametric SS-T1* maps and collagen-sensitive stains.

Figure 6 shows an example of fluorescence image at 40x magnification (myocytes stain green and Cx43 red). All samples contained healthy myocardial tissue, GZ and scar (see Fig. 6a). The connexin Cx43 (red dots) were clustered at the intercalated disks of myocytes. All ROIs analyzed quantitatively (see magnified images in Fig. 6b and 6c) were selected from MCLE-derived healthy, GZ and scar zones. Qualitatively, there was an apparent reduction in Cx43 density in the GZ compared to healthy tissue in both longitudinal *(up)* and transverse *(bottom)* directions. The connexin Cx43 was completely inhibited in the collagenous scar (images not included). Notable also was a clear increased extra-cellular space and cellular disconnection in GZ due to collagenous fibrils separating viable myocytes.

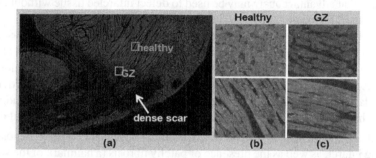

Fig. 6. Qualitative fluorescence microscopy images of Cx43 (see text for details).

Furthermore, as seen in Fig. 7, the quantitative analysis demonstrated a significant reduction ($> 50\%$) in Cx43 density in the ventricular peri-infarct (GZ) in both longitudinal and transverse directions of myocytes, compared to the Cx43 density in the healthy tissue. This reduction may contribute to the electrical uncoupling of myocytes (resulting in slower conduction of impulse) and formation of arrhythmia substrate.

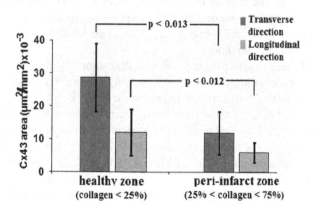

Fig. 7. Quantitative analysis for gap junctions: comparison of Cx43 density in the GZ and healthy tissue, for both longitudinal and transverse directions of myocytes (significant differences denoted by p values < 0.05).

4 Discussion

Accurate characterization of chronic infarct remodelling is an important task in order to identify substrate of potentially lethal ventricular arrhythmia (VT/VF). This clearly motivates the development of pre-clinical experimental frameworks and advanced technologies based on non-invasive imaging methods as MCLE, which can offer rich transmural and structural information to supplement the invasive, sparse and surfacic electrophysiological measurements. Using MCLE, one can generate accurate SS-T_1^* parametric maps that advantageously may be used to detect infarcted tissue without the need to estimate the TI for nulling myocardium as in LGE methods.

Our study is the first to evaluate quantitatively the relation between MR tissue properties based on *in vivo* T_1^* relaxation maps and the density of collagen in chronic fibrosis. Overall, the comparison between MCLE-derived GZ and scar areas vs. histologically-derived GZ and scar areas, demonstrated that 2D SS-T1* parametric maps provided an accurate classification of infarct heterogeneities and identified subtle sub-endocardial lesions at the tissue-blood interface. The T_1–like maps reflected the structural changes in focal fibrosis, showing an increase in extracellular space and collagen deposition in the healing infarct, as well as the presence of patchy fibrosis (a hallmark of the substrate for lethal VT/VF).

We acknowledge that a limitation of this study is the relatively small sample size (i.e., n = 6 animals with chronic myocardial infarction), that is because such *in vivo* preclinical experiments are expensive and difficult to perform, while whole-mount histology is laborious and expensive. However, the 10 histological tissue samples provided sufficient data points for an accurate statistical analysis and to obtain very good correlation between the pixel-by-pixel MCLE-derived GZ/scar areas and collagen-defined GZ/scar areas. The slightly poorer goodness of fit for GZ compared to scar can be explained by the partial volume effect in the peri-infarct area, and was due to the mixture of viable and non-viable myocytes (which contributed to an average value of T1* within the 5 mm slice thickness).

The structural changes in fibrotic areas were accompanied by alterations at molecular level as indicated by inhibition of the Cx43 protein, a major ventricular gap junction that facilitates the flow of ionic current between the cells [3]. Our results showed that Cx43 density was significantly reduced in the ischemic GZ by > 50% in both directions of the myocytes (longitudinal and transverse), which could trigger delayed propagation of activation wave [14]. The closure of connexin Cx43 in the ischemic GZ is important because it reduces the conduction velocity in the arrhythmogenic substrate and facilitates the triggering of VT/VF [15].

To conclude, *in vivo* 2D SS-T_1^* maps enabled accurate MR probing of remodelled myocardium. Such maps can be used to evaluate the changes in structural and molecular characteristics with focal fibrosis, and to distinguish between myocardial tissue categories (i.e., dense scar vs. GZ). Future work will focus on improving the MCLE segmentation algorithm and compare it to other intensity-based segmentation methods. Also, we hope to improve the comparison between tissue categories as defined by MRI and histology, by performing a non-rigid registration between the MCLE images and the whole-mount stained slides. Lastly, as per preliminary results presented in [16], we will continue to improve the *in vivo* spatial resolution of MCLE images and further develop accurate 3D MCLE-based heart models to predict the inducibility of scar-related VT through *in silico* simulations.

Acknowledgement. The authors would like to thank preclinical veterinary technicians at Sunnybrook for help with the infarct creation and MR studies. This work was financially supported by a CIHR grant (MOP #93531).

References

1. Ursell, P.C., Gardner, P.I., Albala, A., Fenoglio, J., Wit, A.L.: Structural and electrophysiological changes in the epicardial border zone of canine myocardial infarcts during infarct healing. Circ. Res. **56**, 436–451 (1985)
2. Bolick, D., Hackel, D., Reimer, K., Ideker, R.: Quantitative analysis of myocardial infarct structure in patients with ventricular tachycardia. Circulation **74**(6), 1266 (1986)
3. Janse, M.J., Wit, A.L.: Electrophysiological mechanisms of ventricular arrhythmias resulting from myocardial ischemia and infarction. Physiol. Rev. **69**(4), 1049–1169 (1989)
4. Bello, D., Fieno, D.S., Kim, R.J., et al.: Infarct morphology identifies patients with substrate for sustained ventricular tachycardia. J. Am. Coll. Cardiol. **45**(7), 1104–1110 (2005)
5. Kim, R.J., Fieno, D.S., Parrish, T.B., Harris, K., Chen, E.L., Simonetti, O., Bundy, J., Finn, J.P., Klocke, F.J., Judd, R.M.: Relationship of MRI delayed contrast enhancement to irreversible injury, infarct age, and contractile function. Circulation **100**, 1992–2002 (1999)
6. Codreanu, A., Odille, F., Aliot, E., et al.: Electro-anatomic characterization of post-infarct scars comparison with 3D myocardial scar reconstruction based on MR imaging. J. Am. Coll. Cardiol. **52**, 839–842 (2008)
7. Wijnmaalen, A., van der Geest, R., van Huls van Taxis, C., Siebelink, H., Kroft, L., Bax, J., Reiber, J., Schalij, M., Zeppenfeld, K.: Head-to-head comparison of contrast-enhanced magnetic resonance imaging and electroanatomical voltage mapping to assess post-infarct scar characteristics in patients with ventricular tachycardias: Real-time image integration and reversed registration. Eur. Heart J. **32**(1), 104–114 (2011)
8. Roes, S.D., Borleffs, C.J., van der Geest, R.J., et al.: Infarct tissue heterogeneity assessed with contrast-enhanced MRI predicts spontaneous ventricular arrhythmia in patients with ischemic cardiomyopathy and implantable cardioverter-defibrillator. Circ. Cardiovasc. Imaging **2**, 183–190 (2009)
9. Detsky, J.S., Paul, G., Dick, A.J., Wright, G.A.: Reproducible classification of infarct heterogeneity using fuzzy clustering on multicontrast delayed enhancement magnetic resonance images. IEEE Trans. Med. Imaging **28**(10), 1606–1614 (2009)
10. Pop, M., Ramanan, V., Yang, F., Zhang, L., Newbigging, S., Ghugre, N., Wright, G.A.: High resolution 3D T1* mapping and quantitative image analysis of the gray zone in chronic fibrosis. IEEE Trans. Biomed. Eng. **61**(12), 2930–2938 (2014)
11. Salerno, M., Janardhanan, R., Jiji, R.S., Brooks, J., Adenaw, N., Mehta, B., Yang, Y., Antkowiak, P., Kramer, C.M., Epstein, F.H.: Comparison of methods for determining the partition coefficient of gadolinium in the myocardium using T1 mapping. J. Magn. Reson. Imaging **38**(1), 217–224 (2013)
12. Pop, M., Ghugre, N.R., Ramanan, V., Morikawa, L., Stanisz, G., Dick, A.J., Wright, G.A.: Quantification of fibrosis in infarcted swine hearts by ex vivo late gadolinium-enhancement and diffusion-weighted MRI methods. Phys. Med. Biol. **58**(15), 5009 (2013)

13. Jansen, J., van Veen, T., de Jong, S., van der Nagel, R., van Stuijvenberg, L., Driessen, H., Labzowsky, R., Oefner, C.M., Bosch, A., Nguyen, T.Q., Goldschmeding, R., Vos, M.A., de Bakker, J.M.T., van Rijen, H.: Reduced Cx43 expression triggers increased fibrosis due to enhanced fibroblast activity. Circ. Arrhythmias Electrophysiol. **5**(2), 380–391 (2012)
14. Kawara, T., Derksen, R., de Groot, J.R., Coronel, R., Tasseron, S., Janse, M.J., de Bakker, M.J.: Activation delay after premature stimulation in chronically diseased myocardium relates to architecture of interstitial fibrosis. Circulation **104**, 3069–3075 (2001)
15. Stevenson, W.G.: Ventricular scars and VT tachycardia. Trans. Am. Clin. Assoc. **120**, 403–412 (2009)
16. Zhang, L., Athavale, P., Pop, M., Wright, G.A.: Multi-contrast reconstruction using compressed sensing with low rank and spatially-varying edge-preserving constraints for high-resolution MR characterization of myocardial infarction. Magn. Reson. Med., September 2016, in press. (doi:10.1002/mrm.26402)

Microstructural Analysis of Cardiac Endomyocardial Biopsies with Synchrotron Radiation-Based X-Ray Phase Contrast Imaging

Hector Dejea[1,2]([envelope]), Patricia Garcia-Canadilla[1,2], Marco Stampanoni[1,3], Monica Zamora[4], Fatima Crispi[4,5], Bart Bijnens[2,6], and Anne Bonnin[1]

[1] Swiss Light Source, Paul Scherrer Institut (PSI), 5232 Villigen, Switzerland
Hector.Dejea@psi.ch
[2] Physense, DTIC, Universitat Pompeu Fabra, 08018 Barcelona, Spain
[3] Institute for Biomedical Engineering, ETH Zürich, 8092 Zurich, Switzerland
[4] Barcelona Center for Maternal-Fetal and Neonatal Medicine (BCNatal),
Hospital Clinic and Hospital Sant Joan de Deu, Barcelona, Spain
[5] Centre for Biomedical Research on Rare Diseases (CIBER-ER),
Hospital Clinic, Barcelona, Spain
[6] Institució Catalana de Recerca i Estudis Avançats (ICREA), Barcelona, Spain

Abstract. Nowadays, unexplained cardiovascular diseases (CVD) and heart transplant response are assessed by qualitative histological analysis of extracted endomyocardial biopsies (EMB), which is a time consuming procedure involving structural damage of the tissue and the analysis in only a few slices of a 3D structure. In this paper we propose synchrotron radiation-based X-ray phase contrast imaging (X-PCI) as a suitable technique for the analysis of different cardiac microstructures, such as collagen matrix, cardiomyocytes and microvasculature, and how they are affected in abnormal conditions. Following an established procedure in clinics, biopsies from Wistar Kyoto rats are extracted, imaged with X-PCI, and processed in order to show that the quantification of the endomysial collagen matrix, cardiomyocytes and microvasculature is possible, thus demonstrating that the intrinsic properties of X-PCI make it a powerful technique for cardiac microstructure imaging and a promising methodology for a faster and more accurate EMB analysis for CVD diagnosis and evaluation.

Keywords: Cardiac microstructure · Cardiac endomyocardial biopsy · X-ray tomography · Phase contrast imaging

1 Introduction

The heart anatomy has been an important research target for centuries, which has led to a good understanding at macroscopic level. Nevertheless, the architecture at individual cell scale within the whole heart as well as its relationship and effects on the contractile mechanism remain still unclear. This lack of knowledge, together with the fact that cardiovascular diseases are the first cause of mortality

© Springer International Publishing AG 2017
M. Pop and G.A. Wright (Eds.): FIMH 2017, LNCS 10263, pp. 23–31, 2017.
DOI: 10.1007/978-3-319-59448-4_3

in the world [1], are the reasons why a big effort from the scientific community is done towards obtaining a detailed multiscale description of the heart structure and function.

The cardiomyocytes are the contractile cells of the heart, with a size of approximately 10–20 μm diameter and 100–120 μm length, arranged in fibre-like structures (myofibres). These structures, in turn, are organized and oriented in a special manner in order to maximize the contractile function of the heart. In addition to the myocytes, a matrix of mainly type I fibrillar collagen (~85%) is present in the extracellular space of the myocardium, maintained by resorption and synthesis processes carried out by cells called fibroblasts. This collagen matrix is characterized by three different layers: endomysium (around cardiomyocytes), perimysium (separates myofibres) and epimysium (around groups of myofibres). The main function of the collagenous matrix is to serve as scaffold for myocyte alignment and to avoid an overstretching of the sarcomeres. Furthermore, it has secondary functions related to the functionality, electrical behaviour and vasomotor reactivity of the myocardium and its structures [2]. All these tissue components may change in shape, size or density under the effects of cardiac disorders, such as myocardial infarction, hypertension or transplant rejection [3,4].

Endomyocardial biopsies (EMB) are often performed in clinics to evaluate rejection of heart transplants as well as to diagnose cardiovascular diseases that can not be assessed by means of diagnostic non-invasive imaging techniques, such as ultrasound or magnetic resonance [5–7]. EMB are usually qualitatively analyzed with histological procedures, which involve sample preparation processes that lead to tissue damage and alteration of the internal structure. In addition, only 2D information from a few slices is used for the analysis of the whole 3D sample.

In order to understand the microstructural organization within the heart with minimum tissue alteration, non-destructive imaging techniques with resolution at micrometre scale are needed. Current imaging techniques provide either high resolution (down to 155 nm pixel size and 1 μm z-step) with a small field of view ($\sim 160 \times 160$ μm^2) and keep altering the sample [8,9], or a whole heart acquisition without enough resolution to resolve individual structures [10,11].

Synchrotron radiation-based X-ray phase contrast imaging (X-PCI) is a recently emerged technique with potential to overcome the aforementioned limitations. In addition to the typical absorption imaging, X-PCI exploits the differences in refractive index between different materials, which improve the contrast while keeping the same X-ray dose. Among the different techniques exploiting the phase contrast effects, propagation-based imaging, used in this study, is of great interest for biomedical soft-tissue research [12]. In a recent study, phase contrast images of cardiac tissue have been obtained with a pixel size of 3.5 μm with the use of this technique [13].

In this study, we propose a high-resolution procedure to assess the detailed cardiac microstructure in EMB biopsies from rat models, thus allowing the segmentation and quantification of cardiac tissue components.

2   Materials and Methods

2.1   Sample Preparation

For this study, three cardiac biopsies of the basal septum, lateral wall and apex of the left ventricle (Fig. 1), with an approximate size of $2 \times 2 \times 4 \, \text{mm}^3$, were extracted from two 12 weeks old male control Wistar Kyoto rats (WKY) and fixed in 4% paraformaldehyde. The biopsies were then placed in thin-walled borosilicate glass tubes of 2 mm of inner diameter using 70% ethanol as medium, and mounted on the sample stage for image acquisition.

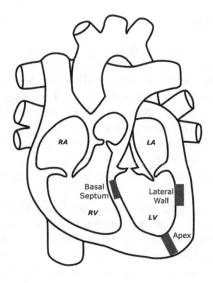

Fig. 1. Extracted biopsies for each of the hearts: basal septum, lateral wall and apex of the left ventricle.

2.2   Data Acquisition

The synchrotron-based X-ray tomography campaign was performed at the TOM-CAT beamline (X02DA) of the Swiss Light Source (Paul Scherrer Institute, Switzerland). Propagation-based X-PCI was achieved with an X-ray beam of 20 keV, a voxel size of 0.65 μm and propagation distance of 20 cm. Since the samples were larger than the field of view ($1.6 \times 1.44 \, \text{mm}^2$), three volumes from each biopsy were imaged using a LuAG:Ce 20 μm scintillator coupled to a PCO.Edge 5.5 CMOS Camera. For each volume, 2501 projections, 20 darks and 50 flats were acquired in approximately 7 min acquisition time. The so called darks are projections taken without beam exposure, used in order to correct for the detector's electronic noise, while flats are direct projections of the beam without sample, so as to correct for non-uniformities of the beam and imperfections in the optical components [14]. A sketch of the experimental setup can be found in Fig. 2.

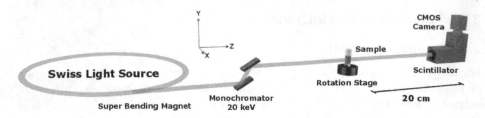

Fig. 2. Sketch of the experimental setup at the TOMCAT beamline at the Swiss Light Source

2.3 Image Processing

The acquired projections were reconstructed using the Gridrec algorithm [15]. Current detector technologies only allow intensity measurements, thus losing the information coded in the phase of the signal. Nevertheless, the phase information can be retrieved by several algorithms incorporating knowledge on wave propagation, such as the single distance phase retrieval method developed by Paganin [16]. Therefore, the data was reconstructed both with and without phase retrieval in order to later be able to fuse information from both phase and intensity images, respectively. The δ/β ratio used in the Paganin algorithm was 56.9. Then, representative $300 \times 300 \times 300$ μm^3 subvolumes were cropped for processing.

For the segmentation task, the chosen tool was the open-source software *Ilastik* [18], based on interactive machine and active learning classifiers. The training process consisted on the iterative labelling of the images for cells, endomysial collagen and background in the three directions (2 slices per direction). Since the imaging conditions were the same for all samples, *Ilastik* was able to segment new input images with just small supervision (a few annotations in 1–2 slices) even if they did not correspond to the training datasets. The phase retrieved images were used to segment cells and background due to the increased contrast given by the Paganin approach. Nevertheless, the contrast of the image areas corresponding to the collagen matrix was reduced in most cases, so the non-phase retrieved images were used to segment it, as they are sharper and show a greater intensity difference compared to the rest of structures. Finally, objects smaller than 150 pixels in 3D were removed to reduce noise, both masks were fused and percentage of endomysial collagen per unit of cell area was computed. Due to acquisition problems, an apical biopsy dataset could not be used for analysis.

Moreover, we were also able to segment part of the cardiac vasculature with the Carving module included in *Ilastik*, based on seeded watershed algorithm where the seeds are given interactively by the user. In this case, the phase retrieved images were used, again due to the higher contrast between the vasculature wall and background.

3   Results

The results obtained are represented for different rats, cardiac areas and volume sizes in order to show representative examples for each of the aforementioned acquisition and image processing procedures.

Two sets of orthogonal views of a representative subvolume are shown in Fig. 3, both with and without phase retrieval. Note how individual cardiomyocytes and its fibre-like arrangement can be distinguished. The brighter intensity spots correspond to the collagen matrix.

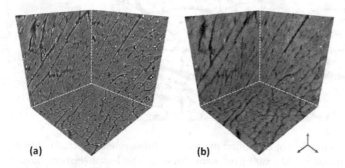

(a)   (b)

Fig. 3. Three orthogonal views (a) without and (b) with phase retrieval, corresponding to a $300 \times 300 \times 300$ μm^3 subvolume of a biopsy of the left ventricle lateral wall of a WKY rat.

Figure 4 shows representative images of the segmentation procedure for the collagen matrix and cells. The 3D representation of the collagen matrix in Fig. 4f shows its organization in fibre-like structures following the direction of the cardiomyocytes. The relative amount of endomysial collagen within the different processed subvolumes for the WKY rats is detailed in Table 1. Finally, Fig. 5 shows the results of the vascular segmentation, where its branching morphology can be clearly observed.

4   Discussion

In this study we demonstrate that X-PCI, specifically in free-space propagation mode, can be used in order to assess the organization of the cardiac tissue at micrometer scale. In order to do so, images with a voxel size of 0.65 μm have been acquired, thus improving the state-of-the-art 3.5 μm for this imaging technique in the study of cardiac tissue [13]. Such improvement in resolution allows to analyse the samples by directly looking at its micrometer level structures, such as cardiomyocytes, endomysium network or microvasculature, which for the best of our knowledge, has never been achieved before with the use of X-ray imaging techniques.

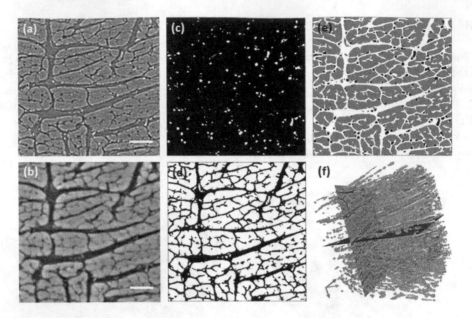

Fig. 4. Representation of the different steps in the segmentation of cardiomyocytes and collagen for a $300 \times 300 \times 300~\mu m^3$ subvolume of the basal septum biopsy of a WKY rat. Scale bar corresponds to 50 μm (a) Non-phase retrieved image. (b) phase retrieved image. (c) collagen segmentation. (d) cardiomyocytes segmentation. (e) resulting fusion of segmentations. (f) 3D rendering of the endomysial collagen segmentation in the entire subvolume.

Table 1. Calculated values for the endomysial collagen percentage in each of the biopsies.

	Apex	LV Wall	Septum
WKY1	4.23%	6.37%	4.60%
WKY2	——	5.14%	1.64%

In Table 1 it is shown that the collagen percentage is higher in the left ventricular wall than in the rest of areas. Moreover, the collagen percentage difference between rats in the left ventricular wall is in a smaller range (~1%) in comparison to the basal septum region (~2%). In the literature [17], endomysial collagen fraction with respect to total area was quantified from confocal microscopy images (0.4 μm voxel size). Left ventricular midwall WKY samples were analysed resulting in a ~2.5% of collagen. In our case, similar but slightly higher values are observed in the left ventricular wall. The reasons for this increase are several. First, the calculation of collagen percentage is done with respect to cardiomyocyte area and not total area, which is more significant taking into account that endomysium is found around individual cells. In addition, the voxel size used

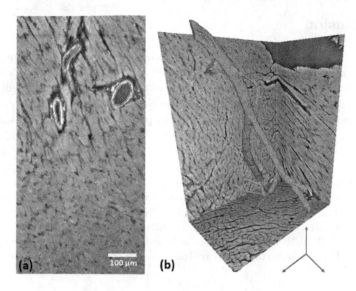

Fig. 5. (a) Selected slice and (b) 3D rendering of the vasculature segmentation in a $680 \times 920 \times 440 \ \mu m^3$ subvolume of the basal septum biopsy of a WKY rat.

in X-PCI is larger than in the confocal microscopy technique used in the compared study, thus introducing a smoothing effect that may increase the amount of collagen quantified. Finally, as a limitation of the methodology, we would like to mention that platelets and collagen have very similar characteristics in the images and, therefore, when platelets are present in very small diameter vessels it is currently very difficult to differentiate them. Therefore, in order to overcome this limitation as much as possible, the analysis was performed in subvolumes without presence of visible vasculature.

The data obtained is of high interest in scientific and clinical terms, as it allows to characterize the tissue by direct observation of its main structural components to better understand how they influence the macroscopic organization and functionality. This means that we can also detect how these microstructures are affected under the influence of certain cardiac diseases or disorders, the so called cardiac remodelling, and thus comprehend in more detail the different abnormalities at all scales.

Nowadays, several diseases involving cardiac remodelling are assessed by histological imaging of extracted EMB. During such techniques, the biopsies undergo a series of destructive procedures that change the properties of the tissue and the final images are usually 2-dimensional, which can be overcome by the non-destructiveness of X-PCI and improved by the 3-dimensional time-efficient nature of the technique. In addition, the proposed image processing framework allows the quantitative evaluation of the acquired X-ray images, thus improving the accuracy of the currently qualitative assessment of histological images.

5 Conclusion

The results show that X-PCI is a very promising technique for the analysis of the cardiac microstructure, as it is a 3-dimensional time-efficient non-destructive technique that allows the detailed observation of the different elements of the tissue at micrometer level. Therefore, the development and translation of X-PCI to a table-top system has the potential to improve the biopsy-based diagnostic procedures thanks to easier, less aggressive and shorter preparation and imaging procedures.

Acknowledgments. This research has received funding from the EU FP7 for research, technological development and demonstration under grant agreement VP2HF (no. 611823) and from the Spanish Ministry of Economy and Competitiveness (gTIN2014-52923-R, the Maria de Maeztu Units of Excellence Programme MDM-2015-0502) and FEDER. P.G.C. wants to acknowledge EMBO for the short-term fellowship to do her research stay at X-ray Tomography group in Paul Scherrer Institut (PSI).

References

1. World Health Organization (WHO): Cardiovascular Diseases (CVDs). http://www.who.int/mediacentre/factsheets/fs317/en/. Accessed 11 May 2016
2. Weber, K.T., Sun, Y., Bhattacharya, S.K., Ahokas, R.A., Gerling, I.C.: Myofibroblast-mediated mechanisms of pathological remodelling of the heart. Nat. Rev. Cardiology **10**(1), 15–26 (2013)
3. Burchfield, J.S., Xie, M., Hill, J.A.: Pathological ventricular remodeling mechanisms: part 1 of 2. Circulation **128**(4), 388–400 (2013)
4. Cohn, J.N., Ferrari, R., Sharpe, N.: Cardiac remodeling–concepts and clinical implications: a consensus paper from an international forum on cardiac remodeling. J. Am. Coll. Cardiol. **35**(3), 569–582 (2000)
5. Dujardin, K.S., Enriquez-Sarano, M., Rossi, A., Bailey, K.R., Seward, J.B.: Echocardiographic assessment of left ventricular remodeling: are left ventricular diameters suitable tools. J. Am. Coll. Cardiol. **30**(6), 1534–1541 (1997)
6. Vogel-Claussen, J., Rochitte, C.E., Wu, K.C., Kamel, I.R., Foo, T.K., Lima, J.A.C., Bluemke, D.A.: Delayed enhancement MR imaging: utility in myocardial assessment 1. Radiographics **26**(3), 795–810 (2006)
7. From, A.M., Maleszewski, J.J., Rihal, C.S.: Current status of endomyocardial biopsy. Mayo Clin. Proc. **86**(11), 1095–1102 (2011)
8. Tobita, K., Garrison, J.B., Liu, L.J., Tinney, J.P., Keller, B.B.: Three-dimensional myober architecture of the embryonic left ventricle during normal development and altered mechanical loads. Anat. Rec. Part A Discoveries Mol. Cell. Evol. Biol. **283**(1), 193–201 (2005)
9. Jouk, P., Mourad, A., Milisic, V., Michalowicz, G., Raoult, A., Caillerie, D., Usson, Y.: Analysis of the fiber architecture of the heart by quantitative polarized light microscopy. Accuracy, limitations and contribution to the study of the fiber architecture of the ventricles during fetal and neonatal life. Eur. J. Cardiothorac. Surg. **31**(5), 915–921 (2007)

10. Schmid, P., Jaermann, T., Boesiger, P., Niederer, P.F., Lunkenheimer, P.P., Cryer, C.W., Anderson, R.H.: Ventricular myocardial architecture as visualised in post-mortem swine hearts using magnetic resonance diffusion tensor imaging. Eur. J. Cardiothorac. Surg. **27**(3), 468–472 (2005)

11. Lombaert, H., Peyrat, J., Croisille, P., Rapacchi, S., Fanton, L., Cheriet, F., Clarysse, P., Magnin, I., Delingette, H., Ayache, N.: Human atlas of the cardiac fiber architecture: study on a healthy population. IEEE Trans. Med. Imaging **31**(7), 1436–1447 (2012)

12. Cloetens, P., Ludwig, W., Baruchel, J., Guigay, J., Pernot-Rejmánková, P., Salomé-Pateyron, M., Schlenker, M., Buffière, J., Maire, E., Peix, G.: Hard x-ray phase imaging using simple propagation of a coherent synchrotron radiation beam. J. Phys. D Appl. Phys. **32**(10A), A145 (1999)

13. Mirea, I., Varray, F., Zhu, Y.M., Fanton, L., Langer, M., Jouk, P.S., Michalowicz, G., Usson, Y., Magnin, I.E.: Very high-resolution imaging of post-mortem human cardiac tissue using X-ray phase contrast tomography. In: van Assen, H., Bovendeerd, P., Delhaas, T. (eds.) FIMH 2015. LNCS, vol. 9126, pp. 172–179. Springer, Cham (2015). doi:10.1007/978-3-319-20309-6_20

14. Seibert, J.A., Boone, J.M., Lindfors, K.K.: Flat-field correction technique for digital detectors. Proc. SPIE Med. Imaging **1998**(3336), 348–354 (1998)

15. Marone, F., Stampanoni, M.: Regridding reconstruction algorithm for real-time tomographic imaging. J. Synchrotron Radiat. **19**(6), 1029–1037 (2012)

16. Paganin, D., Mayo, S.C., Gureyev, T.E., Miller, P.R., Wilkins, S.W.: Simultaneous phase and amplitude extraction from a single defocused image of a homogeneous object. J. Microsc. **206**(1), 33–40 (2002)

17. LeGrice, I.J., Pope, A.J., Sands, G.B., Whalley, G., Doughty, R.N., Smaill, B.H.: Progression of myocardial remodeling and mechanical dysfunction in the spontaneously hypertensive rat. Am. J. Physiol. Heart Circulatory Physiol. **303**(11), H1353–H1365 (2012)

18. Sommer, C., Straehle, C., Köthe, U., Hamprecht, F.A.: Ilastik: interactive learning and segmentation toolkit. In: 2011 IEEE International Symposium on Biomedical Imaging: From Nano to Macro, pp. 230–233 (2011)

Cartan Frames for Heart Wall Fiber Motion

Babak Samari[1], Tristan Aumentado-Armstrong[1], Gustav Strijkers[2],
Martijn Froeling[3], and Kaleem Siddiqi[1(✉)]

[1] School of Computer Science and Centre for Intelligent Machines,
McGill University, Montreal, Canada
siddiqi@cim.mcgill.ca
[2] Academic Medical Center, University of Amsterdam,
Amsterdam, The Netherlands
[3] University Medical Center, Utrecht, Utrecht, The Netherlands

Abstract. Current understanding of heart wall fiber geometry is based
on ex vivo static data obtained through diffusion imaging or histology.
Thus, little is known about the manner in which fibers rotate as the heart
beats. Yet, the geometric organization of moving fibers in the heart wall
is key to its mechanical function and to the distribution of forces in it to
effect efficient, repetitive pumping. We develop a moving frame method
to address this problem, with a spatio-temporal formulation of the associated Cartan matrix. We apply our construction to simulated (canine)
data obtained from a left ventricular mechanics challenge, and to in vivo
human left ventricular data. The method shows promise in providing
Cartan connection parameters to describe spatio-temporal rotations of
fibers, which in turn could benefit subsequent analyses or be used for
diagnostic purposes.

1 Introduction

Mammalian heart wall muscle is comprised of densely packed elongated myocytes
in an extra-cellular matrix [4,11]. The geometry of this packing facilitates efficient pumping to optimize ejection fraction while also providing strength. The
precise manner in which fibers move and rotate with the material medium in
which they are placed during the heart beat cycle is as yet not known. Current
models of heart wall fiber geometry are derived from diffusion imaging of static
ex vivo hearts and from histology. Models do exist for such data, for example
the rule-based model of [1], and the techniques in [7]. However, such methods
lack a temporal dimension to capture fiber rotation as the heart beats. Whereas
there is on-going progress in our community towards in vivo cardiac diffusion
imaging, with the possibility to now acquire full heart fiber orientation data in
beating hearts [6,17], the mathematical tools for analysis lag behind.

The key contribution of the current article is the use of Cartan connection forms [12,16] to model both spatial (within a time sample) and temporal
(between time samples) rotations of frame fields attached to heart wall fibers.
This allows one to capture spatial and temporal geometric signatures within a

© Springer International Publishing AG 2017
M. Pop and G.A. Wright (Eds.): FIMH 2017, LNCS 10263, pp. 32–41, 2017.
DOI: 10.1007/978-3-319-59448-4_4

single consistent framework. We begin by reviewing Cartan connection forms and their use in moving frame methods. Extending a method for the case of static fibers in [13], we add a temporal dimension to the Cartan matrix. We then specialize the model by adopting frames fit to heart wall myofiber orientations recovered from diffusion data. We demonstrate the promise of this method for the recovery of geometric curvature type spatio-temporal signatures for moving myofibers during a heart beat cycle, using both finite element simulation on canine data [2] and human in vivo cardiac diffusion imaging.

2   Materials and Methods

2.1   Cartan Connection Forms

We begin by reviewing the mathematics of Cartan connection forms for the case of an orthonormal unit frame field $F = [f_1, f_2, f_3]^T$ defined in $\mathbb{R}^3$, which is the construction used in [13] to describe the geometry of myofibers in a static heart wall. A covariant derivative of this frame field with respect to a vector v at point p is given by

$$\nabla_v f_i = \omega_{i1}(v) f_1(p) + \omega_{i2}(v) f_2(p) + \omega_{i3}(v) f_3(p), \tag{1}$$

where $i \in \{1, 2, 3\}$, and $\omega_{ij}(v) = \nabla_v f_i \cdot f_j$. [12] shows that these ω_{ij}s satisfy the definition of 1-forms in $\mathbb{R}^3$. For any vector field v in $\mathbb{R}^3$, $\nabla_v f_i = \sum_j \omega_{ij}(v) f_j$. Using the alternating property of 1-forms, $\omega_{11} = \omega_{22} = \omega_{33} = 0$ and $\omega_{21} = -\omega_{12}, \omega_{31} = -\omega_{13}, \omega_{32} = -\omega_{23}$. Let us assume that the frame field F has the following parametrization in the universal Cartesian coordinate system $E = [e_1, e_2, e_3]^T$:

$$f_i = \alpha_{i1} e_1 + \alpha_{i2} e_2 + \alpha_{i3} e_3, \tag{2}$$

where $i \in \{1, 2, 3\}$, and each $\alpha_{ij} = f_i \cdot e_j$ is a real valued function. The matrix $A = [\alpha_{ij}]$ is called the attitude matrix, and $dA = [d\alpha_{ij}]$ is a matrix whose elements are 1-forms. With $\omega = [\omega_{ij}]$ it can be shown that $\omega_{ij} = \sum_k (d\alpha_{ik}) \alpha_{kj}$ [12], so $\omega = dA A^T$. Connection forms describe the rate of change of a frame field $[f_1, f_2, f_3]^T$ in the direction of an arbitrary vector v. The dual 1-form of the frame field $[f_1, f_2, f_3]^T$ is obtained when it is itself parametrized via 1-forms, i.e., for each vector v at p, $\psi_i(v) = v \cdot f_i(p)$. For the sake of simplicity we will use the same notation for the frame and its dual representation, i.e.,

$$f_i(v) = \psi_i(v) = v \cdot f_i(p). \tag{3}$$

From the definition of the 1-form,

$$dx_i(v) = \sum_j v_j \frac{\partial x_i}{\partial x_j}(p) = v_i. \tag{4}$$

Combining Eqs. (3), (4) results in

$$f_i(e_j) = f_i \cdot e_j = \left(\sum_k \alpha_{ik} e_k \right) e_j = \alpha_{ij}. \tag{5}$$

From Eq. (4), (5) we have $f_i = \sum_j \alpha_{ij} dx_j$ which, with $F = [f_i]^T$, can be written in the dual 1-form representation as $F = A[dx_1 \; dx_2 \; dx_3]^T$. The Cartan structural equation is then given by [12]:

$$df_i = \sum_j \omega_{ij} \wedge f_j, \text{ i.e., } dF = \omega F, \quad d\omega_{ij} = \sum_k \omega_{ik} \wedge \omega_{kj}, \text{ i.e., } d\omega = \omega\omega. \quad (6)$$

Since there are three unique connection forms $\omega_{12}(v), \omega_{13}(v), \omega_{23}(v)$, feeding the frame field's unit vectors to them produces nine different connection parameters $c_{ijk} = \omega_{ij}(f_k)$. Here, for the vector v at a point p, $\omega_{ij}(v)$ represents the amount of $f_i(p)$'s turn toward $f_j(p)$ when p moves in the direction of v. The estimation of these connection parameters for frame fields attached to diffusion MRI data of ex vivo hearts was the strategy proposed in [13] to parametrize the static geometry of heart wall myofibers.

2.2   Cartan Forms for Moving Fibers

We assume that for any query time $t \in \mathbb{R}$ we have a stationary state of an orthonormal moving frame field F^t attached to heart wall fibers as in [13], with the corresponding universal coordinate $E^t = [e_1, e_2, e_3]^T$: f_1 is aligned with the fiber orientation, f_3 is given by the component of the normal to the heart wall that is orthogonal to f_1 and f_2 is their cross-product. Our goal is to parametrize the local rotation of F^t in both space and time so we extend the existing 3D coordinate system to add a 4th dimension to represent the time axis. Let $E = [e_1, e_2, e_3, e_4]^T$ represent our extended universal coordinate system, $\mathbb{R}^4$, in-which $e_i \cdot e_j = \delta_{i,j}$ where $\delta_{i,j}$ is the Kronecker delta, and $e_4 = [0, 0, 0, 1]^T$ is a basis vector for the time axis in our 4D representation. Thus a query $< x, y, z, t >$, where $x, y, z, t \in \mathbb{R}$, represents a query $< x, y, z >$ in E^t. We extend the local frames F^t to 4D as follows:

Definition 1. *Let* $F^t = [f_1, f_2, f_3]^T = \begin{bmatrix} f_{1,x}^t & f_{1,y}^t & f_{1,z}^t \\ f_{2,x}^t & f_{2,y}^t & f_{2,z}^t \\ f_{3,x}^t & f_{3,y}^t & f_{3,z}^t \end{bmatrix}$ *describe the frame field*

in $\mathbb{R}^3$ *at time t, then:*

$$\begin{bmatrix} f_{1,x}^t & f_{1,y}^t & f_{1,z}^t & 0 \\ f_{2,x}^t & f_{2,y}^t & f_{2,z}^t & 0 \\ f_{3,x}^t & f_{3,y}^t & f_{3,z}^t & 0 \\ 0 & 0 & 0 & 1 \end{bmatrix} \quad (7)$$

is the 4D extension of the frame fields F^t.

Since we know that $F^t = [f_1, f_2, f_3]^T$ is an orthogonal matrix, it is not hard to show that $F = [f_1, f_2, f_3, f_4]^T$ is also orthogonal, i.e., $\forall i \neq j, f_i \cdot f_j = 0$.

The next step is to extend the existing connection forms from 3D to 4D. Let $F = [f_1, f_2, f_3, f_4]^T$ be an arbitrary frame field on $\mathbb{R}^4$, which, for $i \in \{1, 2, 3, 4\}$, has the following parametrization in the extended universal coordinate system $E = [e_1, e_2, e_3, e_4]^T$:

$$f_i = \alpha_{i1}e_1 + \alpha_{i2}e_2 + \alpha_{i3}e_3 + \alpha_{i4}e_4. \quad (8)$$

A covariant derivative of this frame field with respect to a vector v at point p is given by

$$\nabla_v f_i = \omega_{i1}(v)f_1(p) + \omega_{i2}(v)f_2(p) + \omega_{i3}(v)f_3(p) + \omega_{i4}(v)f_4(p), \qquad (9)$$

where $[\alpha_{ij}] = [f_i \cdot e_j]$ is the attitude matrix in $\mathbb{R}^4$ and $\omega_{ij} = \sum_k (d\alpha_{ik})\alpha_{kj}$ is a 1-form.

Definition 2. *Let $i < j$ and $i, j, k \in \{1, 2, 3, 4\}$. By feeding the frame field's unit vectors ($f_i s$) to $\omega_{ij}(v)$ we have*

$$c_{ijk} = \omega_{ij}(f_k). \qquad (10)$$

Remark: For $j = 4$, $c_{ijk} = \omega_{ij}(f_k) = 0$. For a point p and with $i < j \in \{1, 2, 3\}, \omega_{ij}(f_k)$ represents the amount of $f_i(p)$'s turn toward $f_j(p)$ when taking a step towards $f_k(p)$.

Given the skew-symmetry property of the 4×4 connection form matrix we build $3 \times 4 = 12$ different non-zero and unique c_{ijk}s. We can use these coefficients to estimate the motion of the frame field using first order approximation of the Taylor expansion of the frame field F in the direction of the vector v at point p

$$f_i(v) \simeq f_i(p) + df_i(v). \qquad (11)$$

Finally, by applying the Cartan structural Eq. (6) on the Eq. (11), where $v = \sum_k v_k f_k$, we have

$$f_i(v) \simeq f_i(p) + \sum_j \left(\sum_k v_k c_{ijk} \right) f_j(p). \qquad (12)$$

We now show how to calculate the 1-form coefficients of a given 4D frame. From the definition of the c_{ijk}s in [13] and their first order Taylor expansions, for an arbitrary vector v at point p, and $i, j, k, n \in \{1, 2, 3, 4\}$ we can re-write Eq. (6) as

$$c_{ijk} = \omega_{ij}(f_k) = f_j^T J(f_i) f_k, \qquad (13)$$

where $J(f_i) = [\frac{\partial f_{ij}}{\partial x_k}]$ is a Jacobian matrix. Given a discretized frame field, we then apply Eq. (13) to calculate the connection form coefficients.

We illustrate the above extension by a simulation in Fig. 1, where we consider an initial fiber direction, with an attached frame field, and then apply a specific set of c_{ijk}s to it. Here f_1 is in the direction of the fiber, f_3 is in the in-page direction orthogonal to f_1 and $f_2 = f_3 \times f_1$. The figure (left to right) shows three samples in time of the orientations in the local neighborhood of the fiber, generated with the parameters $c_{123} = 0.5$ radians/voxel, $c_{124} = 0.03$ radians/time-step, and $c_{ijk} = 0$ for all remaining connection parameters. The positive c_{123} value results in a clockwise rotation of fibers in the in-page direction (panel A) and the positive c_{124} value results in an increase in the total in-page rotation of fibers in time (panels B and C).

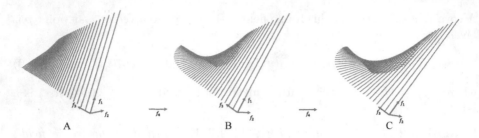

Fig. 1. The application of a specific set of 4D connection parameters c_{ijk} to a frame field attached to a fiber direction. See text for a discussion.

2.3 Beating Heart Fiber Data

We construct two data sets for evaluating our method for moving fibers in the heart wall. The first uses canine heart data from the STACOM 2014 LV mechanics challenge [2]. This includes a hexahedral mesh at the beginning of the beat cycle, the associated local cardiomyocyte fiber orientations from the first principal eigenvector of ex vivo diffusion tensor MRI data, an endocardial pressure curve, and the positions of several reference points on the base of the left ventricle at three points in the beat cycle. We used this data for a finite element simulation of the heart wall using the transversely isotropic Holzapfel-Ogden constitutive equations, which describe a non-linearly elastic incompressible material designed to model the empirical behaviour of the cardiac tissue [8]. We apply to this model a controlled rotation θ of the undeformed fiber orientation at each point in the heart wall, as a linear function of time and transmural distance from the midwall. We then apply the local deformation gradient during the simulation to yield the effective rotation in time of the fibers in the beating heart. The finite element simulation, the material model and the rotation of fibers were implemented in FEBio [10] as a plugin.

The second is from in vivo diffusion imaging of an entire human heart. Data of a single healthy volunteer was acquired on a 3T scanner (Philips, Achieva) using a 32-channel cardiac coil. DWI was performed using a SE sequence with cardiac triggering in free breathing with asymmetric bipolar gradients [6,17] and additional compensation for the slice and readout gradients. Data was acquired with 150 and 220 ms delays after the cardiac trigger and b-values of 0, 10, 20, 30, 50, 100, 200 and 400 s/mm^2 with 6, 3, 3, 3, 3, 3, 3 and 24 gradient-directions, respectively. The imaging parameters were: TE = 62 ms; TR = 14 heart beats; FOV = 280×150 mm^2 (using outer volume suppression using rest slabs); slices = 14; voxel size = $7 \times 2.5 \times 2.5$ mm^3; acquisition matrix = 112×48 SENSE factor = 2.5; partial Fourier = 0.85; EPI bandwidth = 42 Hz/pix; averages = 1; fat suppression = SPAIR; Gmax = 62 mT/m; max slope = 100 mT/m/ms and acquisition time = 13 min. Data processing was done using DTITools for Mathematica 10 and included registration to correct for subject motion and eddy current deformations, noise suppression and tensor calculation using weighted linear least squares estimation [5], with the ventricles segmented manually using

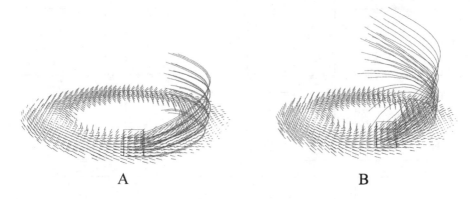

Fig. 2. An illustration of the simulated motion of canine heart wall fibers in the left ventricle at two time points in the systolic phase: 50 ms (A) and 350 ms (B). The fiber orientations are shown in red and partial tractography provides a visualization of fibers in a local neighborhood. In this simulation the total transmural applied rotation (θ) from 0 ms to 400 ms is $20°$. (Color figure online)

ITK-SNAP [18]. The primary eigenvector of the tensors within the left ventricle were extracted and used for subsequent analysis.

3   Results

In our experiments on both the canine data and the in vivo human data, our frame axes f_1, f_2, f_3 are chosen in the manner explained in Sect. 2.2. The results from past fitting of Cartan connection parameters to static ex vivo mammalian heart data in [13] have shown $c_{123}, c_{131}, c_{232}$ to be the most significant parameters, with the others being close to zero. The first of these has to do with the helix angle change in a transmural direction and the latter two are related to sectional curvatures of the heart wall. The addition of the temporal dimension in the present paper now adds the possibility to look at frame axis rotation in time, i.e., c_{ij4} connection parameters. In the canine data with a controlled increase θ in the total epicardium to endocardium rotation we expect the new parameter c_{124} to pick up this effect. Figure 2 shows two time frames of a canine data set, where the total angular change in orientation θ is $20°$. This choice of enforced rotation was motivated in part by prior biological work, which suggests a small but measurable increase in the rate of transmural fiber angle change as systole progresses [3]. The tractography in the visualization shows how the total epicardium to endocardium turn increases in time.

In Fig. 3 we show histograms over the entire left ventricle of the spatio-temporal connection form parameters with the highest magnitudes: c_{123} and c_{124} for two canine data sets. The units of c_{ijk} are radians/voxel when $k \neq 4$ and radians/10 ms when $k = 4$. In our simulations the active contraction begins at around 100 ms, so the red curve for the c_{ij4} terms in the right column of the

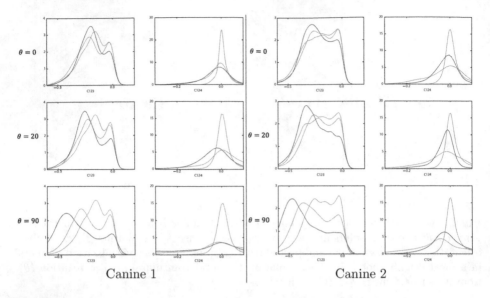

Canine 1   Canine 2

Fig. 3. We focus on c_{123} and c_{124} and show the effect of different choices for the total applied additional transmural rotation (θ): 0° (top), 20° (middle) and 90° (bottom) for two canine data sets. The red, green and blue curves show histograms corresponding to the 50 ms, 200 ms and 350 ms time samples in the systolic phase from 0 ms to 400 ms (Color figure online)

figure is in fact expected to be very small, since it corresponds to 50 ms at which time the fibers are relatively static. We also consider the effect of no additional rotation θ (top row) and the extreme case of $\theta = 90$ (bottom row). Although the latter case is not biologically realistic, this type of variation in controlled rotation of fibers gives us a way to evaluate our method. For both canine data sets we see the clear shift to the left in the c_{123} histograms (red → blue) as θ is increased, reflecting our expectation. The variation in the c_{124} parameters in time is more subtle, because this reflects the instantaneous (time sample to time sample) in plane rotation of the fibers.

Figure 4 illustrates our in vivo data (top) and the Cartan connection parameters (bottom) obtained by fits to the entire left ventricle. In vivo diffusion imaging presently suffers from many challenges, including low spatial resolution, a limit on the range of possible timing and low signal-to-noise compared to ex vivo acquisitions. The data can also contain artifacts which are spatially varying, e.g., results near the apex are typically not reliable. In our full heart sequence the time samples are concentrated at end systole, when the heart wall is thick and fibers are relatively stationary. As such we see the clear role of the non-zero c_{123} term, capturing transmural rotation, but the c_{124} term is small.

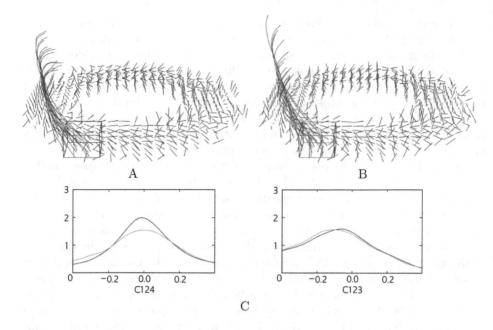

Fig. 4. In vivo heart wall fibers in a human left ventricle at two time points in the systolic phase: (A) 150 ms, (B) 220 ms. (C) Histograms of the Cartan connection parameters c_{124}, c_{123}, obtained via fits to (A) (red) and (B) (green). (Color figure online)

4   Discussion

We have developed a moving frame method for the modeling of fibers in a dynamic, beating heart. The construction of the Cartan matrix for spatio-temporally varying fiber orientations, along with a method for fitting the Cartan connection parameters, shows promise when applied to simulated canine data with controlled rotations to fibers. In our experiments we see that non-linearities in the variation of the (time) sample to (time) sample transmural rotation, as reflected in the c_{124} parameter, can arise. We speculate that such non-linearities may occur in the beat cycle to accommodate the wringing motion qualitatively seen when viewing the beating left ventricle in echocardiography or cardiac MRI. Our experiments on in vivo human data, though proof-of-concept at the present time, demonstrate the ability to recover curvature like signatures in such settings as well.

While our approach is a natural extension of the methods of moving frames in space to space and time, there are a number of theoretical considerations that can be further addressed. In comparison to other models, such as using regression or spline fits between time points, the use of Cartan forms provides an easily interpretable and intuitive set of measurements able to characterize the evolution of the frame fields over time. While it may be possible to obtain equivalent measurements with other models, we anticipate that this would compromise the elegance, extensibility, and simplicity of the approach we have taken.

For the frame fields themselves, one biologically motivated consideration is the use of the other eigenvectors of the diffusion tensor as basis vectors. This would allow a definition of a truly local frame field and permit investigation of the geometry of the other eigenvectors, postulated to represent a laminar structure [9]. A practical challenge here is that such an approach would suffer more acutely from the low spatial and temporal resolution and high noise level of in vivo moving data than our current approach.

As in vivo diffusion imaging advances, Cartan frame spatio-temporal fitting could offer informative geometric descriptors for analysis. As in the strictly spatial case, a number of natural applications of this formalism are apparent. Previous methods for inpainting and denoising potentially damaged measurements in ex vivo static hearts via differential forms [15], as well as atlas construction to allow inter-species comparison using differential geometric measures [14], are both readily extended to the spatiotemporal domain based on the approach described herein. In these and other cases, the statistical properties of connection form values can be adapted for use in learning-based approaches (e.g. for segmentation or other medical imaging tasks). More importantly, they provide a means for the analysis of fiber orientation as the heart beats, which will become increasingly important as in vivo diffusion imaging becomes more common. In clinical applications connection form values are likely to assist in providing signals of abnormal geometry, such as those which might occur in the presence of local cardiac infarcts. Moving beyond diffusion imaging, other modalities may benefit from our approach. For example, with ultrasound we may be able to take advantage of increased temporal resolution, as well as latent fiber information, by utilizing information extracted from the modeling of ex vivo hearts.

Acknowledgments. This work was supported by research and training grants from the Natural Sciences and Engineering Research Council of Canada.

References

1. Bayer, J., Blake, R., Plank, G., Trayanova, N.: A novel rule-based algorithm for assigning myocardial fiber orientation to computational heart models. Ann. Biomed. Eng. **40**(10), 2243–2254 (2012)
2. Camara, O., Mansi, T., Pop, M., Rhode, K., Sermesant, M., Young, A.: LV mechanics challenge. In: STACOM. LNCS, vol. 8896. Springer, Cham (2014)
3. Chen, J., Liu, W., Zhang, H., Lacy, L., Yang, X., Song, S.K., Wickline, S.A., Yu, X.: Regional ventricular wall thickening reflects changes in cardiac fiber and sheet structure during contraction: quantification with diffusion tensor mri. Am. J. Physiol. Heart Circ. Physiol. **289**(5), H1898–H1907 (2005)
4. Franzone, P.C., Guerri, L., Tentoni, S.: Mathematical modeling of the excitation process in myocardial tissue: influence of fiber rotation on wavefront propagation and potential field. Math. Biosci. **101**(2), 155–235 (1990)

5. Froeling, M., Nederveen, A.J., Heijtel, D.F., Lataster, A., Bos, C., Nicolay, K., Maas, M., Drost, M.R., Strijkers, G.J.: Diffusion-tensor MRI reveals the complex muscle architecture of the human forearm. J. Magn. Reson. Imaging **36**(1), 237–248 (2012). http://dx.doi.org/10.1002/jmri.23608
6. Froeling, M., Strijkers, G.J., Nederveen, A.J., Luijten, P.R.: Whole heart DTI using asymmetric bipolar diffusion gradients. J. Cardiovasc. Magn. Reson. **17**(Suppl 1), 15 (2015)
7. Helm, P., Beg, M.F., Miller, M.I., Winslow, R.L.: Measuring and mapping cardiac fiber and laminar architecture using diffusion tensor mr imaging. Ann. N. Y. Acad. Sci. **1047**(1), 296–307 (2005)
8. Holzapfel, G.A., Ogden, R.W.: Constitutive modelling of passive myocardium: a structurally based framework for material characterization. Phil. Trans. R. Soc. A Math. Phys. Eng. Sci. **367**(1902), 3445–3475 (2009)
9. LeGrice, I.J., Smaill, B., Chai, L., Edgar, S., Gavin, J., Hunter, P.J.: Laminar structure of the heart: ventricular myocyte arrangement and connective tissue architecture in the dog. Am. J. Physiol. Heart Circ. Physiol. **269**(2), H571–H582 (1995)
10. Maas, S.A., Ellis, B.J., Ateshian, G.A., Weiss, J.A.: Febio: finite elements for biomechanics. J. Biomech. Eng. **134**(1), 011005 (2012)
11. Nielsen, P., Le Grice, I., Smaill, B., Hunter, P.: Mathematical model of geometry and fibrous structure of the heart. Am. J. Physiol. Heart Circ. Physiol. **260**(4), H1365–H1378 (1991)
12. O'Neill, B.: Elementary Differential Geometry. Academic Press, New York (2006)
13. Piuze, E., Sporring, J., Siddiqi, K.: Maurer-Cartan forms for fields on surfaces: application to heart fiber geometry. IEEE TPAMI **31**(12), 2492–2504 (2015)
14. Piuze, E., Lombaert, H., Sporring, J., Strijkers, G.J., Bakermans, A.J., Siddiqi, K.: Atlases of cardiac fiber differential geometry. In: Ourselin, S., Rueckert, D., Smith, N. (eds.) FIMH 2013. LNCS, vol. 7945, pp. 442–449. Springer, Heidelberg (2013). doi:10.1007/978-3-642-38899-6_52
15. Piuze, E., Sporring, J., Siddiqi, K.: Moving frames for heart fiber reconstruction. In: Ourselin, S., Alexander, D.C., Westin, C.-F., Cardoso, M.J. (eds.) IPMI 2015. LNCS, vol. 9123, pp. 527–539. Springer, Cham (2015). doi:10.1007/978-3-319-19992-4_41
16. Shifrin, T.: Multivariable Mathematics: Linear Algebra, Multivariable Calculus, and Manifolds, vol. 10. John Wiley & Sons Inc. (2005)
17. Stoeck, C.T., von Deuster, C., Genet, M., Atkinson, D., Kozerke, S.: Second order motion compensated spin-echo diffusion tensor imaging of the human heart. J. Cardiovasc. Magn. Reson. **17**(Suppl 1), 81 (2015)
18. Yushkevich, P.A., Piven, J., Hazlett, H.C., Smith, R.G., Ho, S., Gee, J.C., Gerig, G.: User-guided 3D active contour segmentation of anatomical structures: significantly improved efficiency and reliability. NeuroImage **31**(3), 1116–1128 (2006)

Robust Model-Based Registration of Cardiac MR Images for T1 and ECV Mapping

Sofie Tilborghs[1,4]([✉]), Tom Dresselaers[2,4], Piet Claus[3,4], Guido Claessen[3],
Jan Bogaert[2,4], Frederik Maes[1,4], and Paul Suetens[1,4]

[1] Department of Electrical Engineering, ESAT/PSI, KU Leuven, Leuven, Belgium
sofie.tilborghs@kuleuven.be
[2] Department of Imaging and Pathology, Radiology, KU Leuven, Leuven, Belgium
[3] Department of Cardiovascular Sciences, KU Leuven, Leuven, Belgium
[4] Medical Imaging Research Center, UZ Leuven,
Herestraat 49 - 7003, 3000 Leuven, Belgium

Abstract. Quantification of myocardial T1 and extra cellular volume
(ECV) in cardiac MRI provides relevant diagnostic information about
myocardial structure. However, since these maps are pixel-wise calculated from two sequences of T1-weighted images, they are frequently
disturbed by motion artifacts originating from e.g. patient motion, failure in breath-hold or cardiac motion. We propose a new non-rigid registration framework combining a robust data-driven initialization with a
model-based refinement. The data-driven algorithm finds an optimal registration sequence of images for calculation of an initial T1 map. The registration is subsequently refined by exploiting the exponential relaxation
model of T1. Validation using 20 in-vivo data sets showed a decrease in
mean boundary error and an increase in global Dice coefficient.

Keywords: T1 mapping · ECV mapping · Image registration

1 Introduction

Accurate measurement of myocardial T1 and extra cellular volume (ECV) using
cardiac MRI is highly relevant for the diagnosis of diffuse myocardial diseases
such as diffuse fibrosis, amyloidosis and Anderson Fabry disease [1]. Compared
to the conventional late gadolinium enhanced (LGE) images where diagnosis is
based on the subjective assessment of relative contrast differences, T1 and ECV
mapping allow quantitative characterization of the myocardium. In T1-mapping,
a map of the longitudinal relaxation time (T1) is obtained by fitting an exponential curve to each pixel in a sequence of T1-weighted images acquired over multiple heart cycles using the Modified Look-Locker Inversion-Recovery (MOLLI)
protocol [2]. Additionally, an ECV map [3] can be constructed by combining
the T1 maps before (native) and after (enhanced) gadolinium contrast injection.
Despite ECG-gating and imposed breath-hold, the pixel-wise T1 maps are still
often disturbed by motion artifacts (Fig. 1) originating from involuntary patient

© Springer International Publishing AG 2017
M. Pop and G.A. Wright (Eds.): FIMH 2017, LNCS 10263, pp. 42–50, 2017.
DOI: 10.1007/978-3-319-59448-4_5

motion, imperfect breath hold (especially in patients), septal shifts resulting from pressure differences between both ventricles during a breath-hold or timing differences (due to heart-rate related triggering errors or drifts). Retrospective motion correction is thus imperative for correct quantification of T1 in the whole myocardium.

The use of off-the-shelf state-of-the-art data-driven image registration approaches for this application is complicated by the intrinsic complexity of the image data, including contrast inversion, partial volume effects and signal nulling for images acquired near the zero crossing of the T1 relaxation curve [4], such that dedicated approaches are needed. In [5], a local non-rigid registration framework was developed that simultaneously estimates the motion field and the intensity variation to cope with the large variations in contrast. In [6], a groupwise method for quantitative MRI was proposed whereby all images are registered simultaneously to a mean space by minimization of a cost function based on principal component analysis, assuming a non-specified low dimensional signal model. Model-based approaches on the other hand exploit the underlying T1 relaxation model, such that direct registration between images with largely different or inverted contrast can be avoided. In [4], motion-free synthetic images resembling the original contrasts are constructed based on a rough initial T1 estimate to guide the registration. In [7], the error on the exponential curve fitting is used as registration criterion, which is assumed to increases in case of misalignment. A limitation of these model-based algorithms is however that a good initialization for T1 is required, which in practice involves a sufficiently accurate initialization of the registration. Furthermore, relying on the model is complicated by the need for signal polarity restoration when magnitude-reconstructed images are used.

In this paper, we present a model-based registration method for T1 mapping which iteratively minimizes the errors on the T1 curve fit by registering each image to its corresponding model derived from the T1 map. Compared to existing model-based algorithms, we propose a robust data-driven initial registration to avoid large bias in the initial T1 estimate in cases with large motion. Robustness is assured by automatically determining an optimal registration sequence for each image and by exploiting specific knowledge about the acquisition to identify the low-signal images in the sequence that are difficult to register, requiring specific attention. Furthermore, we also integrate the inter-scan registration (native and enhanced) in the framework to obtain motion free ECV maps.

2 Methods

The proposed algorithm combines a data-driven initialization (Sect. 2.1) with a model-based refinement (Sect. 2.2). The data-driven algorithm is required to obtain a robust first estimate for the pixelwise T1 before exploiting the T1-relaxation model. This combined method is applied to both native and enhanced MOLLI scans. A motion free ECV map is additionally obtained by inter-scan registration (Sect. 2.3).

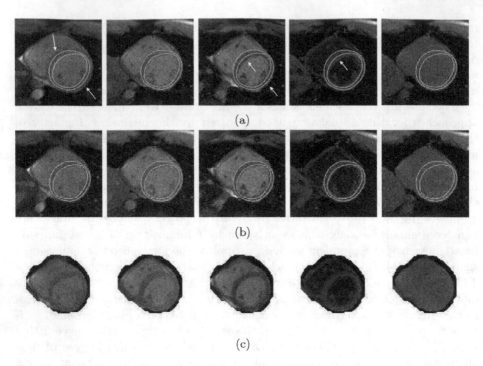

(a)

(b)

(c)

Fig. 1. Five of eight T1-weighted images of a MOLLI scan in order of acquisition before (a) and after (b) model-based registration and the model images I_t^m of the first iteration with heart mask overlaid (c). The endo- and epicardial contours of the second image, which was used as reference image for initialization, are depicted on all other images. The arrows indicate motion.

2.1 Data-Driven Initialization

A rectangular region of interest (ROI) around the heart is manually selected on the first image of each sequence prior to registration. The data-driven algorithm considers all pairwise registrations between all images I_1 to I_n in the sequence simultaneously and applies a global optimization approach to find an optimal reference image and optimal registration sequences in order to minimize registration failure. The optimization is based on a measure $C(a, b)$ for the registration affinity between any two images I_a and I_b which takes into account (1) the consistency $C_c(a, b)$ between their forward $T^{a \to b}$ backward $T^{b \to a}$ rigid registration transformation and (2) the similarity C_s of each image with all other images in the sequence after rigid registration:

$$C(a, b) = (C_c(a, b) + C_c(b, a)) + w \cdot (C_s(a) + C_s(b)), \qquad (1)$$

with $C_c(a, b) = \left\| (T^{b \to a} \cdot T^{a \to b} - \mathbf{1}) \cdot c_a \right\|$, with c_a the center of the image I_a and $C_s(a) = 1/ \sum_i MI(a, i)$ the inverse of the total mutual information of image a with any other image i. The two terms are balanced by the weight w to have a more or less equal contribution to the cost $C(a, b)$. The optimal sequence of pairwise registrations between all images is determined by minimizing the total accumulated cost of each sequence using the Floyd-Warshall algorithm applied to the cost matrix C [8]. The image for which direct registration is selected to be optimal for most of the other images, is selected as reference image. The required pairwise registrations are subsequently refined by non-rigid registration using the mutual information similarity measure, a B-spline parametrization and a rigidity penalty, calculated using $Elastix$ [9]. Each image is then warped to the reference image by concatenating the transformations according to its optimal registration sequence.

Because images with low signal were found to be difficult to register with any other image, they are treated differently to increase registration robustness. These images are automatically detected in the sequence based on their mean signal intensity over the ROI and are initially only aligned rigidly with the reference images, based on the rigid registrations of their adjacent images in acquisition order, assuming the rigid motion component to be continuous.

2.2   Model-Based Registration

Refinement of the registration is achieved by exploiting the exponential model of T1 relaxation. The estimated T1 of a pixel is obtained by fitting the theoretic three parameter curve to the intensities $s(t)$ of the images of the MOLLI scan, acquired at timepoints t after the inversion pulse:

$$s(t) = A - B \cdot e^{\frac{-t}{T_{app}}}, \tag{2}$$

where A and B are dimensionless parameters, T_{app} is the apparent T_1 and the true T_1 is $T_1 = T_{app} * (B \setminus A - 1)$. The optimal parameters are calculated using the Nelder-Mead simplex direct search algorithm [10]. Because magnitude-reconstructed images are used in our experiments, the exponential curve cannot be fitted directly through the image intensities as the polarity of the signals is unknown. Hence, we use a multifitting approach [11] where for each pixel, four curve fittings are performed (none, one, two or three first points inverted). The curve fitting with the lowest squared error is considered to be the correct one.

From the estimated parameters A, B and T_{app} in Eq. 2, ideal model images I_t^m are derived for every timepoint t in the MOLLI sequence with similar contrast as the original images I_t, but with reduced motion artefacts (Fig. 1c). Hence, it is expected that in case of misalignment, the difference between I_t^m and I_t will increase [12]. The model-based motion correction therefore minimizes the sum-of-squared differences $\|I_t - I_t^m\|^2$ between every corresponding original and model image over a region representing the heart. This mask is automatically segmented from the native T1 map obtained after initial data-driven registration (Fig. 1c). After calculating a B-spline motion field for all images, a new T1 map with reduced motion artefacts is obtained. This iterative process of motion correction based on model images and recalculation of the T1 map and the model images is repeated five times.

2.3   Inter-scan Registration

Alignment of the native and enhanced T1 map is achieved by registering the optimal reference images found in the data-driven intra-scan registration. Since these are convenient for pairwise registration with most of the other images, registration between them was found to be sufficiently robust. A non-rigid transformation using a B-spline parametrization and the mutual information similarity criterion with heart mask as explained in Sect. 2.2 are used.

3   Experiments

The registration method was evaluated using the native and enhanced mid-cavity short axis images of 15 young athletes and 5 clinical cases with variable pathology, including cases with clearly apparent motion and cases with no or limited motion. The 2D+time scans were recorded using a 1.5 T MR scanner (Achieva, Philips Healthcare, Best, the Netherlands) and a 6-channel cardiac phased array receiver coil. For the enhanced scans, 0.15 mmol/kg of gadobutrol (Gadovist^{TM}, Bayer Schering) was administered to the subjects. The 5s(3s)3s and the 4s(1s)3s(1s)2s MOLLI scheme [13] were used for the native and enhanced scans respectively. The athlete study was approved by our institutional review board, and informed consent was obtained from all participants.

The endo- and epicardium were manually delineated on every (non-corrected) T1-weighted image by two different, independent observers (athlete data) or one observer (patient data). The segmentations were propagated to the reference image space by applying the calculated transformations to binary images representing the myocardium. For perfect image alignment and assuming no segmentation errors, the overlap of the myocardial segmentations should be optimal. This is assessed using the global Dice similarity coefficient DSC_G [6] which is sensitive to the misregistration of a single image:

$$DSC_G = \frac{n * (S_1 \bigcap S_2 ... \bigcap Sn)}{S_1 + S_2 + ... + Sn}, \tag{3}$$

where S_k is the k_{th} segmentation and n is the number of images in the MOLLI scan(s). Additionally, the mean distance between the endo- and epicardial contours is calculated (Mean Boundary Error, MBE). The results compare the scans before motion correction, after data-driven motion correction (naive, Sect. 2.1, but with no specific treatment for low signal images) and after the proposed method (initialization and refinement). The statistical significance of obtained results is assessed using the two sided Wilcoxon signed rank test with a significance level of 5%.

4   Results

The obtained values for DSC_G and MBE are given in Tables 1 and 2. The results of the athlete data (Table 1) are shown separately for the two observers

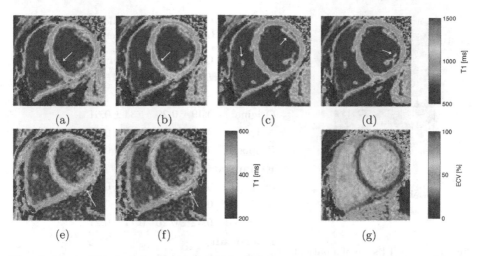

Fig. 2. Native T1 map of an athlete before registration (a), after initialization (b) and after model-based registration in the first (c) and fifth (d) iteration, enhanced T1 map before (e) and after (f) model-based registration and ECV map after registration (g).

to assess the influence of segmentation error on the results. For the athlete data, the model-based approach significantly decreased the MBE and increased the inter-scan DSC_G with respect to the original images. Furthermore, statistical significant improvement of the model-based algorithm compared to the naive data-driven registration was demonstrated for all calculated metrics. For the patient data, a similar trend is observed. Visual inspection of the separate images

Table 1. Mean and standard deviation of DSC_G [%] and MBE [mm] before registration, after naive intensity based registration and after the proposed model-based registration for the subjects of the athlete study for the segmentations of observer 1 and 2.

		Observer 1		Observer 2	
		DSC_G	MBE	DSC_G	MBE
Native	Original	0.53 ± 0.19	1.16 ± 0.48	0.64 ± 0.17	1.20 ± 0.53
	Naive	0.47 ± 0.13	0.97 ± 0.30	0.55 ± 0.11	0.99 ± 0.25
	Proposed	0.57 ± 0.11	0.74 ± 0.14	0.63 ± 0.08	0.94 ± 0.21
Enhanced	Original	0.48 ± 0.27	1.40 ± 1.29	0.56 ± 0.24	1.51 ± 1.38
	Naive	0.25 ± 0.21	1.37 ± 0.68	0.37 ± 0.19	1.27 ± 0.69
	Proposed	0.50 ± 0.20	0.77 ± 0.36	0.57 ± 0.16	0.97 ± 0.41
Inter-scan	Original	0.12 ± 0.10	3.75 ± 1.81	0.25 ± 0.16	3.74 ± 1.96
	Naive	0.19 ± 0.14	1.42 ± 0.44	0.29 ± 0.18	1.56 ± 0.40
	Proposed	0.39 ± 0.10	1.08 ± 0.25	0.47 ± 0.10	1.27 ± 0.21

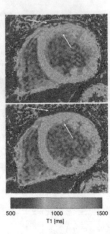

Fig. 3. Native T1 scan of a patient before (top) and after (bottom) registration. Clear improvement is shown at the arrows.

Table 2. Mean and standard deviation of DSC_G [%] and MBE [mm] for 5 patients.

	Observer 1	
	DSC_G	MBE
Native		
Original	0.49 ± 0.26	1.34 ± 0.61
Naive	0.42 ± 0.31	1.42 ± 0.93
Proposed	0.51 ± 0.21	0.88 ± 0.34
Enhanced		
Original	0.52 ± 0.14	1.18 ± 0.24
Naive	0.51 ± 0.17	0.81 ± 0.18
Proposed	0.52 ± 0.20	0.77 ± 0.27
Inter-scan		
Original	0.16 ± 0.12	4.99 ± 3.39
Naive	0.32 ± 0.24	1.92 ± 1.94
Proposed	0.40 ± 0.24	1.54 ± 0.44

showed no unexpected deformation (Fig. 1b) and improvement of the T1 maps is illustrated in Figs. 2 and 3. The mean septal T1 and ECV of the athletes consistently decreased for every subject after registration with on average respectively 1048 ± 38 ms and $36 \pm 8\%$ before registration and 1007 ± 22 ms and $26 \pm 2\%$ after registration. This indicates that the myocardial T1 and ECV values are less contaminated by the higher T1 and ECV of the the blood pool, thus confirming improved alignment of the images. Computation using non-optimized Matlab code on a recent desktop PC required between 50 ± 17 s for the data-driven initialization (7 to 14 images per scan), 61 ± 8 s for one iteration of the model-based registration and 3–4 s for inter-scan registration. Overall computation time to obtain two registered T1 maps and a registered ECV map equaled 13.53 ± 2.06 min.

5   Discussion

The proposed model-based algorithm with exclusion of low-signal images in the initialization, shows to be more robust compared to the more naive data-driven approach. This can be appreciated from Tables 1 and 2 where the global DSC, which is sensitive to the misregistration of a single image, decreases after naive registration whereas the MBE, which averages over all images, also decreases. An example of a low signal image where intensity-based registration will fail because of the relatively large influence of noise compared to other contrasts, is the fourth image of Fig. 1. To avoid misregistration, this image was only given a rigid initialization based on the registration of the adjacent images. It is clear from Fig. 1b that the model-based registration is subsequently able to correct the motion in this image. A second type of low signal images that results in misregistration presents a bright blood pool and signal nulling of the myocardium. Consequently, the myocardium-lung interface disappears and the endocardium of this image is mapped on the epicardium of the reference image in this region.

Our initial experiments showed the importance of good initialization for model-based registration. In case of misregistration or large motion, the motion artefacts of the T1 map are propagated to the model images, which affects the registration. Alternatively, motion artefacts can be avoided if the initial T1 map is calculated from few images suitable for intensity-based registration. This however results in a noisy T1 map again affecting the model images and thus the registration.

The lack of ground truth is a persistent issue in medical image analysis. The numeric validation in this paper was derived from manual segmentations, but large inter-subject variability is clear from the differences between the two observers in Table 1. Phantom data offers only a limited solution because of the large complexity and variability of in-vivo data. Indeed, application of the naive approach on a simple phantom (data not shown) revealed near-pixel accuracy whereas the in-vivo data clearly shows the limitations of this method.

In future work, we will further validate the performance of our method on pathological data to assess clinical applicability. Furthermore, our method was not optimized for computation time. Hence, we expect large possibilities for improvement here.

6 Conclusion

In this paper, a new registration strategy for the intra- and inter-scan registration of MOLLI images was presented, which combines a model-based registration approach with a robust T1 initialization obtained by a registration approach based on the automatic selection of optimal reference image and optimal registration sequences. Validation using 20 in-vivo data sets showed a significant decrease in mean boundary error and an increase in global DSC. The impact and clinical relevance of improved motion correction on regional T1 and ECV quantification for assessment of myocardial pathology needs to be further investigated.

References

1. Schelbert, E.B., Messroghli, D.R.: State of the art: clinical applications of cardiac T1 mapping. Radiology **278**(3), 658–676 (2016)
2. Messroghli, D.R., Radjenovic, A., Kozerke, S., Higgins, D.M., Sivananthan, M.U., Ridgway, J.P.: Modified Look-Locker inversion recovery (MOLLI) for high-resolution T1 mapping of the heart. Magn. Reson. Med. **52**, 141–146 (2004)
3. Kellman, P., Arai, A.E., Xue, H.: T1 and extracellular volume mapping in the heart: estimation of error maps and the influence of noise on precision. J. Cardiovasc. Magn. Reson. **15**(1), 56 (2013)

4. Xue, H., Shah, S., Greiser, A., Guetter, C., Littmann, A., Jolly, P.-P., Arai, A.E., Zuehlsdorff, S., Guehring, J., Kellman, P.: Motion correction for myocardial T1 mapping using image registration with synthetic image estimation. Magn. Reson. Med. **57**, 1644–1655 (2012)
5. Roujol, S., Foppa, M., Weingartner, S., Manning, W.J., Nezefat, R.: Adaptive registration of varying contrast-weighted images for improved tissue characterization (ARCTIC): Application to T1 mapping. Magn. Reson. Med. **73**, 1469–1482 (2015)
6. Huizinga, W., Poot, D., Guyader, J.-M., Klaassen, R., van Coolen, B., Kranenburg, M., van Geuns, R., Uitterdijk, A., Polfliet, M., Vandemeulebroucke, J., Leemans, A., Niessen, W., Klein, S.: PCA-based groupwise image registration for quantitative MRI. Med. Image Anal. **29**, 65–78 (2016)
7. Van De Giessen, M., Tao, Q., Van Der Geest, R.J., Lelieveldt, B.P.F.: Model-based alignment of look-locker MRI sequences for calibrated myocardial scar tissue quantification. In: 10th IEEE International Symposium on Biomedical Imaging From Nano to Macro, pp. 1038–1041 (2013)
8. Floyd, R.W.: Algorithm 97: shortest path. Commun. ACM **5**(6), 345 (1962)
9. Klein, S., Staring, M., Murphy, K., Viergever, M., Pluim, J.: A toolbox for intensity-based medical image registration. IEEE Trans. Med. Imaging **29**(1), 196–205 (2010)
10. Lagarias, J.C., Reeds, J.A., Wright, M.H., Wright, P.E.: Convergence properties of the Nelder-Mead simplex method in low dimensions. SIAM J. Optim. **9**(1), 112–147 (1998)
11. Nekolla, S., Gneiting, T., Syha, J., Deichmann, R., Haase, A.: T1 Maps by K-space reduced snapshot-FLASH MRI. J. Comput. Assist. Tomogr. **16**, 327–332 (1992)
12. Mewton, N., Liu, C.Y., Croisille, P., Bluemke, D., Lima, J.A.C.: Assessment of myocardial fibrosis with cardiovascular magnetic resonance. J. Am. Coll. Cardiol. **57**(8), 891–903 (2001)
13. Kellman, P., Hansen, M.S.: T1-mapping in the heart: accuracy and precision. J. Cardiovasc. Magn. Reson. **16**, 2 (2014)

Improving Understanding of Long-Term Cardiac Functional Remodelling via Cross-Sectional Analysis of Polyaffine Motion Parameters

Kristin McLeod[1,2(✉)], Maxime Sermesant[3], and Xavier Pennec[3]

[1] Simula Research Laboratory, Centre for Cardiological Innovation, Oslo, Norway
kristin@simula.no
[2] KardioMe s.r.o, Bratislava, Slovakia
[3] Université Côte d'Azur, Inria, Nice, France

Abstract. Changes in cardiac motion dynamics occur as a direct result of alterations in structure, hemodynamics, and electrical activation. Abnormal ventricular motion compromises long-term sustainability of heart function. While motion abnormalities are reasonably well documented and have been identified for many conditions, the remodelling process that occurs as a condition progresses is not well understood. Thanks to the recent development of a method to quantify full ventricular motion (as opposed to 1D abstractions of the motion) with few comparable parameters, population-based statistical analysis is possible. A method for describing functional remodelling is proposed by performing statistical cross-sectional analysis of spatio-temporally aligned subject-specific polyaffine motion parameters. The proposed method is applied to pathological and control datasets to compare functional remodelling occurring as a process of disease as opposed to a process of ageing.

1 Introduction

Cardiac function can be altered under disease states, reducing the long-term ability of the heart to pump efficiently. While there has been effort dedicated towards understanding typical functional abnormalities occurring for different diseases, the functional remodelling processes that occur over time are poorly understood. A good understanding of functional remodelling is key in optimising treatment to, for example, predict long-term functional remodelling for decision making of cardiac implants, or to determine the best timing to intervene for valve replacement surgery.

1.1 State-of-the-art in Cardiac Motion Tracking

Cardiac motion tracking via non-rigid registration has been used in recent years to quantify 3D motion of the heart. Quantitative measures of motion can be used to detect abnormal regions and to provide more detailed measures beyond

© Springer International Publishing AG 2017
M. Pop and G.A. Wright (Eds.): FIMH 2017, LNCS 10263, pp. 51–59, 2017.
DOI: 10.1007/978-3-319-59448-4_6

simple abstractions of function such as volumes or ejection fraction. An overview of recent methods is given in [1]. Population-based analysis of motion has been used to quantify changes over time in cardiac function by performing group-wise analysis of strain derived from tagged magnetic resonance imaging [2] or by studying spatio-temporally aligned displacements [3]. While informative, these methods only give a snapshot into the function of the heart and do not provide insight into how the function remodels over time.

1.2   State-of-the-art in Cardiac Remodelling

Functional changes have been studied by, for example, comparing strain values before and shortly after myocardial infarction [4], and pre- and post-surgery for aortic stenosis patients [5]. However, few studies have investigated long-term functional changes, even for 1D measures of strain [6]. Analysis of changes in the heart over time is an ongoing area of research, which has traditionally been addressed using longitudinal patient data. In light of the fact that longitudinal studies require long-term follow-up of the same patients, potentially over years, the prospect of using the results of such studies to guide therapy planning remains low. In contrast, cross-sectional analysis of patients within the same population has the ability to describe long-term changes without the need to wait years for a patient to age. Long-term cardiac structural remodelling was analysed by Mansi et al. [7] to study the long-term shape remodelling that occurs in the right ventricle of Tetralogy of Fallot patients in response to increased load caused by regurgitated blood from the lungs. This study was performed only on the end diastolic phase of the cardiac cycle, and is therefore not able to capture functional remodelling that (most likely) occurs over time in these patients.

1.3   Aim and Paper Organisation

In the present work we propose to analyse long-term remodelling of cardiac function using a method that is able to represent the full 3D dynamics of the heart. The proposed method makes use of a recent method to track cardiac motion by describing motion regionally as affine transformations, and fusing these to a global smooth transformation [8]. In contrast to previous work, where the motion was analysed only at a single time point, the present work builds upon recent effort in cardiac motion tracking and population-based analysis to analyse changes in motion over time. Here, we are interested in studying motion over two time scales: the 3D dynamics during the heart beat (motion parameters), and the remodelling over years (changes in these motion parameters). The contributions of the present work are:

1. Derivation of a long-term functional model using image registration, polyaffine projection, and cross-sectional statistical analysis
2. Quantitative analysis of the accuracy of predicting age given motion parameters using leave-one-out numerical validation
3. Qualitative analysis of the accuracy of predicting motion given age by analysing motion sequences and image-derived parameters

The remainder of the paper is organised as follows: Sect. 2 summarises the techniques used to track the motion of the left ventricle and project this onto a polyaffine space, in which the polyaffine motion parameters can be analysed. Sect. 3 describes the proposed method to perform cross-sectional analysis of the polyaffine motion parameters. Sect. 4 describes experiments performed to quantitatively and qualitatively evaluate the proposed method, and in Sect. 5 the methods are evaluated, discussed, and future perspectives are given.

2   Background: Polyaffine Motion Tracking

A global (affine) description of motion is not sufficient to characterise the motion of the heart. On the other hand, voxel-wise descriptors most likely over-represent the motion when considering that the heart consists of connected tissue which moves in an elastic manner. Describing the motion at a regional level can provide a suitable compromise between these two, to sufficiently describe the motion while representing the motion with as few parameters as possible. Polyaffine projection allows motion to be described at a regional level while maintaining a globally smooth transformation by means of weight functions that smooth the transformation between regions. Moreover, for the purpose of estimating the remodelling over time, it is important to have a robust yet informative description of the motion. We believe that the Polyaffine transformations provide a good trade-off.

Polyaffine projection from a dense deformation field was proposed by Seiler et al. [9] via log-affine matrices M_i described in each region i that are fused to a global deformation field by $v(x) = \sum_i^N \omega_i(x) M_i x$. $\omega(x)$ are weight parameters in the N regions, describing the weight of the transformation of each region of a given voxel x written in homogeneous coordinates. As described in [9], the polyaffine projection can be formulated as a linear least squares problem by solving $\nabla C_M = 0$, where C is described by the following formula:

$$ C(M) = \int_\Pi \| \sum_i \omega_i(x) \cdot M_i \tilde{x} - \boldsymbol{v}(x) \|^2 \, dx. $$

The polyaffine projection technique used in this work was implemented following the method described in [8], which was developed specifically for motion tracking of the left ventricle of the heart.

3   Methods: Regression of Polyaffine Motion Parameters

The compact parameterisation of motion parameters using the polyaffine formulation described in the previous section enables cross-sectional analysis of motion. Common methods for cross-sectional analysis include principal component analysis (PCA) and partial least squares (PLS). PLS has the advantage over PCA of computing directly the components most related to the output variable. In both cases, the regression model is performed by computing the motion

parameters from age since the converse requires prediction of many output parameters: motion described by 5916 parameters per subject (12 affine parameters
× 17 regions × 29 time frames), given few input parameters: age described by a
single value (in years). Following the method for computing shape features from
age proposed by Mansi et al. [7], canonical correlation analysis (CCA) can be
used to reverse the relationship in order to obtain a model of motion given age.
The proposed pipeline is summarised in Fig. 1.

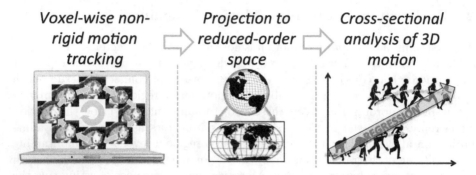

Fig. 1. Proposed pipeline from image to motion model. Non-rigid motion tracking is
performed to describe the voxel-wise motion in the heart. This is then projected to
a polyaffine space in order to represent the motion with few parameters, consistently
defined from one subject to another. Finally, cross-sectional analysis is performed on
the motion parameters to formulate a generative model of functional remodelling.

4   Experiments: Polyaffine Motion Parameter Regression

Two types of experiments were performed to validate the proposed method. The
first experiment was a quantitative validation of the accuracy of predicting age
given the motion parameters. While this is not a clinically meaningful validation
(since the age of patients is generally known), it serves as a means to validate the
regression since it is straightforward to compute the accuracy when predicting a
single output value. The second experiment was a qualitative assessment of the
prediction of motion given age, which is difficult to validate quantitatively given
the large number of predicted parameters, but can be analysed qualitatively by
visualising the output to determine if it matches physiological motion dynamics.

4.1   Testing Datasets

The motion parameters used in the experiments were taken directly from the
previous work of [8], which consists of two datasets; a healthy control dataset,
and a pathological group of Tetralogy of Fallot patients. The healthy control

subjects from the openly available STACOM 2011 cardiac motion tracking challenge dataset [1] ($n = 15$, 3 female, mean age $\pm$ SD $= 28 \pm 5$) were used since these have already been used to validate the polyaffine cardiac motion tracking. The pathological subjects were a dataset of Tetralogy of Fallot patients ($n = 10$, 5 female, mean age $\pm$ SD $= 21 \pm 7$), which were already used to analyse the polyaffine motion parameters in [10]. The Tetralogy of Fallot cohort forms a suitable testing set for this problem since there are known functional changes over time in these patients in response to poor pulmonary valve function or in the absence of a pulmonary valve. Furthermore, modelling functional remodelling in this population complements the previous analysis of shape remodelling performed by Mansi et al. [7]. In both cohorts, only left ventricular motion was tracked due to the absence of open testing sets to validate bi-ventricle motion tracking methods. The motion tracking results from these experiments were consistent with those presented in [8].

4.2   Statistical Analysis

Following the techniques described in [7] to predict shape remodelling, partial least squares (PLS) was applied to each vectorised motion array to reduce the parameter space to a vector of PLS component loadings, from which CCA between the loading vector (representing motion) and age was computed. CCA was performed iteratively on the number of components, to find the minimal number of components required to obtain a statistically significant combination of PLS components with respect to the dependent variable. Statistical significance was computed using the Bartlett-Lawley hypothesis test. Since we were interested in modelling motion (and not age), the important variable to explain is Y (vector of Polyaffine motion parameters from M), rather than X (age). PLS captured more than 90% of the variability in Y with only 2 components for the ToF group, and 7 components for the control group.

The canonical correlations are plotted for each component from PLS of each group in Fig. 2. The canonical correlations for PLS are ordered in magnitude by design since PLS is maximising the variance in Y and covariance in X and Y.

Quantitative Validation of the Prediction of Age Given Motion. The testing error of predicting age given motion was computed by performing leave-one-out validation (training on $n - 1$ subjects, and predicting the age of the nth subject. In all error computations, the number of CCA components for the full training set were used. The training error of the Tetralogy of Fallot group was 0.30 years and the testing error was 5.85 years. For the control group, the respective training and testing errors were 0.47 years and 6.81 years. For the ToF group, the yielded errors were below the population standard deviation of age (6.49). For the control group, however, the testing errors were above the population standard deviation of 5.0 years.

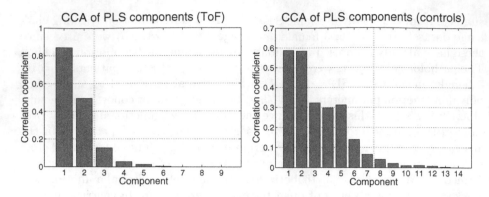

Fig. 2. Canonical correlations for PLS for the Tetralogy of Fallot group (left) and the control group (right). The red line indicates the minimum number of components needed for each group to obtain statistically significant combinations of the modes. Fewer components were required in the more heterogeneous pathological group (i.e. the ToF group) than relatively homogeneous control group. (Color figure online)

Qualitative Analysis of Prediction of Motion Given Age. Quantitative validation of the motion parameters computed from age would require longitudinal data, which is a challenge to obtain given the long time intervals required between time points. Furthermore, quantitative validation voxel-wise is challenging due to the fact that the heart remodels structurally over time in addition to the functional remodelling. Since there is currently no method to identify how/where voxels move over time as a result of structural remodelling compared to those that move as a result of functional remodelling, quantitative validation of the motion estimation is not possible. Therefore, only qualitative analysis of motion via visualisation of the motion sequences was performed. Figure 3 shows the predictive (healthy or pathological) deformation models applied to a single magnetic resonance image using the motion visualisation method described in [8], shown here at two phases of the cardiac cycle to compare the remodelling of motion in the Tetralogy of Fallot group to the control group. As expected, the control group shows little changes in motion through ageing, since the population is healthy the motion should remain stable over time. In contrast, the Tetralogy of Fallot group show large differences in motion over time, particularly in the free wall.

As a preliminary means of quantitative validation, the predicted motion sequences for the Tetralogy of Fallot group were segmented to obtain the left ventricular volumes, plotted in Fig. 4. As the curves show, the minimum of the curves increase with age, which ultimately decreases ejection fraction; a commonly used measure of function by cardiologists. Ejection fraction is known to decrease with age after initial repair in these patients [11].

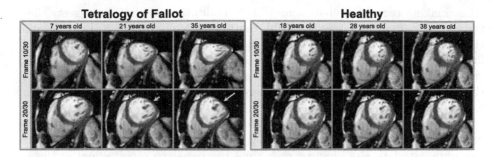

Fig. 3. Snapshots of the sequences shown at two phases of the cardiac cycle: frame 10 (top row) and frame 20 (bottom row) of a total of 30 frames for the Tetralogy of Fallot group (left) and healthy control group (right). Three ages are shown for each group at −2 standard deviations of the age, the mean age, and +2 standard deviations of age. Note that the motion in the control group remains stable over the ages, since very little motion changes occur over normal ageing. In contrast, noticeable changes in the motion of the Tetralogy of Fallot group are visible, as highlighted by the yellow arrows. (Color figure online)

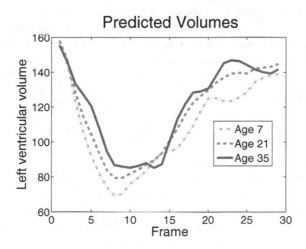

Fig. 4. Segmented volumes computed from sequences predicted from the Tetralogy of Fallot group model. Ejection fraction (used as a measure of cardiac function) decreased over time, consistent with clinical findings in this patient group.

5   Discussion and Perspectives

The experiments suggest that the proposed methods are promising for computing the long-term functional changes in the left ventricle. The computation of age from motion parameters was more accurate in the Tetralogy of Fallot group than the healthy control group, which is as expected since the control group consists of a range of ages in which the motion parameters should be reasonably stable (thus making regression challenging). The qualitative validation suggests that the methods are able to predict feasible motion dynamics from age alone, despite the small sizes of the datasets. These findings should be, however, confirmed in a larger population in which longitudinal data is available to quantitatively validate the prediction of motion from age. Additionally, with larger datasets the models could be sub-grouped to ensure that the change due to the evolution is much larger than the cross-sectional variability.

An important limitation of the present work is that no longitudinal data was used for the regression. As mentioned by Eng et al. [12], cross-sectional analysis can underestimate size changes over time compared to longitudinal analysis. Since longitudinal data was not available at the time of this study they were not used, therefore future studies will focus on a combined longitudinal and cross-sectional study design.

This work presents only functional changes without consideration of structure, despite the fact that these are inherently coupled and could be more informative when used in combination. A very interesting future work will address coupling shape (for example using the methods in [7]) with function.

In the present work, regression was performed on matricised tensors. However, the regression could be performed directly on the tensors. Performing regression directly on the tensor may be an advantage for retaining the structure of the tensor in the regression, and to reduce the number of parameters required to describe the generative motion models.

6   Conclusions

In this work we propose a method to analyse motion changes in a population over time using a non-rigid registration algorithm in conjunction with polyaffine projection and statistical analysis. The proposed method was validated quantitatively and qualitatively on a control dataset and on a dataset of Tetralogy of Fallot patients and complements the work of [7] to allow future potential work on joint shape and motion evolution modelling. This is, to our knowledge, the first method that is able to predict long-term functional remodelling at a level that retains the full complexity of motion dynamics.

Acknowledgements. This project was carried out as a part of the Centre for Cardiological Innovation (CCI), Norway, funded by the Research Council of Norway. The authors would like to thank Tommaso Mansi for providing the code for computing CCA.

References

1. Tobon-Gomez, C., De-Craene, M., et al.: Benchmarking framework for myocardial tracking and deformation algorithms: an open access database. Med. Im. Anal. **17**(6), 632–648 (2013)
2. Qian, Z., Liu, Q., Metaxas, D.N., Axel, L.: Identifying regional cardiac abnormalities from myocardial strains using nontracking-based strain estimation and spatio-temporal tensor analysis. IEEE Trans. Med. Im. **30**(12), 2017–2029 (2011)

3. Rougon, N.F., Petitjean, C., Preteux, F.J.: Building and using a statistical 3D motion atlas for analyzing myocardial contraction in MRI. In: Proceedings SPIE 5370, Medical Imaging 2004: Image Processing, pp. 253–264 (2004)
4. Gerber, B.L., Rochitte, C.E., Melin, J.A., McVeigh, E.R., Bluemke, D.A., Wu, K.C., Becker, L.C., Lima, J.A.: Microvascular obstruction and left ventricular remodeling early after acute myocardial infarction. Circulation 101(23), 2734–2741 (2000)
5. Bauer, F., Eltchaninoff, H., Tron, C., Lesault, P.F., Agatiello, C., Nercolini, D., Derumeaux, G., Cribier, A.: Acute improvement in global and regional left ventricular systolic function after percutaneous heart valve implantation in patients with symptomatic aortic stenosis. Circulation 110(11), 1473–1476 (2004)
6. Kuznetsova, T., Thijs, L., Knez, J., Cauwenberghs, N., Petit, T., Gu, Y.M., Zhang, Z., Staessen, J.A.: Longitudinal changes in left ventricular diastolic function in a general population. Circ. Cardiovasc. Imaging 8(4) (2015)
7. Mansi, T., Voigt, I., Leonardi, B., Pennec, X., Durrleman, S., Sermesant, M., Delingette, H., Taylor, A.M., Boudjemline, Y., Pongiglione, G., Ayache, N.: A statistical model for quantification and prediction of cardiac remodelling: Application to tetralogy of fallot. IEEE Trans. Med. Im. 9(30), 1605–1616 (2011)
8. McLeod, K., Sermesant, M., Beerbaum, P., Pennec, X.: Spatio-temporal tensor decomposition of a polyaffine motion model for a better analysis of pathological left ventricular dynamics. IEEE Trans. Med. Im. (2015)
9. Seiler, C., Pennec, X., Reyes, M.: Capturing the multiscale anatomical shape variability with polyaffine transformation trees. Med. Image Anal. 16(7), 1371–1384 (2012)
10. McLeod, K., Sermesant, M., Beerbaum, P., Pennec, X.: Descriptive and intuitive population-based cardiac motion analysis via sparsity constrained tensor decomposition. In: Navab, N., Hornegger, J., Wells, W.M., Frangi, A.F. (eds.) MICCAI 2015. LNCS, vol. 9351, pp. 419–426. Springer, Cham (2015). doi:10.1007/978-3-319-24574-4_50
11. Waien, S.A., Liu, P.P., Ross, B.L., Williams, W.G., Webb, G.D., Mclaughlin, P.R.: Serial follow-up of adults with repaired tetralogy of fallot. J. Am. Coll. Cardiol. 20(2), 295–300 (1992)
12. Eng, J., McClelland, R.L., Gomes, A.S., Hundley, W.G., Cheng, S., Wu, C.O., Carr, J.J., Shea, S., Bluemke, D.A., Lima, J.A.: Adverse left ventricular remodeling and age assessed with cardiac mr imaging: the multi-ethnic study of atherosclerosis. Radiology 278(3), 714–722 (2015)

Advanced Cardiac Image Analysis Tools for Diagnostic and Interventions

Multi-cycle Reconstruction of Cardiac MRI for the Analysis of Inter-ventricular Septum Motion During Free Breathing

Teodora Chitiboi[1(✉)], Rebecca Ramb[1], Li Feng[1], Eve Piekarski[2], Lennart Tautz[3], Anja Hennemuth[3], and Leon Axel[1]

[1] NYU School of Medicine, Center for Biomedical Imaging, New York, USA
Teodora.Chitiboi@nyumc.org
[2] Nuclear Medicine Ward, Pitié Salpétrière Hospital, Paris, France
[3] Fraunhofer MEVIS, Bremen, Germany

Abstract. Small variations in left-ventricular preload due to respiration produce measurable changes in cardiac function in normal subjects. We show that this mechanism is altered in patients with reduced ejection fraction (EF), hypertrophy, or volume-loaded right ventricle (RV). We propose a multi-dimensional retrospective image reconstruction, based on an adaptive, soft classification of data into respiratory and cardiac phases, to study these effects.

Keywords: Cardiac function · Preload · Respiration

1 Introduction

Free-breathing cine MRI provides novel information on cardiac dynamics, extending routine cardiac function analysis to patients unable to hold their breath or suffering from arrhythmia. The recently developed XD-GRASP [1] (eXtra-Dimension-Golden-Angle-Radial-Sparse-Parallel) reconstruction technique enables the independent analysis of cardiac and respiratory motion within the same measurement, which can bring new insights into the normal and abnormal variability of cardiac function.

Breathing induces changes in the intrathoracic pressure, with associated variations in venous return. This causes small modulations in the interventricular septum position, including changes of the left ventricle (LV) diameter, reflecting normal preload variations in the corresponding cardiac cycles.

In this work, we investigate the relation between cardiac function and normal LV preload variations induced by respiration, in normal subjects and patients. We applied the XD-GRASP [1] technique to reconstruct the acquired data in a multi-dimensional image space composed of the spatial dimensions (X, Y) and two physiological dimensions (the cardiac and respiratory cycle phases – Fig. 1). Using this data, we tracked the septum and measured its displacement over the two temporal dimensions to compute surrogate measures for preload and associated cardiac function response. This was applied to a small series of normal subjects and patients. We found significantly larger changes in diameter during the respiratory cycle for normal subjects compared to

© Springer International Publishing AG 2017
M. Pop and G.A. Wright (Eds.): FIMH 2017, LNCS 10263, pp. 63–72, 2017.
DOI: 10.1007/978-3-319-59448-4_7

patients. We also found a significant correlation between preload and cardiac function for normal subjects, which was reduced for patients.

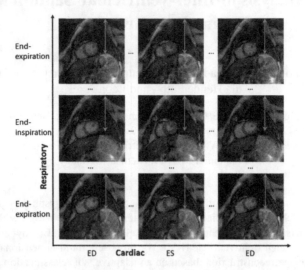

Fig. 1. Multi-dimensional cardiac MRI with two physiological dimensions: cardiac (horizontal) and respiratory (vertical – arrows show displacement during inspiration and expiration).

In the following section, we describe the image reconstruction approach that enabled us to evenly sample the respiratory physiological dimension, with a focus on data sorting and classification. In the remainder of the paper, we present the septum displacement analysis and its results.

2 Related Work

To reduce motion artifacts due to respiration, routine clinical MRI is typically performed during breath-hold [2]. However, not only is breath-holding sometimes too physically demanding for patients, but potentially important information about the influence of the respiratory cycle is lost. The presence of respiratory-synchronous fluctuations in stroke volume has been documented [3] and a quantitative assessment of the respiratory-related motion of the heart followed [4]. For the latter, subjects were imaged at different subjective respiratory positions, during several distinct measurements. A recent study found significant effects of free-breathing, compared to breath-held acquisition, on ventricular mechanics in patients with Duchenne muscular dystrophy and normal subjects [5]. However, artifacts due to respiratory motion still affected image quality. Common methods to reduce artifacts during free breathing include using a navigator [6] or respiratory bellows [7]. Nevertheless, data are conventionally still acquired for only a single respiratory state.

Recent rapid continuous non-Cartesian MRI techniques allow free-respiratory cine imaging [8]. However, they are still affected by motion artifacts, and respiratory states need to be determined separately in postprocessing. The recent XD-GRASP technique [1] enables the independent analysis of cardiac and respiratory physiological dimensions. This is achieved by populating a multi-dimensional image space by reconstruction from the continuously acquired radial spokes (incremented with the golden-angle scheme), based on separately determined cardiac and respiratory motion phases. This allows us to quantify the effect of respiration over different cardiac phases. The presented approach extends initial work on the analysis of septal motion [9] using adaptive data sorting, with automatic myocardium segmentation, and is applied to routine clinical data.

3   Methods

We considered 35 subjects: 11 normal (N), 9 with reduced EF (rEF), without left bundle branch block or infarction, 7 with volume-loaded RV (vlRV), 5 with asymmetric septal hypertrophy (aHyp), and 3 with diffuse hypertrophy (dHyp). We acquired single-slice images with the XD-GRASP technique, using a 2D radial golden-angle increment bSSFP readout with TR/TE = 2.8/1.4 ms, spatial resolution = 2×2 mm^2 and 8 mm slice thickness on a 1.5T MRI scanner (Avanto, Siemens Healthcare, Erlangen, Germany). Images were acquired in a short-axis mid-ventricle position under free-breathing for an acquisition time of 23 s. Typically, between 6–9 respiratory cycles are sampled over around 20 consecutive hearts beats.

3.1   Multi-dimensional Reconstruction

Respiratory and cardiac motion-related time-phase curves were separately extracted from the image data during post-processing. The acquired data were resorted into 9 cardiac $\times$ 9 respiratory states. The 4D dataset shown in Fig. 1 was reconstructed using a compressed sensing approach which employs sparsity priors along the cardiac and respiratory dimensions in an iterative reconstruction scheme [1].

Detecting Cardiac and Respiratory Phases
The continuous acquisition scheme allows for reconstruction of various temporal resolutions retrospectively. First, a low temporal resolution reconstruction R0 is derived using the KWIC algorithm [10] in a sliding window approach. Although this results in an overly smooth image series, it mitigates the majority of undersampling artifacts. The frequency interval H = 0.9–1.7 Hz was used to represent the range of regular heartbeats; a lower interval R = 0.2–0.4 Hz was used for the respiratory cycles.

To extract cardiac motion, a mask was automatically constructed around the heart region, detected by first selecting the frequency responses in the heartbeat interval H, then summing the forward difference, and finally performing automatic thresholding [11]. The averaged temporal intensity curve over the cardiac mask represents the cardiac signal (Fig. 2-left). After some temporal filtering, the local maxima represent ED and local minima ES frames. Cardiac cycles are automatically segmented and corresponding k-space data are labeled based on their relative position within the cycles.

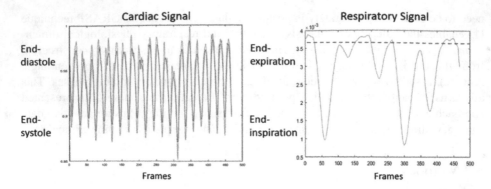

Fig. 2. Left – cardiac signal before and after smoothing. End-systole time-points marked (*). Right – respiratory signal. End-expiration time-points have less variation than end-inspiration.

To extract respiratory motion, R0 is further averaged over the temporal domain to reduce noise. The 2D image space is partitioned into partly overlapping rectangles of 15×15 pixels, over which average temporal frequency variations are computed. The sub-image block with the highest frequency response in the respiratory interval R, over the maximum frequency response in the cardiac interval H, represents the respiratory signal (Fig. 2-right).

The inspiration and expiration motion effects observed over the heart and diaphragm regions are different [12]. Therefore, it is desirable to distinguish between inspiration and expiration intervals, by mirroring the latter along the Y axis (Fig. 3-left). In practice, however, the sub-image block that produces the respiratory signal is usually located close to the diaphragm, where the presence of bright vessels or fat often causes large intensity variations, especially in the presence of through-plane motion. This makes it challenging to automatically distinguish whether local maxima represent end-expiration (Eexp) or end-inspiration (Einsp) phases by solely analyzing the temporal histogram. However, considering physiological mechanisms, we observed that, while inspiration

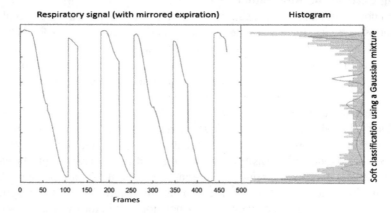

Fig. 3. Left – respiratory signal with mirrored expiration phases. Right – histogram of respiratory signal with overlaid GMM soft classification. Observe the different class sizes.

can be varying in depth, Eexp phases are more similar (Fig. 2). Therefore, it can be derived that the set of local extrema (either minima or maxima) with the smallest standard deviation likely represents Eexp phases.

Adaptive Data Classification in Multidimensional Space

Based on the cardiac and respiratory motion information, the data was classified into 9×9 sets of cardio-respiratory phases. A relatively low number of bins was chosen, due to the short acquisition time, to account for the high undersampling, especially of the respiratory dimension. Since the analysis was focused on differences between systole and diastole, the low number of bins was sufficient.

To normalize cardiac cycle duration, we considered a fixed relative systolic time per patient, computed as the average duration between the occurrence of a local maximum of the cardiac signal and the subsequent minimum, and a variable-length diastole. Isolated arrhythmic cycles can be automatically detected based on their atypical length, and excluded from the reconstruction. As the acquired data is evenly distributed over the cardiac dimension, we performed a hard classification of the data into 9 equal bins corresponding to cardiac phases.

The respiratory time curves (with mirrored expiration phases) show that the data are not normally distributed, since inspiration and expiration often differ in duration and dynamics (Fig. 3). Particularly during irregular respiration, only few respiratory phases may be well-represented. Hard classification into equal bins therefore results in data

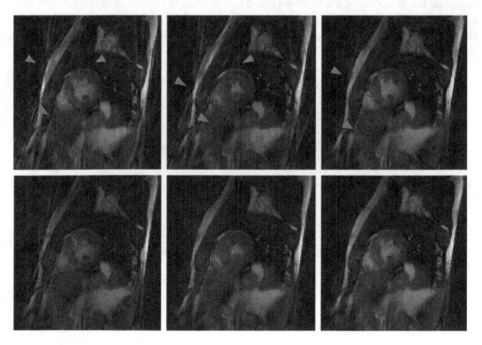

Fig. 4. Comparison between reconstructions using a hard classification into equal size respiratory bins (top), with associated artifacts (arrowheads), and using soft binning with variable bin size after a GMM classification (bottom) for different cardiac and respiratory phases.

subsets with heterogeneous motion states and, hence, may cause artifacts in the reconstruction (Fig. 4-top). An alternative is to seek an optimal classification into a fixed number of classes, based on standard clustering approaches, e.g., k-means. However, with such approaches, classes often contain too few k-space samples to allow a reconstruction with acceptable image quality, even using compressed sensing. In our approach, we propose using a Gaussian mixture model (GMM) [13] which produces a probabilistic (soft) classification along the respiratory dimension (Fig. 3-right). Data in similar respiratory phases are grouped together, but class sizes can be very different. The resulting probabilities are then used to weight the contribution of the data to their assigned class, and are used as weights in a soft-gated reconstruction scheme, as presented in [14]. Similarly, more information can be added to classes with few members in a strictly binary classification sense, by sharing the data from other classes in order of decreasing probability until a minimum threshold is met.

3.2 Septum Motion Analysis

We analyzed the variation of cardiac function with respiration by measuring the LV diameter perpendicular to the septum midpoint $d_{c,r}$ over the cardiac (c) and respiratory (r) phases. To ensure a consistent diameter computation, the myocardium was automatically segmented using an object-based approach [15], and manually corrected, as needed. The RV and the interventricular sulci were also automatically detected, to compute the midpoint of the septum (Fig. 5). The relative change in diameter between end-systole (d_{ES}) and end-diastole (d_{ED}) was computed for each respiratory phase, as a measure of the functional response (a 1D measure of cardiac function) $\nabla d_r = (d_{ED,r} - d_{ES,r})/d_{ED,r}$.

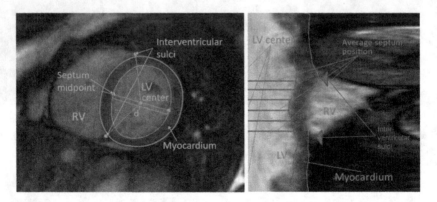

Fig. 5. Segmented myocardium in a short-axis slice. The interventricular sulci are automatically detected in a polar coordinates representation and the septum midpoint and corresponding diameter (d) are computed.

4   Results and Discussion

The LV diameters d_{ED} and d_{ES} (Fig. 6-top), and the relative changes in diameter ∇dr (Fig. 6-bottom) were plotted over the respiratory cycle, for N, rEF, vlRV, aHyp and dHyp. First, we observe that the relative position of the septum is modulated during respiration by the normally varying preload, due to changes in intrathoracic pressure. The LV diameter at end-diastole d_{ED} varies most during the respiratory cycle, while the end-systolic diameter d_{ES} remains relatively constant. d_{ED} varies significantly with respiration for N ($p < 0.001$), providing a measure of LV preload (the blood volume in the LV at the beginning of the next beat). However, for an already dilated LV, as in the case of patients with reduced EF, or a thick and stiffer myocardium, different respiratory phases are associated with smaller changes in diameter, suggesting less variation in LV preload.

A second observation is that the relative change in LV diameter ∇d_r decreases during inspiration and increases during expiration for N. We found a significant correlation between ∇d_r (a measure of cardiac function) and respiratory phase for N ($r = 0.45$, $p < 0.001$), but a lower correlation for rEF ($r = 0.35$, $p = 0.001$) and for vlRV ($r = 0.35$, $p = 0.002$), and the two were not significantly correlated for hypertrophy patients.

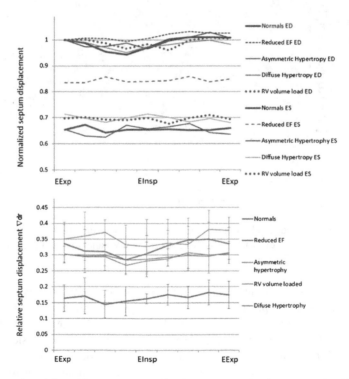

Fig. 6. $d_{ES,r}$ (red) and $d_{ED,r}$ (blue), normalized by $d_{ED,r}$ (Top), and ∇d_r (Bottom) for the respiratory cycle from end-expiration (EExp) to end-inspiration (EInsp) and back to EExp For the normal, reduced EF, asymmetric hypertrophy, diffuse hypertrophy, and volume loaded RV subgroups. Note the greater variation in the normal subjects. (Color figure online)

Assuming that though-plane motion is not significant at the mid-ventricular level, ∇d_r over the respiratory cycle could be interpreted as a measure of the LV functional response to the different respiratory states. The relative difference in ∇d_r between Eexp and Einsp phases, measuring the cardiac function response to the different respiratory states, was significantly larger for N compared to the set of all patients (5.8 ± 3.6% vs. 2.5 ± 3.4%, p = 0.035).

Finally, we can plot the functional response ∇d_r as a function of preload variation d_{ED}. In Fig. 7, d_{ED} is normalized to the average d_{ED} over the respiratory cycle. The slope of the relationship between ∇d_r (i.e. between a 1D EF measure) and the corresponding preload is greater for N, reduced for rEF and vlRV, and almost flat for the hypertrophic subgroups. This slope could be considered an equivalent to the Frank-Starling relationship, which is found to be altered for patients with reduced contractility, with a hypertrophic myocardium, or with volume-loaded RV.

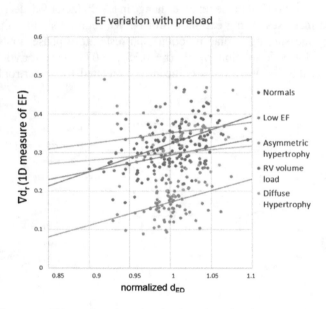

Fig. 7. 1D EF measure ∇d_r as a function of preload, measured as d_{ED} (normalized to the average d_{ED} during the respiratory cycle), for N, rEF, aHyp, vlRV, and dHyp subject groups. The corresponding 1D regressions are shown.

Our hypothesis is that some patients with preserved EF may show reduced correlation between ∇d_r and d_{ED}, presumably due to hypertrophy, while for others this correlation is partially preserved despite a lower EF, possibly indicating an earlier stage of disease. This will be further investigated in a larger group of patients. Besides the small number of subjects, another limitation of our work is the restriction of image acquisition and analysis to a single mid-ventricular slice, as small variations in the slice position may affect the measurements. In the future, we plan to perform a 3D analysis of the septum relative displacement for a full stack of slices. The short acquisition time may also be a limitation when not enough respiratory cycles are recorded.

Limitations

Besides the relatively small number of subjects, an important limitation of this study is the restriction to a single mid-ventricular slice. In this work, we assumed that though-plane motion has a minimal impact on functional measurements, based on the constant wall thickness. However, it is possible that for other patients with local septal hypertrophy, though-plane motion may have a significant effect on the results, as a local bulge in the septum may move in and out of the plane. Furthermore, the relatively short acquisition time may not offer a sufficient sampling of different respiratory states, especially when breathing is not consistent. In the future, we will acquire longer datasets and also consider different breathing patterns, such as Cheyne–Stokes respiration, which may have a different impact on cardiac function.

5 Conclusions

Free-breathing cardiac cine MRI provides novel information on cardiac dynamics. We demonstrated a new method for analyzing the influence of preload on cardiac function and function reserve, using the normal variation induced by respiration. For this, we devised a multi-dimensional reconstruction approach, using an adaptive soft classification of data into respiratory phases. The analysis revealed significant differences between normal subjects and patients with reduced EF and hypertrophy.

References

1. Feng, L., Axel, L., Chandarana, H., Block, K.T., Sodickson, D.K., Otazo, R.: XD-GRASP: golden-angle radial MRI with reconstruction of extra motion-state dimensions using compressed sensing. Magn. Reson. Med. **75**, 775–788 (2016)
2. Paling, M.R., Brookeman, J.R.: Respiration artifacts in MR imaging: reduction by breath holding. J. Comput. Assist. Tomogr. **10**, 1080–1082 (1986)
3. Toska, K., Eriksen, M.: Respiration-synchronous fluctuations in stroke volume, heart rate and arterial pressure in humans. J. Physiol. **472**, 501 (1993)
4. McLeish, K., Hill, D.L., Atkinson, D., Blackall, J.M., Razavi, R.: A study of the motion and deformation of the heart due to respiration. IEEE Trans. Med. Imaging **21**, 1142–1150 (2002)
5. Reyhan, M.L., Wang, Z., Kim, H.J., Halnon, N.J., Finn, J.P., Ennis, D.B.: Effect of free-breathing on left ventricular rotational mechanics in healthy subjects and patients with duchenne muscular dystrophy. Magn. Reson. Med. **77**(2), 864–869 (2016)
6. Wang, Y., Rossman, P.J., Grimm, R.C., Riederer, S.J., Ehman, R.L.: Navigator-echo-based real-time respiratory gating and triggering for reduction of respiration effects in three-dimensional coronary MR angiography. Radiology **198**, 55–60 (1996)
7. Ehman, R.L., McNamara, M., Pallack, M., Hricak, H., Higgins, C.: Magnetic resonance imaging with respiratory gating: techniques and advantages. Am. J. Roentgenol. **143**, 1175–1182 (1984)
8. Uecker, M., Zhang, S., Voit, D., Karaus, A., Merboldt, K.-D., Frahm, J.: Real-time MRI at a resolution of 20 ms. NMR Biomed. **23**, 986–994 (2010)
9. Tautz, L., Feng, L., Otazo, R., Hennemuth, A., Axel, L.: Analysis of cardiac interventricular septum motion in different respiratory states. In: International Society for Optics and Photonics SPIE Medical Imaging, p. 97880X (2016)

10. Kim, K.W., Lee, J.M., Jeon, Y.S., Kang, S.E., Baek, J.H., Han, J.K., Choi, B.I., Bang, Y.J., Kiefer, B., Block, K.T., Ji, H.: Free-breathing dynamic contrast-enhanced MRI of the abdomen and chest using a radial gradient echo sequence with K-space weighted image contrast (KWIC). Eur. Radiol. **23**(5), 1352–1360 (2013)
11. Otsu, N.: A threshold selection method from gray-level histograms. IEEE Trans. Syst. Man Cybern. Syst. **9**, 62–66 (1979)
12. Nehrke, K., Bornert, P., Manke, D., Bock, J.C.: Free-breathing cardiac MR imaging: study of implications of respiratory motion—initial results 1. Radiology **220**, 810–815 (2001)
13. Banfield, J.D., Raftery, A.E.: Model-based Gaussian and non-Gaussian clustering. Biometrics **49**, 803–821 (1993)
14. Johnson, K.M., Block, W.F., Reeder, S., Samsonov, A.: Improved least squares MR image reconstruction using estimates of k-Space data consistency. Magn. Reson. Med. **67**, 1600–1608 (2012)
15. Chitiboi, T., Hennemuth, A., Tautz, L., Huellebrand, M., Frahm, J., Linsen, L., et al.: Context-based segmentation and analysis of multi-cycle real-time cardiac MRI. In: IEEE 11th International Symposium on Biomedical Imaging (ISBI), pp. 943–946 (2014)

Learning-Based Heart Coverage Estimation for Short-Axis Cine Cardiac MR Images

Giacomo Tarroni[1]([✉]), Ozan Oktay[1], Wenjia Bai[1], Andreas Schuh[1],
Hideaki Suzuki[2], Jonathan Passerat-Palmbach[1], Ben Glocker[1],
Antonio de Marvao[3], Declan O'Regan[3], Stuart Cook[3], and Daniel Rueckert[1]

[1] BioMedIA, Department of Computing, Imperial College London, London, UK
g.tarroni@imperial.ac.uk
[2] Restorative Neurosciences, Imperial College London, London, UK
[3] MRC London Institute of Medical Sciences, Imperial College London, London, UK

Abstract. The correct acquisition of short axis (SA) cine cardiac MR image stacks requires the imaging of the full cardiac anatomy between the apex and the mitral valve plane via multiple 2D slices. While in the clinical practice the SA stacks are usually checked qualitatively to ensure full heart coverage, visual inspection can become infeasible for large amounts of imaging data that is routinely acquired, e.g. in population studies such as the UK Biobank (UKBB). Accordingly, we propose a learning-based technique for the fully-automated estimation of the heart coverage for SA image stacks. The technique relies on the identification of cardiac landmarks (i.e. the apex and the mitral valve sides) on two chamber view long axis images and on the comparison of the landmarks' positions to the volume covered by the SA stack. Landmark detection is performed using a hybrid random forest approach integrating both regression and structured classification models. The technique was applied on 3000 cases from the UKBB and compared to visual assessment. The obtained results (error rate = 2.3%, sens. = 73%, spec. = 90%) indicate that the proposed technique is able to correctly detect the vast majority of the cases with insufficient coverage, suggesting that it could be used as a fully-automated quality control step for CMR SA image stacks.

Keywords: Quality control · Cardiac MR · Landmark detection · Heart coverage

1 Introduction

Nowadays, cardiovascular magnetic resonance (CMR) imaging offers a wide variety of different applications for the anatomical and functional assessment of the heart. Unfortunately, the success of a CMR acquisition relies on the ability of the MR operator to correctly tune the acquisition parameters to the subject being scanned [1]. Moreover, CMR can be affected by a long list of imaging artefacts (caused for instance by respiratory and cardiac motion, flow artefacts and

© Springer International Publishing AG 2017
M. Pop and G.A. Wright (Eds.): FIMH 2017, LNCS 10263, pp. 73–82, 2017.
DOI: 10.1007/978-3-319-59448-4_8

magnetic field inhomogeneities) [2]. As a consequence, a quality control step is required to assess the usability of the acquired images. In clinical practice this step is usually performed by visual inspection.

In recent years, several initiatives for the acquisition of large-scale population studies have been launched. For example, the UK Biobank (UKBB) study will collect CMR images from 100,000 subjects using several protocols, including cine CMR long and short axis views of the ventricles [3]. At the time of submission the acquisition is ongoing, with more than 10,000 subjects already scanned. Together with this trend towards the implementation of open access large-scale multi-centre imaging datasets, the need for fast, automated and reliable quality control techniques for CMR images has become evident. Due to the very high throughput demanded from the acquisition pipeline, the visual inspection of the obtained images would be in most cases infeasible, but also highly subjective. On the other hand, failure to correctly identify corrupted or unusable images might affect the results of automated analyses performed on the dataset, with undesirable effects.

Several research efforts have been dedicated to the automated identification of quality metrics from MR images. Most of these efforts have focused on the automated estimation of noise levels [4,5]. However, many aspects related to the usability of the acquired images are inherently modality-specific. As far as brain MR imaging is concerned, several automated pipelines for quality control have been proposed [6]. However, to our knowledge, no comprehensive automated quality control pipelines have been proposed so far for CMR images. One crucial aspect of CMR acquisitions, which is essentially not present in brain imaging, is that most of the applications require the MR operator to manually or semi-automatically select the imaging planes with respect to specific anatomical landmarks. In particular, the acquisition of the cine short axis (SA) image stacks relies on the identification of the direction of the left ventricular (LV) long axis (LA), which is the line going from the apex to the centre of the mitral valve: the correct planning will generate a SA stack encompassing both those landmarks with slices perpendicular to the LV LA. If this selection is incorrect, the acquired SA stack may include an insufficient number of SA slices to fully cover the LV. As a consequence, any functional analysis performed on the stack (e.g. ventricular volumes estimation) may be compromised.

There are mainly two possible approaches to address this problem: the first one involves trying to identify the apical and basal slice from the SA stack [7], while the second one aims at automatically detecting specific landmarks from the acquired images. Automatic landmark localization in medical images has been an active area of research for decades. While the earlier techniques relied on image registration against annotated atlases [8], most of the recently proposed methods involve the learning of a model from an annotated dataset [9,10]. These methods establish an association between image content (together with potential prior information) and landmark locations, which can then be exploited at testing time on previously unseen images. In this paper we present a learning-based technique for the fully-automated estimation of the heart coverage for SA image stacks.

The proposed technique relies on the identification of cardiac landmarks (i.e. the apex and the mitral valve sides) from vertical (two chamber view) LA images (LA 2CH, Fig. 1A), and on the comparison of the landmarks' coordinates to the volume covered by the SA stack. Landmark detection is performed using a hybrid random forest approach integrating both regression and structured classification models. The technique was applied on a set of 3000 CMR images from the UKBB study and compared to the visual assessment of the heart coverage.

2 Methods

A decision tree consists in the combination of split and leaf nodes arranged in a tree-like structure [11]. Decision trees route a sample $x \in \mathcal{X}$ (in our case an image patch) by recursively branching left or right at each split node j until a leaf node k is reached. Each leaf node is associated with a posterior distribution $p(y|x)$ for the output variable $y \in \mathcal{Y}$. Each split node j is associated with a binary split function $h(x, \theta_j) \in \{0, 1\}$, defined by the set of parameters θ_j: if the outcome is 0 the node sends x to the left, otherwise to the right. In most cases, h is a decision *stump*, i.e. a single feature dimension n of x is compared with a threshold τ: $\theta = (n, \tau)$ and $h(x, \theta) = [x(n) < \tau]$. A random forest is an ensemble of T independent decision trees: during testing, given a sample patch x, the predictions of the different trees are combined into a single output using an ensemble model. During training, at each node the goal is to find the set of parameters θ_j which maximizes a previously defined *information gain* I_j, whose construction depends on the task at hand (e.g. classification, regression). Importantly, different types of nodes (maximizing different information gains) can be interleaved within a single tree structure. As in previous approaches [12,13], in the present technique structured classification nodes (aiming at the detection of the boundaries of an object close to the desired landmarks, in our case the LV myocardial contours) and regression nodes (aiming at landmark localization) are combined. In particular, in the proposed framework the landmark localization is conditioned on the results of the detection of the myocardium, potentially improving the localization accuracy [12]. The reason behind this result is that joint training of structured classification and regression nodes requires additional ground-truth information about myocardial position and shape, which can be implicitly incorporated in the model. The actual landmark detection will finally be performed by the creation of Hough vote maps for each landmark.

Structured Classification. Structured classification extends the concept of classification by using structured labels for $\mathcal{Y}$ in lieu of integer labels. In our case, each label $y \in \mathcal{Y}$ (associated to the image patch x) consists of a myocardial edge map outlining the endo- and epicardial LV contours within x. To train a structured classification node it is necessary to find a way to cluster structured labels at each split node into two subgroups depending on some similarity measure between them. This can be done in two steps [14]. First, $\mathcal{Y}$ is mapped to an intermediate space $\mathcal{Z}$ in which a distance between maps can be computed: more specifically, each edge map y is associated with a binary vector z encoding

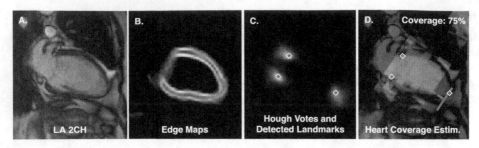

Fig. 1. Steps of the proposed technique. A. Two chamber view LA image. B. Myocardial edge maps obtained from the structured learning part of the model. C. Hough Vote maps obtained from the regression part of the model and detected landmarks' position. D. Heart coverage estimation and visualization of the portion of LV LA not covered by the SA stack. Note: the Hough Votes map is conditioned on the obtained edge maps.

whether every pair of pixels belong to a simply-connected region in the map or not. Then, at each split node the edge maps y are mapped into a binary set of labels $c \in C = \{0, 1\}$. More in detail, this second step is achieved by reducing the dimensionality - randomly sampling K elements from the associated z vectors and then performing PCA on the resulting vectors - and then by applying binary quantization to the identified principal component [14]. The information gain usually defined for standard classification nodes can thus be adopted: $I_j = H(S_j) - \sum_{i \in \{0,1\}} (|S_j^i|/|S_j|) \cdot H(S_j^i)$, where i is the class identifier, S_j, S_j^0 and S_j^1 are respectively the training set (comprising of samples x and associated labels y) arriving at node j, leaving the node to the left and leaving the node to the right, and $H(S)$ is defined as the Shannon entropy [14]. As a consequence, at each split node similar edge maps will be assigned to the same binary label c. The learned class posterior distributions for myocardial edge maps (Fig. 1B) are stored in the leaf nodes and will be used as weighting terms in the generation of the Hough vote maps.

Regression and Hough Vote Maps. To train regression nodes, it is necessary to add to each sample patch x an additional label $\mathcal{D} = (d^1, d^2, \ldots, d^L)$, where d^l represents for each of the L landmarks the N-dimensional displacement vector from the patch centre to the landmark location. The information gain for regression nodes is defined as $I_j = log|\Lambda(S_j)| - \sum_{i \in \{0,1\}} (|S_j^i|/|S_j|) \cdot log|\Lambda(S_j^i)|$, where $|\Lambda(S)|$ is the determinant of the full covariance matrix defined by the $N \cdot L$ landmark displacement vector components [11]. Minimizing this information gain encourages the clustering of samples in two groups with similar landmark displacement vectors. Moreover, by using the full covariance matrix, the spatial relationships between the different landmarks are taken into account and implicitly used to train the regression model. The regression information is stored at each leaf node k using a parametric model following a $N \cdot L$-dimensional multivariate normal distribution with $\overline{d_k^l}$ and Σ_k^l mean and covariance matrices, respectively. At testing time, for each landmark, Hough vote maps are generated by summing up the regression posterior distributions obtained from each tree for each patch (minus normalization factors). Following the assumption that pixels belonging to the myocardial contours are more informative for myocardial landmark detection than background ones, the posterior distributions for structured classification are used for each patch as weighting factor during the generation of the Hough vote maps [13], effectively restricting voting rights only to pixels on myocardial contours. Finally, the location of the landmarks are determined by identifying the pixel with the highest value on each of the L Hough vote maps (Fig. 1C).

Input Features and Tree Structure. Each patch x is represented by C number of channels, which are multi-resolution image intensity, histogram of gradients (HoG) and gradient magnitude. The multi-resolution intensity channels are constructed with a two-scale Gaussian image pyramid, with the first layer downsampled by a factor of 2. Smoothed gradient magnitude and orientation are computed using oriented Gaussian derivative kernels. Orientation histograms are computed using soft-binning, where bin weights are determined by the gradient magnitude. Pair-wise differences features are also computed from the single channels [13]. As far as the alternation of the different types of split nodes is concerned, during training a structured classification node or a regression node are randomly selected for each node of the tree with equal probability.

Heart Coverage Estimation. The described hybrid random forest can be used to identify the apex (l^1) and the mitral valve sides (l^2 and l^3) from a LA 2CH image. Using the orientation matrix extracted from the DICOM headers of the acquired SA and LA image stacks, it is possible to define the coordinates of the three landmarks (identified in the coordinate system of the LA 2CH image) in the coordinate system of the SA stack itself ($\hat{l}^{1,2,3}$). Finally, the coverage of the SA stack is defined as the relative portion of the distance between $\hat{l}^1$ and the average between $\hat{l}^2$ and $\hat{l}^3$ encompassed by the stack (Fig. 1D).

3   Experiments and Results

Image Acquisition. Training and testing of the proposed technique were carried out on two datasets acquired at different sites. This setting represents a realistic scenario often found in practice, in which new test data might come from a different site than the one used to acquire the training dataset. For training, CMR imaging was performed on 473 healthy subjects using a 1.5T Philips Achieva system equipped with a 32 element cardiac phased-array coil (33 mT/m and 160 T/m/s gradient system). 2D cine balanced steady-state free precession (b-SSFP) LA 2CH images were acquired with in-plane spatial resolution 1.48×1.48 mm and slice thickness 8 mm. For testing, images from 3000 subjects were randomly extracted from the UKBB study, whose CMR infrastructure features a 1.5T Siemens MAGNETOM Aera system equipped with a 18 channels anterior body surface coil (45 mT/m and 200 T/m/s gradient system).

2D cine b-SSFP LA 2CH images were acquired with in-plane spatial resolution 1.8×1.8 mm and slice thickness 6 mm. 2D cine b-SSFP SA image stacks were also acquired with in-plane spatial resolution 1.8×1.8 mm, slice thickness 8 mm and slice gap 2 mm [3]. On both datasets, only end-diastolic frames were considered.

Performance Evaluation. The training dataset underwent intensity normalization and data augmentation (4 rotations randomly selected in the $\pm 30°$ range). Manual annotation was performed to provide LV myocardial segmentations and landmarks locations. A model was then generated for the proposed technique with image patch size 64×64 px, labels size (structured classification) 12×12 px, number of trees $T = 8$, maximum tree depth = 64, total number of image samples = 8 million, number of samples for $\mathcal{Z}$-undersampling $K = 256$. To provide references for performance evaluation, the testing dataset was visually inspected: the coverage was deemed insufficient when the SA stack was short of at least one slice (in either the basal or apical direction) to provide a full LV coverage. To assess the accuracy of the landmark detection alone, the landmarks were also manually identified on a subset of 30 cases. The automated technique was then applied to the whole testing set (after bias correction using N4ITK [15], intensity normalization and spatial re-sampling to match the training set specifications) and the relative coverage value was estimated for each

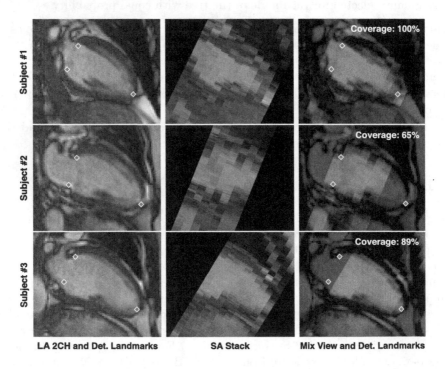

Fig. 2. Results obtained in 3 different subjects using the proposed technique.

subject (the results obtained in three cases are qualitatively displayed in Fig. 2). Time required for landmarks detection and coverage estimation was around 1 s per subject without resorting to parallel computing. In the subset of 30 manually annotated images, the reference landmarks were compared to the automatically detected ones by measuring the Euclidean distance between the two sets of points. Similarly, the LA length values were extracted from the automatically detected landmarks and compared to the values computed from the reference ones. Following the rationale presented for visual inspection and considering that the SA stacks consisted on average of 10 slices, a coverage value of less than 90% was considered insufficient. Finally, the results of visual assessment and automated analysis were compared by means of standard statistical metrics. Of note, when at least one of the three relative distances between the landmarks was over 2 standard deviations greater or smaller than the respective mean distance value ($\left| \left| l^i - l^j \right| - \mu^{ij} \right| > 2\sigma^{ij}$, where $i,j \in \{1,2,3\}$ and mean μ^{ij} and standard deviation σ^{ij} of the distances between the respective landmarks were computed on 200 images randomly extracted from the testing dataset), the automated technique was considered unreliable and the specific case excluded from the statistical comparison. This mostly allowed the exclusion of LA 2CH images with too high noise levels, wrong acquisition planning or wrong file naming among the different sequences of the CMR protocol.

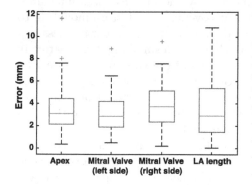

Table 1. Average errors (in mm) for landmark localization and LA length estimation.

	Error (mean ± std)
Apex	3.8 ± 2.8
Mitral V. (left side)	3.2 ± 1.8
Mitral V. (right side)	3.8 ± 2.1
LA length	3.6 ± 2.6

Fig. 3. Error distributions for landmark localization and LA length estimation.

The results for landmark detection and LA length estimation accuracy are reported in Table 1, while the error distributions are shown in in Fig. 3. All of the average errors are small compared to the reconstructed slice thickness (10 mm), suggesting the reliability of the proposed approach. Our technique seems to outperform other published studies in landmark detection from LA images: for instance, Lu et al. [9] reported an average error in mitral valve detection of 5.1 ± 6.8 mm, while Mahapatra [10] reported 5.67 ± 5.83 mm. As far as landmark detection is concerned, the present technique can be considered an extension of the work of Oktay et al. [13]: importantly, while their technique addressed the automated initialization of image segmentation and registration tasks, in the present approach landmark detection is performed as a preliminary step for slice coverage estimation. Moreover, we excluded *stratification nodes* from the implementation of our forest based on the empirical observation that the variations in pose and shape of the heart within the LA 2CH images in our training dataset were not large enough to justify the associated increase in complexity and computational time.

Table 2. Confusion matrix for the heart coverage assessment. N are cases with a sufficient coverage, P with an insufficient one.

		Proposed technique	
		P	N
Visual	P	60 (TP)	46 (FP)
Assessment	N	22 (FN)	2783 (TN)

The computed confusion matrix for the 2911 evaluated cases is presented in Table 2. Some of the metrics that can be extracted include error rate = 2.3%, sensitivity = 73%, specificity = 98%, area under the curve (ROC analysis) = 0.98%. The interpretation of these results is hindered by the massive class imbalance between cases with sufficient and insufficient coverage, the latter amounting to a 2.8% of the total. While this number could lead to question the usefulness of a slice coverage check, it is important to note that even a small number of wrongly imaged subjects could adversely affect the results of the subsequent automated analyses. Moreover, this estimate has been obtained from a small subset of cases from the UKBB, and might vary considerably in a different subset or a different dataset. It could be argued that another way to perform the coverage estimation might be to simply count the number of acquired SA slices: however, while a very low slice count is undoubtedly sign of insufficient coverage, the slice number varies according to the heart size, and even a number of slices within the normal range is not necessarily a reliable indicator of sufficient coverage (e.g. Subject #3 in Fig. 2). By applying the proposed automated technique, it was possible to correctly detect the 73% of the cases with insufficient coverage, and to thus improve the percentage of undetected wrongly imaged subjects from 2.8% to 0.76%. This comes at the price of having to visually check an additional 1.6% of cases that actually featured a sufficient coverage. Notably, most of the 46 FP cases actually had a sub-optimal coverage, but not of the amount required to consider them as wrongly imaged when following the rule adopted during visual inspection.

4   Conclusion

In this study, a technique for automated heart coverage estimation for SA cine stacks has been presented and tested on 3000 cases extracted from the UKBB study. The technique relies on the localization of cardiac landmarks on two chamber view LA images using a hybrid random forest approach integrating both regression and structured classification models. The presented results show its capability in identifying the cases of insufficient heart coverage, allowing either the execution of a second acquisition or the exclusion of the case from the dataset. In conclusion, the proposed technique could potentially be adopted as part of a comprehensive quality control pipeline for CMR images, which is particularly needed for the analysis of large datasets.

Acknowledgements. This research has been conducted using the UK Biobank Resource [3] under Application Number 18545. The first author benefits from a Marie Skłodowska-Curie Fellowship.

References

1. Zhuo, J., Gullapalli, R.P.: MR artifacts, safety, and quality control. RadioGraphics **26**(1), 275–297 (2006)
2. Ferreira, P.F., Gatehouse, P.D., Mohiaddin, R.H., Firmin, D.N.: Cardiovascular magnetic resonance artefacts. J. Cardiovasc. Magn. Reson. **15**(1), 41 (2013)
3. Petersen, S.E., Matthews, P.M., Francis, J.M., Robson, M.D., Zemrak, F., Boubertakh, R., Young, A.A., Hudson, S., Weale, P., Garratt, S., Collins, R., Piechnik, S., Neubauer, S.: UK Biobank's cardiovascular magnetic resonance protocol. J. Cardiovasc. Magn. Reson. **18**(1), 8 (2016). Official journal of the Society for Cardiovascular Magnetic Resonance
4. Coupé, P., Manjón, J.V., Gedamu, E., Arnold, D., Robles, M., Collins, D.L.: Robust Rician noise estimation for MR images. Med. Image Anal. **14**(4), 483–493 (2010)
5. Maximov, I.I., Farrher, E., Grinberg, F., Jon Shah, N.: Spatially variable Rician noise in magnetic resonance imaging. Med. Image Anal. **16**(2), 536–548 (2012)
6. Gedamu, E.L., Collins, D.L., Arnold, D.L.: Automated quality control of brain MR images. J. Magn. Reson. Imag. **28**(2), 308–319 (2008)
7. Paknezhad, M., Marchesseau, S., Brown, M.S.: Automatic basal slice detection for cardiac analysis. J. Med. Imag. **3**(3), 034004 (2016)
8. Gass, T., Szekely, G., Goksel, O.: Multi-atlas segmentation and landmark localization in images with large field of view. In: Menze, B., Langs, G., Montillo, A., Kelm, M., Müller, H., Zhang, S., Cai, W.T., Metaxas, D. (eds.) MCV 2014. LNCS, vol. 8848, pp. 171–180. Springer, Cham (2014). doi:10.1007/978-3-319-13972-2_16
9. Lu, X., Georgescu, B., Jolly, M.-P., Guehring, J., Young, A., Cowan, B., Littmann, A., Comaniciu, D.: Cardiac anchoring in MRI through context modeling. In: Jiang, T., Navab, N., Pluim, J.P.W., Viergever, M.A. (eds.) MICCAI 2010. LNCS, vol. 6361, pp. 383–390. Springer, Heidelberg (2010). doi:10.1007/978-3-642-15705-9_47
10. Mahapatra, D.: Landmark detection in Cardiac MRI using learned local image statistics. In: Camara, O., Mansi, T., Pop, M., Rhode, K., Sermesant, M., Young, A. (eds.) STACOM 2012. LNCS, vol. 7746, pp. 115–124. Springer, Heidelberg (2013). doi:10.1007/978-3-642-36961-2_14
11. Criminisi, A., Shotton, J., Konukoglu, E.: Decision forests: a unified framework for classification, regression, density estimation, manifold learning and semi-supervised learning. Found. Trends® Comput. Graph. Vis. **7**(2–3), 81–227 (2011)

12. Gall, J., Yao, A., Razavi, N., Van Gool, L., Lempitsky, V.: Hough forests for object detection, tracking, and action recognition. IEEE Trans. Pattern Anal. Mach. Intell. **33**(11), 2188–2202 (2011)
13. Oktay, O., Bai, W., Guerrero, R., Rajchl, M., de Marvao, A., O'Regan, D.P., Cook, S.A., Heinrich, M.P., Glocker, B., Rueckert, D.: Stratified decision forests for accurate anatomical landmark localization in cardiac images. IEEE Trans. Med. Imag. **36**(1), 332–342 (2017)
14. Dollar, P., Zitnick, C.L.: Fast edge detection using structured forests. IEEE Trans. Pattern Anal. Mach. Intell. **37**(8), 1558–1570 (2015)
15. Tustison, N.J., Avants, B.B., Cook, P.A., Zheng, Y., Egan, A., Yushkevich, P.A., Gee, J.C.: N4ITK: improved N3 bias correction. IEEE Trans. Med. Imag. **29**(6), 1310–1320 (2010)

Using Atlas Prior with Graph Cut Methods for Right Ventricle Segmentation from Cardiac MRI

Shusil Dangi[1(✉)] and Cristian A. Linte[1,2]

[1] Chester F. Carlson Center for Imaging Science, Rochester, USA
sxd7257@rit.edu
[2] Biomedical Engineering, Rochester Institute of Technology, Rochester, NY, USA

Abstract. Right ventricle segmentation helps quantify many functional parameters of the heart and construct anatomical models for intervention planning. Here we propose a fast and accurate graph cut segmentation algorithm to extract the right ventricle from cine cardiac MRI sequences. A shape prior obtained by propagating the right ventricle label from an average atlas via affine registration is incorporated into the graph energy. The optimal segmentation obtained from the graph cut is iteratively refined to produce the final right ventricle blood pool segmentation. We evaluate our segmentation results against gold-standard expert manual segmentation of 16 cine MRI datasets available through the MICCAI 2012 Cardiac MR Right Ventricle Segmentation Challenge. Our method achieved an average Dice Index 0.83, a Jaccard Index 0.75, Mean absolute distance of 5.50 mm, and a Hausdorff distance of 10.00 mm.

1 Introduction

Magnetic Resonance Imaging (MRI) has been a preferred imaging modality for non-invasive cardiac diagnosis and planning of cardiac interventions thanks to its high image quality, good tissue contrast, and lack of exposure to ionizing radiation. MRI is commonly employed in clinical practice for quantifying important Right Ventricular (RV) parameters such as mass, volume, and ejection fraction, as well as to generate faithful anatomical models used to plan and provide visualization during minimally invasive interventions. Moreover, since the MR image acquisition technology has evolved to the extent that it enables the acquisition of peri-operative (just in time) images of the patient within the interventional suite moments before the procedure, the need for fast and accurate segmentation approaches has been increasing.

The complex anatomy of the RV and presence of trabeculations (with similar gray level as the surrounding myocardium), amplified by indistinct borders and partial volume effects, make the segmentation of RV a challenging task [1]. RV segmentation is performed manually in clinical practice, requiring about 15 min per patient per cardiac phase, and is prone to inter and intra-expert variability [2]. These limitations suggests a need for fast, accurate, and robust semi- or fully-automatic RV segmentation algorithms.

© Springer International Publishing AG 2017
M. Pop and G.A. Wright (Eds.): FIMH 2017, LNCS 10263, pp. 83–94, 2017.
DOI: 10.1007/978-3-319-59448-4_9

Various techniques for segmentation of the heart chambers have been proposed in the literature [1], including edge and region-based approaches [3,4], deformable models [5], active shape and appearance models [6–8], and atlas-based approaches [9–11]. Many methods are based on a joint segmentation of both ventricles [5,12–14], such that their geometry information, relative position of ventricles, and similar intensity values characteristic to their blood pool cavities can be exploited.

Of these techniques, multi-atlas based approaches [9–11] have been shown to produce good segmentation results, nevertheless at the expense of computationally intensive and time consuming non-rigid image registration and label fusion steps. On the other hand, combinatorial optimization based graph-cut techniques [3] were also shown to serve as powerful tools for image segmentation; these techniques are fast and guarantee convergence within a known factor of the global minimum for a special class of functions (i.e., regular functions) [15]. In the attempt to further refine these techniques, adding shape constraints to the graph cut framework has been shown to significantly improve cardiac image segmentation results [6,16,17]. While these techniques were extensively exploited for left ventricle (LV) segmentation, their use to extract the RV anatomy from cine cardiac MRI images has been minimal.

Here we describe the integration of atlas shape priors and graph cut techniques to segment the end diastolic RV blood pool from clinical quality cardiac MR images and evaluate the performance of these segmentation tools using 16 image datasets accompanied by ground truth manual delineations of the RV.

2   Methodology

2.1   Data Preprocessing

This study employed 16 cardiac cine MRI datasets available through the MICCAI 2012 Cardiac MR Right Ventricle Segmentation Challenge (RVSC)[1]. Expert manual segmentations of the RV images were provided with the dataset and served as the gold-standard for evaluating the proposed segmentation technique.

We first select a reference patient volume with good contrast and preferred LV-RV orientation. The DICOM Image Orientation Patient (IPP) field is used to rotate the patient volumes about the z-axis to roughly align the LV-RV orientation of each dataset with that of the selected reference patient. We then find the region of interest (ROI) (in the xy-plane) enclosing the left and right ventricles using the method described in [18]. To restrict the ROI along the z-axis, we require a manual input that indicates the start and end slices of the RV anatomy. The image volumes are then cropped according to the above ROI, and the intensity range of each slice is normalized prior to the subsequent steps.

2.2   Atlas Generation

We affinely register all cropped 3D volumes to the selected reference patient using the intensity based Nelder-Meade downhill simplex algorithm [19] available in

---

[1] http://www.litislab.eu/rvsc.

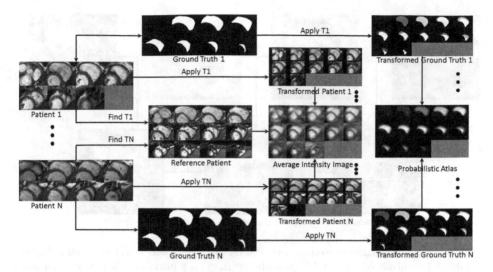

Fig. 1. All patient images are affinely registered to the reference patient, and the obtained optimum transformations are applied to the corresponding ground truth images. An average intensity image is obtained by averaging the intensities of all transformed patient images, while the averaging of the transformed ground truth images yields a probabilistic atlas.

SimpleITK. The resulting affine transforms are applied to the respective ground truths datasets. Averaging the transformed volumes as well as the transformed ground truths yields an average appearance atlas and a blood pool (BP) probabilistic atlas, respectively Fig. 1.

2.3 RV Blood Pool Segmentation Using Iterative Graph Cuts

To exploit the 3D cardiac geometry information, we use the BP segmentation of a given slice to initialize the BP ROI in the neighboring slices. As such, we first segment the mid-slice, followed by its neighboring slices, and proceed accordingly, until the complete volume is segmented.

Blood Pool Probabilistic Map: The average appearance atlas is registered to a test volume (Fig. 2a) using intensity based 3D affine registration. The resulting affine transform is used to transfer the BP probabilistic label to the test data. The obtained label is normalized (0–1) per slice to generate a BP probabilistic map, which helps in BP initialization and serves as a shape constraint for subsequent graph cut segmentation.

Blood Pool Initialization: The probabilistic map is thresholded to obtain an initial BP ROI (Fig. 2b). All the basal slices (including and above the mid-slice) are thresholded at a very small value (0.01) to account for the large BP size, whereas, the threshold is determined automatically, as a mean value for the

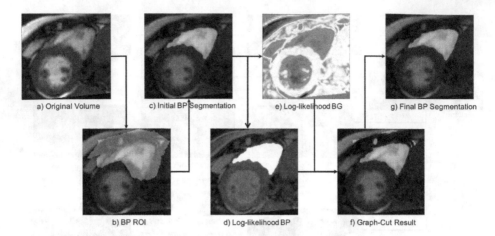

Fig. 2. BP Segmentation procedure: (a) original MR image slice, (b) the initial BP ROI obtained by thresholding the BP probability map, (c) initial BP segment obtained by otsu thresholding and cleaning; (d) BP log-likelihood and (e) BG log-likelihood obtained from intensity distribution model; (f) the graph-cut segmentation result; (g) final BP segmentation for current iteration obtained after post-processing.

probabilistic map, for the apical slices to restrict the BP size to a smaller region. Otsu thresholding [20] applied to the obtained BP ROI yields a set of potentially bright BP segments. A single connected component closest to the center of the BP ROI is used as initial BP segmentation shown in Fig. 2c.

Intensity Distribution Model: The BP intensity model is obtained by fitting the intensity values within the initial BP segmentation to a Gaussian Mixture Model (GMM) comprising one Gaussian. The remaining intensity values inside the initial BP ROI are fitted to a two-Gaussian GMM to yield the background (BG) intensity model. Figure 2d and e show the resulting BP and BG log-likelihood map, respectively. Although Rician distribution [21] has been found more appropriate for modeling intensity noise in MR images, signal-to-noise ratio for our MR data is high, hence, we rely on a simpler model using Gaussian approximation.

Graph-Cut Segmentation: A four-neighborhood graph is constructed with each pixel representing a node. Two special terminal nodes representing two classes — the source blood pool (BP), and the sink background (BG) — are added to the graph. The segmentation is formulated as an energy minimization problem over the space of optimal labelings f:

$$E(f) = \sum_{p \in \mathcal{P}} D_p(f_p) + \sum_{\{p,q\} \in \mathcal{N}} V_{p,q}(f_p, f_q), \qquad (1)$$

where the first term represents the data energy that reduces the disagreement between the labeling f_p given the observed data at every pixel $p \in P$, and the

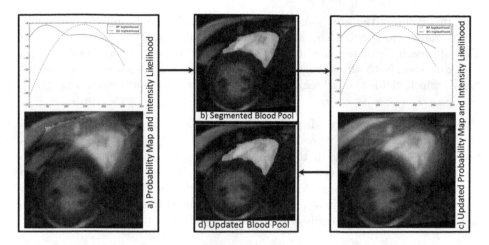

Fig. 3. (a) Probability map and intensity distribution model for current iteration, (b) BP segmentation obtained from graph cut using (a),(c) updated probability map and intensity distribution model obtained using (b), (d) new BP segmentation obtained from graph cut using (c).

second term represents the smoothness energy that forces pixels p and q defined by a set of interacting pair $\mathcal{N}$ (in our case, the neighboring pixels) towards the same label.

The data energy term encoded as terminal link (t-link) between each node to source (or sink) is assigned based on the log likelihood computed from the intensity distribution model:

$$D_p(f_p) = -lnPr(I_p|f_p) \qquad (2)$$

where $Pr(I_p|f_p)$ is the likelihood of observing the intensity I_p given that pixel p belongs to class f_p. The log-likelihood for BP and BG are shown in (Fig. 2d) and (Fig. 2e), respectively.

The smoothness energy term is computed over the links between neighboring nodes (n-links) and is assigned as a weighted sum of intensity similarity between the pixels and average probability of the pixels belonging to BP based on BP probabilistic atlas:

$$V_{p,q}(f_p, f_q) = \begin{cases} w_I * exp\left(-\frac{|I_p - I_q|}{\tau}\right) + w_A * exp\left(-(P_A(p) + P_A(q))\right) & \text{if } f_p = f_q \\ 0 & \text{if } f_p \neq f_q \end{cases} \qquad (3)$$

where, τ is the iteration number, w_I and w_A are weights for the intensity similarity term and atlas prior term, respectively, and $P_A(.)$ is the probability of a pixel belonging to BP obtained from the atlas. The intensity similarity term changes during each iteration such that the pixels with higher intensity difference can be assigned the same label to allow the expansion of BP as the iteration proceeds.

After defining the graph energy configuration, the minimum cut equivalent to the maximum flow is identified via the expansion algorithm described in [22]. This approach yields the labeling that minimizes the global energy of the graph and corresponds to the optimal segmentation. Figure 2f shows the graph-cut result, which yields the BP segmentation after some post processing Fig. 2g.

Blood Pool Probability Map Refinement: The mean value of the BP probabilistic map is used as a threshold it to obtain the BP region. The signed distance map corresponding to the extracted BP region is affinely registered to the signed distance map generated from the boundary of the BP segmentation obtained from graph-cuts, such that, the sum of squared differences between the two distance maps is minimized. The optimum affine transform, when applied to the BP probability map, transforms it according to the latest BP segmentation.

Iterative Refinement: We employed the iterative refinement technique as described in [23]. The latest BP segmentation obtained from the graph cut is used to update the intensity distribution model for BP/BG. Very high intensity likelihood $Pr(I_p|f_p)$ is assigned to the pixels labeled as BP in the current iteration, such that they don't change their labels. The refined BP probability map is used to impose shape constraint into the graph-cut framework in the form of smoothness energy. An updated BP segmentation is obtained via another graph cut operating on the new graph energy configuration. This iterative process is repeated until the dice coefficient between two latest BP segmentations exceed 99%. On average, the

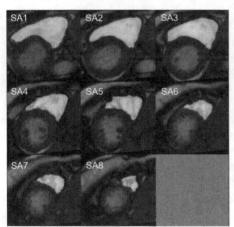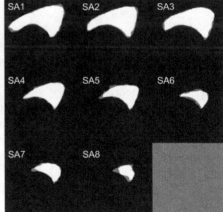

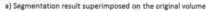

a) Segmentation result superimposed on the original volume   b) Our segmentation result versus the Gold-standard segmentation

Fig. 4. (a) Final BP segmentation of all slices of a patient dataset (shown in blue) superimposed with the patient volume (shown in red); (b) Final BP segmentation assessed against the provided gold-standard manual segmentation; white regions represent true positives, red regions represent false negatives, and blue regions represent false positives. (Color figure online)

process requires about three iterations for convergence, hence the maximum number of iterations is restricted to 10. The latest segmentation result yields the final BP segmentation. Figure 3 illustrates the iterative refinement process.

3   Results

Segmentation results for a patient dataset are overlaid onto each slice of the patient volume and shown in Fig. 4a. Figure 4b shows the visual comparison of our segmentation result against the provided manual segmentation.

Table 1. Evaluation of our segmentation results against the provided gold-standard manual segmentation for the basal and apical slices according to Dice Index, Jaccard Index, Mean Absolute Distance, and Hausdorff Distance

Assessment metric	Basal slices	Apical slices	All slices
Dice index	0.89 ± 0.12	0.62 ± 0.32	0.83 ± 0.22
Jaccard index	0.82 ± 0.15	0.52 ± 0.30	0.75 ± 0.23
Mean absolute distance (mm)	3.73 ± 4.16	11.11 ± 15.05	5.50 ± 8.69
Hausdorff distance (mm)	8.25 ± 7.42	15.57 ± 12.57	10.00 ± 9.40

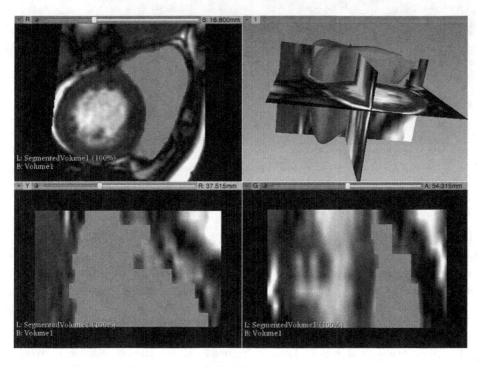

Fig. 5. Axial, Sagittal, Coronal, and the 3D model (counter-clockwise from top left) of the segmented RV blood pool overlayed onto the MRI image.

We evaluated the obtained RV blood pool segmentations against the provided expert manual segmentations for 16 datasets according to the following metrics: Dice Metric (DM), Jaccard Metric (JM), Mean Absolute Distance (MAD), and Hausdorff Distance (HD) [1]. The metrics are summarized in Table 1 for all slices together, as well as for the apical-slices (last two slices) and remaining slices separately.

The collated results for RVSC are reported on [1]. Inter-expert variability study in [1] shows the variability in DM and HD are 0.90 ± 0.10 and 5.02 ± 2.87 mm, respectively. The best reported mean DM and HD in the collated studies were 0.75 and 11.00 for the fully automatic methods evaluated and 0.805 and 8.26 for the semi-automatic techniques, respectively. As evident from Table 1, the proposed algorithm performs better than the best fully automatic method and is comparable to the best semi-automatic method reported in [1]. However, it should be noted that our method was evaluated on the training dataset (compared against the provided ground truth segmentation) whereas the collated results were reported for the test dataset. Nevertheless, there is still a room for improvement to reach the performance equivalent to the inter-expert variability.

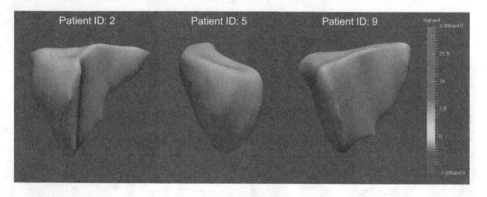

Fig. 6. Shown for three cases: Signed distance error between the segmented RV blood pool and the corresponding gold standard manual segmentation overlayed onto the 3D model of the latter.

Axial, Sagittal, Coronal, and the 3D reconstruction of the segmented RV blood pool overlayed onto the MRI volume is shown in Fig. 5. The signed distance error computed between the 3D models of the segmented RV blood pool and the provided expert manual segmentation is overlaid onto the latter and shown for three different cases in Fig. 6.

We computed the End Diastole (ED) RV blood pool volume for 16 datasets and compared it against the volume computed from the provided expert manual segmentation as shown in Table 2. Figure 7a shows the correlation between the RV volume reconstructed from our automated segmentation and the volume

Table 2. Comparing the end diastole RV blood pool volume estimated by the proposed algorithm against the ground truth volume for 16 cine MRI images

Assessment metric	Ground truth	Proposed algorithm	Difference
ED Volume (ml)	129.7 ± 38.1	122.0 ± 29.0	7.8 ± 16.6

reconstructed from the manual expert segmentation. As illustrated, a linear correlation (defined by $y = 1.2x$) exists, where y is the ground truth RV volume and x is the segmented RV volume in mL, respectively. Moreover, the linear R^2 value is 0.83, showing a strong correlation between the automatically segmented and ground truth RV volume.

The Bland-Altman plot in Fig. 7b shows the difference between the ground truth and automatically segmented volume for all 16 datasets. Although the estimated volume is linearly correlated to the ground truth volume, our automatic segmentation algorithm slightly under-estimated the RV volume by 7.8±16.6 mL on average. Specifically, the volume was under-estimated for 5 of the patients (1, 5, 10, 11, 14) and over-estimated it for the remaining 11 patients. Lastly, with the exception of one single dataset (patient 10), who is clearly an outlier as it exhibited a difference between the automatically segmented and ground truth volume of more than 3 SDs from the mean (i.e., a z-score larger than 3), all other cases exhibited a volume difference within 2 SDs from the mean.

The proposed algorithm was implemented in Python and required 76 secs on average to propagate atlas label and segment RV blood pool from cine MRI volumes (12 ± 1 axial slices) on a Intel$^{\circledR}$ Xenon$^{\circledR}$ 3.60 GHz 32 GB RAM PC, therefore posing a great potential for peri-operative segmentation of RV without delaying the procedure workflow. The atlas is precomputed, in 450 s, by co-registering and averaging the training images.

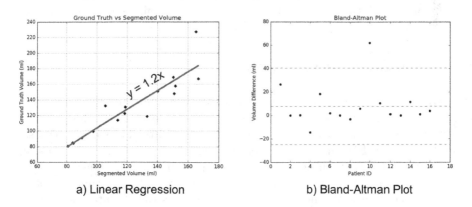

a) Linear Regression   b) Bland-Altman Plot

Fig. 7. (a) Linear regression result for ground truth volume against the volume estimated by the algorithm. The best fit line is $y = 1.2x$, with r-squared value of 0.83; (b) Difference between the ground truth volume and volume estimated by the algorithm plotted for all 16 cases. The mean difference (7.8 ml) along with the 95% confidence interval (± 1.96SD: -24.8 to 40.4 ml) are shown as the dotted lines.

4   Discussion

We observed that the segmentation results for apical slices with smaller BP region are consistently worse than that for the basal slices. Although they have very little impact on volume computation, they could be a limiting factor on other fields such as studies of the fiber structure. Slice-wise refinement process might be hurting the performance of our segmentation algorithm due to the misleading cues on apical slices, which suggests a need for special processing of these slices. An alternative solution for better segmentation of apical slices would be to perform a complete 3D segmentation, which requires a robust algorithm to align the cine MRI slices into a complete 3D volume to correct for stair-step artifacts caused by inherent motion between subsequent slices prior to segmentation.

Since the expert annotated ground truth segmentations of the right ventricle are only available in end diastole and end systole, only two atlases could be constructed – a diastole and a systole atlas. We have in fact attempted to generate a systole atlas based on the same techniques used to generate the diastole atlas, however the apical slices are compromised due to the wrapping of the right ventricle around the left ventricle, making it difficult to identify the blood pool from the myocardium in this region and hence compromising the accuracy of the atlas and its robustness across all datasets.

Nevertheless, a more feasible approach would be to take advantage of the temporal information and correlation of the subsequent frames in the cardiac cycle and use the frame-to-frame motion to "animate" the single phase diastolic right ventricle segmentation extracted using the diastole right ventricle atlas faithfully generated as described in this paper. The proposed method will follow the approach described in [24]. Briefly, we will utilize non-rigid image registration to extract the frame-to-frame motion from the sequence of multi-phase cardiac images depicting the right ventricle and then apply the non-rigid displacement field to propagate the segmented right ventricle diastolic surface through the cardiac cycle.

5   Conclusion and Future Work

We proposed a fast and automatic segmentation method using atlas prior in the graph cut framework with iterative refinement to segment the RV blood pool from cardiac cine MRI images. Quantitative results of our blood pool segmentation in the ED phase are better than the fully-automatic methods and comparable to the semi-automatic methods reported in the challenge.

We plan to segment the RV myocardium and extend the evaluation of our algorithm to end diastole and end systole phases of all the 48 datasets (16 training, 32 testing). Furthermore, we will be studying the effect of weighting factor variability on different terms of graph energy towards final segmentation result and lastly, we will also be exploring other options to improve the segmentation results for apical slices.

References

1. Petitjean, C., Zuluaga, M.A., Bai, W., Dacher, J.N., Grosgeorge, D., Caudron, J., Ruan, S., Ayed, I.B., Cardoso, M.J., Chen, H.C., Jimenez-Carretero, D., Ledesma-Carbayo, M.J., Davatzikos, C., Doshi, J., Erus, G., Maier, O.M., Nambakhsh, C.M., Ou, Y., Ourselin, S., Peng, C.W., Peters, N.S., Peters, T.M., Rajchl, M., Rueckert, D., Santos, A., Shi, W., Wang, C.W., Wang, H., Yuan, J.: Right ventricle segmentation from cardiac MRI: a collation study. Med. Image Anal. **19**(1), 187–202 (2015)
2. Caudron, J., Fares, J., Lefebvre, V., Vivier, P.H., Petitjean, C., Dacher, J.N.: Cardiac MRI assessment of right ventricular function in acquired heart disease: factors of variability. Acad. Radiol. **19**(8), 991–1002 (2012)
3. Maier, O.M.O., Jiménez, D., Santos, A., Ledesma-Carbayo, M.J.: Segmentation of RV in 4d cardiac MR volumes using region-merging graph cuts. In: 2012 Computing in Cardiology, pp. 697–700 (2012)
4. Wang, C.W., Peng, C.W., Chen, H.C.: A simple and fully automatic right ventricle segmentation method for 4-dimensional cardiac MR images. In: Workshop in Medical Image Computing and Computer Assisted Intervention, pp. 1–8 (2012)
5. Grosgeorge, D., Petitjean, C., Caudron, J., Fares, J., Dacher, J.N.: Automatic cardiac ventricle segmentation in MR images: a validation study. Int. J. Comput. Assist. Radiol. Surg. **6**(5), 573–581 (2011)
6. Grosgeorge, D., Petitjean, C., Dacher, J.N., Ruan, S.: Graph cut segmentation with a statistical shape model in cardiac MRI. Comput. Vis. Image Underst. **117**(9), 1027–1035 (2013)
7. Abi-Nahed, J., Jolly, M.-P., Yang, G.-Z.: Robust active shape models: a robust, generic and simple automatic segmentation tool. In: Larsen, R., Nielsen, M., Sporring, J. (eds.) MICCAI 2006. LNCS, vol. 4191, pp. 1–8. Springer, Heidelberg (2006). doi:10.1007/11866763_1
8. ElBaz, M.S., Fahmy, A.S.: Active shape model with inter-profile modeling paradigm for cardiac right ventricle segmentation. In: Ayache, N., Delingette, H., Golland, P., Mori, K. (eds.) MICCAI 2012. LNCS, vol. 7510, pp. 691–698. Springer, Heidelberg (2012). doi:10.1007/978-3-642-33415-3_85
9. Bai, W., Shi, W., Wang, H., Peters, N.S., Rueckert, D.: Multiatlas based segmentation with local label fusion for right ventricle MR images. Image **6**, 9 (2012)
10. Ou, Y., Doshi, J., Erus, G., Davatzikos, C.: Multi-atlas segmentation of the cardiac MR right ventricle. In: Proceedings of 3D Cardiovascular Imaging: A MICCAI Segmentation Challenge (2012)
11. Zuluaga, M., Cardoso, M., Ourselin, S.: Automatic right ventricle segmentation using multi-label fusion in cardiac MRI. In: Medical Image Computing and Computer-Assisted Intervention, Workshop on RV Segmentation Challenge in Cardiac MRI (2012)
12. Peters, J., Ecabert, O., Meyer, C., Schramm, H., Kneser, R., Groth, A., Weese, J.: Automatic whole heart segmentation in static magnetic resonance image volumes. In: Ayache, N., Ourselin, S., Maeder, A. (eds.) MICCAI 2007. LNCS, vol. 4792, pp. 402–410. Springer, Heidelberg (2007). doi:10.1007/978-3-540-75759-7_49
13. Zuluaga, M.A., Cardoso, M.J., Modat, M., Ourselin, S.: Multi-atlas propagation whole heart segmentation from MRI and CTA using a local normalised correlation coefficient criterion. In: Ourselin, S., Rueckert, D., Smith, N. (eds.) FIMH 2013. LNCS, vol. 7945, pp. 174–181. Springer, Heidelberg (2013). doi:10.1007/978-3-642-38899-6_21

14. Bai, W., Shi, W., O'Regan, D.P., Tong, T., Wang, H., Jamil-Copley, S., Peters, N.S., Rueckert, D.: A probabilistic patch-based label fusion model for multi-atlas segmentation with registration refinement: Application to cardiac MR images. IEEE Trans. Med. Imaging **32**(7), 1302–1315 (2013)
15. Kolmogorov, V., Zabin, R.: What energy functions can be minimized via graph cuts? IEEE Trans. Pattern Anal. Mach. Intell. **26**(2), 147–159 (2004)
16. Freedman, D., Zhang, T.: Interactive graph cut based segmentation with shape priors. In: Proceedings - 2005 IEEE Computer Society Conference on Computer Vision and Pattern Recognition, CVPR 2005 I, pp. 755–762 (2005)
17. Mahapatra, D.: Cardiac image segmentation from cine cardiac MRI using graph cuts and shape priors. J. Digit. Imaging **26**(4), 721–730 (2013)
18. Ben-Zikri, Y.K., Linte, C.A.: A robust automated left ventricle region of interest localization technique using a cardiac cine MRI atlas. In: Proceedings of the SPIE Medical Imaging 2016: Image-Guided Procedures, Robotic Interventions, and Modeling, vol. 9786 (2016). 9786-27-12
19. Nelder, J.A., Mead, R.: A simplex method for function minimization. Comput. J. **7**(4), 308–313 (1965)
20. Otsu, N.: A threshold selection method from gray-level histograms. IEEE Trans. Syst. Man Cybern. **9**(1), 62–66 (1979)
21. Gudbjartsson, H., Patz, S.: The rician distribution of noisy MRI data. Magn. Resonance Med. **34**(6), 910–914 (1995)
22. Boykov, Y., Veksler, O., Zabih, R.: Fast approximate energy minimization via graph cuts. IEEE Trans PAMI **23**, 1222–39 (2001)
23. Dangi, S., Cahill, N., Linte, C.A.: Integrating atlas and graph cut methods for left ventricle segmentation from cardiac cine MRI. In: Mansi, T., McLeod, K., Pop, M., Rhode, K., Sermesant, M., Young, A. (eds.) STACOM 2016. LNCS, vol. 10124, pp. 76–86. Springer, Cham (2017). doi:10.1007/978-3-319-52718-5_9
24. Dangi, S., Ben-Zikri, Y.K., Lamash, Y., Schwarz, K.Q., Linte, C.A.: Automatic LV feature detection and blood-pool tracking from multi-plane TEE time series. In: van Assen, H., Bovendeerd, P., Delhaas, T. (eds.) FIMH 2015. LNCS, vol. 9126, pp. 29–39. Springer, Cham (2015). doi:10.1007/978-3-319-20309-6_4

Image Segmentation and Modeling of the Pediatric Tricuspid Valve in Hypoplastic Left Heart Syndrome

Alison M. Pouch[1](✉), Ahmed H. Aly[2], Andras Lasso[3], Alexander V. Nguyen[4],
Adam B. Scanlan[4], Francis X. McGowan[4], Gabor Fichtinger[3], Robert C. Gorman[5],
Joseph H. Gorman III[5], Paul A. Yushkevich[1], and Matthew A. Jolley[4]

[1] Department of Radiology, University of Pennsylvania, Philadelphia, PA, USA
pouch@seas.upenn.edu
[2] Department of Bioengineering, University of Pennsylvania, Philadelphia, PA, USA
[3] Laboratory for Percutaneous Surgery, Queen's University, Kingston, Canada
[4] Department of Anesthesiology and Critical Care Medicine,
Children's Hospital of Philadelphia, University of Pennsylvania
Perelman School of Medicine, Philadelphia, PA, USA
[5] Gorman Cardiovascular Research Group, Department of Surgery,
Perelman School of Medicine, University of Pennsylvania, Philadelphia, PA, USA

Abstract. Hypoplastic left heart syndrome (HLHS) is a single-ventricle congenital heart disease that is fatal if left unpalliated. In HLHS patients, the tricuspid valve is the only functioning atrioventricular valve, and its competence is therefore critical. This work demonstrates the first automated strategy for segmentation, modeling, and morphometry of the tricuspid valve in transthoracic 3D echocardiographic (3DE) images of pediatric patients with HLHS. After initial landmark placement, the automated segmentation step uses multi-atlas label fusion and the modeling approach uses deformable modeling with medial axis representation to produce patient-specific models of the tricuspid valve that can be comprehensively and quantitatively assessed. In a group of 16 pediatric patients, valve segmentation and modeling attains an accuracy (mean boundary displacement) of 0.8 ± 0.2 mm relative to manual tracing and shows consistency in annular and leaflet measurements. In the future, such image-based tools have the potential to improve understanding and evaluation of tricuspid valve morphology in HLHS and guide strategies for patient care.

Keywords: Medial axis representation · Multi-atlas segmentation · Tricuspid valve · Hypoplastic left heart syndrome · 3D echocardiography

1 Introduction

Hypoplastic left heart syndrome (HLHS) is a congenital heart disease characterized by underdevelopment of the left heart, including atresia of the mitral and/or aortic valves and an undersized left ventricle that is incapable of supporting systemic circulation. Although the condition constitutes 2–3% of all congenital heart disease [1, 2], HLHS, if left unpalliated, would account for 25–40% of all neonatal cardiac deaths. Palliative surgical intervention is often performed in three stages culminating in the Fontan

© Springer International Publishing AG 2017
M. Pop and G.A. Wright (Eds.): FIMH 2017, LNCS 10263, pp. 95–105, 2017.
DOI: 10.1007/978-3-319-59448-4_10

procedure, which puts systemic and pulmonary circulation in series with a single functioning right ventricle. At all stages, the tricuspid valve is the only functioning atrioventricular valve, and the development of tricuspid regurgitation is understandably associated with morbidity and mortality [3–5]. Accurate assessment of tricuspid valve morphology in HLHS is therefore essential for determining the mechanism of regurgitation and potential techniques for surgical repair [6, 7]. Three-dimensional echocardiography (3DE) captures the complex geometry of the valve in real-time and reduces the subjectivity in image interpretation associated with conventional 2D echocardiography [7–9]. Moreover, 3DE can be used to quantitate morphological abnormalities that may be predictive of clinical outcome in HLHS patients even before the first stage of surgical palliation is performed [5]. Despite a potential relationship between valve morphology and outcome, there is currently no commercial system for semi- or fully automated 3D modeling of tricuspid valves and, in particular, congenitally abnormal valves such as those in HLHS. Although manually cropped volume renderings displayed on the 3DE scanner are valuable for qualitative assessment, they do not automatically generate optimal visualizations or facilitate automatic valve quantification.

Investigation of tricuspid valve morphology in the HLHS population necessitates the development of semi- or fully-automated 3DE image segmentation methods. Valve analysis in this setting, however, presents several challenges relative to valve assessment in the adult population: pediatric patients are more prone to motion than adults, making high resolution EKG gated acquisitions difficult; the images must be acquired transthoracically due to the lack of availability of pediatric transesophageal probes; and there may be visible structural changes in the images introduced by multiple palliative surgeries. To overcome these challenges, we present a strategy that integrates robust image segmentation and shape modeling techniques. Shown in Fig. 1, the tricuspid valve analysis pipeline begins with identification of landmarks that provide initialization for subsequent fully automatic steps. With these landmarks, the input target image is segmented by multi-atlas label fusion (MALF), which uses deformable registration to warp a set of expert-labeled example 3DE images called "atlases" to the target image, and it combines the warped atlas segmentations into a consensus voxel-wise label map of the tricuspid leaflets. Next, in order to make consistent quantitative measurements, a deformable model of the tricuspid valve is warped to capture the geometry of the valve in the label map. The combination of label fusion and deformable modeling has a number of benefits that have been demonstrated in adult mitral and aortic valve applications [10, 11]: it exploits knowledge of tricuspid valve image appearance through the use of expert-labeled atlases; the deformable template has immutable topology, thereby preventing extraneous holes or artifacts in the output model; and the valve is represented volumetrically, as a structure with locally varying thickness. The goal of this work is to demonstrate the first semi-automated image segmentation and geometric modeling of the tricuspid valve in 3DE images from pediatric patients with HLHS. Three-dimensional tricuspid valve models allow interactive 3D visualization and quantitative evaluation of valve morphology, which could improve understanding of tricuspid valve disease in HLHS and potentially translate into enhanced surgical repair.

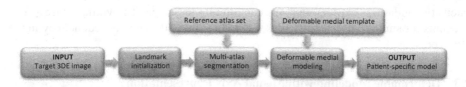

Fig. 1. Tricuspid valve image analysis pipeline.

2 Materials and Methods

2.1 Image Acquisition and Data Sets

Transthoracic 3DE images of the tricuspid valve were acquired from 16 pediatric patients with HLHS aged 4 days to 15 years. Fifteen of these patients had previously undergone the Fontan procedure (stage 3 of surgical palliation). Investigation was approved by the Institutional Review Board at the Children's Hospital of Philadelphia. The images were acquired with the iE33 platform (Philips Medical Systems, Andover, MA) in full volume or 3D zoom mode using an X7 or X5 probe. For each subject, a 3DE image of the closed tricuspid valve in mid-systole (the frame midway between valve opening and closing) was selected for analysis. Systole was chosen since it is the phase of the cardiac cycle during which atrioventricular valves are closed and resist the pressure of ventricular contraction. Abnormal annular shape and leaflet profile during systole has been associated with disease in adult mitral valves [12] and tricuspid valves in HLHS [5, 7]. The images were anonymized and exported in Cartesian format with an approximate resolution of $0.7 \times 0.7 \times 0.5$ mm^3. Each 3DE image was manually traced to create an atlas with the three leaflets (anterior, posterior, and septal) assigned a separate label. In addition, five landmarks were identified in each image for the purpose of initializing multi-atlas segmentation: the mid-septal (MS) annular point, anteroseptal (AS) commissure, the anteroposterior (AP) commissure, the posteroseptal (PS) commissure, and the coaptation point of the three leaflets. Manual tracings of all images were completed by two observers in the "Segments" module of the 3D Slicer application [13]. The proposed segmentation framework was evaluated in a leave-one-out manner: each of the 16 images was segmented using the remaining 15 images and their manual segmentations as atlases for MALF. The resulting segmentations were compared to the manual segmentations to establish the accuracy of the automated method.

2.2 Multi-atlas Segmentation with Joint Label Fusion

With a set of user-identified landmarks, the first step of fully automated tricuspid valve image analysis is multi-atlas joint label fusion. Briefly, a collection of atlases (i.e., 3DE images and labels for the tricuspid leaflets) is registered to a target image, first with a landmark-guided affine transformation and then diffeomorphic deformable registration. The candidate segmentations generated by each atlas are combined to create a consensus label map using the weighted voting method detailed in [14]. Each atlas contributes to the final consensus label map according to locally varying weights determined by intensity similarities between that atlas and the target image. The voting process also accounts for similarities between atlases to reduce bias attributed to redundancy in the atlas set.

2.3 Deformable Modeling with Medial Axis Representation

MALF generates a label map of the tricuspid valve in the 3DE image, but it does not guarantee the correct topology of the segmentation or provide a straightforward means of computing morphological measurements. To capture the shape of the tricuspid valve in a manner that preserves leaflet topology and enables automated morphometry, deformable modeling with medial axis representation is used to describe valve geometry in the label map generated by label fusion.

Medial axis representation and deformable model fitting. Deformable medial modeling represents an object's geometry in terms of its medial axis, which Blum has defined as a locus of the centers of maximal inscribed balls (MIBs) that lie inside the object and cannot be made any larger without crossing the object boundary [15]. The center of each MIB is associated with a radius R, the distance between that point on the skeleton and the object boundary (Fig. 2). Deformable medial modeling determines the medial axis of a structure through *inverse skeletonization*, wherein the medial axis of the object is first explicitly defined as a continuous parametric manifold $\mathbf{m}:\Omega \rightarrow \mathbb{R}^3, \Omega \in \mathbb{R}^2$ with an associated parametric *thickness* scalar field $R:\Omega \rightarrow \mathbb{R}^+$. The boundary of the object is then derived analytically as described in [16]. Given a label map of an object, a pre-defined template $\{\mathbf{m}, R\}$ is deformed through optimization of a Bayesian objective function with respect to a set of coefficients that parameterize $\mathbf{m}$ and R. The objective function maximizes overlap between the medial template and label image, while enforcing medial constraints and model smoothness through regularization penalties described in [10]. The regularization penalties help to ensure smoothness of the medial edge and prevent leaflet overlap during model deformation. The result is a fitted, patient-specific medial model of the valve. In this work, the tricuspid valve template is a single, non-branching medial manifold (Fig. 2b).

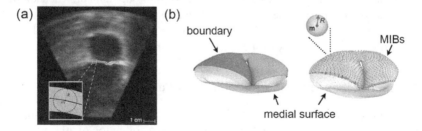

Fig. 2. (a) Image of the tricuspid valve with a 2D medial geometry diagram showing the cross-section of a maximally inscribed ball (MIB) in a leaflet. (b) 3D medial axis diagram of the tricuspid valve showing the medial surface and the cutaway boundary (left) and the MIBs whose centers form the medial surface (right). $\mathbf{m}$ refers to a point on the medial axis, and R is the radius of the MIB.

Template generation. To obtain a deformable medial template of the tricuspid valve, one of the 16 manual segmentations was selected at random and a triangulated mesh of the segmentation's skeleton was created using the procedure described in [11]. To reduce potential bias associated with a single-subject template, the single-subject template was fitted to all the subject's multi-atlas segmentations, and generalized Procrustes analysis was used to compute a new "average" template. This mean shape, shown in Fig. 2b, served as the final medial template and was deformed to all subjects' MALF results to obtain patient-specific models. While this template generation was not completely unbiased, creation of the mean template served to reduce bias introduced by individual valve geometries.

2.4 Tricuspid Valve Morphometry

To demonstrate automated tricuspid valve morphometry, annular and leaflet measurements were computed from medial models that were deformed to the MALF results and from models that were deformed directly to the manual segmentations:

- Surface area: sum of triangle areas on the atrial side of the anterior, posterior, or septal leaflet
- Bending angle: angle between the normal vectors of the anterior and posterior planes. The anterior plane is defined as the plane that best fits the anterior annular contour and the septal annular contour between the AS commissure and MS point. The posterior plane is defined as the plane that best fits the posterior annular contour and the septal annular contour between the PS commissure and MS point (Fig. 3).

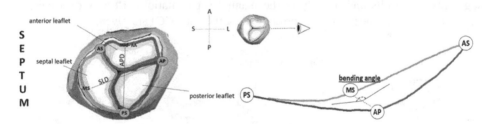

Fig. 3. Schematic of the tricuspid valve (left) and diagram of the annular bending angle measurement as viewed from the AP to MS (right). (A – anterior, P – posterior, S – septal, L – lateral, AP – anteroposterior commissure, AS – anteroseptal commissure, PS – posteroseptal commissure, MS – mid-septum, mid-AA – mid-anterior annulus, SLD – septal-lateral diameter, APD – antero-posterior diameter)

- Annular area: area enclosed by the projection of the annulus onto the annular best-fit plane
- Annular circumference: sum of distances between consecutive points on the annulus
- Septal-lateral diameter (SLD): distance between the MS annular point and the AP commissure, defined as the point on anterior annulus nearest to the posterior annulus

- Antero-posterior diameter (APD): distance between the mid-anterior annular (mid-AA) point and the PS commissure, defined as the point on the posterior annulus nearest to the septal annulus
- Regional annular height: the perpendicular distance from the annulus to the annular best-fit plane, as a function of rotation angle about the annulus
- Global annular height: difference between the maximum and minimum annular height

In addition to annular bending angle, regional annular height was computed as a function of rotation angle around the valve, beginning at the junction of the anterior and septal leaflets. Regional annular height has been shown to be a meaningful descriptor of adult mitral and tricuspid valve geometry [12, 17] and is complementary to annular bending angle in quantifying annular curvature.

3   Results

Examples of manual segmentations, MALF results, and model fittings are shown in Fig. 4 for three pediatric patients with HLHS. The symmetric mean boundary error (MBE) between the manual segmentations and the deformable medial model fitted to the multi-atlas segmentation results was 0.8 ± 0.2 mm. For reference, the MBE between the manual segmentations and the deformable models fitted directly to the manual segmentations was 0.3 ± 0.1 mm, and the MBE between the manual and multi-atlas segmentations without model fitting was 0.7 ± 0.2 mm. Table 1 lists and compares morphological measurements derived from medial models fitted to the multi-atlas segmentation results and directly to the manual segmentations. Results of a paired Student's t-test and the intraclass correlation coefficient (ICC) are given.

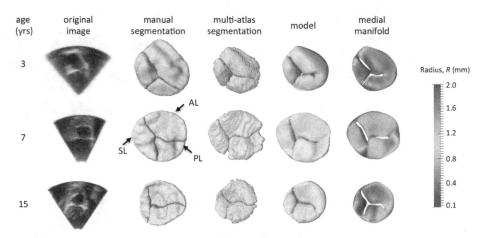

Fig. 4. Tricuspid valve segmentation results for three patients. The medial surface is shown with the radius function in color (rightmost column).

Table 1. Measurements derived from the deformable models fitted to the multi-atlas (automated) and manual segmentations. The p-value, bias, and intraclass correlation (ICC) relate to the comparison of automated and manual measurements, and r_{BSA} is the Pearson correlation between the percent error in the automated measurement and the patient's BSA.

Measurement	Automated segmentation	Manual segmentation	p-value	Bias	ICC	r_{BSA}
Atrial anterior leaflet surface area (mm^2)	407.6 ± 129.2	436.5 ± 159.6	0.577	−28.9	0.75	−0.27
Atrial posterior leaflet surface area (mm^2)	259.8 ± 102.1	325.7 ± 155.5	0.167	−65.9	0.67	−0.12
Atrial septal leaflet surface area (mm^2)	210.0 ± 88.3	290.5 ± 114.0	0.033	−80.5	0.52	−0.17
Bending angle (degrees)	165.3 ± 10.4	168.1 ± 8.13	0.401	−2.8	0.78	−0.05
Annular area (mm^2)	716.2 ± 245.6	819.1 ± 294.1	0.292	−102.8	0.84	0.00
Annular circumference (mm)	96.4 ± 17.1	102.4 ± 19.9	0.369	−6.0	0.86	−0.08
Septal-lateral diameter (mm)	30.2 ± 5.7	32.3 ± 7.1	0.359	−2.1	0.89	−0.43
Anterior-posterior diameter (mm)	29.5 ± 5.8	30.8 ± 6.6	0.567	−1.3	0.88	−0.06
Global annular height (mm)	3.7 ± 1.5	3.4 ± 1.3	0.613	0.3	0.69	−0.12

To assess segmentation and measurement accuracy relative to valve size, MBE and percent error in the measurements in Table 1 were evaluated in terms of body surface area (BSA), which has a known relationship with valve size: valve diameter scales linearly with BSA$^{1/2}$ and valve area scales linearly with BSA [18]. In Fig. 5, MBE is plotted as a function of BSA$^{1/2}$, showing no significant linear relationship (Pearson correlation $r = 0.21$, $p = 0.4$). The percent error in linear measurements such as antero-lateral diameter was evaluated as a function of BSA$^{1/2}$, and the percent error in area measurements such as leaflet surface area was evaluated as a function of BSA. No statistically significant linear relationship between measurement accuracy and BSA$^{1/2}$ or BSA was observed (Table 1, Pearson correlation $|r| < 0.43$, $p > 0.1$).

In Fig. 6, regional annular height is plotted as a function of rotational angle about the annulus. Here, regional annular height is shown for each individual subject (black curves), and the mean regional annular height for the population is shown in red.

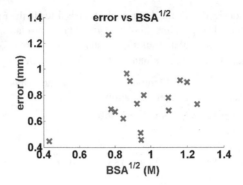

Fig. 5. Mean boundary error as a function of the square root of body surface area (BSA).

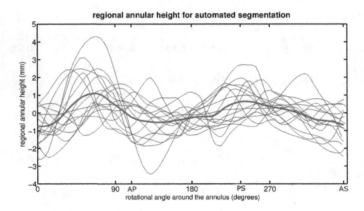

Fig. 6. Annular height as a function of rotation angle around the annulus. Black curves correspond to individual patients, and red is the mean regional height. (AP – anteroposterior commissure, PS – posteroseptal commissure, AS – anteroseptal commissure at 0°) (Color figure online)

4   Discussion

The basic three-dimensional structure of the tricuspid valve in HLHS has been shown to be associated with valve competence and with patient survival [5, 7]. The ability to rapidly, automatically, and comprehensively characterize tricuspid valve morphology from readily available transthoracic 3DE is critical to further understanding of valve dysfunction and translation to clinical application. This study is the first to demonstrate a semi-automated method for image segmentation, geometric modeling, and morphometry of the tricuspid valve in patients with HLHS. Further, this work demonstrates preliminary evidence that the adaptation and evolution of image processing algorithms that have effectively modeled mitral valves in transesophageal 3DE in the adult population are applicable to transthoracic images in congenital heart disease.

Accuracy of automatic segmentation and quantification is essential to reliable valve modeling. MBE between the manual segmentation and the medial models fitted to the results of multi-atlas segmentation (0.8 ± 0.2 mm) was on the order of 1–2 voxels. The advantage of a medial representation over multi-atlas segmentation alone is that deformable modeling enables consistent morphometry in a patient population. Medial axis representation, in particular, distinguishes the atrial and ventricular surfaces of the leaflets and facilitates measurements such as locally varying thickness (derived from the parameter R). The manual versus semi-automatic segmentation comparison in this study is on par with studies of mitral and aortic valve segmentation in adult transesophageal 3DE images, wherein MBEs of 1.54 ± 1.17 mm [19], 0.59 ± 0.49 mm [20], and 0.6 ± 0.2 mm [10] have been reported. In this study, no linear relationship was observed between segmentation accuracy and $BSA^{1/2}$ (Fig. 5), suggesting that the segmentation algorithm is applicable to tricuspid valves of a wide size range (53–128 mm in circumference) in the HLHS population. Since the majority of the subjects in this study had a BSA lower than 0.6 M^2, a larger cohort of patients ranging from neonates to adults will need to be assessed in the future to evaluate broad applicability.

In addition to image segmentation, this work demonstrates an evolving tool for evaluation of tricuspid valve morphology in HLHS (Table 1, Figs. 5 and 6). When comparing measurements derived from deformable models fitted to the manual and multi-atlas segmentations, no significant difference ($p < 0.05$) was observed except for measurement of septal leaflet surface area. Except for this measurement, the intraclass correlations between manual and automatically derived measurements were good to excellent (Table 1). Illustrating individual and mean regional annular height, Fig. 6 demonstrates the ability to characterize localized tricuspid annular geometry from medial models. When compared to normal adult mitral annuli, which have an average height of 7 ± 2 mm and are characterized by a distinctive saddle shape [12], both the individual and average regional annular height contours of the HLHS patients in Fig. 6 suggest that the tricuspid annulus is relatively flat at mid-systole, but exhibits substantial variability between patients. As a measure of annular flatness, bending angle in prestage 1 palliation has been associated with tricuspid valve failure at medium-term follow-up of HLHS [5]. The bending angles computed in this study (Table 1) are comparable to those reported in the literature for HLHS patients [7].

While this study is a proof of concept of segmentation and modeling of the tricuspid valve in pediatric transthoracic 3DE, future work will focus on expanding atlases and application across a large age range, making these tools fully automated, and extending the current leaflet quantification metrics. Segmentation at multiple frames in the cardiac cycle, as well as evaluation of intra- and inter-observer variability in automated valve analysis, are important extensions of this work. Future application of this evolving methodology to larger HLHS populations may facilitate valvular shape characterization, associate morphological metrics with clinical outcomes, and eventually translate into clinical decision making.

Acknowledgement. This work was supported by grant numbers EB017255 and HL103723 from the National Institutes of Health, as well as the Children's of Hospital of Philadelphia Department of Anesthesia and Critical Care Medicine, Cancer Care Ontario with funds provided by the Ontario Ministry of Health and Long-Term Care, the Natural Sciences and Engineering Research Council of Canada (NSERC), and the Neuroimage Analysis Center supported by the National Institute of Biomedical Imaging and Bioengineering (P41 EB015902).

References

1. Gordon, B.M., Rodriguez, S., Lee, M., Chang, R.K.: Decreasing number of deaths of infants with hypoplastic left heart syndrome. J. Pediatr. **153**(3), 354 (2008)
2. Reller, M.D., Strickland, M.J., Riehle-Colarusso, T., Mahle, W.T., Correa, A.: Prevalence of congenital heart defects in metropolitan Atlanta, 1998–2005. J. Pediatr. **153**(6), 807 (2008)
3. Barber, G., Helton, J.G., Aglira, B.A., Chin, A.J., Murphy, J.D., Pigott, J.D., Norwood, W.I.: The significance of tricuspid regurgitation in hypoplastic left-heart syndrome. Am. Heart J. **116**, 1563–1567 (1988)
4. Elmi, M., Hickey, E.J., Williams, W.G., Van Arsdell, G., Caldarone, C.A., McCrindle, B.W.: Long-term tricuspid valve function after Norwood operation. J. Thorac. Cardiovasc. Surg. **142**, 1341–1347 (2011). e4
5. Kutty, S., Colen, T., Thompson, R.B., Tham, E., Li, L., Vijarnsorn, C., Polak, A., Truong, D.T., Danford, D.A., Smallhorn, J.F., Khoo, N.S.: Tricuspid regurgitation in hypoplastic left heart syndrome. Circ. Cardiovasc. Imaging **7**, 765–772 (2014)
6. Bharucha, T., Honjo, O., Seller, N., Atlin, C., Redington, A., Caldarone, C.A., van Arsdell, G., Mertens, L.: Mechanisms of tricuspid valve regurgitation in hypoplastic left heart syndrome: a case-matched echocardiographic-surgical comparison study. Eur. Heart J. Cardiovasc. Imaging **14**, 135–141 (2013)
7. Takahashi, K., Mackie, A.S., Rebeyka, I.M., Ross, D.B., Robertson, M., Dyck, J.D., Inage, A., Smallhorn, J.F.: Two-dimensional versus transthoracic real-time three-dimensional echocardiography in the evaluation of the mechanisms and sites of atrioventricular valve regurgitation in a congenital heart disease population. J. Am. Soc. Echocardiogr. **23**, 726–734 (2010)
8. Badano, L.P., Agricola, E., Perez de Isla, L., Gianfagna, P., Zamorano, J.L.: Evaluation of the tricuspid valve morphology and function by transthoracic real-time three-dimensional echocardiography. Eur. J. Echocardiogr. **10**, 477–484 (2009)
9. Anwar, A.M., Geleijnse, M.L., Soliman, O.I., McGhie, J.S., Frowijn, R., Nemes, A., van den Bosch, A.E., Galema, T.W., Ten Cate, F.J.: Assessment of normal tricuspid valve anatomy in adults by real-time three-dimensional echocardiography. Int. J. Cardiovasc. Imaging **23**, 717–724 (2007)
10. Pouch, A.M., Wang, H., Takabe, M., Jackson, B.M., Gorman 3rd, J.H., Gorman, R.C., Yushkevich, P.A., Sehgal, C.M.: Fully automatic segmentation of the mitral leaflets in 3D transesophageal echocardiographic images using multi-atlas joint label fusion and deformable medial modeling. Med. Image Anal. **18**, 118–129 (2014)
11. Pouch, A.M., Tian, S., Takebe, M., Yuan, J., Gorman, R., Cheung, A.T., Wang, H., Jackson, B.M., Gorman, J.H., Gorman, R.C., Yushkevich, P.A.: Medially constrained deformable modeling for segmentation of branching medial structures: application to aortic valve segmentation and morphometry. Med. Image Anal. **26**, 217–231 (2015)

12. Jassar, A.S., Vergnat, M., Jackson, B.M., McGarvey, J., Cheung, A.T., Ferrari, G., Woo, Y.J., Acker, M.A., Gorman, R.C., Gorman III, J.H.: Regional annular geometry in patients with mitral regurgitation: Implications for annuloplasty ring selection. Ann. Thorac. Surg. **97**(1), 64–70 (2014)
13. Fedorov, A., Beichel, R., Kalpathy-Cramer, J., Finet, J., Fillion-Robin, J.-C., Pujol, S., Bauer, C., Jennings, D., Fennessy, F.M., Sonka, M., Buatti, J., Aylward, S.R., Miller, J.V., Pieper, S., Kikinis, R.: 3D slicer as an image computing platform for the quantitative imaging network. Magn. Reson. Imaging **30**(9), 1323–1341 (2012)
14. Wang, H., Suh, J.W., Das, S., Pluta, J., Craige, C., Yushkevich, P.: Multi-atlas segmentation with joint label fusion. IEEE Trans. Pattern Anal. Mach. Intell. **35**(3), 611–623 (2013)
15. Blum, H.: A transformation for extracting new descriptors of shape. In: Wathen-Dunn, W. (ed.) Models for the Perception of Speech and Visual Form, pp. 362–380. MIT Press, Cambridge (1967)
16. Yushkevich, P.A., Zhang, H., Gee, J.C.: Continuous medial representation for anatomical structures. IEEE Trans. Med. Imaging **25**(12), 1547–1564 (2006)
17. Fukuda, S., Saracino, G., Matsumura, Y., Daimon, M., Tran, H., Greenberg, N.L., Hozumi, T., Yoshikawa, J., Thomas, J.D., Shiota, T.: Three-dimensional geometry of the tricuspid annulus in healthy subjects and in patients with functional tricuspid regurgitation. Circulation **114**(suppl I), I-492–I-498 (2006)
18. Sluysmans, T., Colan, D.: Theoretical and empirical derivation of cardiovascular allometric relationships in children. J. Appl. Physiol. **99**, 445–457 (2005)
19. Ionasec, R.I., Voigt, I., Georgescu, B., Wang, Y., Houle, H., Vega-Higuera, F., Navab, N., Comaniciu, D.: Patient-specific modeling and quantification of the aortic and mitral valves from 4-D cardiac CT and TEE. IEEE Trans. Med. Imaging **29**, 1636–1651 (2010)
20. Schneider, R.J., Tenenholtz, N.A., Perrin, D.P., Marx, G.R., del Nido, P.J., Howe, R.D.: Patient-specific mitral leaflet segmentation from 4D ultrasound. Med. Image Comput. Comput. Assist. Interv. **14**, 520–527 (2011)

Strain-Based Parameters for Infarct Localization: Evaluation via a Learning Algorithm on a Synthetic Database of Pathological Hearts

Gerardo Kenny Rumindo[1]([✉]), Nicolas Duchateau[1], Pierre Croisille[1],
Jacques Ohayon[2], and Patrick Clarysse[1]

[1] Univ.Lyon, INSA-Lyon, Université Claude Bernard Lyon 1, UJM-Saint Etienne,
CNRS, Inserm, CREATIS UMR 5220, U1206, Lyon, France
Kenny.Rumindo@creatis.insa-lyon.fr
[2] University of Savoie Mont Blanc, Polytech Annecy-Chambéry,
Laboratory TIMC-IMAG/DyCTiM2, UGA, CNRS, Grenoble, France

Abstract. Localization of infarcted regions is essential to determine the most appropriate treatment for patients with cardiac ischemia. Myocardial strain partially reflects the location of infarcted regions, which demonstrated potential use in clinical practice. However, strain patterns are complex and simple thresholding is not sufficient to locate the infarcts. Besides, many strain-based parameters exist and their sensitivities to myocardial infarcts have not been directly investigated. In our study, we propose to evaluate nine strain-based parameters to locate infarcted regions. For this purpose, we designed a large database (n = 200) of synthetic pathological finite-element heart models from 5 real healthy left ventricle geometries. The infarcts were incorporated with random location, shape and degree of severity. In addition, we used a state-of-the-art learning algorithm to link deformation patterns and infarct location. Based on our evaluation, we propose to sort the strain-based parameters into three groups according to their performances in locating infarcts.

Keywords: Finite-element model · Myocardial infarct · Myocardial strain · Infarct diagnosis · Machine learning

1 Introduction

The clinical value of determining myocardial viability to help the physicians decide on the best treatment for patients with cardiac ischemia has been established [5]. Late Gadolinium Enhancement - an MR imaging method - is generally accepted as the gold standard to assess myocardial viability and therefore to locate myocardial infarcts. However, it requires contrast injection, is costly and not available for all patients. Myocardial strains - extracted from cardiac MR [14] or echocardiography [3] - have also been used to identify dysfunctional

© Springer International Publishing AG 2017
M. Pop and G.A. Wright (Eds.): FIMH 2017, LNCS 10263, pp. 106–114, 2017.
DOI: 10.1007/978-3-319-59448-4_11

regions of the heart [1]. However, due to the complex strain patterns, simple thresholding is not sufficient and further processing is required [12]. Learning-based algorithms have been investigated for better detecting [9] and locating [2,4] infarcts from strain data. These methods are promising and can be generalized to different modalities and subjects, in addition to showing high accuracy of infarct detection.

Several strain-based parameters have been investigated for infarct localization. However, many more exist and their potential has not been evaluated. In our study, we propose to evaluate the following nine strain-based parameters; principal strain, effective strain, fractional anisotropy [13], three local directional strains (radial, circumferential, longitudinal), and three stretch-dependent invariants (in fiber, cross-fiber, and sheet-normal directions). They were assessed in terms of their localization performance (sensitivity and specificity) given a learning algorithm to locate the infarcts. The algorithm uses a regression to find the transfer function between the tested parameters and the infarct location [4]. To do so, we designed a large database of synthetically-generated pathological cases that incorporates infarcts with different locations, shapes, sizes, and degrees of severity. Notably, we demonstrated that by properly designing and exploiting a large and sufficiently varied database of pathological cases, we were able to really push the limits of each strain parameter used in the detection algorithm and to better estimate their localization performances.

2   Materials and Methods

Five LV meshes (3552 hexahedral elements) from healthy volunteers were obtained from an open access source [6]. A finite-element model of each healthy LV in the diastolic (filling) phase was simulated. Afterwards, 40 pathological cases were generated based on each LV geometry - resulting in 200 pathological cases - and the nine strain-based parameters were computed. Finally, the learning algorithm was applied to evaluate the localization performance of each parameter.

2.1   Database of Pathological Cases

Simulation of Healthy Cases. The diastolic simulation for the healthy LV model follows the principles detailed in [6,11]. Rule-based fiber orientation was incorporated into the LV geometry by first defining a pseudoprolate spheroidal coordinate system that consists of radial (orthogonal to the LV surface), circumferential and longitudinal (tangential to the LV surface) directions. Subsequently, an orthogonal fiber coordinate system was defined, which is comprised of fiber, sheet and sheet-normal directions. The fibers are oriented with an elevation angle distribution of $-60°$ to $+60°$ from the epicardium to the endocardium [7].

The constitutive law was the transversely isotropic Fung-type law [6] (Eq. 1). The strain energy density function is divided into two parts; the isochoric term, which is based on the Green-Lagrange strain tensor $\mathbf{E}$, and the volumetric term:

$$\mathbf{W} = \overbrace{\frac{C}{2}(e^Q - 1)}^{\text{Isochoric term}} + \overbrace{\frac{1}{D}(\frac{J^2-1}{2} - ln\ J)}^{\text{Volumetric term}}$$

$$\underbrace{Q = b_f E_{ff}^2 + b_t E_{ss}^2 + b_t E_{nn}^2 + b_t(E_{sn}^2 + E_{ns}^2) + b_{fs}(E_{fs}^2 + E_{sf}^2 + E_{fn}^2 + E_{nf}^2)}$$

Exponential terms of the isochoric part; $\boldsymbol{E}$: Green-Lagrange strain tensor

(1)

where C, b_f, b_t and b_{fs} are the material parameters to be personalized for each subject; J is the determinant of the deformation gradient tensor $\mathbf{F}$; whereas D was set to 0.001 to enforce quasi-incompressibility. The subscripts f, s and n denote the fiber, sheet and sheet-normal directions, respectively. Active tension was not taken into account as we only simulated the diastolic phase in this study. The models were solved using the finite-element software ABAQUS[1], and the constitutive law was implemented in ABAQUS user material subroutine UMAT.

The material parameters were personalized for each subject based on the end-diastolic pressure-volume relationships described in [8]. The beginning- and end-diastolic volumes of each subject were known; the corresponding pressures were set to 0 and 9 mmHg, respectively.

Simulation of Pathological Cases. Forty synthetically-generated cases were generated from each of the 5 healthy LV models, resulting in 200 cases. Infarct regions were incorporated through a binary value at each mesh element. These regions were defined as follows: first, the intra-ventricular junction and apex of each LV were manually selected from the MR images, which enabled us to define the left anterior descending (LAD) coronary artery territory on each LV. The infarcts had truncated ellipsoidal or spheroidal shapes with arbitrary sizes ranging from 0.5–99.7 ml, whose center points were constrained to be on the endocardial surface and within the LAD territory. This territory was chosen due to its high prevalence [10]. The material properties of the infarct were set to be stiffer compared to the healthy ones, and they were assigned in a uniformly-distributed manner by changing the material parameter C (Eq. 1) from 1 up to 2.50 times of the personalized healthy values. The resulting infarcts have an average myocardial mass of 22.9 ± 21.3 g and an average volume of 21.6 ± 20.1 ml, corresponding to $15.6 \pm 13.2\%$ of the myocardium total volume.

2.2 Strain-Based Parameters

Nine strain-based parameters were evaluated in this study (Table 1). Starting from the deformation gradient tensor $\mathbf{F}$ extracted from ABAQUS simulations, the right Cauchy stretch tensor $\mathbf{C}$ and the Green-Lagrange strain tensor $\mathbf{E}$ were calculated: $\mathbf{C} = \mathbf{F}^T.\mathbf{F}$; $\mathbf{E} = \frac{1}{2}(\mathbf{C} - \boldsymbol{I})$; where $\boldsymbol{I}$ is the identity tensor.

---

[1] http://www.3ds.com/products-services/simulia/products/abaqus/.

Table 1. Strain-based parameters evaluated in this study

Parameters	Formulae/Comments	
Radial strain	With respect to pseudoprolate coordinate system	
Circ. strain		
Long. strain		
Fiber invariant	$\mathbf{f_0}.(\mathbf{C}\mathbf{f_0})$ $\mathbf{f_0}$: fiber direction	
Cross-fiber invariant	$\mathbf{s_0}.(\mathbf{C}\mathbf{s_0})$ $\mathbf{s_0}$: cross-fiber direction	
Sheet-normal invariant	$\mathbf{n_0}.(\mathbf{C}\mathbf{n_0})$ $\mathbf{n_0}$: sheet-normal direction	
Principal strain	The max positive Eigenvalue of $\mathbf{E}$	
Effective strain	$\sqrt{\dfrac{2}{9}[(E_{xx} - E_{yy})^2 + (E_{yy} - E_{zz})^2 + (E_{xx} - E_{zz})^2 + 6(E_{xy}^2 + E_{yz}^2 + E_{xz}^2)]}$	
Fractional anisotropy	$\sqrt{\dfrac{3}{2}\dfrac{(\lambda_1-\lambda)^2+(\lambda_2-\lambda)^2+(\lambda_3-\lambda)^2}{\lambda_1^2+\lambda_2^2+\lambda_3^2}}$	$\lambda = (\lambda_1 + \lambda_2 + \lambda_3)/3$
		$\lambda_1, \lambda_2, \lambda_3$: Eigenvalues of $\mathbf{E}$

2.3   Infarct Localization

Data Alignment. Due to the use of different geometries, the strain parameters were not directly comparable and spatial correspondence should be obtained. Thus, normalized local coordinates were computed on each geometry corresponding to the pseudoprolate coordinate system explained in Sect. 2.1. Then, a reference geometry was computed by Procrustes analysis with similarity transformation. Local coordinates were also computed for this reference geometry. Finally, local data (the strain parameters and the infarct binary labels) were transported to the reference geometry using the correspondence of the local coordinates. The data transportation involved interpolation, addressed by ridge regression with

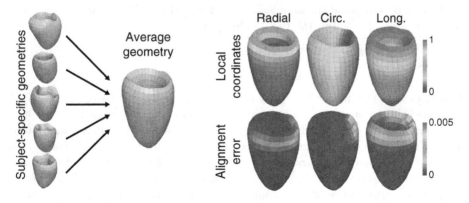

Fig. 1. Left: The five LV and the calculated reference. Right: The local normalized pseudoprolate coordinates (radial, circumferential, longitudinal) and the average alignment error due to the interpolation of the local coordinates onto the reference.

a Gaussian kernel, with a balanced weight between similarity and regularization terms. The interpolation resulted in negligible errors. The whole process is summarized in Fig. 1.

Regression. The aligned strain parameters and infarct location were treated as column vectors of length equal to the number of elements for the reference mesh. The link between each strain parameters and infarct location was learned via kernel ridge regression, inspired by the algorithm described in [4]. Direct regression was preferred over going through an intermediate space of reduced dimensionality, whose purpose is mainly for uncertainty modeling without substantially affecting the localization performance.

Each healthy LV geometry was used to generate 40 out of the total 200 pathological cases, which in turn might present some biases in the detection algorithm due to a substantial amount of relatively similar cases, thereby limiting the difference in deformation patterns and in the distribution of the evaluated parameters. To avoid this bias when a case was investigated, all other cases derived from the same geometry were not included in the training set. The training set was randomly constructed from the remaining cases with a certain percentage, which was in the range of 5–40% of the total population. The localization performance of each parameter was then tested against the size of the training set. The limit of 40% was set since it was observed that the algorithm was able to accurately detect the infarcts with this percentage of population as the training set.

The regression output consisted of a non-binary scalar value at each mesh element due to the linear combination of infarct locations. A binary localization of the infarct was obtained after thresholding this output by an appropriate value, as inspired by the two-step process described in [4]: (1) ROC analysis was first performed for each case to obtain a set of best individual thresholds. However, this analysis relies on the ground truth infarct location for each case, which is unrealistic in real-world clinical application, thus (2) the same optimal threshold was applied to all cases, whose values were chosen as the median of the set of the best individual thresholds. Thus, we ended up with a single consensual threshold, which is used in the rest of this paper.

3   Results

Figure 2 depicts the infarct localization for pathological cases based on principal strain and sheet-normal invariant, which were observed as one of the best and worst parameters, respectively. The degree of severity of the infarct was based on how much stiffer it was in comparison to the healthy ones, with respect to the value of the material parameter C. The principal strain is able to localize the infarcts substantially better than the sheet-normal invariant. The infarct localization performance was evaluated in terms of sensitivity and specificity after thresholding the regression output by the consensual value as explained in Sect. 2.3. The results were summarized in terms of the distance to the ideal localization performance, namely a sensitivity and a specificity equal to 1.

Figure 3 summarizes the performance of the nine strain-based parameters, with the color bar depicting the distance to an ideal localization. The horizontal axis represents a combination of the degree of severity and the size of the infarct normalized from 0 to 1, where 0 signifies smaller infarcts and/or infarcts with

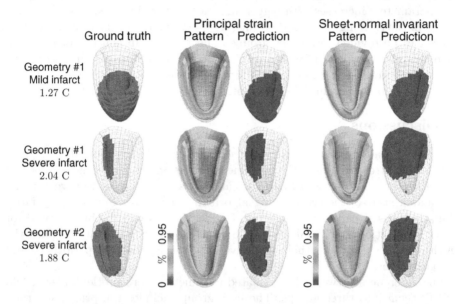

Fig. 2. Infarct localization for some pathological cases based on principal strain and sheet-normal invariant; amongst the best and the worst parameters, respectively.

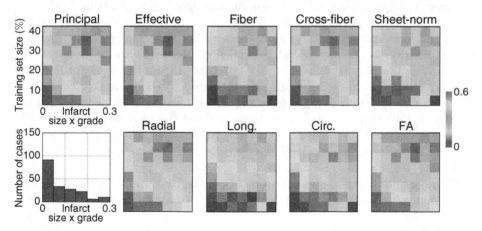

Fig. 3. Performance of the strain-based parameters (median over the tested cases). The color bar depicts the distance to an ideal localization. Horizontal axis: a normalized combination of the degree of severity and the size of the infarct. Vertical axis: the percentage of the population used as the training set. The histogram shows the number of cases that correspond to the infarct size-severity measure on the horizontal axis. (Color figure online)

closer material properties to the healthy ones; whereas the vertical axis represents the percentage of the population included in the training set. Lower values on the color bar signify better accuracy. In addition, the histogram presents the distribution of the number of pathological cases that correspond to each infarct size-severity bin on the horizontal axis. As expected, the localization performance of all parameters increases with bigger and more severe infarcts, as well as with the size of the training set. Based on this, the performance of the evaluated parameters can be divided into three groups. The best group consists of principal strain, effective strain, and fractional anisotropy. The second-best group consists of cross-fiber invariant and radial strain; whereas the last group consists of fiber and sheet-normal invariant, circumferential and longitudinal strain.

4   Discussions

Our study details a novel strategy for evaluating different strain-based parameters for infarct diagnosis. We combined finite-element modeling to generate a large synthetic database of pathological hearts based on healthy volunteers' data and a state-of-the-art learning algorithm to localize myocardial infarct. Nine strain-based parameters, several of which have yet to be investigated for infarct localization, were evaluated. The large database allowed us to push the algorithm to its limit and thoroughly evaluate the localization performance of each parameter.

The differences between the evaluated parameters were subtle, but we could identify them into three groups. The best group includes the principal strain, effective strain, and fractional anisotropy. Their performances were very similar. The second group shows slightly inferior, yet still good performance. It notably comprises the radial strain, which is readily available in clinical setting. It is interesting that the best parameters are those that are direction-independent, which was somewhat expected as they may be able to better highlight and extract the general distribution of myocardial deformation. However, the invariant in the fiber direction performed worse than the radial strain and the invariant in cross-fiber direction. The reason for this might be due to the boundary conditions applied for the diastolic-filling simulation, which forced the elements to deform radially as the volume of the cavity increases and the incompressibility of the element is enforced. This should be investigated further. In addition, the combination between parameters were not investigated and are also left for future work.

Our synthetic database was derived only from five LV geometries. Although these geometries come from MRI examinations of healthy volunteers - thereby ensuring that the personalized healthy material parameters are physiological - it is of interest in the future to be able to generate a database with larger variability in terms of geometries and subsequently deformation patterns. It is also well known that geometrical alterations of the LV might occur in chronic ischemic patients, i.e. the thickening and thinning of ventricular wall around the remote and infarcted myocardium, respectively. Since our database was derived from healthy volunteers, this effect was not taken into account.

It is certainly essential to evaluate the infarct detection performance of the strain-based parameters on real clinical data of ischemic patients. However, the complexity in acquiring and processing strain data from images with clinical routine quality is still very challenging. Thus, investigation on clinical data is reserved for future studies. Additionally, the exclusion of clinical data allowed us to evaluate the strain-based parameters on a fully-controlled setting. Our study was also limited by the regression approach used for the learning, which relies on a global distance between deformation patterns and linearly combines the infarct candidates. Although the learning algorithm was simply used in our study as a state-of-the-art infarct detection tool to test the performance of the strain-based parameters, further work using different regression approaches or learning algorithms needs to be explored.

5   Conclusions

We have proposed a thorough evaluation of various strain-based parameters in locating myocardial infarct. We took advantage of finite-element modeling to generate a large database of pathological hearts and test the limits of a given localization algorithm against a variety of infarct configurations. Although the strain-based parameters only showed slightly different performance towards locating infarcts, we were able to divide them into three groups, which we showed to be coherent with physiological interpretations.

Acknowledgement. GK Rumindo is supported by the European Commission H2020 Marie Sklodowska-Curie European Training Network VPH-CaSE (www.vph-case.eu), grant agreement No 642612. This work was performed within the LABEX PRIMES (ANR-11-LABX-0063) of Université de Lyon, within the program "Investissements d'Avenir" (ANR-11-IDEX-0007) operated by the French National Research Agency (ANR); and the IMPULSION project from the Programme Avenir Lyon - St. Etienne.

References

1. Antoni, M.L., Mollema, S.A., Delgado, V., Atary, J.Z., Borleffs, C.J., Boersma, E., Holman, E.R., van der Wall, E.E., Schalij, M.J., Bax, J.J.: Prognostic importance of strain and strain rate after acute myocardial infarction. Eur. Heart J. **31**, 1640–1647 (2010)
2. Bleton, H., Margeta, J., Lombaert, H., Delingette, H., Ayache, N.: Myocardial infarct localization using neighbourhood approximation forests. In: Camara, O., Mansi, T., Pop, M., Rhode, K., Sermesant, M., Young, A. (eds.) STACOM 2015. LNCS, vol. 9534, pp. 108–116. Springer, Cham (2016). doi:10.1007/978-3-319-28712-6_12
3. Dandel, M., Lehmkuhl, H., Knosalla, C., Suramelashvili, N., Hetzer, R.: Strain and strain rate imaging by echocardiography - basic concepts and clinical applicability. Curr. Cardiol. Rev. **5**(2), 133–148 (2009)
4. Duchateau, N., De Craene, M., Allain, P., Saloux, E., Sermesant, M.: Infarct localization from myocardial deformation: prediction and uncertainty quantification by regression from a low-dimensional space. IEEE Trans. Med. Imaging **35**(10), 2340–2353 (2016)

5. Flachskampf, F.A., Schmid, M., Rost, C., Achenbach, S., DeMaria, A.N., Daniel, W.G.: Cardiac imaging after myocardial infarction. Eur. Heart J. **32**(3), 272–283 (2011)
6. Genet, M., Lee, L.C., Nguyen, R., Haraldsson, H., Acevedo-Bolton, G., Zhang, Z., Ge, L., Ordovas, K., Kozerke, S., Guccione, J.M.: Distribution of normal human left ventricular myofiber stress at end diastole and end systole: a target for in silico design of heart failure treatments. J. Appl. Physiol. **117**, 142–152 (2014)
7. Holzapfel, G.A., Ogden, R.W.: Constitutive modelling of passive myocardium: a structurally-based framework for material characterization. Phil. Trans. R. Soc. A **367**, 3445–3475 (2009)
8. Klotz, S., Hay, H., Dickstein, M.L., Yi, G.H., Wang, J., Maurer, M.S., Kass, D.A., Burkhoff, D.: Single-beat estimation of end-diastolic pressure-volume relationship: a novel method with potential for noninvasive application. Am. J. Physiol. Heart Circ. Physiol. **291**, H403–H412 (2006)
9. Medrano-Garcia, P., Zhang, X., Suinesiaputra, A., Cowan, B., Young, A.A.: Statistical shape modelling of the left ventricle: myocardial infarct classification challenge. In: MICCAI - STACOM 2015
10. Ortiz-Pérez, J.T., Rodriguez, J., Meyers, S.N., Lee, D.C., Davidson, C., Wu, E.: Correspondence between the 17-segment model and coronary arteryal anatomy using contrast-enhanced cardiac magnetic resonance imaging. JACC Cardiovasc. Imaging **1**(3), 282–293 (2008)
11. Rumindo, G.K., Ohayon, J., Viallon, M., Stuber, M., Croisille, P., Clarysse, P.: Comparison of different strain-based parameters to identify human left ventricular myocardial infarct during diastole: a 3D finite element study. In: CMBBE 2016, Tel Aviv, Israel
12. Sjøli, B., Ørn, S., Grenne, B., Ihlen, H., Edvardsen, T., Brunvand, H.: Diagnostic capability and reproducibility of strain by Doppler and by speckle tracking in patients with acute myocardial infarction. JACC Caridovasc. Imaging **2**, 24–33 (2009)
13. Soleimanifard, S., Abd-Elmoniem, K.Z., Agarwal, H.K., Tomas, M.S., Sasano, T., Vonken, E., Youssef, A., Abraham, M.R., Abraham, T.P., Prince, J.L.: Identification of myocardial infarction using three-dimensional strain tensor fractional anisotropy. In: Proceedings of the IEEE International Symposium on Biomedical Imaging, pp. 468–471 (2010)
14. Wang, H., Amini, A.A.: Cardiac motion and deformation recovery from MRI: a review. IEEE Trans. Med. Imaging **31**(2), 487–503 (2012)

Towards Cognition-Guided Patient-Specific FEM-Based Cardiac Surgery Simulation

Nicolai Schoch$^{(\boxtimes)}$ ⓘ and Vincent Heuveline ⓘ

Engineering Mathematics and Computing Lab (EMCL),
Heidelberg University, Heidelberg, Germany
nicolai.schoch@iwr.uni-heidelberg.de

Abstract. Biomechanical surgery simulation can provide surgeons with useful ancillary information for intervention planning, diagnosis and therapy. The simulation therefore most importantly needs to be patient-specific, surgical knowledge-based and comprehensive in terms of the underlying simulation model and the patient's data. Moreover, the simulation setup and execution should be largely automated and integrated into the surgical treatment workflow. However, this still rarely holds and simulation-based surgery support is not yet commonly established in the clinic. In this work, we address this problem in the context of cardiac surgery, and present the setup and results of a prototypic cognition-guided, patient-specific FEM-based cardiac surgery simulation system. We have designed a semantic data infrastructure and implemented cognitive software components that autonomously interact with the medical data via a common ontology. Using this setup, we enable the creation of knowledge-based, patient-specific surgery simulation scenarios for mitral valve reconstruction surgery, that are executed by means of the FEM simulation software HiFlow³. The obtained simulation results are provided to the surgeon in order to support surgical decision making.

Keywords: Cognition-guidance · Surgical information processing · FEM surgery simulation · Biomechanical modeling and simulation workflow · Cardiac surgery · Mitral valve reconstruction

1   Introduction

For a successful surgical operation, surgeons are required to take into account an enormous number of different, patient-individual impact factors. Moreover, in addition to the available medical patient data, they have to make use of their experience and surgical expert knowledge in order to then define a patient-specifically suitable holistic surgical treatment strategy and to properly conduct surgery. This poses a large potential for computer assistance in surgery: Big clinical data bases have risen recently, and semantic technologies as well as intelligent information processing algorithms have paved the way for a more comprehensive, holistic, patient-individual surgical treatment planning [3]. These advances have, however, not yet been exploited in the context of FEM-based surgery simulation,

© Springer International Publishing AG 2017
M. Pop and G.A. Wright (Eds.): FIMH 2017, LNCS 10263, pp. 115–126, 2017.
DOI: 10.1007/978-3-319-59448-4_12

even though they seem very promising for solving some of the most prominent issues that have so far prevented surgery simulation tools from commonly being employed in the operation room (OR) [12]. Surgery simulation can give surgeons the opportunity not only to plan but to simulate, too, some steps of an intervention and to forecast relevant or potentially critical surgical situations. Yet, they are only helpful if they integrate surgical knowledge and work on the respective patient's medical data. Moreover, in order for an easy usage and for their in-OR-applicability they need to be largely automated and integrated into surgical treatment workflows [12].

In this work, we particularly look at minimally-invasive *mitral valve reconstruction* (MVR) surgery by means of annuloplasty, which is to re-establish the functionality of an incompetent mitral valve (MV) in the heart through implantation of an artificial ring prosthesis [4]. We aim at supporting MVR by providing surgeons with a set of patient-individual biomechanical FEM-based MVR surgery simulation scenarios that may be appropriate for the treatment of the respective patient's disease. These surgery simulation scenarios are to enable the surgeons *before* the actual operation to virtually assess the simulated behavior of the MV *after* the implantation of a specific ring prosthesis, i.e., after MVR. With this, we aim at supporting the actual ring selection process. However, as indicated above, such surgery simulations are really beneficial to surgeons only if they are patient-specific, surgical expert knowledge-based, and if their setup and execution is automated and integrated into the surgical treatment workflow.

There are many research groups that deal with numerical simulation for cardiac surgery assistance and for MVR support. We point out the pioneering work of Mansi et al. [11], who first presented an integrated framework for the FE-based modeling of the MV biomechanics in the context of mitral clip intervention planning. Further, we refer to three summarizing reviews to provide an overview: In 2013, Votta et al. [22] summarize topical works towards patient-specific cardiac valve simulations. Chandran and Kim [5] report in 2014 on FEM- and FSI-based computational MV evaluation and potential clinical applications. Early in 2016, Morgan et al. [12] conclude with a discussion of the evolution of FEM-based modeling and simulation of MVR in particular. They argue that currently available automated processes to generate FE models from tomographic images and to thereon-based run the respective simulation scenarios are often incomplete and prohibitively expensive. Just very recently, Zhang and colleagues [23] published new results in the context of a series of works that integrate into their established FEM-based simulation system methods for data assimilation to calibrate simulation scenario setups patient-individually using 3D TEE images.

However, to the best of the authors' knowledge, the cognitive integration of the respective simulation frameworks into established clinical and surgical treatment workflows has not yet been solved, and only a handful of works (see above) start to address the associated questions of data integration, information processing workflows, and cognition-guidance in the OR.

In the works of Schoch et al. [16–19], an adequate solution approach for a such cognition-guided, patient-specific, simulation-enhanced MVR surgery assistance system has been proposed, and partial components, specific technical system features and interim results have been presented. In the work at hands, the different component parts are put together for the first time, and preliminary results that were obtained from this prototypic system are presented and discussed.

In the following, in Sect. 2, we first present the MVR simulation application and the underlying mathematical model formulation. We then look at the biomechanical modeling workflow, at the end of which the simulation application is executed. This will lead to focus on the cognitive data and software architecture that eventually allows for cognition-guidance and for the comprehensive, automated medical information processing in order for an enhanced in-OR-usability of surgery simulation. We emphasize that our work prioritizes the cognition-guidance aspect over the elaboration of the biomechanical model itself. In Sect. 3, we present first results obtained from the prototypic system, and evaluate the system's functionality and performance. Finally, in Sect. 4, we conclude with a discussion of the presented work and analyze future intended work.

2   Materials and Methods

Modeling and Simulation of the MV Behavior. To describe and simulate the deformation behavior of the MV, we build on the theory of elasticity. We extend and further specify the work of Schoch et al. [16], which proposes a basic model for MVR surgery simulation. We start with the *boundary value problem* (BVP) formulation for *contact problems in elasticity* that reads

$$- \operatorname{div} \boldsymbol{\sigma}(\boldsymbol{u}) + \rho\, \ddot{\boldsymbol{u}} = \rho\, \boldsymbol{g} \quad \text{on } \Omega_R\ , \tag{1}$$

$$\boldsymbol{u} = \boldsymbol{u}_D \quad \text{on } \Gamma_D\ , \tag{2}$$

$$\boldsymbol{\sigma}(\boldsymbol{u})\, \boldsymbol{n} = \boldsymbol{t}_N \quad \text{on } \Gamma_N\ , \tag{3}$$

$$\boldsymbol{\sigma}(\boldsymbol{u})\, \boldsymbol{n} = \boldsymbol{p}_C \quad \text{on } \Gamma_C\ . \tag{4}$$

The balance Eq. (1) accounts for the conservation laws, considering stiffness and inertia, and governs the deformation of the soft body $\Omega_R \subset \mathrm{I\!R}^3$. It contains the stress tensor $\boldsymbol{\sigma}$ with respect to the MV body's displacement $\boldsymbol{u}$, the material density ρ, and the body force $\boldsymbol{g}$. The boundary conditions (BCs) state the constraints for fixation or displacement (2) on Γ_D (*Dirichlet BC*), as well as for blood pressure onto the leaflet surfaces and for the pulling effects of the chordae tendinae (3) on Γ_N (*Neumann BC*), and take account for the event of contact between the two MV leaflets (4) on Γ_C (*Contact BC*). Contact is implemented by means of a penalty term $\boldsymbol{p}_C$, that prevents leaflet interpenetration.

To numerically solve this problem with the Finite Element Method (FEM), we derive – according to the standard procedure as, e.g., in [1] – the *variational formulation* as shown in Fig. 1. The illustration also indicates how the required patient-specific features are represented in the model: We allow for considering patient-specific MV morphologies and material properties, integrate the patient-individual blood pressure and the pulling chordae, account for contact, and model the annuloplasty ring implantation.

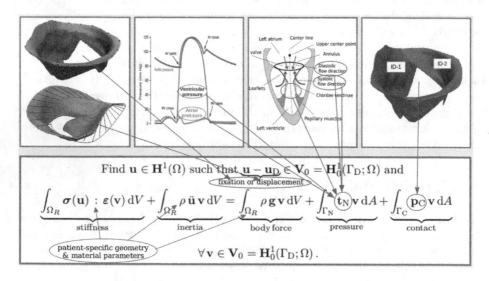

Fig. 1. Illustration of the specifications of the variational formulation of contact in elasticity for describing the MV behavior and MVR annuloplasty surgery.

Following the conforming Galerkin approach, we apply the *FEM* for space discretization, and make use of the implicit *Newmark* time integration scheme for time discretization, as it is reported to be stable, efficient and robust with respect to oscillations and shocks [1] during the cardiac cycle. To describe the MV behavior we thus obtain the system

$$M\,\ddot{u}(t) + D\,\dot{u}(t) + K\,u(t) = f\,,\tag{5}$$

with stiffness matrix K, damping matrix D (Rayleigh Damping), mass matrix M, and load vector f to account for body forces and the BCs.

For an efficient assembly and computation of the respective equation systems in every time step, we make use of the open-source C++ FEM software toolkit HiFlow[3] [10]. We optimized the performance of the simulation application by means of massive parallelization of assembly and contact search. Moreover, through specifically defined simulation interfaces, a flexible manipulation and control of the simulation scenario setup is enabled, and we allow for all available patient-specific medical data and for surgical expert knowledge to be integrated after appropriate simulation preprocessing (see below).

The Biomechanical Modeling Workflow and Cognition-Guidance. The above mentioned simulation interfaces have been designed to be compatible with the *Medical Simulation Markup Language* (MSML) [21], which we employ for the simulation preprocessing. The MSML not only describes the biomechanical modeling workflow, but also acts as a middleware between all steps from MV image segmentation, via 3D mesh generation, to the definition of BC data structures, and, finally, the execution of the simulation using the desired physics engine. A set of dedicated MSML-based MVR simulation preprocessing algorithms [18] guarantees the comprehensive analysis and the further processing of medical patient data as well as of surgically relevant MVR expert knowledge in order for the fully automated specification of appropriate MVR simulation scenarios, that can directly be executed using the above HiFlow^3-based MVR simulation application.

The thus designed simulation-enhanced MVR surgery assistance workflow (see Fig. 2) makes use of morphologic patient data (MV segmentations) that were obtained through Engelhardt et al. [7], and of surgically relevant patient parameter data (health histories and surgery indication records) that comes from Schoch et al. [19]. All data is stored and semantically annotated under consideration of Linked Data principles [2] in a semantic knowledge base [9], which constitutes the core of the overall cognitive system architecture and of

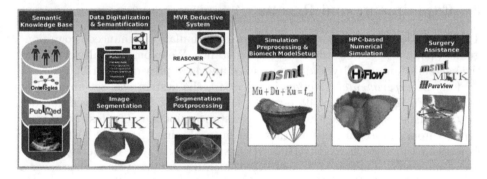

Fig. 2. Illustration of the overall pipeline for cognition-guided, patient-specific, simulation-enhanced MVR surgery assistance. The proposed system first processes a patient's medical data via a deductive reasoning system that implements surgical knowledge and surgical guidelines [19] and via image segmentation tools [7], in order to thus obtain all required input for subsequently merging the information in the biomechanical modeling step. The MSML-based modeling step then sets up simulation scenarios according to the patient's processed geometric data and to the chosen surgical strategy, i.e., considering which ring type and size had been proposed by the MVR deductive system, and how the ring is suggested to be implanted onto the natural MV. The thereby obtained comprehensive, patient-specific, knowledge-based biomechanical models are then used for simulation and executed by the HiFlow^3-based MVR simulation application. Finally, the MVR surgery simulation results are visualized and provided to the surgeons, in order to assist them in a better and more holistic decision making with respect to the ring implantation strategy.

its semantic data infrastructure. Analogously, we have extended all tools and algorithms by means of a semantic description, too. For every tool or algorithm, a such description wraps the tool and specifies its required input and produced output data as well as its functionalities (as first described by Philipp et al. [13]). Abiding by one common ontology, we guarantee that the tools' descriptions are compatible with the semantic data annotation scheme of the underlying semantic knowledge base [9,19]. This way, we also facilitate that all tools and algorithms in the assistance workflow can intelligently and autonomously interact with each other and with the available data. Moreover, we have all tools and algorithms executed through a *data-driven* rule engine, meaning that every application is run as soon as the appropriate input data (as defined in the semantic description) is available. Finally, having equipped all applications with RESTful Web APIs, and exposing them via the Web, we have the entire workflow executed *fully automatically*, and on possibly different computer systems, from OR workstation, via reasoning server, to High-Performance Computing cluster. Our system thus facilitates the *comprehensive* processing of *heterogeneous* medical patient data (both images and parameter data) from miscellaneous data and information sources. The obtained pipeline of *cognitive* applications, see Fig. 2, thus allows for setting up *patient-individual and surgical knowledge-based* MVR simulation scenarios.

3   Experiments and Results

This work describes a prototypic system for simulation-enhanced, cognition-guided, patient-specific cardiac surgery assistance. We experimentally evaluated the system's functionality and performance and here present first results.

For the *experimental functional evaluation*, the system was set-up as illustrated in Fig. 2, and a set of 49 test cases has artificially been produced in order for the validation of the entire surgery assistance pipeline. The test cases were derived from MV segmentations that had been obtained from 4D TEE images from two patients with severity III of MV insufficiency [7]. We deliberately make use of multiple segmentation frames over the entire cardiac cycle to thus cover a broad spectrum of possible image data input variations for validation purposes. Along with the MV segmentations, all surgically relevant parameter data was digitalized and semantified for both patients. Using the deductive system for annuloplasty ring selection [19], a set of potentially suitable ring prostheses was then obtained for both patients individually: The respective rings were derived according to a semantified machine representation of MVR surgery guidelines by Carpentier [4] and Fedak [8] and based on the above mentioned patient parameter and image data. Moreover, each ring is geometrically represented for subsequent integration into 3D models. Resulting from this setup, all of the thus derived 49 MVR surgery scenarios could successfully be further processed via the MSML-based simulation preprocessing step and via the HiFlow^3-based simulation application, through the entire data-driven MVR surgery assistance pipeline as shown in Fig. 2. We consequently obtained simulation results for 49 patient-individual MVR scenarios with respectively different ring prostheses.

After simulation, the simulation results are visualized and provided to the surgeon to thus enable the virtual patient- and ring-specific assessment of potential MVR surgery outcomes *before* the actual operation. See Fig. 3 for an example output of the MVR assistance system, which allows to assess patient- and ring-specifically (**1**) the MV leaflets' deformation behavior, (**2**) the MV coaptation functionality, and (**3**) the stress behavior during the post-operative cardiac cycle.

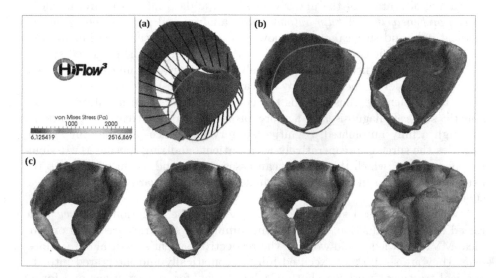

Fig. 3. Visualization of MVR surgery simulation results that were obtained from the cognition-guided, patient-specific MVR surgery assistance system prototype, showing by the example of one test patient (**a**) the patient- and ring-specific biomechanical model of the MVR ring implantation process, that constitutes the basis for the subsequent simulation, (**b**) the virtual annuloplasty ring implantation process itself, and (**c**) the MV closing behavior during a post-surgical cardiac cycle. In addition to the actual deformation behavior, the colors encode the arising von Mises stress distribution. (Color figure online)

Looking at the *performance* of the presented prototypic MVR assistance system (as in Fig. 2), we separately evaluated the deductive and MSML-based simulation preprocessing components on the one hand, and the simulation component on the other hand. For the deductive annuloplasty ring selection system and the MSML-based simulation preprocessing component, we employ Intel Core i7-4600U notebook PCs with 8 GB RAM, which successfully and fully automatically yields results after 72 ± 4 s. The simulation application is executed on a High-Performance Computing cluster with 128 cores, distributed over 16 nodes, each with 64 GB RAM. Space discretization with approx. 120 000 Tet4 cells and 70 000 DoFs corresponds to the original TEE-based imaging resolution, such that the usage of the HiFlow[3]-based CG solver along with its Symmetric Gauss-Seidel and Algebraic MultiGrid preconditioners yields simulation results (as shown in Fig. 3) in just under three minutes of computation time.

4   Discussion

In this work, we present, for the first time, a prototypic system for cognition-guided, patient-specific, simulation-enhanced MVR surgery assistance. This system prototype integrates a biomechanical simulation for MVR surgery and a set of dedicated MVR-related cognitive software components as part of a data-driven pipeline into a semantic surgery assistance system architecture.

Taking advantage of the underlying semantic data infrastructure, it allows to *comprehensively* and *fully automatically* analyze and process heterogeneous medical data and surgical expert knowledge in order for the situation-adapted and patient-specific simulation of MVR surgery scenarios. The proposed system hence allows – in addition to the previously available medical image and parameter data – to infer further surgically relevant simulation-based knowledge for the respective patient. It thus extends the current standard and facilitates a more holistic surgical diagnosis and a more profound surgical treatment planning: Through a fully automated, cognitive surgery assistance pipeline, our system provides the surgeon preoperatively with patient- and ring-specific MVR simulation scenarios, which the surgeon can assess in 3D and use in addition to the usual medical data in order to then make a decision towards the best surgical strategy.

With the proposed system, we hence address the cognition-guidance issue raised in the introduction. The resulting simulation scenarios are based on surgical MVR knowledge and work on the respective patient's medical data. Moreover, they are set-up and executed fully automatically and integrated into the surgical treatment workflow, both of which is key for an easy usage and for an enhanced in-OR-applicability of simulation-based surgery assistance.

Compared to other state-of-the-art works on simulation-based surgery assistance systems (see again Sect. 1), in this work, we do not present a further advanced biomechanical model, but aim at investigating on *cognition-guidance* and *patient-specifity* for simulation-enhanced surgery assistance. Consequently, we have put particular focus on

- a generic, flexible and arbitrarily extensible MVR simulation application setup, with competent interfaces which facilitate the consideration of patient-specific medical data and of surgical MVR knowledge,
- the integration of this simulation into a semantic system architecture,
- cognition-guidance and cognitive data integration, through a semantic representation of all data and its analysis and further processing by means of software components that handle all information with respect to the same common ontology, and
- the complete automation of the entire simulation-enhanced surgery assistance workflow through the underlying data-driven information processing pipeline setup, as shown in Fig. 2.

We demonstrated how surgically relevant and *beneficial insights* can hence be gained from the respective simulation results without requiring any interaction by staff in the operation room. Our system allows to patient- and ring-specifically assess (**1**) the simulated MV deformation behavior, (**2**) the coaptation functionality, and (**3**) the stress distribution, and thus provides the surgeon with means to distinguish optimal from suboptimal rings *before* the actual operation.

However, we point out that the current *system prototype* is still subjected to essential *limitations*: The initial imaging and segmentation step [7] and the deductive system for ring selection [19] do not yet allow us to fully analyze the valve's geometry and other surgically relevant impact factors (e.g., the chordae tendinae or patient-individual material properties). Looking at our elasticity model that describes the behavior of the MV soft tissue [16], we make use of predefined sets of material parameter values (taken from [11,15]), since biomechanical tissue properties cannot directly be reconstructed from images. In taking such average value sets for healthy or sick, and for old or young patient populations, we however only achieve an approximated (quasi-)representation of the actual patient-specific parameters. Moreover, our elasticity model can be further enhanced, e.g., towards the models desribed in [6,12,14] or [22]. Finally, as both patient data and simulation model are subjected to inaccuracies and uncertainties, there is a need for methods for Uncertainty Quantification in order to better quantify the reliability of the obtained simulation results.

Knowing about these limitations, we have *not yet clinically evaluated* the prototypic system, and so far focused on the *general feasibility* of a cognition-guided, simulation-based cardiac surgery assistance system. In this regard, concerning the data and information processing in the proposed MVR surgery assistance system, we point out that we can guarantee for the system's *internal robustness and compatibility*, as well as for an *error-free processing of data* through the whole pipeline: the entire data flow is *semantically controlled* by the underlying semantic system architecture, and supervised by means of *I/O and functionality matching* between the respective workflow steps, fully automatically. Also, all workflow steps and underlying applications have been *evaluated separately* (as recorded in the references). In particular, we investigated on the deductive system for annuloplasty ring selection, which is reported in [19]: it has been described that accepted surgical guidelines for annuloplasty ring selection (by Carpentier [4] and Fedak [8]) could semantically be represented in order to allow for an appropriate patient-specific exclusion of all non-optimal rings. Subsequent simulation of corresponding MVR surgery scenarios that integrate the patient's data and the selected, potentially suitable ring(s) hence extends the current *standard* for surgical decision making in MVR: Our system provides the surgeon with the corresponding patient- and ring-specific MVR simulation scenarios, which can visually be assessed in order for an enhanced surgical decision making.

Beyond investigation on the general feasibility of our prototypic system for cognition-guided, simulation-based MVR surgery assistance, we emphasize again that the above limitations are the reason why we are missing a solid clinical evaluation, so far. Yet, the described benefits and the system's functionality that is based on a cognitive, data-driven, fully-automated information processing pipeline architecture indicate the system's *applicability* and *usability*.

Future work is to enhance several parts of the prototype and to eliminate the above limitations in order to then allow for a reasonable clinical evaluation. We especially intend to integrate into the simulation workflow a framework for *Data Assimilation* as proposed, e.g., by Zhang et al. [23], in order to patient-individually calibrate the simulation setup during simulation using our cognitive data integration architecture. We particularly expect data assimilation algorithms to be suitable for better estimating the real values of the material parameters and of the chordae force distribution. Through combining the simulation setup with data assimilation methods, we intend to let the afore-mentioned average material parameters converge towards the respective patients' real tissue properties. Beyond that, a current work in progress [20] aims at establishing a *simulation ontology* that is connected to an *anatomy and surgery ontology*, through which we plan to make our results comparable and understandable for different work groups and the community. Thanks to the cognition-guided overall setup of our proposed system prototype and its data-driven, modular pipeline architecture, we expect our framework to be perfectly suitable for these further enhancements and for subsequent clinical evaluation.

To conclude, the proposed prototypic cardiac surgery assistance system connects, for the first time, a patient-specific biomechanical FEM surgery simulation and its results with a comprehensive, semantic information management architecture. It thus enables cognition-guided, *knowledge-based and patient-individual* surgery assistance and decision support: It provides the surgeon fully automatically with suitably selected MVR surgery simulation scenarios, which can be assessed and used to assist surgical planning before the operation. With the suggested extensions and improvements, the proposed work can be expected to significantly support MVR surgery in the future.

Acknowledgments. This work was carried out with the support of the *German Research Foundation* (DFG) in the framework of the Collaborative Research Center SFB/TRR 125 *'Cognition-Guided Surgery'*. We particularly thank our colleagues Sandy Engelhardt, Ivo Wolf (Institute of Informatics, University of Applied Science, Mannheim, Germany) and Raffaele de Simone (Department of Cardiac Surgery, University Hospital Heidelberg, Heidelberg, Germany) in the context of cardiac surgery and medical imaging, for the fruitful cooperation and for valuable explanations and discussions concerning our work. Also, we thank Patrick Philipp and York Sure-Vetter (Institute of Applied Informatics and Formal Description Methods (AIFB), Karlsruhe Institute of Technology, Karlsruhe, Germany) for the help and experience with respect to the cognitive semantic software and data infrastructure. We performed all simulations on the *bwUniCluster*, funded by the Ministry of Science, Research and the Arts Baden-Wuerttemberg and the Universities of the State of Baden-Wuerttemberg, Germany, within the framework program *bwHPC*.

References

1. Bathe, K.-J.: Finite Element Procedures. Prentice Hall, Englewood Cliffs (1996)
2. Berners-Lee, T.: Linked Data. W3C. Design Issues, 27 July 2006
3. Biem, A., Butrico, M., Feblowitz, M., Klinger, T., Malitsky, Y., Ng, K., Perer, A., Reddy, C., Riabov, A., Samulowitz, H., Sow, D., Tesauro, G., Turaga, D.: Towards cognitive automation of data science. In: Proceedings of AAAI Conference on Artificial Intelligence 2015 (2015)
4. Carpentier, A.: Cardiac valve surgery the 'French' correction. J. Thorac. Cardiovasc. Surg. **86**, 323–337 (1983)
5. Chandran, K., Kim, H.: Computational mitral valve evaluation and potential clinical applications. Ann. Biomed. Eng. **43**(6), 1348–1362 (2014)
6. Choi, A., Rim, Y., Mun, J.S., Kim, H.: A novel finite element-based patient-specific mitral valve repair: virtual ring annuloplasty. Biomed. Mater. Eng. **24**(1), 341–347 (2014)
7. Engelhardt, S., Lichtenberg, N., Al-Maisary, S., Simone, R., Rauch, H., Roggenbach, J., Müller, S., Karck, M., Meinzer, H.-P., Wolf, I.: Towards automatic assessment of the mitral valve coaptation zone from 4D ultrasound. In: van Assen, H., Bovendeerd, P., Delhaas, T. (eds.) FIMH 2015. LNCS, vol. 9126, pp. 137–145. Springer, Cham (2015). doi:10.1007/978-3-319-20309-6_16
8. Fedak, P.W.M., McCarthy, P.M., Bonow, R.O.: Evolving concepts and technologies in mitral valve repair. Circulation **117**(7), 963–974 (2008)
9. Fetzer, A., Metzger, J., Katic, D., Mrz, K., Wagner, M., Philipp, P., Engelhardt, S., Weller, T., Zelzer, S., Franz, A.M., Schoch, N., Heuveline, V., Maleshkova, M., Rettinger, A., Speidel, S., Wolf, I., Kenngott, H., Mehrabi, A., Mller, B., Maier-Hein, L., Meinzer, H.-P., Nolden, M.: Towards an open-source semantic data infrastructure for integrating clinical and scientific data in cognition-guided surgery. In: Proceedings of SPIE 9789 Medical Imaging 2016, vol. 9789, pp. 978900–978908 (2016)
10. Augustin, W., Baumann, M., Gengenbach, T., Hahn, T., Helfrich-Schkarbanenko, A., Heuveline, V., Ketelaer, E., Lukarski, D., Nestler, A., Ritterbusch, S., Ronnas, S., Schick, M., Schmidtobreick, M., Subramanian, C., Weiss, J.-P., Wilhelm, F., Wlotzka, M.: HiFlow^3- a hardware-aware parallel finite element package. In: Brunst, H., Müller, M., Nagel, W., Resch, M. (eds.) Tools for High Performance Computing, pp. 139–151. Springer, Heidelberg (2012)
11. Mansi, T., Voigt, I., Georgescu, B., Zheng, X., Mengue, E.A., Hackl, M., Ionasec, R.I., Noack, T., Seeburger, J., Comaniciu, D.: An integrated framework for finite element modeling of mitral valve biomechanics from medical images. J. Med. Image. Anal. **16**(7), 1330–1346 (2012)
12. Morgan, A.E., Pantoja, J.L., Weinsaft, J., Grossi, E., Guccione, J.M., Ge, L., Ratcliffe, M.: Finite Element modeling of mitral valve repair. J. Biomech. Eng. **138**(2), 021009 (2016)
13. Philipp, P., Maleshkova, M., Katic, D., Weber, C., Goetz, M., Rettinger, A., Speidel, S., Kaempgen, B., Nolden, M., Wekerle, A.-L., Dillmann, R., Kenngott, H., Mueller, B., Studer, R.: Toward cognitive pipelines of medical assistance algorithms. Int. J. CARS **11**(9), 1743–1753 (2015)
14. Prot, V., Skallerud, B.: Nonlinear solid finite element analysis of mitral valves with heterogeneous leaflet layers. J. Comput. Mech. **43**, 353–368 (2009)
15. Prot, V., Skallerud, B., Sommer, G., Holzapfel, G.A.: On modelling and analysis of healthy and pathological human mitral valves: two case studies. J. Mech. Behav. Biomed. Mater. **3**, 167–177 (2010)

16. Schoch, N., Engelhardt, S., De Simone, R., Wolf, I., Heuveline, V.: High performance computing for cognition-guided cardiac surgery: soft tissue simulation for mitral valve reconstruction in knowledge-based surgery assistance. In: Proceedings of High Performance Scientific Computing (HPSC) (2015, in press)
17. Schoch, N., Engelhardt, S., Zimmermann, N., Speidel, S., de Simone, R., Wolf, I., Heuveline, V.: Integration of a biomechanical simulation for mitral valve reconstruction into a knowledge-based surgery assistance system. In: Proceedings of SPIE 9415 Medical Imaging 2015, vol. 9415, pp. 941502–941502-7 (2015)
18. Schoch, N., Kissler, F., Stoll, M., Engelhardt, S., de Simone, R., Wolf, I., Bendl, R., Heuveline, V.: Comprehensive patient-specific information preprocessing for cardiac surgery simulations. Int. J. CARS **11**(6), 1051–1059 (2016). (Special Issue IPCAI)
19. Schoch, N., Philipp, P., Weller, T., Engelhardt, S., Volovyk, M., Fetzer, A., Nolden, M., de Simone, R., Wolf, I., Maleshkova, M., Rettinger, A., Studer, R., Heuveline, V.: Cognitive tools pipeline for assistance of mitral valve surgery. In: Proceedings of SPIE 9786 Medical Imaging 2016, 9786: 978603–978603-8 (2016)
20. Schoch, N., Speidel, S., Sure-Vetter, Y., Heuveline, V.: Towards semantic simulation for patient-specific surgery assistance. In: Online-Proceedings of Surgical Data Science 2016 (2016)
21. Suwelack, S., Stoll, M., Schalck, S., Schoch, N., Dillmann, R., Berndl, R., Heuveline, V., Speidel, S.: The medical simulation markup language - simplifying the biomechanical modeling workflow. J. Stud. Health. Techn. Inf. **196**, 394–400 (2014)
22. Votta, E., Le, T.B., Stevanella, M., Fusini, L., Caiani, E.G., Redaelli, A., Sotiropoulos, F.: Toward patient-specific simulations of cardiac valves: state-of-the-art and future directions. J. Biomech. **46**, 217–228 (2013)
23. Zhang, F., Kanik, J., Mansi, T., Voigt, I., Sharma, P., Ionasec, R.I., Subrahmanyan, L., Lin, B.A., Sugeng, L., Yuh, D., Comaniciu, D., Duncan, J.: Towards patient-specific modeling of mitral valve repair: 3D transesophageal echocardiography-derived parameter estimation. J. Med. Image Anal. **35**, 599–609 (2017)

FastVentricle: Cardiac Segmentation with ENet

Jesse Lieman-Sifry$^{(\boxtimes)}$, Matthieu Le, Felix Lau, Sean Sall, and Daniel Golden

Machine Learning, San Francisco, USA
{jesse,matthieu,felix,sean,dan}@arterys.com

Abstract. Cardiac Magnetic Resonance (CMR) imaging is commonly used to assess cardiac structure and function. One disadvantage of CMR is that postprocessing of exams is tedious. Without automation, precise assessment of cardiac function via CMR typically requires an annotator to spend tens of minutes per case manually contouring ventricular structures. Automatic contouring can lower the required time per patient by generating contour suggestions that can be lightly modified by the annotator. Fully convolutional networks (FCNs), a variant of convolutional neural networks, have been used to rapidly advance the state-of-the-art in automated segmentation, which makes FCNs a natural choice for ventricular segmentation. However, FCNs are limited by their computational cost, which increases the monetary cost and degrades the user experience of production systems. To combat this shortcoming, we have developed the FastVentricle architecture, an FCN architecture for ventricular segmentation based on the recently developed ENet architecture. FastVentricle is 4× faster and runs with 6× less memory than the previous state-of-the-art ventricular segmentation architecture while still maintaining excellent clinical accuracy.

1 Introduction

Patients with known or suspected cardiovascular disease often receive a cardiac MRI to evaluate cardiac function. These scans are annotated with ventricular contours in order to calculate cardiac volumes at end systole (ES) and end diastole (ED); from the cardiac volumes, relevant diagnostic quantities such as ejection fraction and myocardial mass can be calculated. Manual contouring can take upwards of 30 min per case, so radiologists often use automation tools to help speed up the process.

Active contour models [1] are a heuristic-based approach to segmentation that have been utilized previously for segmentation of the ventricles [2,3] with optional use of a ventricle shape prior [4,5]. However, active contour-based methods not only perform poorly on images with low contrast, they are also sensitive to initialization and hyperparameter values. We encourage the interested reader to refer to recent review papers [6,7] as a jumping-off point for further insight on the usage of these (and many other) non-deep learning approaches for cardiac segmentation.

© Springer International Publishing AG 2017
M. Pop and G.A. Wright (Eds.): FIMH 2017, LNCS 10263, pp. 127–138, 2017.
DOI: 10.1007/978-3-319-59448-4_13

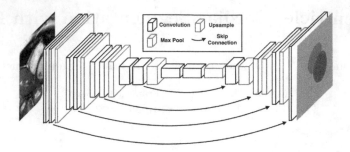

Fig. 1. Schematic representation of a fully convolutional encoder-decoder architecture with skip connections that utilizes a smaller expanding path than contracting path.

Deep learning methods for segmentation have recently defined state-of-the-art with the use of fully convolutional networks (FCNs) [8]. Simple FCN architectures similar to that described by [8] have been utilized for cardiac segmentation [9]. The general idea behind FCNs is to use a downsampling path to learn relevant features at a variety of spatial scales followed by an upsampling path to combine the features for pixelwise prediction (see Fig. 1). DeconvNet [10] pioneered the use of a symmetric contracting-expanding architecture for more detailed segmentations, at the cost of longer training and inference time, and the need for larger computational resources. UNet [11], originally developed for use in the biomedical community where there are often fewer training images and even finer resolution is required, added the use of skip connections between the contracting and expanding paths to preserve details.

A UNet variant, DeepVentricle, has previously been used for cardiac segmentation [12] and has received FDA clearance for clinical usage [13]. Both UNet and DeepVentricle are FCNs with symmetric downsampling and upsampling paths, skip connections, two convolution operations before each pooling or upsampling operation, and double/half the number of filters as in the previous layer following each pool/upsample, respectively. A key difference is that UNet utilizes valid padding, resulting in segmentation maps that are smaller than the input image, whereas DeepVentricle uses same padding.

One disadvantage of fully symmetric architectures in which there is a one-to-one correspondence between downsampling and upsampling layers is that they can be slow, especially for large input images. ENet, an alternative FCN design, is an asymmetrical architecture optimized for speed [14]. ENet utilizes early downsampling to reduce the input size using only a few feature maps. This reduces both training and inference time, given that much of the network's computational load takes place when the image is at full resolution, and has minimal effect on accuracy since much of the visual information at this stage is redundant. Furthermore, the ENet authors show that the primary purpose of the expanding path in FCNs is to upsample and fine-tune the details learned by the contracting path rather than to learn complicated upsampling features; hence, ENet utilizes an expanding path that is smaller than its contracting path. ENet also makes use of bottleneck modules, which are convolutions with a small receptive field that are applied in order to project the feature maps into a lower dimensional space in which larger kernels can be applied [15]. Finally, throughout the network, ENet leverages a diversity of low cost convolution operations. In addition to the more-expensive $n \times n$ convolutions, ENet also uses cheaper asymmetric ($1 \times n$ and $n \times 1$) convolutions and dilated convolutions [16].

In this paper, we present FastVentricle, an ENet variation with skip connections for segmentation of the LV endocardium (LV endo), LV epicardium (LV epi), and RV endocardium (RV endo), and compare it to DeepVentricle, the architecture previously cleared by the FDA for clinical use. We show that inference with FastVentricle requires significantly less time and memory than inference with DeepVentricle while achieving statistically indistinguishable volume accuracy.

2   Methods

Training Data. We use a database of 1143 short-axis cine Steady State Free Precession (SSFP) scans annotated as part of standard clinical care to train and validate our model. We split the data chronologically with 80% of studies used for training, 10% for validation, and 10% as a hold out set. We curate the hold out set to only include contours from trusted annotators, i.e. licensed radiologists or technologists. The hold out set is used for all plots and tables in this paper. Annotated contour types include LV endo, LV epi and RV endo. Scans are annotated at ED and ES. Contours were annotated with different frequencies; 96% (1097) of scans have LV endo contours, 22% (247) have LV epi contours and 85% (966) have RV endo contours.

Data Preprocessing. We normalize all MRIs such that the 1st and 99th percentile of pixel intensities of a batch of images fall at -0.5 and 0.5, i.e. their "usable range" falls between -0.5 and 0.5. We crop and resize the images such that the ventricle contours take up a larger percentage of the image; the actual crop and resize factors are hyperparameters. Cropping the image increases the fraction of the image that is taken up by the foreground (ventricle) class, making it easier to resolve fine details and helping the model converge.

Training. We use the Keras [17] deep learning package with TensorFlow [18] as the backend to implement and train all of our models. We modify the standard per-pixel cross-entropy loss to account for missing ground truth annotations in our dataset. We discard the component of the loss that is calculated on images for which ground truth is missing; we only backpropagate the component of the loss for which ground truth is known. This allows us to train on our full training dataset, including series with missing contours. Network weights are updated per the Adam rule [19]. We train and test the models using NVIDIA Maxwell Titan X GPUs. We augment our data during training by flipping, shearing, shifting, zooming, and rotating the image/label pairs. To compare different models, we use relative absolute volume error (RAVE). Using a relative metric ensures

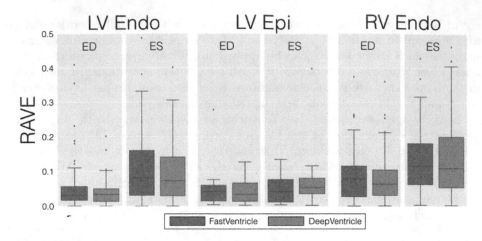

Fig. 2. Boxplots comparing the relative absolute volume error (RAVE) between FastVentricle and DeepVentricle for each of LV endo, LV epi, and RV endo at ED (left panels) and ES (right panels). The line at the center of each box denotes the median RAVE, the ends of the box show 25% (Q1) and 75% (Q3) of the distribution. Whiskers are placed at the first value within 1.5 interquartile ranges past the first and third quartiles.

that equal weight is given to pediatric and adult hearts. We focus on the volume error, as opposed to the Sørensen–Dice index or a similar overlap metric, because ventricular volumes are the clinical endpoint used to diagnose disease and determine clinical care. RAVE is defined as RAVE $= |V_{pred} - V_{truth}|/V_{truth}$, where V_{truth} is the ground truth volume and V_{pred} is the volume computed from the predicted 2D segmentation masks. Volumes are calculated from segmentation masks using a frustum approximation.

Hyperparameter Search. We fine tune the DeepVentricle and FastVentricle network architectures using random hyperparameter search [20]. In practice, for each of the DeepVentricle and FastVentricle architectures, we: (i) run models with random sets of hyperparameters for 20 epochs (i.e. 20 passes over the training set), (ii) select from the resulting corpus of models the three models with the highest validation set accuracy, (iii) select the final model from the three candidates based on lowest average RAVE on the validation set. The hyperparameters of the DeepVentricle architecture include the use of batch normalization, the number of convolution layers, the number of initial filters, and the number of pooling layers. The hyperparameters of the FastVentricle architecture include the kernel size for asymmetric convolutions, the number of times Section 2 of the network is repeated, the number of initial bottleneck modules, the number of initial filters, the projection ratio, and whether to use skip connections. These connections run from the contracting path (the initial block and Section 1) to the equivalent image size on the expanding path (Section 5 and Section 4, respectively). Refer to [14] for nomenclature of the Sections and detailed descriptions of the hyperparameters. For both architectures, the hyperparameters also include the batch size, learning rate, dropout probability, crop fraction, image size, and the strength of all data augmentation parameters. We trained 50 DeepVentricle models and 20 FastVentricle models for our hyperparameter search, with the scope of the search limited by computational limitations and time constraints. We note that skip connections are used in the 5 best FastVentricle architectures in terms of validation set accuracy, demonstrating their usefulness for this problem.

3   Results

Volume Error Analysis. Figure 2 presents box plots of the RAVE comparing DeepVentricle and FastVentricle for each combination of ventricular structure (LV endo, LV epi, RV endo) and phase (ES, ED). Within our hold out set, we find that the performance of the models are very similar across structures and phases (see Table 1 for median RAVEs as well as corresponding sample sizes). For both models, segmentations at ED tend to be better than at ES, as the chambers at ES are smaller and dark-colored papillary muscles tend to blend with the myocardium when the heart is contracted. RV endo is the most difficult of the structures due to its more complex shape. Furthermore, we find that model performance at the apex and center of the ventricle is often better than at the base, as it is generally ambiguous from just the basal slice where the valve plane (separating ventricle from atrium) is. Although trained on only ES and ED annotations, we are able to perform visually pleasing inference on all time points. Figure 3 shows examples of network predictions on different slices and time points for studies for both DeepVentricle and FastVentricle.

We additionally make Bland-Altman plots [21] for the volume error for each combination of ventricular structure and phase (Fig. 4). These plots are used to

Table 1. Median RAVEs, U statistics and p-values from the Wilcoxon-Mann-Whitney test, and corresponding sample sizes for our comparison of DeepVentricle and FastVentricle for every combination of phase and ventricular anatomy on our hold out set. We observe no statistically significant difference between the RAVE distributions of DeepVentricle and FastVentricle.

	LV Endo (ED)	LV Endo (ES)	LV Epi (ED)	LV Epi (ES)	RV Endo (ED)	RV Endo (ES)
DeepVentricle median RAVE	0.033	0.073	0.036	0.049	0.064	0.109
FastVentricle median RAVE	0.031	0.081	0.044	0.043	0.080	0.116
U statistic	4678	4967	190	160	4623	4142
p-value	0.86	0.35	0.79	0.56	0.28	0.80
Sample size	96	96	19	19	112	112

show the agreement between the "gold standard" method (i.e. expert manual annotations) and our automated DeepVentricle and FastVentricle methods. If the 95% limits of agreement (mean $\pm1.96SD$) are within the range that would not make a clinical difference, the new method can be used in place of the old with no adverse effects.

Statistical Analysis. We measure the statistical significance of the difference between DeepVentricle and FastVentricle's RAVE distributions for each combination of phase and anatomy for which we have ground truth annotations. We use the Wilcoxon-Mann-Whitney test[1] to assess the null hypothesis H_0 that the distribution of DeepVentricle and FastVentricle's RAVE are equal. Table 1 displays the results. We find that there is no statistical evidence to claim one model as the best, since the lowest measured p-value is 0.28.

Computational Complexity and Inference Speed. To be clinically and commercially viable, any automated algorithm must be faster than manual annotations, and lightweight enough to deploy easily. As seen in Table 2, we find that FastVentricle is roughly 4× faster than DeepVentricle and uses 6× less memory for inference. Because the model contains more layers, FastVentricle takes longer to initialize before being ready to perform inference. However, in a production setting, the model only needs to be initialized once when provisioning the server, so this additional cost is incidental.

Table 2. Model volume error, speed, and computational complexity for DeepVentricle and FastVentricle. Inference time per sample and GPU memory required for inference calculated with a batch size of 16.

	DeepVentricle	FastVentricle
Average RAVE	0.100	**0.098**
Median RAVE	**0.057**	0.065
Inference GPU time per sample (ms)	31	**7**
Initialization GPU time (s)	**1.3**	13.3
Number of parameters	19,249,059	**755,529**
GPU memory required for inference (MB)	1,800	**270**
Size of the weight file (MB)	220	**10**

Internal Representation. Neural networks are infamous for being black boxes, i.e., it is very difficult to "look inside" and understand why a certain prediction is being made. This is especially troublesome in the medical setting, as doctors prefer to use tools that they can understand. We follow the results of [23] to visualize the features that FastVentricle is "looking for" when performing inference.

---

[1] Using the SciPy 0.17.0 implementation with default parameters https://docs.scipy.org/doc/scipy/reference/generated/scipy.stats.mannwhitneyu.html.

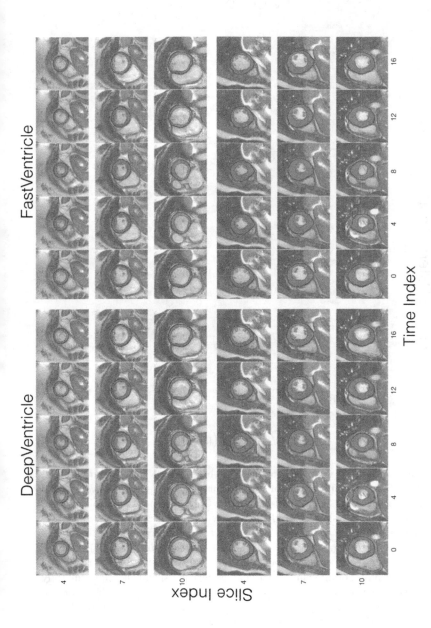

Fig. 3. DeepVentricle and FastVentricle predictions for a healthy patient (top) and on a patient with hypertrophic cardiomyopathy (HCM, bottom). RV endo is outlined in red, LV endo in green, and LV epi in blue. DeepVentricle's average RAVE is 0.080 and 0.095 for the healthy and HCM patients, respectively, and FastVentricle's average RAVE is 0.057 and 0.053 for each patient, respectively. The x-axis of the grid corresponds to time indices sampled throughout the cardiac cycle and the y-axis corresponds to slice indices sampled from apex (low slice index) to base (high slice index). (Color figure online)

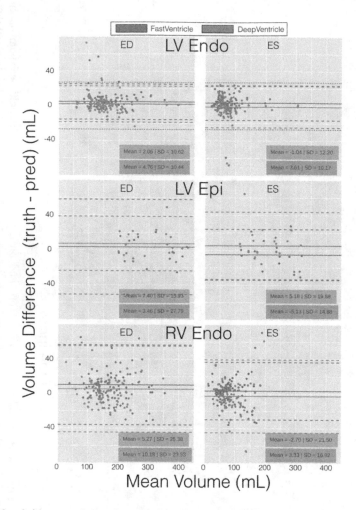

Fig. 4. Bland-Altman plots of ventricle volumes for both DeepVentricle (green) and FastVentricle (blue) for each combination of ventricular structure and phase vs. ground truth annotations. Solid colored lines denote the mean difference between ground truth and predicted volumes, and dashed colored lines show the mean $\pm 1.96SD$, where SD is the standard deviation of the difference. The mean can be interpreted as the bias of the new method (automatic segmentation) vs. the old (doctor's annotations), and the dashed lines are the 95% limits of agreement between the two methods. For reference, [22] find that the 95% limits of agreement between expert physicians is approximately ± 27 mL for LV endo measurements; we display this value as a dotted black line on the LV endo plots. Our LV endo volume 95% limits of agreement is within that of the 95% limits of agreement between expert annotators for both DeepVentricle and FastVentricle. Information on the 95% limits of agreement between expert annotators is unavailable for LV epi and RV endo. (Color figure online)

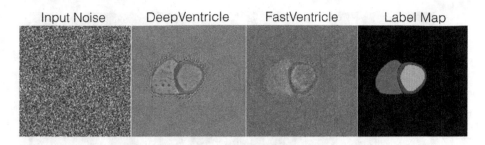

Input Noise   DeepVentricle   FastVentricle   Label Map

Fig. 5. A random input (left) is optimized using gradient descent for DeepVentricle and FastVentricle (middle) to fit the label map (right, RV endo in red, LV endo in cyan, LV epi in blue). The generated image has many qualities that the network is "looking for" when making its predictions, such as high contrast between endocardium and epicardium and the presence of papillary muscles. (Color figure online)

Beginning with random noise as a model input and a real segmentation mask as the target, we perform backpropagation to update the pixel values in the input image such that the loss is minimized. Figure 5 shows the result of such an optimization for DeepVentricle and FastVentricle. We find that, as a doctor would, the model is confident in its predictions when the endocardium is light and the contrast with the epicardium is high. The model seems to have learned to ignore the anatomy surrounding the heart. We also note that the optimized input for DeepVentricle is less noisy than that for FastVentricle, probably because the former model is larger and utilizes skip connections at the full resolution of the input image. DeepVentricle also seems to "imagine" structures which look like papillary muscles inside the ventricles.

4   Discussion

Performance. Though accuracy is the most important property of a model when making clinical decisions, speed of algorithm execution is also critical for maintaining positive user experience and minimizing infrastructure costs. Within the bounds of our experiments, we find no statistically significant difference between the accuracy of DeepVentricle and that of its 4× faster cousin, FastVentricle. This suggests that FastVentricle can replace DeepVentricle in a clinical setting with no detrimental effects.

Both DeepVentricle and FastVentricle have a median RAVE of between 5 and 7%. To put this in context, [22] investigated the reproducibility of LV endo volume measurements from CMR over 15 studies with 7 expert readers and found the interrater standard deviation to be, on average, 14mL, i.e. approximately 10% of the volume being measured. Although there are occasional cases on which our algorithm performs poorly, for example the congenital case displayed in Fig. 6, we auto-detect inference results with noisy contours when using DeepVentricle or FastVentricle for clinical use and opt not to display them to the user. Note that this quality detection algorithm has not been run for any plots

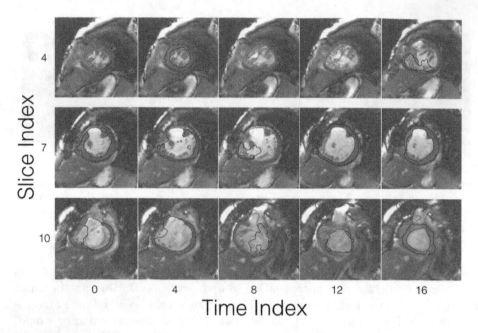

Fig. 6. One main failure mode is on patients with congenital defects, for which the anatomy is often ambiguous. In clinical use, a post-processing algorithm that auto-detects poor inference results would report "cannot find reasonable segmentation" for this study.

in this paper; all presented results contain the full hold out set for completeness. For all cases, our production application allows radiologists to view and modify any erroneous contours before completing the report.

5   Conclusion

We show that a new ENet-based FCN with skip connections, FastVentricle, can be used for quick and efficient segmentation of cardiac anatomy. Trained on a sparsely annotated database, our algorithm provides LV endo, LV epi, and RV endo contours to clinicians for the purpose of calculating important diagnostic quantities such as ejection fraction and myocardial mass. FastVentricle is 4× faster and runs with 6× less memory than the previous state-of-the-art.

We experimented with a fully convolutional 3D network to improve the consistency of basal slice segmentations, but found that the memory requirements of this model limited the resolution of the input images such that the results did not outperform 2D segmentation. A more careful consideration of additional spatial information, perhaps with a 2.5D network incorporating long axis views, could potentially improve performance. Finally, as with any deep learning-based model, additional training cases could further improve performance.

References

1. Kass, M., Witkin, A., Terzopoulos, D.: Snakes: active contour models. Int. J. Comput. Vis. **1**, 321–331 (1988)
2. Choa, J., Benkeserb, P.J.: Cardiac segmentation by a velocity-aided active contour model. Comput. Med. Imag. Graph. **30**, 31–41 (2006)
3. Zhu, W., et al.: A geodesic-active-contour-based variational model for short-axis cardiac MRI segmentation. Int. J. Comput. Math. **90**(1), 124–139 (2013)
4. Pluempitiwiriyawej, C., et al.: STACS: new active contour scheme for cardiac MR image segmentation. IEEE Trans. Med. Imag. **24**, 593–603 (2005)
5. Schwarz, T., Heimann, T., Wolf, I., Meinzer, H.: 3d heart segmentation and volumetry using deformable shape models. In: Computers in Cardiology, pp. 741–744. IEEE (2007)
6. Petitjean, C., Dacher, J.N.: A review of segmentation methods in short axis cardiac MR images. Med. Image Anal. **15**(2), 169–184 (2011)
7. Peng, P., Lekadir, K., Gooya, A., Shao, L., Petersen, S.E., Frangi, A.F.: A review of heart chamber segmentation for structural and functional analysis using cardiac magnetic resonance imaging. Magn. Reson. Mater. Phys. Biol. Med. **29**(2), 155–195 (2016)
8. Long, J., Shelhamer, E., Darrell, T.: Fully convolutional networks for semantic segmentation. In: Proceedings of the IEEE CVPR, pp. 3431–3440 (2015)
9. Tran, P.V.: A fully convolutional neural network for cardiac segmentation in short-axis MRI. arXiv preprint (2016). arXiv:1604.00494
10. Noh, H., Hong, S., Han, B.: Learning deconvolution network for semantic segmentation. In: Proceedings of the IEEE ICCV, pp. 1520–1528 (2015)
11. Ronneberger, O., Fischer, P., Brox, T.: U-Net: convolutional networks for biomedical image segmentation. In: Navab, N., Hornegger, J., Wells, W.M., Frangi, A.F. (eds.) MICCAI 2015. LNCS, vol. 9351, pp. 234–241. Springer, Cham (2015). doi:10.1007/978-3-319-24574-4_28
12. Lau, H.K., et al.: DeepVentricle: automated cardiac MRI ventricle segmentation using deep learning. In: Conference on Machine Intelligence in Medical Imaging (2016)
13. Food and Drug Administration: Arterys cardio dl. http://www.accessdata.fda.gov/cdrh_docs/pdf16/K163253.pdf
14. Paszke, A., Chaurasia, A., et al.: Enet: a deep neural network architecture for real-time semantic segmentation. arXiv preprint (2016). arXiv:1606.02147
15. He, K., Zhang, X., Ren, S., Sun, J.: Deep residual learning for image recognition. In: Proceedings of the IEEE CVPR, pp. 770–778 (2016)
16. Yu, F., Koltun, V.: Multi-scale context aggregation by dilated convolutions. arXiv preprint (2015). arXiv:1511.07122
17. Chollet, F.: Keras (2015). https://github.com/fchollet/keras
18. Abadi, M., et al.: Tensorflow: large-scale machine learning on heterogeneous distributed systems. arXiv preprint (2016). arXiv:1603.04467
19. Kingma, D., Ba, J.: Adam: a method for stochastic optimization. arXiv preprint (2014). arXiv:1412.6980
20. Bergstra, J., Bengio, Y.: Random search for hyper-parameter optimization. J. Mach. Learn. Res. **13**, 281–305 (2012)
21. Bland, J.M., Altman, D.: Statistical methods for assessing agreement between two methods of clinical measurement. Lancet **327**(8476), 307–310 (1986)

22. Suinesiaputra, A., et al.: Quantification of LV function and mass by cardiovascular magnetic resonance: multi-center variability and consensus contours. J. Cardiovas. Magn. Reson. **17**(1), 63 (2015)
23. Mordvintsev, A., et al.: Deep Dream (2015). https://research.googleblog.com/2015/06/inceptionism-going-deeper-into-neural.html. Accessed 17 Jan 2017

Slice-to-Volume Image Registration Models for MRI-Guided Cardiac Procedures

L.W. Lorraine Ma and Mehran Ebrahimi$^{(\boxtimes)}$

Faculty of Science, University of Ontario Institute of Technology,
2000 Simcoe Street North, Oshawa, ON L1H 7K4, Canada
{lok.ma,mehran.ebrahimi}@uoit.ca

Abstract. A mathematical formulation for intensity-based slice-to-volume registration is proposed. The approach is flexible and accommodates various regularization schemes, similarity measures, and optimizers. The framework is evaluated by registering 2D and 3D cardiac magnetic resonance (MR) images obtained *in vivo*, aimed at real-time MR-guided applications. Rigid-body and affine transformations are used to validate the parametric model. Target registration error (TRE), Jaccard, and Dice indices are used to evaluate the algorithm and demonstrate the accuracy of the registration scheme on both simulated and clinical data. Registration with the affine model appeared to be more robust than with the rigid model in controlled cases. By simply extending the rigid model to an affine model, alignment of the cardiac region generally improved, without the need for complex dissimilarity measures or regularizers.

1 Introduction

Recently, there has been increased interest in using magnetic resonance imaging (MRI) for image-guided procedures that have traditionally been guided by X-ray imaging. In some patients with a history of myocardial infarction (MI), electrical activity in the heart may be disrupted by substrate formed from a previous MI, triggering arrhythmia. Treatment options include catheter ablation, where the offending substrate is surgically ablated to correct the arrhythmia. Catheter ablation is traditionally guided by X-ray fluoroscopy, however, possible concerns over radiation exposure have led to MRI being proposed as an alternative. Besides being a non-invasive imaging modality, other advantages of MRI include superior soft tissue contrast to better image anatomical features in and around the heart, the ability to capture depth information without multiple projections and to easily adjust the positions of imaging planes to access areas of interest. MRI-guided procedures necessitate fast imaging techniques to capture images in real-time. Fortunately, MR sequences for real-time 2D visualization exist and have been used to guide cardiac procedures [1].

The short acquisition time required by real-time MRI means that the 2D images lack in quality compared to slices of cine MR volumes that are acquired

© Springer International Publishing AG 2017
M. Pop and G.A. Wright (Eds.): FIMH 2017, LNCS 10263, pp. 139–151, 2017.
DOI: 10.1007/978-3-319-59448-4_14

prior to surgical intervention without the constraints of producing images at real-time frame rates and are therefore less noisy. 3D pre-operative images provide detailed anatomical information while 2D intra-operative images provide live positional updates. Ideally, one can register the pre- and intra-operative images together to combine the advantages of both.

Various methods to perform slice-to-volume registration have been proposed. Of interest to us is the application of slice-to-volume registration to MRI-guided procedures. Registration between pre- and intra-operative images has been studied for non-cardiac applications in humans [2], but on a more relevant note, work has been done *in vivo* on swine to register 2D intra-operative cardiac MR image slices to pre-operative MR image volumes [3]. Xu et al. also present registration of high-quality pre-operative MR image volumes to live cardiac MR images on human volunteers *in vivo* with applications to MRI-guided radiofrequency ablation of substrate in the heart, but also focuses on registration incorporating rigid-body transformations [4].

In most of the studies mentioned above, rigid-body registration was employed. While rigid registration is generally employed to reduce computational cost and to speed up the registration process, it risks oversimplifying the displacement of body tissues, which are generally not rigid. The highly deformable nature of the heart and displacement at various stages of the breathing cycle make registration of the cardiac region more challenging. Deformable registration may be more accurate, but is computationally much more expensive.

Work involving slice-to-volume registration has been recent and not nearly as numerous as projective 2D-3D registration, especially with respect to applications in MRI-guided procedures. In addition, there seems to be a lack of a precise model in the literature, in contrast to 2D-2D or 3D-3D registration [5].

We propose a general mathematical framework for slice-to-volume registration which can accommodate parametric and non-parametric transformation models. A rigid transformation model can be used in this framework, but the user can easily adapt a different parametric transformation model.

We will demonstrate this framework on parametric models, specifically, using this framework to extend existing 2D-3D rigid registration to affine registration. Although the number of parameters in an affine parametric model (12 parameters) is twice the number of parameters in a rigid model (6 parameters), the figure dwarfs in comparison to the number of parameters dealt with in deformable registration, and thus is still a computationally inexpensive method that accounts for some non-rigid deformations. The intensity-based registration framework is flexible and can accommodate various models and parameters. We demonstrate by registering high-resolution 3D MR images to noisier 2D real-time MR images, using rigid and affine parametric models, and investigate the ill-posedness of 2D-3D registration as an inverse problem.

2   Model

Consider the registration problem of a 3D 'template' image $\mathcal{T}$ to a 2D 'reference' image $\mathcal{R}$, where $\mathcal{R}$ is a realization of $\mathcal{T}$ deformed via a transformations y and sliced at a certain location z. The reference and template images are represented by mappings $\mathcal{R} : \Omega \subset \mathbb{R}^2 \to \mathbb{R}$ and $\mathcal{T} : \Omega \times \mathcal{Z} \subset \mathbb{R}^3 \to \mathbb{R}$ of compact support. Considering a slice location z, the goal is to find the transformation $y : \mathbb{R}^3 \to \mathbb{R}^3$ such that $\mathcal{L}_z(\mathcal{T}[y])$ is similar to $\mathcal{R}$, in which $\mathcal{T}[y]$ is the transformed template image and $\mathcal{L}_z : \mathbb{L}^2(\Omega \times \mathcal{Z}) \to \mathbb{L}^2(\Omega)$ is the slicing operator at level $z \in \mathcal{Z} \subset \mathbb{R}$, where $\mathcal{L}_z(\mathcal{T}(x^1, x^2, x^3)) := \mathcal{T}(x^1, x^2, z)$ for $(x^1, x^2, x^3) \in \mathbb{R}^3$. A formulation of the 2D-3D image registration of a template image $\mathcal{T}$ to a reference image $\mathcal{R}$ can be written as the following problem.

2D-3D Image Registration Problem: Given two images $\mathcal{R} : \Omega \subset \mathbb{R}^2 \to \mathbb{R}$ and $\mathcal{T} : \Omega \times \mathcal{Z} \subset \mathbb{R}^3 \to \mathbb{R}$ and an arbitrary given slice location $z \in \mathbb{R}$, find a transformation $y : \mathbb{R}^3 \to \mathbb{R}^3$ that minimizes the objective functional

$$\mathcal{J}[y] := \mathcal{D}[\mathcal{L}_z(\mathcal{T}[y]), \mathcal{R}] + \mathcal{S}[y - y^{\text{ref}}]. \tag{1}$$

Here, $\mathcal{D}$ is a distance that measures the dissimilarity of $\mathcal{L}_z(\mathcal{T}[y])$ and $\mathcal{R}$, and $\mathcal{S}$ is a regularization expression on the transformation y that penalizes transformations "away" from y^{ref}.

2.1   Parametric 2D-3D Registration

It is possible that y can be parametrized via parameters w. For example if y is an affine transformation, the transformation on a point $x = (x^1, x^2, x^3)$ can be expressed as

$$y(w; x) = \begin{pmatrix} w_1 & w_2 & w_3 \\ w_5 & w_6 & w_7 \\ w_9 & w_{10} & w_{11} \end{pmatrix} \begin{pmatrix} x^1 \\ x^2 \\ x^3 \end{pmatrix} + \begin{pmatrix} w_4 \\ w_8 \\ w_{12} \end{pmatrix}.$$

In general, for the parametric registration problem we equivalently aim to minimize

$$\mathcal{J}[w] := \mathcal{D}[\mathcal{L}_z(\mathcal{T}[y(w)]), \mathcal{R}] + \mathcal{S}[w - w^{\text{ref}}]. \tag{2}$$

Here we assume sum of squared distances (SSD) is the dissimilarity measure $\mathcal{D}$

$$\mathcal{D}[\mathcal{L}_z(\mathcal{T}), \mathcal{R}] = \mathcal{D}^{\text{SSD}}[\mathcal{L}_z(\mathcal{T}), \mathcal{R}] := \frac{1}{2} \int_\Omega (\mathcal{L}_z(\mathcal{T}(x)) - \mathcal{R}(x))^2 \, dx.$$

Furthermore, the regularization functional $\mathcal{S}$ can be defined as

$$\mathcal{S}[w - w^{\text{ref}}] := \frac{1}{2} \times (w - w^{\text{ref}})^T \mathbf{M} (w - w^{\text{ref}}) \tag{3}$$

for a symmetric positive definite weight matrix $\mathbf{M}$ that acts as a regularizer (see [5]). If no regularization is imposed on w, for any pair of given images $\mathcal{R}$ and $\mathcal{T}$ the above model is ill-posed. Therefore, to yield a unique w, we require a regularizer $\mathcal{S}$. The following theorem proves this claim.

Theorem 1. *Consider a given* z. *Any two affine transformations* w^A *and* w^B *that satisfy the following conditions yield* $\mathcal{L}_z(\mathcal{T}[y(w^A; x)]) = \mathcal{L}_z(\mathcal{T}[y(w^B; x)])$:

$$\begin{pmatrix} w_1^A \\ w_5^A \\ w_9^A \end{pmatrix} = \begin{pmatrix} w_1^B \\ w_5^B \\ w_9^B \end{pmatrix}, \quad \begin{pmatrix} w_2^A \\ w_6^A \\ w_{10}^A \end{pmatrix} = \begin{pmatrix} w_2^B \\ w_6^B \\ w_{10}^B \end{pmatrix}, \quad \begin{pmatrix} w_3^A - w_3^B \\ w_7^A - w_7^B \\ w_{11}^A - w_{11}^B \end{pmatrix} z + \begin{pmatrix} w_4^A - w_4^B \\ w_8^A - w_8^B \\ w_{12}^A - w_{12}^B \end{pmatrix} = \begin{pmatrix} 0 \\ 0 \\ 0 \end{pmatrix}.$$

This suggests that if no regularization is imposed, the first two columns of w^A and w^B have to match. In addition, for any given third columns of w^A and w^B, a given z, and a given fourth column of w^A, we can always compute the fourth column of w^B that yields the same sliced result. This suggests that if we impose no regularization, the parameters of w have to be reduced to 9 instead of 12. In practice, since we typically have information about the reference w^{ref}, we impose regularization and keep the number of parameters as 12 in the parametric affine case. Furthermore, regardless of how many parameters we choose for w, the registration problem may be ill-posed in theory due to the intensities of images $\mathcal{R}$ and $\mathcal{T}$. For example, if $\mathcal{R}$ is image of a disk in 2D and $\mathcal{T}$ is image of a sphere in 3D, the problem yields infinitely many solutions since infinitely many cross-sections of a sphere can yield a disk. Due to the structure of the employed input images, this does not happen in practice. That being said, we regularize the affine transformation w in all cases.

2.2   Discretization

Here we employ a discretize-then-optimize paradigm (see the FAIR software [5] for details) to minimize the functional in Eq. (2).

Discretizing Ω into n pixels and $\mathcal{Z}$ into l pixels, we can define grids $\mathbf{x}_R = [x_k^1, x_k^2]_{k=1,\dots,n}$ and $\mathbf{x}_T = [x_j^1, x_j^2, x_j^3]_{j=1,\dots,n \times l}$ relating to $\mathcal{R}$ and $\mathcal{T}$, respectively, to be the discretizations of Ω and $\Omega \times \mathcal{Z}$. Furthermore, $\mathbf{y} \approx y(\mathbf{w}, \mathbf{x}_T)$, $\mathbf{w} = w$, the cell-centered-discretized images are $T \approx \mathcal{T}(\mathbf{x}_T)$ and $R \approx \mathcal{R}(\mathbf{x}_R)$ (containing nl and n pixels, respectively), and discretization of the operators $\mathcal{D}$ and $\mathcal{S}$ are represented by D and S (see [5]). For a given z, the discretization of the operator $\mathcal{L}_z$, denoted by L_z can be computed as

$$L_z = I_{n \times nl} := I_{n \times n} \otimes \overbrace{[0, \dots, 0, \underbrace{1}_{\lceil l(z+\omega)/2\omega \rceil \text{-th component}}, 0, \dots, 0]}^{1 \times l \text{ size}} \qquad (4)$$

in which we have assumed $\mathcal{Z}$ is the interval $(-\omega, \omega)$. The discretized problem is now to minimize the functional

$$J[\mathbf{w}] := D[L_z(T(\mathbf{y}(\mathbf{w}))), R] + S(\mathbf{w} - \mathbf{w}^{\text{ref}}). \qquad (5)$$

2.3   Optimization

We compute the derivative and Hessian of J denoted by dJ and H_J respectively in a Gauss-Newton approach described in [6]. For simplicity, we allow ourselves to interchangeably refer to derivatives of real-valued functions as Jacobians as well. The Hessian and Jacobian of the regularization S are denoted respectively as dS and H_S. To proceed, we represent the Jacobian of the objective function J as $dJ := \frac{\partial J}{\partial w}$. Now define $L := L_z(T(y(w)))$ and $r := L - R$. Choosing the SSD distance measure and defining $\Psi(r) := \frac{1}{2}r^T r = D[L_z(T(y(w))), R]$ yields $J[w] = \Psi + S(w - w^{\text{ref}})$. Hence using the chain rule

$$\frac{\partial J}{\partial w} = \left(\frac{\partial \Psi}{\partial r}\right)\left(\frac{\partial r}{\partial L}\right)\left(\frac{\partial L}{\partial T}\right)\left(\frac{\partial T}{\partial y}\right)\left(\frac{\partial y}{\partial w}\right) + \left(\frac{\partial S}{\partial w}\right)$$

$$= r^T \times I_{n \times n} \times I_{n \times nl} \times dT \times dy \quad + \quad dS$$

$$= r^T \times I_{n \times n} \times I_{n \times nl} \times dT \times dy \quad + (w - w^{\text{ref}})^T \mathbf{M}$$

in which $dT := \frac{\partial T}{\partial y}$ represents the derivative of the interpolant and $dy := \frac{\partial y}{\partial w}$ is the derivative of the transformation y with respect to w. Derivatives dy and dT are both available in FAIR [5]. Finally, the Hessian of J denoted by H_J can be approximated as

$$H_J = d^2\Psi + H_S \approx dr^T dr + H_S = dr^T dr + \mathbf{M}, \tag{6}$$

where

$$dr = \left(\frac{\partial r}{\partial L}\right)\left(\frac{\partial L}{\partial T}\right)\left(\frac{\partial T}{\partial y}\right)\left(\frac{\partial y}{\partial w}\right) = I_{n \times n} \times I_{n \times nl} \times dT \times dy = I_{n \times nl} \times dT \times dy. \tag{7}$$

In practice, to speed up the computations, matrix-free implementation of the algorithm can be applied. We also consider different discrete representations of the image registration problem, and address the discrete problems sequentially in the so-called multi-level approach.

3   Experiments and Results

3.1   Data

3D pre-procedural and 2D real-time cardiac MRI were acquired from 6 volunteers using a 1.5T MRI scanner (GE Healthcare, Waukesha, WI).

3.1.1   Prior 3D (Cine) Images

Each pre-procedural 3D volume consists of a stack of 12 to 14 short-axis (SAX) slices of the heart with a resolution of $1.37 \times 1.37 \times 8\,\text{mm}^3$ and a field of view (FOV) of $350 \times 350\,\text{mm}^2$. The images were acquired at end-expiration breath-hold with an electrocardiogram (ECG) triggered GE FIESTA pulse sequence and only end-diastolic images were used.

3.1.2   Real-Time Images

2D real-time images were acquired at the same slice locations as in the pre-procedural scans, but under free-breathing conditions. The images were obtained continuously with a fast spiral balanced steady state free precession (bSSFP) sequence at a frame rate of 8 fps, an in-plane resolution of 2.2×2.2 mm^2, slice thickness 8 mm, and a FOV of 350×350 mm^2. The images were ECG-gated and only images acquired end-diastole were used in the following experiments. It should be noted that stacked images do not produce meaningful volumes as there is no synchronization between different slices.

3.2   Validation of Results

If registration between two (non-identical) images is successful, a slice obtained from transforming the 3D template image with the transformation parameters obtained from registration and then slicing at a predetermined slice location would yield a 2D image similar to the 2D reference image. While a look at the end-result images can give us a subjective impression of whether registration was successful and transformation parameters returned have aligned objects in the image well, no 'ground truth' is available in general. We can, however, evaluate end-result images for their purpose in application.

The images in question are cardiac MR images, where the region of interest is the left ventricle (LV). One way of measuring how well two images have been aligned by registration is to measure how much the LVs in the template and reference images overlap before and after registration. Overlap can be quantified by the Dice coefficient and the Jaccard index, which are, respectively, defined as defined as: $\mathrm{Dice}(A, B) = \frac{2|A \cap B|}{|A| + |B|}$ and $\mathrm{Jaccard}(A, B) = \frac{|A \cap B|}{|A \cup B|}$.

The LV also contains papillary muscles which can be used as landmarks. Alignment of the landmarks can be quantified by computing the distance between corresponding landmarks in the reference and template images, before and after registration. This quantity, called the target registration error (TRE), is the l^2-normed distance between landmarks in the template image and the corresponding landmarks in the reference image.

The LV and landmarks in the cine volume were manually selected. The endocardium of the LV was outlined for each slice, and the in-plane segmentations stacked to form a 3D segmentation mask. To obtain a 2D segmentation mask of L after registration, the 3D segmentation mask is transformed using the parameters obtained from registration, and then sliced. In the real-time images, the LV and landmarks were also segmented, by an expert. The coordinates for the landmarks in the image are 2D, but knowing the location where the slice was taken from allows us to append an approximate third coordinate to the landmarks.

3.3   Cine/Cine Controlled Experiments

Before demonstrating the affine model on registration between a cine (pre-operative) image and a real-time (intra-operative) image, we first perform controlled experiments. Controlled registration experiments were performed between a 3D cine volume (template) and a 2D image (reference) that is a slice of a transformed version of the 3D volume. Since that initial transformation is known, ground truth is available. For all experiments following, domains $\Omega = (-175, 175) \times (-175, 175)\,\text{mm}^2$ and $\mathcal{Z} = (-48, 48)\,\text{mm}$, and discretizations $n = 128^2 = 16384$ and $l = 12$. For affine registration, we will also assume the regularizer $\mathbf{M}$ is a diagonal 12×12 matrix with unit entries on the main diagonal except for locations 3, 7, and 11 (third column of matrix) where entries are 10^6, i.e., large. If $\mathbf{w}^{\text{ref}}$ is chosen to be the identity transformation, the regularizer ensures the computed parameters $[\mathbf{w}_3, \mathbf{w}_7, \mathbf{w}_{11}]$ to be close to $[0, 0, 1]$; see Theorem 1. Linear interpolation and an Armijo line search [6] were used in the multi-level Gauss-Newton optimization framework.

3.3.1   Affine Initial Transformation

If the initial transformation applied to generate the reference image is affine, successful affine registration should produce a transformed template slice that aligns with the reference image. Recall that the motivation behind using an affine model as opposed to a rigid was to more accurately represent the deformable nature of organs in the body. To demonstrate that the rigid model does indeed fail when the nature of the deformation applied to the reference image R is not rigid, we individually perturbed each entry of the identity transform $\mathbf{w} = [1, 0, 0, 0, 0, 1, 0, 0, 0, 0, 1, 0]$, and applied the perturbed set of parameters in the initial transformation to obtain R. For each R that was obtained, rigid and affine registration was performed. Due to Theorem 1, perturbing entries w_3, w_7, and w_{11} is equivalent (in terms of producing the same template image slice) to perturbing w_4, w_8, and w_{12} but scaled by a factor of z, the location of the slicing operator, so only 9 entries of $\mathbf{w}$ need to be perturbed; w_4, w_8, and w_{12} were not perturbed. For each of the 9 entries, an ϵ between -0.5 to 0.5 was added to the entry to produce a set of initial parameters used to obtain R.

Figures 1, 2, and 3 show the effects of perturbing the entries of w_2 on the TRE before registration, after rigid registration, and after affine registration. As expected, affine registration improves results over rigid registration. Similar results were found for entries w_1, w_5, w_6, w_9, and w_{10} (first two columns of matrix), but for the sake of brevity, no figures will be shown for those entries. In all figures, the box represents the 25th–75th percentile, and the line in the box marks the median. Figures 4, 5, and 6 show the effects of perturbing the entries of w_3 on the TRE before registration, after rigid registration, and after affine registration. Due to Theorem 1, perturbations in w_3, w_7, and w_{11} can be compensated for when the image is reduced from 3D to 2D by changing the values of w_4, w_8, and w_{12}, which are translation parameters and thus forms a rigid transformation, if no other shear terms are present. Rigid registration was therefore comparable to affine transformation for perturbations on w_3, as seen in Figs. 5 and 6. The same was found for entries w_7 and w_{11} (third column of matrix), but for the sake of brevity, no figures will be shown for those entries.

Recall that an indicator of good alignment is a simultaneously large Jaccard index and small TRE. Although not shown, LV overlap was quantified with the Jaccard index as well. For perturbations on w_2 (and $w_1, w_5, w_6, w_9, w_{10}$ as well), affine registration generally performed well, increasing Jaccard indices and reducing TREs. Rigid registration did not improve results; Jaccard indices after rigid registration became more varied and generally appear to worsen. For perturbations on w_3 (and w_7, w_{11}), the results of rigid registration were comparable to affine registration. This can be explained by Theorem 1 – variations in w_3, w_7, and w_{11} can be compensated for by changing the values of w_4, w_8, and w_{12} to obtain the same 2D slice of a 3D volume.

3.4   Real-Time/Cine Experiments

No initial 3D transformation was applied to obtain the reference image as was done on the test cases, since the reference images here are 2D real-time images. It is also not meaningful to perform a 2D transformation on a real-time image to obtain the reference image for registration, since a modified image no longer represents a clinical setting. Because the slice locations in the real-time and cine cardiac MRI are already rather aligned initially in the z-direction, registration between images from same the slice prescription would align things mostly within the xy-plane, and give little indication of how well the algorithm works when the images are taken at different slice locations. Performing registration between different slices would be a better indicator of how well the algorithm improves alignment in the z-direction. For the following example (Fig. 7), the real-time slice was taken at spatial location $z = -4$ mm while the slicing operation was applied on the template image at $z = -36$ mm, so the initial slice of the 3D template is at $z = -36$ mm. To register the images successfully, the registration algorithm must return transformation parameters that translate the template image by approximately 32 mm (the physical distance between the spatial locations of the reference image and slicing operator) in the z-direction, along with appropriate alignments in the x- and y-directions. Figure 7 and Table 1 show the results of one experiment. The affine model appears to produce slightly better results for this experiment, due to its ability to deform, apparent in the LV overlap after registration (Fig. 7g). In most clinical applications, initial misalignment will not be as large and the two images registered will be slices in close proximity to one another. Affine and rigid registration was performed on real-time images from 6 data sets, each contributing 1 cine image and between 17 to 29 real-time images, to a total of 143 real-time images across 6 data sets. Each real-time image was registered to the cine image of the same subject at the same slice location and cardiac phase. Although the slice prescriptions are identical, there may be small motion normal to the image plane. The results are listed in Table 2. With the exception of Data Set 3 and Data Set 5, rigid registration improves or leaves results unchanged. Affine registration improves results for all data sets except Data Set 3. For Data Set 3, rigid registration returned values worse than what was initially given and affine registration performed even worse. This was due to local deformation within the cardiac region, consistent throughout the

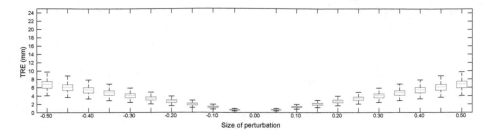

Fig. 1. The TRE as a function of the perturbation on w_2 before registration for all data sets. Reference image obtained by an affine transformation that is the identity transformation except for the addition of the perturbation to w_2.

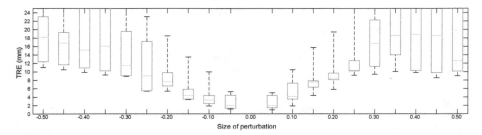

Fig. 2. The TRE as a function of the perturbation on w_2 after rigid registration for all data sets. Reference image obtained by an affine transformation that is the identity transformation except for the addition of the perturbation to w_2.

data available for Data Set 3. Since the body cavity is considerably larger than the cardiac region and comprises most of the content in each image, the algorithm accounted for the body cavity, not the heart, thus the LV becomes more misaligned after registration. For Data Set 5, rigid registration returned slightly worse values than what the algorithm initially started with, but affine registration produced values that were a slight improvement over the initial data. From the values for the rest of the data sets, however, affine registration returns better results in general compared to rigid registration. The TRE was not calculated in this set of experiments because the images are from the same slice locations, image resolution in the z-direction (the direction normal to the short-axis slices) is much coarser than the in-plane resolution, i.e. slice thickness is larger than pixel size; because there is no ground truth available for us to obtain more precise landmark locations, thus z-direction uncertainty would dominate and render the results meaningless.

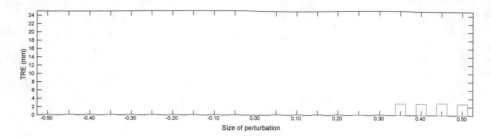

Fig. 3. The TRE as a function of the perturbation on w_2 after affine registration for all data sets. Reference image obtained by an affine transformation that is the identity transformation except for the addition of the perturbation to w_2.

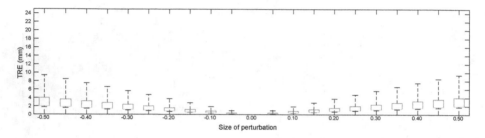

Fig. 4. The TRE as a function of the perturbation on w_3 before registration for all data sets. Reference image obtained by an affine transformation that is the identity transformation except for the addition of the perturbation to w_3.

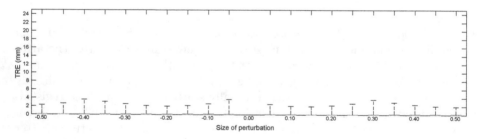

Fig. 5. The TRE as a function of the perturbation on w_3 after rigid registration for all data sets. Reference image obtained by an affine transformation that is the identity transformation except for the addition of the perturbation to w_3.

Table 1. Jaccard indices and Dice coefficients of left ventricle overlap before and after registration between a 3D cine volume and a real-time image taken from $z = -4$ mm. Slicing operations performed at $z = -36$ mm.

	Jaccard	Dice	TRE (mm)
Before registration	0.67	0.80	32.8 ± 0.2
After rigid registration	0.71	0.83	6.6 ± 0.8
After affine registration	0.87	0.93	4.5 ± 0.1

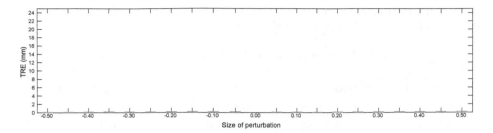

Fig. 6. The TRE as a function of the perturbation on w_3 after affine registration for all data sets. Reference image obtained by an affine transformation that is the identity transformation except for the addition of the perturbation to w_3.

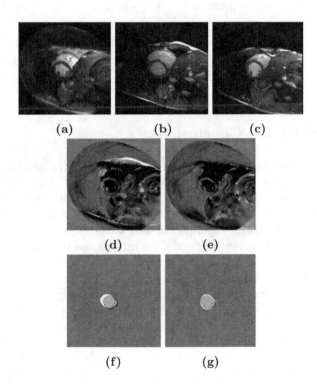

(a)   (b)   (c)

(d)   (e)

(f)   (g)

Fig. 7. Results of affine registration between a 3D cine image and a 2D real-time image on the same subject as in the controlled experiment, with an initial misalignment of approximately 32 mm in the z-direction (through the image plane). (a) Reference image R. (b), (c) Template slice L before and after registration. (d), (e) Difference between the reference image and template slice ($L - R$) before and after registration. (f), (g) Segmentation masks showing left ventricle overlap before and after registration, with in-plane reference image landmarks ($\times$) and out-of-plane template image landmarks projected onto image ($+$).

Table 2. LV overlap before registration, after rigid registration, and after affine registration between a pre-operative 3D cine volume and a noisier, lower-resolution intra-operative 2D real-time image, as in a clinical setting.

	Data set	Before registration	After rigid registration	After affine registration
Jaccard	1	0.86 ± 0.06	0.87 ± 0.07	0.92 ± 0.02
Dice		0.92 ± 0.04	0.93 ± 0.04	0.96 ± 0.01
Jaccard	2	0.75 ± 0.02	0.86 ± 0.02	0.87 ± 0.02
Dice		0.86 ± 0.02	0.93 ± 0.01	0.93 ± 0.01
Jaccard	3	0.77 ± 0.06	0.47 ± 0.13	0.21 ± 0.08
Dice		0.87 ± 0.04	0.63 ± 0.13	0.34 ± 0.12
Jaccard	4	0.49 ± 0.08	0.66 ± 0.05	0.73 ± 0.08
Dice		0.65 ± 0.07	0.80 ± 0.04	0.84 ± 0.06
Jaccard	5	0.80 ± 0.04	0.73 ± 0.03	0.83 ± 0.04
Dice		0.89 ± 0.03	0.84 ± 0.02	0.91 ± 0.02
Jaccard	6	0.76 ± 0.09	0.76 ± 0.09	0.80 ± 0.09
Dice		0.86 ± 0.06	0.86 ± 0.06	0.88 ± 0.06

4 Discussions and Conclusions

In controlled experiments where the reference image is a transformed and sliced version of the template image, it was demonstrated that rigid registration did not sufficiently account for deformations that are affine in nature. We can conclude that the affine model performs better than, or is at least comparable to, the rigid model for controlled experiments, but at the expense of extra computational time. In registration between real-time images and cine images of the same slice location and cardiac phase, affine registration generally performed better than rigid registration, presumably due to its greater flexibility over the rigid model, allowing it to deform the cine image to more closely match the real-time image. We can conclude that between images of the same modality, the proposed multi-level parametric 2D-3D registration scheme can align images well for mis-alignments within reasonable limits encountered in clinical applications, such as motion due to respiration. Despite different acquisition methods in the real-time and prior cine MR images, the registration algorithm improved alignment with the SSD dissimilarity measure.

Affine registration was found to be a generally more robust model than rigid registration in this framework. This suggests that in attempting to improve results for applications employing 2D-3D rigid registration with the SSD, one can first consider simply expanding the transformation model to an affine one before considering more complex dissimilarity measures and regularizers. The advantage of the affine model is its simplicity, allowing more accurate registration at a small cost. For multi-modality registration where intensities of the template and reference images differ more drastically, one can consider using other dissimilarity measures and/or optimizers [5,7] that can fit well within the context of the general proposed model.

Acknowledgments. This research was supported in part by a Natural Sciences and Engineering Research Council of Canada (NSERC) Discovery Grant for M. Ebrahimi. We would like to thank Drs. Graham Wright and Robert Xu of Sunnybrook Research Institute, Toronto, Canada, for valuable discussions and providing the MR data.

References

1. Pushparajah, K., Tzifa, A., Razavi, R.: Cardiac MRI catheterization: a 10-year single institution experience and review. Intervent. Cardiol. **6**, 335–346 (2014)
2. Helen, X., Lasso, A., Fedorov, A., Tuncali, K., Tempany, C., Fichtinger, G.: Multi-slice-to-volume registration for MRI-guided transperineal prostate biopsy. Int. J. CARS **10**(1), 563–572 (2015)
3. Smolíková, R., Wachowiak, M.P., Drangova, M.: Registration of fast cine cardiac MR slices to 3D preprocedural images: toward real-time registration for MRI-guided procedures. In: Proceedings of SPIE, vol. 5370, pp. 1195–1205, May 2004
4. Xu, R., Wright, G.A.: Registration of real-time and prior imaging data with applications to MR guided cardiac interventions. In: Camara, O., Mansi, T., Pop, M., Rhode, K., Sermesant, M., Young, A. (eds.) STACOM 2014. LNCS, vol. 8896, pp. 265–274. Springer, Cham (2015). doi:10.1007/978-3-319-14678-2_28
5. Modersitzki, J.: FAIR: Flexible Algorithms for Image Registration. SIAM, Philadelphia (2009)
6. Nocedal, J., Wright, S.J.: Numerical Optimization, 2nd edn. Springer, New York (2006)
7. Ardeshir Goshtasby, A.: 2-D and 3-D Image Registration. Wiley Press, New York (2005)

Random Forest Based Left Ventricle Segmentation in LGE-MRI

Tanja Kurzendorfer[1]([⊠]), Christoph Forman[2],
Alexander Brost[3], and Andreas Maier[1]

[1] Pattern Recognition Lab,
Friedrich-Alexander University Erlangen-Nuremberg, Erlangen, Germany
tanja.kurzendorfer@fau.de
[2] Siemens Healthcare GmbH, Erlangen, Germany
[3] Siemens Healthcare GmbH, Forchheim, Germany

Abstract. The leading cause of death worldwide is ischaemic heart disease. Late gadolinium enhanced magnetic resonance imaging (LGE-MRI) is the clinical gold standard to visualize regions of myocardial scarring. However, the challenge arises in the segmentation of the myocardial border, as the transition of scar tissue and blood pool can be very smooth, because the contrast agent accumulates in the damaged tissue and leads to various enhancements. In this work, a random forest based boundary detection approach is combined with a scar exclusion criterion. The final endocardial and epicardial border is found with the help of dynamic programming, which finds the distance weighted minimum through the boundary cost array. The segmentation method is evaluated using a 5-fold cross validation on 100 clinical LGE-MRI data sets. The Dice coefficient resulted in an overlap of 0.83 for the endocardium as well as for the epicardium.

1 Introduction

The leading cause of death worldwide is ischaemic heart disease [1]. For diagnosis in clinical routine cardiac magnetic resonance imaging is used, as it can provide information on morphology, tissue characterization, blood flow or perfusion [2,3]. The clinical gold standard for the assessment of myocardial viability is late gadolinium enhanced magnetic resonance imaging (LGE-MRI) [4]. The enhancement of the damaged tissue is based on the different contrast agent accumulation within the tissue, which is based on T_1 weighted imaging [5]. Therefore, necrotic tissue has high signal intensity, whereas the boundaries of the myocardium are hardly enhanced. Consequently, the challenge is the accurate and reliable segmentation of the myocardium for further tissue analysis. As the quantification of the myocardial scar is needed for diagnosis, therapy planning and patient prognosis.

Most segmentation approaches for LGE-MRI require the prior delineation of the myocardium in Cine-MRI data of the same patient which are then propagated to the LGE-MRI [6–9]. However, this contour propagation has several

© Springer International Publishing AG 2017
M. Pop and G.A. Wright (Eds.): FIMH 2017, LNCS 10263, pp. 152–160, 2017.
DOI: 10.1007/978-3-319-59448-4_15

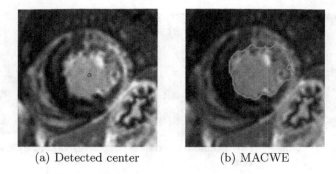

(a) Detected center   (b) MACWE

Fig. 1. (a) Detected center of the left ventricle using circular Hough transforms and circularity constraints. (b) Result of the morphological active contours without edges approach (MACWE).

limitations. The cardiac phases from the Cine-MRI and the LGE-MRI may not accurately match. Inter-slice shifts from multiple breath holds can arise. The global position of the heart may change due to patient movement as contrast has to be injected and the acquisition is done 10 to 20 min after injection. Although these shifts may appear minor, they can lead to significant errors in the scar quantification.

Thus, we propose a random forest based segmentation approach for 2-D LGE-MRI, which is independent of Cine MRI. The major contribution of this approach is, that steerable features are extracted in polar space for the endo-cardial and epicardial boundary respectively. These features are used to train two random forest classifiers, which results in two boundary probability maps for the endocardium and epicardium, respectively. For the endocardium an additional scar exclusion step is added. The final segmentation result is obtained by a dynamic programming approach in polar space.

2   Materials and Methods

The segmentation of the left ventricle can be divided into several steps. First, the left ventricle is detected using a combination of circular Hough transforms, Otsu thresholding and circularity measures. In the second step, a region of interest is identified using morphological active contours. In the third step, potential endocardial boundary positions are detected by casting rays in a cylindrical fashion. The boundary probability is estimated using a random forest classifier. In addition, potential scar areas are excluded from the boundary probabilities. In the final step, the optimal contour is obtained by applying a minimal cost path search to the boundary cost array in polar space.

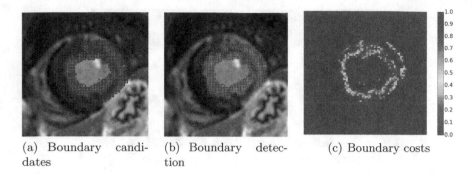

(a) Boundary candi-   (b) Boundary detec-   (c) Boundary costs
dates   tion

Fig. 2. (a) Potential boundary candidates, extracted using ray casting. (b) Boundary detection result obtained from the trained random forest classifier. (c) Boundary cost map in Cartesian coordinates.

2.1 Left Ventricle Detection

The left ventricle is detected in the mid slice of the 2-D LGE-MRI stack. First, the Canny edge detector is used to extract the edges from the image [10]. In the next step, circular Hough transforms are applied [11]. The radii of the circular Hough transforms were in range of 17 mm to 35 mm with a step size of 2 mm due to performance, which was defined according to the anatomical information in literature [12]. The most prominent candidate is selected as potential left ventricle blood pool candidate. To verify this position, an additional roundness measure is applied. Therefore, Otsu's thresholding is applied to the whole slice, to convert the image into a binary mask [13]. Objects that are smaller than a predefined threshold $\theta_o = 25$ are removed. The threshold was defined heuristically. From the remaining objects the eccentricity, i.e. the roundness is estimated $R = \sqrt{\frac{a^2 - b^2}{a^2}}$, where a is the semi-major axis and b is the semi-minor axis of the object. If the object is circular, $R = 0$. If the center points c_1 and c_2 of the roundest object and the result of the circular Hough transform are within θ_c, where $\theta_c = \sqrt{(c_1 - c_2)^2}$, the left ventricle has been accurately detected. Otherwise, the user is asked to verify the center of the left ventricle. The result of the left ventricle detection is shown in Fig. 1(a).

2.2 Endocardial Boundary Estimation

After the left ventricle is detected in the center slice of the MRI stack, this information is used for the boundary detection of the endocardium. The mid-slice is a good slice to start with the segmentation, as the result can be used to propagate in basal and apical direction. To get a rough estimate of the blood pool outline, a morphological active contours without edges (MACWE) approach is applied [14]. This approach alone is not sufficient to get the outline of the blood pool as in LGE-MRI the transition between blood pool and myocardial scar can be very smooth, see Fig. 1(b). However, it gives us a rough outline of the blood pool. This outline can be used to extract potential endocardial boundary candidates using circular ray casting. Therefore, the image is converted to polar coordinates. Boundary candidates are then selected for N equidistant points along R rays, as depicted in Fig. 2(a) in Cartesian coordinates. Each potential boundary candidate is then classified using a trained random forest classifier. The result of the classification is illustrated in Fig. 2(b) as cost map.

Boundary Map Generation: The performance of any classifier is limited by the discriminative power of the features used for training. Steerable features were used [15], as they are computationally efficient and can capture the orientation and scale. In total 16 features were extracted for each boundary candidate, based on local intensity and gradient, which result in a feature vector $x \in \mathbb{R}^{1 \times 16}$, that is used for training and detection. For a given boundary candidate $p(x, y)$ with the intensity I and the gradient $g = (g_x, g_y)$, the following features are extracted: I, $\sqrt{I}$, $\sqrt[3]{I}$, I^2, I^3, $\log I$, $||g||$, $\sqrt{||g||}$, $\sqrt[3]{||g||}$, $||g||^2$, $||g||^3$, $\log ||g||$, g_x, g_y, $\sqrt{g_x^2 + g_y^2}$, $\text{div}(g)$. Note, that all the features are extracted in polar space, which is the steerable space. The center position in Cartesian space, i.e. origin in polar space, has not influence on the classification result.

The training of the random forest is based on ground truth annotations from which positive as well as negative samples are extracted. For the training pathologic as well as healthy subjects are used, to generate a broad range for the training data base. After the training, the classifier can predict the endocardial boundary probability $p_{\text{endo}}(x) \in [0, 1]$. The endocardial boundary probability can be interpreted as costs c, where $c = 1 - p_{\text{endo}}$. If the boundary probability is very high, the costs are close to 0.

To improve the detection of the boundaries from scarred myocardium, an additional scar exclusion step is added. Given the mean intensity of the blood pool μ_{bp} and the standard deviation σ_{bp}, the scar threshold θ_{st} is defined as $\theta_{st} = \mu_{bp} + \sigma_{bp}$. All the pixels above this threshold and outside of the blood pool are defined as potential scar candidates, see Fig. 3(c). The scar map is generated from the scar candidates, where all pixels with increasing radius from potential scar candidates are labeled with 1, as depicted in Fig. 3(d). If a boundary probability overlaps with the scar map, the boundary potentials are impaired, see Fig. 3(e).

Segmentation: In the next step, the final segmentation result of the endocardial contour has to be obtained from the endocardial cost map. Therefore, a dynamic programming approach is used in polar space, to compute the optimal endocardial contour from one end to the other end of the polar image [16]. The minimal cost path (MCP) search is used [17], which finds the distance weighted minimum path through the cost array. The cost path is calculated as the sum of the costs for each move and weighted by the length of the path. The result of the MCP is shown in Fig. 3(f). After the optimal path is found, the contour is transferred back to Cartesian coordinates, see Fig. 4(a). The convex hull is calculated from the contour, as papillary muscles close to the endocardial border might be included, as visualized in Fig. 4(b).

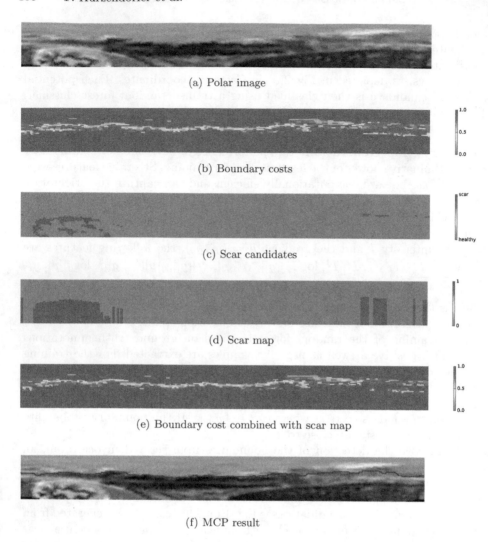

(a) Polar image

(b) Boundary costs

(c) Scar candidates

(d) Scar map

(e) Boundary cost combined with scar map

(f) MCP result

Fig. 3. (a) Mid slice image after polar transformation. (b) Boundary cost map obtained from the trained random forest classifier in polar coordinates. (c) Potential scar candidates which have an intensity value greater than θ_{st} and are not within the blood pool. (d) Scar map, where all scar candidates with increasing radius are labeled with 1. (e) Final endocardial boundary cost map, resulting from the combined boundary detection result and the scar map. (f) Final result of the minimal cost path (MCP) search in polar coordinates.

After the contour is refined in the mid-slice, the information is used for the boundary detection in apical and basal direction. The center is propagated to the succeeding slices and the MACWE is initialized. The boundary detection using the random forest classifier, the scar map generation, and the MCP is repeated for all succeeding slices until the base and apex is reached.

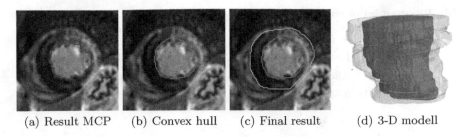

(a) Result MCP   (b) Convex hull   (c) Final result   (d) 3-D modell

Fig. 4. (a) Result of the minimal cost path search in Cartesian coordinates. (b) Convex hull of the final result. (c) The final result of the boundary estimation for the endocardium in red and the epicardium in yellow. (d) 3-D model of the endocardial and epicardial contour in red and yellow, respectively. (Color figure online)

2.3 Epicardial Boundary Estimation

After the endocardial contour is found, the epicardial contour can be estimated. The segmentation starts again with the mid-slice and the result of the refined endocardial contour is used as an initialization for the boundary detection.

Again a random forest classifier is trained for the epicardial boundary detection using the same 16 features as for the endocardial border estimation, resulting in an epicardial boundary probability $p(x)_{epi}$. The result of the epicardial boundary detection is used as cost array for the minimal cost path search. The MCP is applied in polar coordinates for the same reasons as mentioned before. The MCP finds the distance weighted minimal path from the left to the right end of the polar image. The result is then transferred back to Cartesian coordinates and the convex hull is taken. The result is depicted in Fig. 4(c).

The endocardial contour estimation is repeated till the apex and base is reached. Afterwards the contours are extracted as 3-D surface models using the marching cubes algorithm [18]. The output is a list of vertices and faces which are saved in the *.stl or *.obj file format. Figure 4(d) shows an example of a 3-D surface mesh, where the endocardium is visualized in red and the epicardium in yellow.

3   Evaluation and Results

The automatic segmentation of the left ventricle's endo- and epi-cardium was evaluated on 100 clinical LGE-MRI data sets. The inversion recovery 2-D LGE-MRI sequences were acquired with a Siemens MAGNETOM Area 1.5 T scanner (Siemens Healthcare GmbH, Erlangen, Germany). The slice thickness was 8–10 mm, with a pixel size of $(1.59\text{–}2.08 \times 1.59\text{–}2.08)\,\text{mm}^2$ and the spacing between the slices was set to 10 mm. Each data set contained between 10 and 13 short axis slices. Gold standard annotations of the LV endo- and epi-cardium were provided by two clinical experts. The annotations were preformed using MITK [19]. The observers were asked to outline the endocardial and epicardial contour separately.

158   T. Kurzendorfer et al.

Table 1. Segmentation results for the endocardium (Endo), epicardium (Epi) and inter observer variability using the Dice coefficient and the mean surface distance.

	Endo	Epi	Inter Endo	Inter Epi
Dice	0.83 ± 0.08	0.83 ± 0.08	0.95 ± 0.06	0.96 ± 0.05
MSD	3.55 ± 2.08	4.12 ± 2.11	0.89 ± 1.14	0.93 ± 1.13

Given the gold standard annotations, the segmentation was evaluated using different measures, the volumetric Dice coefficient (DC) and the mean surface distance (MSD). Furthermore, the inter-observability was evaluated. The evaluation itself was performed by a 5-fold cross validation, i.e. 20 sequences were excluded from the training of the random forest classifier and used for testing. In Table 1 the average values and the standard deviation of the computed metrics are presented for the endocardium and epicardium. In Fig. 5 the qualitative results of the segmentation are presented. The first row depicts the raw data from base to apex. The second row shows the gold standard annotation of one clinical expert, where the endocardial contour is orange and the epicardial contour green. The last row illustrates the result of the proposed segmentation algorithm, where the endocardium is red and the epicardium yellow.

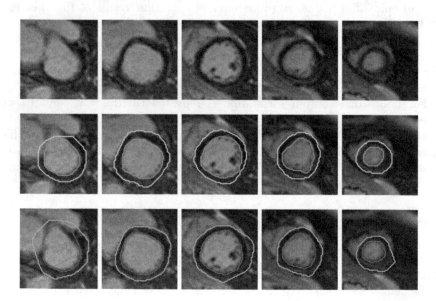

Fig. 5. Comparison of the segmentation result for one data set. From top to bottom: native slices without any contours, gold-standard annotation from clinical expert, and segmentation result of the proposed method. (Color figure online)

The proposed approach was implemented in Python (single threaded, no optimization) and needs less than 10 s for the entire segmentation on a computer equipped with an Intel i7 2.8 GHz CPU and 16 GB of RAM.

4 Discussion and Conclusion

The presented work solely uses LGE-MRI data for the segmentation of the left ventricle, compared to most work reported in literature, which make use of Cine MRI and propagate the contours [6–9]. Albà et al. [20] computed directly the contours from LGE-MRI. Our results are in the same range of the reported errors in literature. However, a direct comparison to the method is not possible, as the data sets differ.

The proposed method achieved a DC of 0.83 for the endocardium and epicardium. The biggest differences occur in the basal region, as the delineation of the left ventricular outflow tract is not always clear. The poor performance of the MSD is mainly due to the larger error in the apex and the left ventricular outflow tract. However, the results in the mid-cavity are convincing, which can be seen in Fig. 5. It is expected, that incorporating a model will directly improve the segmentation result.

In the course of this work, it has been shown that rather simple features can be used for the boundary detection of the endocardium and epicardium. In combination with a minimal cost path search, accurate and consistent results can be achieved. The clear benefit of the method is the independence of registration to Cine MRI and the speed.

Disclaimer: The methods and information presented in this paper are based on research and are not commercially available.

References

1. Mendis, S.: Global Status Report on Noncommunicable Diseases 2014. World Health Organization, Geneva (2014)
2. Petitjean, C., Dacher, J.N.: A review of segmentation methods in short axis cardiac MR images. Med. Image Anal. **15**(2), 169–184 (2011)
3. Suinesiaputra, A., Cowan, B.R., Al-Agamy, A.O., Elattar, M.A., Ayache, N., Fahmy, A.S., et al.: A collaborative resource to build consensus for automated left ventricular segmentation of cardiac MR images. Med. Image Anal. **18**(1), 50–62 (2014)
4. Rashid, S., Rapacchi, S., Shivkumar, K., Plotnik, A., Finn, P., Hu, P.: Modified wideband 3D late gadolinium enhancement (LGE) MRI for patients with implantable cardiac devices. J. Cardiovasc. Magn. Reson. **17**(Suppl 1), Q26 (2015)
5. Kellman, P., Arai, A.: Cardiac imaging techniques for physicians: late enhancement. J. Magn. Reson. Imaging **36**(3), 529–542 (2012)
6. Ciofolo, C., Fradkin, M., Mory, B., Hautvast, G., Breeuwer, M.: Automatic myocardium segmentation in late-enhancement MRI. In: 2008 Proceedings of the 5th IEEE International Symposium on Biomedical Imaging: From Nano to Macro, ISBI 2008, pp. 225–228. IEEE (2008)

7. Dikici, E., O'Donnell, T., Setser, R., White, R.D.: Quantification of delayed enhancement MR images. In: Barillot, C., Haynor, D.R., Hellier, P. (eds.) MICCAI 2004. LNCS, vol. 3216, pp. 250–257. Springer, Heidelberg (2004). doi:10.1007/978-3-540-30135-6_31

8. Wei, D., Sun, Y., Chai, P., Low, A., Ong, S.H.: Myocardial segmentation of late gadolinium enhanced MR images by propagation of contours from cine MR images. In: Fichtinger, G., Martel, A., Peters, T. (eds.) MICCAI 2011. LNCS, vol. 6893, pp. 428–435. Springer, Heidelberg (2011). doi:10.1007/978-3-642-23626-6_53

9. Tao, Q., Piers, S., Lamb, H., van der Geest, R.: Automated left ventricle segmentation in late gadolinium-enhanced MRI for objective myocardial scar assessment. J. Magn. Reson. Imaging 42(2), 390–399 (2015)

10. Canny, J.: A computational approach to edge detection. IEEE Trans. Pattern Anal. Mach. Intell. 8(6), 679–698 (1986)

11. Duda, R.O., Hart, P.E.: Use of the Hough transformation to detect lines and curves in pictures. Commun. ACM 15(1), 11–15 (1972)

12. Lang, R.M., Bierig, M., Devereux, R.B., Flachskampf, F.A., Foster, E., Pellikka, P.A., et al.: Recommendations for chamber quantification. Eur. Heart J. Cardiovasc. Imaging 7(2), 79–108 (2006)

13. Otsu, N.: A threshold selection method from gray-level histograms. Automatica 11(285–296), 23–27 (1979)

14. Marquez-Neila, P., Baumela, L., Alvarez, L.: A morphological approach to curvature-based evolution of curves and surfaces. IEEE Trans. Pattern Anal. Mach. Intell. 36(1), 2–17 (2014)

15. Zheng, Y., Barbu, A., Georgescu, B., Scheuering, M., Comaniciu, D.: Four-chamber heart modeling and automatic segmentation for 3-D cardiac CT volumes using marginal space learning and steerable features. IEEE Trans. Med. Imaging 27(11), 1668–1681 (2008)

16. Qian, X., Lin, Y., Zhao, Y., Wang, J., Liu, J., Zhuang, X.: Segmentation of myocardium from cardiac MR images using a novel dynamic programming based segmentation method. Med. Phys. 42(3), 1424–1435 (2015)

17. Dijkstra, E.: A note on two problems in connexion with graphs. Numer. Math. 1(1), 269–271 (1959)

18. Lorensen, W., Cline, H.: Marching cubes: a high resolution 3D surface construction algorithm. In: ACM Siggraph Computer Graphics, vol. 21, pp. 163–169. ACM (1987)

19. Wolf, I., Vetter, M., Wegner, I., Böttger, T., Nolden, M., Schöbinger, M., et al.: The medical imaging interaction toolkit. Med. Image Anal. 9(6), 594–604 (2005)

20. Albà, X., i Ventura, F., Rosa, M., Lekadir, K., Tobon-Gomez, C., Hoogendoorn, C., et al.: Automatic cardiac LV segmentation in MRI using modified graph cuts with smoothness and interslice constraints. Magn. Reson. Med. 72(6), 1775–1784 (2014)

A Multiple Kernel Learning Framework to Investigate the Relationship Between Ventricular Fibrillation and First Myocardial Infarction

Maciej Marciniak[1](✉), Hermenegild Arevalo[1], Jacob Tfelt-Hansen[2],
Kiril A. Ahtarovski[2], Thomas Jespersen[3], Reza Jabbari[2], Charlotte Glinge[2],
Niels Vejlstrup[2], Thomas Engstrom[2], Mary M. Maleckar[1],
and Kristin McLeod[1,4]

[1] Simula Research Laboratory,
Centre for Cardiological Innovation, Oslo, Norway
maciej.mar92@gmail.com
[2] Department of Cardiology, Rigshospitalet, Copenhagen, Denmark
[3] Department of Biomedical Sciences,
University of Copenhagen, Copenhagen, Denmark
[4] KardioMe s.r.o., Bratislava, Slovakia

Abstract. Myocardial infarction results in changes in the structure and tissue deformation of the ventricles. In some cases, the development of the disease may trigger an arrhythmic event, which is a major cause of death within the first twenty four hours after the infarction. Advanced analysis methods are increasingly used in order to discover particular characteristics of the myocardial infarction development that lead to the occurrence of arrhythmias. However, such methods usually consider only a single feature or combine separate analyses from multiple features in the analytical process. In an attempt to address this, we propose to use cardiac magnetic resonance imaging to extract data on the shape of the ventricles and volume and location of the infarct zone, and to combine them within one analytical model through a multiple kernel learning framework. The proposed method was applied to a cohort of 46 myocardial infarction patients. The location, rather than the volume, of the infarct region was found to be correlated with arrhythmic events and the proposed combination of kernels yielded excellent accuracy (100%) in distinguishing between patients that did and did not present at the hospital with ventricular fibrillation.

1 Introduction

An ST-elevation myocardial infarction (STEMI) leads to alterations in loading conditions which cause distinctive patterns of remodelling in infarcted and non-infarcted zones of the myocardium. Such deformations can include dilatation, hypertrophy, and regional thinning. To an extent, the magnitude of remodelling can be anticipated by the location, volume, and transmurality of the infarct

© Springer International Publishing AG 2017
M. Pop and G.A. Wright (Eds.): FIMH 2017, LNCS 10263, pp. 161–171, 2017.
DOI: 10.1007/978-3-319-59448-4_16

region [1]. In 2–5% of cases of STEMI, ventricular fibrillation (VF) occurs before, during or immediately after the primary percutaneous coronary intervention [2].

Cardiac Magnetic Resonance (CMR) imaging provides a non-invasive way to gain insights into the structure and tissue characteristics of the heart. CMR imaging followed by image processing can provide quantitative information on the overall cardiac shape, level of fibrosis, presence of scar, and tissue deformation, which are increasingly used in assessing cardiac disease state and identifying precursors of arrhythmic disorders. Numerous applications for statistical analysis of such features, used either individually or separately, have been widely described. However, combining the extracted features into one model could provide more effective and accurate analysis than predictions made on a single characteristic, or many characteristics considered independently.

Multiple kernel learning (MKL) is a novel approach in machine learning which is becoming increasingly popular in analyzing data where the similarity metrics are not clearly specified or the data sets come from different sources. Instead of taking into account results from detached models (as in ensemble methods), MKL allows the use of multiple sources of information in one classification model. MKL methods have been used recently to, for example, predict responders of cardiac resynchronisation therapy [3].

In this study we sought to understand why, and in which patients, the first symptom of STEMI is VF, as these patients are at greater risk of sudden cardiac death. MKL algorithms were used to distinguish between first STEMI patients presenting at the hospital with VF and patients presenting with non-arrhythmic symptoms, based on features extracted from CMR data. Personalised three-dimensional (3D) shape models, which required image segmentation and 3D mesh generation, were used to acquire information about the overall shape of the ventricles. Additionally, since CMR imaging enables tissue characterisation through contrast (e.g. gadolinium) enhancement imaging [4], segmentation of the infarct region was performed to obtain data on the volume and location of the ischemic zone in the left ventricle (LV). Based on the shape and infarct region information, an MKL classifier was trained and tested on a data set of 46 patients with the first STEMI.

2   Event-Based Classification from Shape and Infarct Data

We propose a framework to distinguish between patients presenting with different symptoms from image-based features related to ventricular shape and tissue characterisation. The process of data acquisition is divided into two parts, with the shape features described in Sect. 2.1 and the infarct features shown in Sect. 2.2. Construction of kernels used in the learning process is illustrated in Sect. 2.3. A MKL method (described in Sect. 2.4) is used to account for the heterogeneity of the features for a single classification task. The proposed pipeline is summarised in Fig. 1.

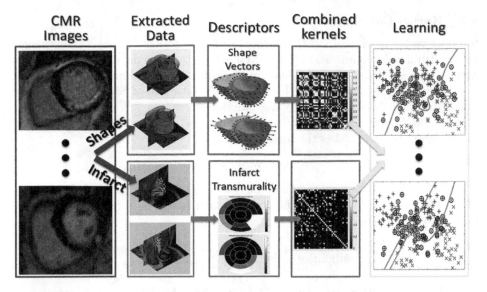

Fig. 1. The proposed pipeline to go from CMR images to classification models. CMR images are semi-automatically segmented to extract the infarct and shape information. From the created models, the relevant data is obtained: shape descriptors and infarct location and volume. Finally, different combinations of kernels are constructed and the best combination is used in the MKL algorithm to perform classification.

2.1 Shape Features

The ventricular shapes are represented by surfaces; one each for the LV endocardium and epicardium and right ventricle (RV) endocardium. We propose to use an atlas-based approach to describe the shapes as deformations of an atlas, to avoid the need to parameterise the shapes point-wise (which is challenging due to the lack of identifiable landmarks). The atlas is computed from a group of (spatially aligned) subject surfaces using a forward approach, as described in [5]. Using the same framework, the deformation of every subject to the atlas is computed using a large deformation diffeomorphic metric mapping (LDDMM) method. The deformation of a new subject (not used in the atlas construction) to the atlas can be computed similarly using the LDDMM framework. The resulting shape features are a matrix of displacement vectors for the atlas shape to each subject.

2.2 Infarct Volume and Location

LV infarct region is identified from contrast-enhanced gadolinium images and represented by either a global measure (volume/percentage of infarct in the LV myocardium) or at a regional level (weighted infarct transmurality (WIT) per region). Global infarct is represented by a scalar for each subject and regional infarct is represented by a vector with the number of dimensions equal to the number of regions.

2.3 Kernel Based Functions

Support vector machine (SVM) is a classification algorithm based on the theory of structural risk minimization [6]. SVM assumes that the data set consists of N identically distributed, independent input samples $\mathbf{x}_i$ (M-dimensional vectors) and corresponding outputs $y_i \in \{-1, 1\}$, $i \in N$. The aim of SVM is to find the linear discriminant with the mapping function $\mathbf{\Phi} \colon \mathbb{R}^M \to \mathbb{R}^L$, which maximizes the margin between samples differing in outcome in the feature space. This discriminant function takes the form:

$$f(\mathbf{x}) = \langle \mathbf{w}, \mathbf{\Phi}(\mathbf{x}) \rangle + b,$$

where $\mathbf{w} \in \mathbb{R}^L$ is the vector of weights and b is the bias term of the hyperplane separating the samples. The cost function takes the form of a quadratic optimization problem:

$$\text{minimize } \frac{1}{2}\|\mathbf{w}\|_2^2 + C\sum_{i=1}^{N}\xi_i,$$

$$\text{with respect to } \mathbf{w} \in \mathbb{R}^L, \ \xi \in \mathbb{R}_+^N, \ b \in \mathbb{R},$$

$$\text{subject to } y_i(\langle \mathbf{w}, \mathbf{\Phi}(\mathbf{x}_i) \rangle + b) \geq 1 - \xi_i \ \ \forall i,$$

where C denotes a trade-off between the classification error and the simplicity of the margin, and ξ stands for the vector of slack variables ('softness' of margin). This problem is not solved directly; instead the dual formulation of an optimization in training the support vector machine classifier is tackled, in the form of the Lagrangian dual function [6]:

$$\text{maximize } \sum_{i=1}^{N}\alpha_i - \frac{1}{2}\sum_{i=1}^{N}\sum_{j=1}^{N}\alpha_i\alpha_j y_i y_j \langle \mathbf{\Phi}(\mathbf{x}_i), \mathbf{\Phi}(\mathbf{x}_j) \rangle,$$

$$\text{with respect to } \alpha \in \mathbb{R}_+^N,$$

$$\text{subject to } \sum_{i=1}^{N}\alpha_i y_i = 0,$$

$$C \geq \alpha_i \geq 0 \ \ \forall i,$$

where $k(\mathbf{x}_i, \mathbf{x}_j) = \langle \mathbf{\Phi}(\mathbf{x}_i), \mathbf{\Phi}(\mathbf{x}_j) \rangle$ is the *kernel function* and α is the vector containing dual variables, controlling separation constraints. With this problem formulation, the discriminant function becomes

$$f(\mathbf{x}) = \sum_{i=1}^{N}\alpha_i y_i k(\mathbf{x}_i, \mathbf{x}) + b.$$

In machine learning, the kernel function is understood as a measure of similarity and determines the distribution of similarities of points around a given point in multi-dimensional space. There are a few kernel functions commonly used in literature, including the linear kernel: $k_L(\mathbf{x}_i, \mathbf{x}_j) = \langle \mathbf{x}_i, \mathbf{x}_j \rangle$ and the polynomial kernel: $k_P(\mathbf{x}_i, \mathbf{x}_j) = (\langle \mathbf{x}_i, \mathbf{x}_j \rangle + p)^q, q \in \mathbb{N}$.

2.4   Classification via a Multiple Kernel Learning Framework

In modern machine learning methods, it is suggested to combine multiple kernel functions and their corresponding parameters into one model, in order to utilize data from various information sources, features of which may be compared using different optimal similarity measures [7]. Such kernel functions are integrated with a combination function, which assigns weights to each predefined kernel:

$$k_\eta(\mathbf{x}_i, \mathbf{x}_j) = f_\eta(\{k_m(\mathbf{x}_i^m, \mathbf{x}_j^m)\}_{m=1}^P | \eta), \tag{1}$$

where the combination function f_η can be linear or non-linear and η is a set of parameters that weighs the importance of every kernel. Each kernel function $\{[k_m \colon \mathbb{R}^D \times \mathbb{R}^D \to \mathbb{R}\}_{m=1}^P$ takes P representations of D-dimensional feature vectors of the separately transformed data instances $\mathbf{x}_i = \{\mathbf{x}_i^m\}_{m=1}^P$. The combination function (Eq. 1) is then used in SVM classification.

3   Arrhythmic Classification of Myocardial Infarction Patients

3.1   Data Preparation

Myocardial Infarction Patient Data Set: We illustrate the proposed method on a data set of patients with first ST-elevation myocardial infarction from the GEVAMI study in Denmark [8]. Forty six patients with varying degrees of infarction were considered (from 0.07% to 24.45%) with mean age $\pm$ standard deviation $= 56 \pm 6$, 34 male. Contrast-enhanced CMR images were acquired for all patients using a gadolinium-based contrast agent. All patients were scanned within three months after primary percutaneous stent intervention. Patients were divided into two groups: a group of patients presenting with ventricular fibrillation (the case group, n = 21), and a group of patients presenting with other non-arrhythmic symptoms (the control group, n = 25).

Segmentation and Pre-processing: The left ventricle endocardium and epicardium, right ventricle endocardium, and left ventricle infarct were segmented using the Segment software from Medviso [9]. Triangulated surface meshes for the ventricles were created using the pipeline described in [10]. Using this pipeline, the point clouds exported from Segment were aligned in space for all subjects (removing differences of pose and orientation), and surface meshes were generated from the aligned point clouds.

Infarct Volume: The left ventricle infarct volumes were computed via Simpson's rule [11]. These volumes for each group are plotted in Fig. 2 (left). To check for the utility of the volume of ischemic zone in the left ventricle, we considered the null hypothesis that the occurrence of arrhythmia has no effect on the size of infarct. I.e. we checked for H_0: $\mu_1 = \mu_2$ versus H_A: $\mu_1 \neq \mu_2$, where H_0 is the null hypothesis, H_A is the alternative, and μ_1, μ_2 are the mean ratios of the ischemic zone in the first and second group. The numbers of patients in both groups are similar (21 patients with arrhythmia and 25 controls) and the variances are alike, therefore a Two Sample t-test was conducted. Confidence level of 95% is considered, thus we reject the null hypothesis for p-value < 0.05.

Figure 2 (left) clearly shows similarities between the groups, especially with regard to the first (0.82% difference), second (0.07% difference) and third quartile (0.17% difference), which might indicate a lack of relevance of this feature. The conducted Two Sample t-test (with t-statistic equal to -0.67 and 44 degrees of freedom) results in a p-value > 0.5, which confirms this statement. We cannot reject the null hypothesis, therefore the information about the percentage of volume of the left ventricle myocardium taken by ischemic zone was not included in the classification, as it would decrease the performance of the model, rather than improve it, by introducing noise.

Infarct Location: The impact of infarct zone location on the outcome was checked in the same manner as the volume of the infarct region in the LV. With a Two Sample t-test we analysed the null hypothesis that occurrence of the arrhythmia does not influence the WIT percentage in each of the 17 segments. What is more, the test for association between the individual features and the event, using Pearson's product-moment correlation coefficients has been calculated. The coefficients are visualised on a 17-Segment AHA plot (Fig. 2).

Out of the 17 segments included in the correlation test, two exhibit statistical significance: the mid-anterolateral segment (no. 12, correlation coef: -0.3313, p-value $= 0.0245$) and the apical inferior segment (no. 15, correlation coef: 0.2976, p-value $= 0.0445$). However, there is a significant difference in positive and negative correlation between the inferior and anterior walls, taken as a whole (correlation coef: 0.2168, p-value $= 0.0379$), therefore information about the infarct location was kept in the model.

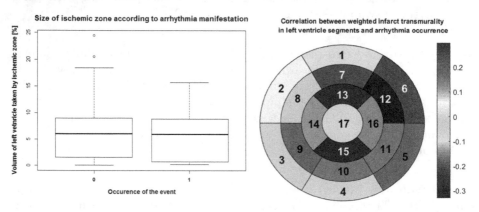

Fig. 2. Left: Variation of global infarct volume for the control group (left boxplot) and the arrhythmia group (right boxplot). Right: Heatmap presenting the correlation coefficients between regional infarct burden and arrhythmias, depicted in a 17-Segment AHA bulls-eye plot.

3.2   Kernel Function Integration

In the process of combining the kernels, alignment-based MKL algorithms were compared. Models were constructed using the MATLAB package described in [7]. Combinations of linear and polynomial kernels (up to the fifth degree) were tested. In order to find the optimal kernel weights, three methods were analysed and used. The first one finds solution to a QCQP problem stated in [12], due to its functional form referred to as 'conic':

$$\text{maximize} \quad \sum_{m=1}^{P} \eta_m \langle \mathbf{K}_m^{tra}, \mathbf{yy^T} \rangle_F,$$

$$\text{with respect to} \quad \eta \in \mathbb{R}_+^P,$$

$$\text{subject to} \quad \sum_{m=1}^{P} \sum_{h=1}^{P} \eta_m \eta_h \langle \mathbf{K}_m, \mathbf{K}_h, \rangle_F \leq 1,$$

where $\langle \mathbf{K}_A, \mathbf{K}_B \rangle_F = \sum_{i=1}^{N} \sum_{j=1}^{N} k_1(\mathbf{x}_i^1, \mathbf{x}_j^1) k_2(\mathbf{x}_i^2, \mathbf{x}_j^2)$. The trace of the acquired kernel matrix is denoted by $\mathbf{K}_m^{tra}$, and $\mathbf{yy}^T$ is the ideal kernel matrix in a classification problem with two classes. The second method solves the QP problem described in [13] ('convex'):

$$\text{minimize} \quad \sum_{m=1}^{P} \sum_{h=1}^{P} \eta_m \eta_h \langle \mathbf{K}_m, \mathbf{K}_h, \rangle_F - 2 \sum_{m=1}^{P} \eta_m \langle \mathbf{K}_m, \mathbf{yy^T} \rangle_F,$$

$$\text{with respect to} \quad \eta \in \mathbb{R}_+^P,$$

$$\text{subject to} \quad \sum_{m=1}^{P} \eta_m = 1.$$

Both of these parametrized combination functions are learnt by solving an optimization problem. The target functions use a similarity metric to select the combination function parameters which maximizes the similarity between the optimal kernel matrix and combined kernel matrix. The last function is based on a heuristic approach proposed in [14] ('ratio'):

$$\eta_m = \frac{A(\mathbf{K}_m, \mathbf{yy}^T)}{\sum_{h=1}^{P} A(\mathbf{K}_h, \mathbf{yy}^T)} \quad \forall m,$$

where $A()$ denotes *kernel alignment* [15]. This approach also uses a parametrized combination function, but the parameters are found by examining alignment acquired from every kernel individually as a measure of performance. The results for each respective algorithm are summarised in Fig. 3.

3.3   Methodology and Parameters

The experiments were performed as follows: the entire data set was randomly divided into a training set (41 samples) and a testing set (5 samples).

Average test accuracy over all values of hyperparameter C for all combinations of kernels [%]				
Combination	Conic	Convex	Ratio	Overall
'L' + 'p2'	60	60	80	66,6667
'L' + 'p3'	96,6667	96,6667	80	91,1111
'L' + 'p4'	80	76,6667	80	78,8889
'L' + 'p5'	80	80	80	80
'p2' + 'p3'	60	60	60	60
'p2' + 'p4'	60	60	60	60
'p2' + 'p5'	60	60	60	60
'p3' + 'p4'	40	40	93,3333	57,7778
'p3' + 'p5'	93,3333	93,3333	93,3333	93,3333
'p4' + 'p5'	40	40	56,6667	45,5556
'L' + 'p2' + 'p3'	60	60	96,6667	72,2222
'L' + 'p2' + 'p4'	60	56,6667	80	65,5556
'L' + 'p2' + 'p5'	60	56,6667	63,3333	60
'L' + 'p3' + 'p4'	80	76,6667	80	78,8889
'L' + 'p3' + 'p5'	80	80	80	80
'L' + 'p4' + 'p5'	80	76,6667	80	78,8889
'p2' + 'p3' + 'p4'	60	60	60	60
'p2' + 'p3'+ 'p5'	60	60	80	66,6667
'p2' + p4'+'p5'	60	60	60	60
'p3' + 'p4' + 'p5'	40	40	90	56,6667
'L' + 'p2' + 'p3' + 'p4'	60	56,6667	96,6667	71,1111
'L' + 'p2' + 'p3' + 'p5'	60	56,6667	96,6667	71,1111
'L' + 'p2' + 'p4' + 'p5'	60	56,6667	80	65,5556
'L' + 'p3' + 'p4' + 'p5'	80	76,6667	80	78,8889
'p2' + 'p3' + 'p4' + 'p5'	60	60	76,6667	65,5556
'L' + 'p2' + 'p3' + 'p4' + 'p5'	60	56,6667	80	65,5556

Average test accuracy over all combinations of kernels for all values of hyperparameter C [%]				
C	Conic	Convex	Ratio	Overall
0.1	40	40	40	40
0.3	40	40	40	40
1	63,0769	55,3846	70,7692	63,0769
3	65,3846	65,3846	79,2308	70
10	65,3846	65,3846	79,2308	70
30	65,3846	65,3846	79,2308	70
100	65,3846	65,3846	79,2308	70
300	65,3846	65,3846	79,2308	70
1000	53,0769	73,0769	70,7692	65,641

(a)

Test accuracy of the ('l' + 'p3') kernel combination for all values of hyperparameter C [%]				
C	Conic	Convex	Ratio	Overall
0.1	40	40	40	40
0.3	40	40	40	40
1	80	80	80	80
3	100	100	80	93,33
10	100	100	80	93,33
30	100	100	80	93,33
100	100	100	80	93,33
300	100	100	80	93,33
1000	40	60	60	53,33

(b)

Fig. 3. Left: Accuracy of the models with regard to the combination of kernels, according to 'conic', 'convex' and 'ratio' algorithms and overall, given all values of hyperparameter C. Right: Accuracy of (a) all models and (b) the most accurate model (with linear and polynomial kernel of the third degree) with regard to different values of hyperparameter C according to 'conic', 'convex' and 'ratio' algorithms, and overall.

The division is necessary in order to test how well do the algorithms perform on the unseen samples. Validation was performed in the form of a grid search with each combination of parameters taken into consideration in a separate model. Hyperparameters are trained via 10-fold stratified cross-validation. The models learned from the training data and its accuracy were examined on the testing set. The models were tuned in terms of the following parameters:

- the hyperparameter C (values 0.1, 0.3, 1, 3, 10, 30, 100, 300, 1000);
- the number of kernels (2, 3, 4 and 5 kernels);
- the combinations of kernels (linear ('L') and polynomial kernels of the second ('p2'), third ('p3'), fourth ('p4') and fifth ('p5') degree;
- the algorithm, i.e. method of combining kernels (conic, convex, and ratio).

The performance of the models with all C values and for the three combining kernel methods described in the previous section are summarized in Fig. 3, along with the overall accuracy of the given kernel combinations. In all cases, the combination of the polynomial kernel of third and fifth degree provided very good accuracy (93.3%). When conic and convex algorithms were taken into account, combining linear and third degree polynomial kernels performed even better, with 96.6% accuracy. In the case of the ratio algorithm, there were three other well-performing combinations: {'L' + 'p2' + 'p3'}, {'L' + 'p2' + 'p3' + 'p4'} and {'L' + 'p2' + 'p3' + 'p5'}, all with 96.6% accuracy. The optimal model in this study was the one combining the linear kernel and polynomial kernel of the third degree with the conic algorithm, with the value for regularization parameter C anywhere between 3 and 300. The reason for this model being the optimal is the simplicity of the model (which decreases the probability of overfitting and misclassifying new samples), and its accuracy measured on the testing set.

4 Discussion

The experiments suggest that the proposed method is useful for identifying patients according to whether or not they were hospitalised as a result of VF. In this study, images from within 3 months of intervention were analysed, which is a potential limitation due to the remodelling that can occur over time. Images acquired closer to the intervention may be more discriminant.

In this study we summarised the results of all VF patients, not individually looking at each patient. With more patients and a wider range of data the exact regions and combinations of regions can be analysed in depth by for example subgrouping patients according to the infarct region. Moreover, since shape data processing can be a lengthy procedure, in the future studies combining other sources of data can be considered, such as ECG recordings and accounting for heterogeneity in the infarct regions.

The results suggest that the proposed method can be useful in distinguishing between arrhythmic and non-arrhythmic patients, despite the fact that the scar is treated as homogenous. Infarct burden in the lateral wall was found to be the most correlated with VF; a result which implies that occlusion in the left coronary artery (i.e. LCX, which supplies blood to the lateral wall) is more associated with a risk of out-of-hospital VF than the right coronary artery (which supplies blood to the inferior wall of the LV). The accuracy was highest when both shape and infarct region information was included, highlighting the need for methods such as MKL that can combine different types of features.

5 Conclusion

In this study we proposed a method to use a MKL classification algorithm in order to differentiate patients according to the symptoms they presented with at the hospital (arrhythmic vs. non-arrhythmic) based on bi-ventricular shape features and volume and location of infarct in the left ventricle. The MKL algorithm with the combination of a linear kernel and polynomial kernel of the third degree provided excellent classification results when linear combination methods are used. In addition, the location, rather than volume, of infarct was found to be correlated to arrhythmia. The results suggest that shape remodelling and scar location may be predictors of arrhythmic events, though the reasons for this remain unknown. Therefore in future studies, the mechanisms behind why the different infarct locations did or did not lead to arrhythmia will be investigated.

Acknowledgements. This project was partially carried out in the Centre for Cardiological Innovation (CCI), Norway funded by the Norwegian Research Council, and partially funded by the Novo Nordisk foundation.

References

1. Pfeffer, M.A., Braunwald, E.: Ventricular remodeling after myocardial infarction. experimental observations and clinical implications. Circulation **81**(4), 1161–1172 (1990)
2. Gorenek, B., Lundqvist, C.B., Terradellas, J.B., Camm, A.J., Hindricks, G., Huber, K., Kirchhof, P., Kuck, K.H., Kudaiberdieva, G., Lin, T., et al.: Cardiac arrhythmias in acute coronary syndromes: position paper from the joint ehra, acca, and eapci task force. Europace **16**, 1655–1673 (2014). euu208
3. Peressutti, D., Sinclair, M., Bai, W., Jackson, T., Ruijsink, J., Nordsletten, D., Asner, L., Hadjicharalambous, M., Rinaldi, C.A., Rueckert, D., et al.: A framework for combining a motion atlas with non-motion information to learn clinically useful biomarkers: application to cardiac resynchronisation therapy response prediction. Med. Image Anal. **35**, 669–684 (2017)
4. Ismail, T.F., Prasad, S.K., Pennell, D.J.: Prognostic importance of late gadolinium enhancement cardiovascular magnetic resonance in cardiomyopathy. Heart **98**(6), 438–442 (2012)
5. Durrleman, S., Pennec, X., Trouvé, A., Ayache, N.: Statistical models of sets of curves and surfaces based on currents. Med. Image Anal. **13**(5), 793–808 (2009)
6. Vapnik, V.: The Nature of Statistical Learning Theory. Wiley, New York (1998)
7. Gönen, M., Alpaydın, E.: Multiple kernel learning algorithms. J. Mach. Learn. Res. **12**, 2211–2268 (2011)
8. Jabbari, R., Engstrøm, T., Glinge, C., Risgaard, B., Jabbari, J., Winkel, B.G., Terkelsen, C.J., Tilsted, H.H., Jensen, L.O., Hougaard, M., et al.: Incidence and risk factors of ventricular fibrillation before primary angioplasty in patients with first st-elevation myocardial infarction: a nationwide study in Denmark. J. Am. Heart Assoc. **4**(1), e001399 (2015)
9. Heiberg, E., Sjögren, J., Ugander, M., Carlsson, M., Engblom, H., Arheden, H.: Design and validation of segment-freely available software for cardiovascular image analysis. BMC Med. Imaging **10**(1), 1 (2010)
10. Marciniak, M., et al.: From CMR image to patient-specific simulation and population-based analysis: tutorial for an openly available image-processing pipeline. In: Mansi, T., McLeod, K., Pop, M., Rhode, K., Sermesant, M., Young, A. (eds.) STACOM 2016. LNCS, vol. 10124, pp. 106–117. Springer, Cham (2017). doi:10.1007/978-3-319-52718-5_12
11. Hergan, K., Schuster, A., Frühwald, J., Mair, M., Burger, R., Töpker, M.: Comparison of left and right ventricular volume measurement using the Simpson's method and the area length method. Eur. J. Radiol. **65**(2), 270–278 (2008)
12. Lanckriet, G.R., Cristianini, N., Bartlett, P., Ghaoui, L.E., Jordan, M.I.: Learning the kernel matrix with semidefinite programming. J. Mach. Learn. Res. **5**, 27–72 (2004)

13. He, J., Chang, S.F., Xie, L.: Fast kernel learning for spatial pyramid matching. In: 2008 IEEE Conference on Computer Vision and Pattern Recognition, CVPR 2008, pp. 1–7. IEEE (2008)
14. Qiu, S., Lane, T.: A framework for multiple kernel support vector regression and its applications to sirna efficacy prediction. IEEE/ACM Trans. Comput. Biol. Bioinf. **6**(2), 190–199 (2009)
15. Cristianini, N., Elisseeff, A., Shawe-Taylor, J., Kandola, J.: On kernel-target alignment. In: Advances in Neural Information Processing Systems (2001)

Real-Time Guiding Catheter and Guidewire Detection for Congenital Cardiovascular Interventions

YingLiang Ma[1,2(✉)], Mazen Alhrishy[3], Maria Panayiotou[3], Srinivas Ananth Narayan[4], Ansab Fazili[3], Peter Mountney[5], and Kawal S. Rhode[3]

[1] School of Computing, Electronics and Mathematics, Coventry University, Coventry, UK
[2] School of Computing and Digital Technology, Birmingham City University, Birmingham, UK
y.ma@bcu.ac.uk
[3] Division of Imaging Sciences and Biomedical Engineering,
King's College London, London, UK
[4] Department of Cardiology, Guy's and St. Thomas' Hospitals NHS Foundation Trust,
London, UK
[5] Medical Imaging Technologies, Siemens Healthineers, Princeton, NJ, USA

Abstract. Guiding catheters and guidewires are used extensively in pediatric cardiac catheterization procedures for congenital heart diseases (CHD). Detecting their positions in fluoroscopic X-ray images is important for several clinical applications, such as visibility enhancement for low dose X-ray images, and co-registration between 2D and 3D imaging modalities. As guiding catheters are made from thin plastic tubes, they can be deformed by cardiac and breathing motions. Therefore, detection is the essential step before automatic tracking of guiding catheters in live X-ray fluoroscopic images. However, there are several wire-like artifacts existing in X-ray images, which makes developing a real-time robust detection method very challenging. To solve those challenges in real-time, a localized machine learning algorithm is built to distinguish between guiding catheters and artifacts. As the machine learning algorithm is only applied to potential wire-like objects, which are obtained from vessel enhancement filters, the detection method is fast enough to be used in real-time applications. The other challenge is the low contrast between guiding catheters and background, as the majority of X-ray images are low dose. Therefore, the guiding catheter might be detected as a discontinuous curve object, such as a few disconnected line blocks from the vessel enhancement filter. A minimum energy method is developed to trace the whole wire object. Finally, the proposed methods are tested on 1102 images which are from 8 image sequences acquired from 3 clinical cases. Results show an accuracy of 0.87 ± 0.53 mm which is measured as the error distances between the detected object and the manually annotated object. The success rate of detection is 83.4%.

1 Introduction

According to the American Heart Association, Congenital heart diseases (CHD) affect an estimated 0.5% to 1% of all live births. They are responsible for up to 40% of all deaths from congenital anomalies and account for 3.0–7.5% of all infant deaths. CHD

© Springer International Publishing AG 2017
M. Pop and G.A. Wright (Eds.): FIMH 2017, LNCS 10263, pp. 172–182, 2017.
DOI: 10.1007/978-3-319-59448-4_17

has been traditionally treated with surgery. However, with remarkable advances of interventional cardiology, the majority of congenital heart problems can be treated using cardiac intervention procedures. The procedures are currently guided using X-ray fluoroscopy, and generally involve guiding catheters and guidewires, which can be then used to deploy interventional devices. Real-time guiding catheter and guidewire detection, is essential for many image guided applications, such as visibility enhancement for low dose X-ray images, and co-registration between 2D and 3D imaging modalities. As intervention procedures are mainly for young patients with CHD, radiographer has to use the minimum amount of X-ray radiation to guide the procedure. Therefore, the majority of X-ray fluoroscopic images are low dose screening images, with little contrast between guidewires and background. Low dose screening images are of low quality, and are generally acquired to aid navigation (For higher quality, high dose cine images can be acquired, which are well contrasted and less noisy than low dose screening images). Furthermore, guiding catheters are often used to give essential backup support, to assist in pushing interventional equipment over guidewires to the area to be treated. However, the guiding catheters for congenital cardiovascular interventions (CCI) is often made from plastic materials, which are less likely to injure vessels or other structures inside the vascular compartment. However, as these materials are not radiodense, there will be very little contrast for guiding catheters on the X-ray image when compared to the guidewires. Those two challenges have made the detection of catheters more difficult.

Conventional methods using region features, such as pixel intensity, texture and histogram, cannot track guiding catheters or guidewires well [1]. Another category of detection methods, such as active contours and level sets [2, 3], can be easily distracted by image artifacts and other wire-like objects (see Fig. 1). Ma et al. [12] have developed a catheter detection method based on blob detection. However, this method only works on catheters with electrodes and will not detect guiding catheters or wires. Beyar et al. [4] designed a guidewire detection method by using combination of a filter based method, and the Hough transform, to extract wire-like objects and fit them with polynomial curves. This method would likely fail in our X-ray images, as there is no classification between image artifacts and real wire-like object. Similarly, Baert et al. [5], used image subtraction and template matching to enhance guidewires, but only detected a part of the guidewire. More recently, Barbu et al. [6],Wang et al. [7], and Navab et al. [8], had developed learning-based methods for guidewire tracking. Based on a database of manual annotations for guidewires, Barbu et al., used a marginal space learning method to track the target object. But overall, they only achieved tracking speed of one frame-per-second (fps). Wang et al., utilized a probabilistic framework, and tracking speed of 2 fps was achieved. All tracking methods need manual or semi-automatic initialization of guidewire models. Navab et al. applied a machine learning approach on random generated deformable models to extract guidewires. However, this was not tested on cases where guidewires have a sudden and large deformation movement.

To achieve real-time speed, and to overcome challenges of low contrast and image artifacts, in this paper, a localized machine learning algorithm is developed. It is used to classify wire-like objects into two categories: wire objects, and image artifacts. As guiding catheters or guidewires have low contrast against the background in X-ray

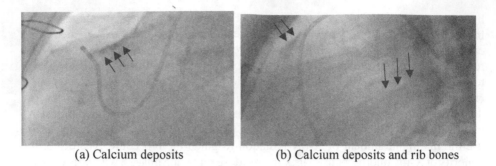

(a) Calcium deposits   (b) Calcium deposits and rib bones

Fig. 1. Image artifacts (indicated by red arrows) in X-ray images. (Color figure online)

images, they could be represented as several disconnected wire objects. Therefore, an energy minimum algorithm is developed to find an optimal wire object.

2   Methods

During CCIs, the first step is to push a guiding catheter toward the area to be treated, which sometimes involves pushing the catheter into a narrow blood vessel (Fig. 2(a)). Then, a guidewire is pushed through the guiding catheter (Fig. 2(b)). Finally, a treatment device, such as a balloon or stent, is push along guidewire inside the guide catheter to the target area and deployed.

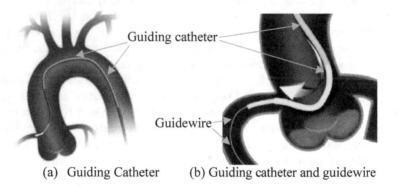

(a)   Guiding Catheter   (b) Guiding catheter and guidewire

Fig. 2. The Clinical workflow for guiding catheter and guidewire.

Guidewires used in CCIs are different from the guidewires shown in [7], as they have to support interventional devices. Therefore, guidewires are thicker and less flexible than examples shown in [7]. This makes the energy minimum method suitable for finding both guiding catheters and guidewires in CCIs.

2.1   Image Pre-processing

A Multiscale vessel enhancement filter [9] is used to enhance the visibility of wire-like structures in the X-ray images. It is based on the idea of approximating wire-like objects, such as tubular or cylindrical structures [9]. In order to classify wire-likes structures, the vessel filter algorithm finds the local coordinate system aligned with the wire and use the curvature in these directions (x or y axis) to classify different structures. This involves the following 5 steps. Step 1, the algorithm performs Gaussian smoothing by convolving the 2D input image with a Gaussian kernel of the appropriate scale s. The Gaussian kernel at position $X = (x, y)$, and scale s is defined as follows:

$$G(X, s) = \frac{1}{2\pi s^2}\exp(-\frac{||X||^2}{2s^2})$$

The smoothed image $L(X, s)$ is computed as $L(X, s) = L(X) * G(X, s)$, where $*$ is the convolution operator. Step 2, the algorithm finds an orthogonal coordinate system aligned with the local features in the image by forming and decomposing the 2×2 Hessian matrix at every image pixel. The Hessian matrix $H_{X,s}$ consists of second order derivatives that contain information about the local curvature. $H_{X,s}$ is defined such as:

$$H_{X,s} = \begin{vmatrix} L_{xx}(X, s) & L_{xy}(X, s) \\ L_{yx}(X, s) & L_{yy}(X, s) \end{vmatrix}$$

Where, $L_{xy}(X, s) = \frac{\partial}{\partial x}(\frac{\partial}{\partial y}L(X, s))$, and the other terms are defined similarly. Step 3, the algorithm performs eigenvalue decomposition for Hessian matrix $H_{X,s}$. Step 4, the algorithm uses the eigenvalues and computes the vessel classification. As $H_{X,s}$ is a 2×2 matrix, there are two eigenvectors and eigenvalues at every image pixel. To quantify any local structures in the image, the eigenvalues for each pixel are arranged in increasing order such that $||\lambda_1|| < ||\lambda_2||$. The ratio differentiates between wire-like structures and blob-like structures, and is given by $R = \lambda_1/\lambda_2$. If $R \approx 1$, detected structures will be blob-like structures. Otherwise, they will be wire-like structures. Step 5, is to apply the vessel filter repeatedly using different Gaussian scales to take into account different vessel sizes within the 2D image.

The multiscale parameter s is one of the important parameters in the vessel enhancement filter. If s is set too high, guiding wires or catheters will be filtered out. If the range of s is set too large, it will slow down the vessel enhancement filter dramatically. In order to get the optimal result for enhancing guiding catheters, the multiscale s of Gaussian kernel should be centred at the average radius of target objects [9]. To calculate the average radius, radii of several guiding catheters and guidewires used in clinical cases are measured. To convert them into image pixel space, R_{dicom} pixel to mm ratio are obtained from X-ray Dicom image header. Magnification factor M of X-ray system is also estimated, which is based on $M = D_{det}/D_{pat}$ (D_{det} is the distance from X-ray source to the detector, and D_{pat} is the distance from X-ray source to the patient). The real pixel to mm ratio is defined as: $R_{xray} = R_{dicom}/M$. The final multiscale s is in the range of $2 \le s \le 6$ (measured in image pixels).

2.2 Object Extraction

In order to reduce computational complexity, and achieve real-time detection speed, the image, after applying the vessel enhancement filter, is binarized using Otsu's method [10] with an image mask. It is well known that X-ray image formation is governed by Beer-Lambert law [13]. So the contrast of an object depends on the background. In order to minimize the contrast variation, an image mask which covers the target objects (guidewires or guiding catheters) is used to remove the background from the calculation of Otsu's method so that the contrast of target objects is directly related to their radio density. The image mask is created from an image after applying the vessel enhancement filter, which involves downsampling, dilation and thresholding using the average intensity. Figure 3 gives an example. Final step of binarization is to use Otsu' method to calculate the thresholding level only within the area of the image mask. Otsu' method is a non-parametrized and adaptive algorithm as it automatically determines the thresholding level based on minimizing the intra-class variance. Otsu' method has been used together with vessel enhancement filter for coronary sinus segmentation on X-ray image [11].

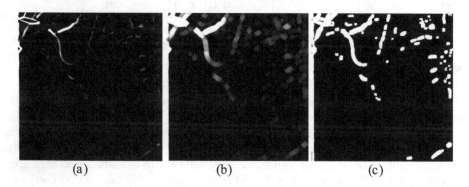

(a)   (b)   (c)

Fig. 3. (a) Image after applying the vessel enhancement filter. (b) After downsampling and dilation. (c) After thresholding using the average intensity. It is the image mask used by Otsu' method.

After binarization, a standard contour detection algorithm is applied to the binarized image to get contours of wire-like objects. Because wire-like objects, such as guiding catheters or guidewires, have different levels of contrast in X-ray images, they might be detected as a completed contour (Fig. 4a), or as several blocks of contours (Fig. 4b). Therefore, centerlines need to be computed among blocks of contours as the final results of detection method.

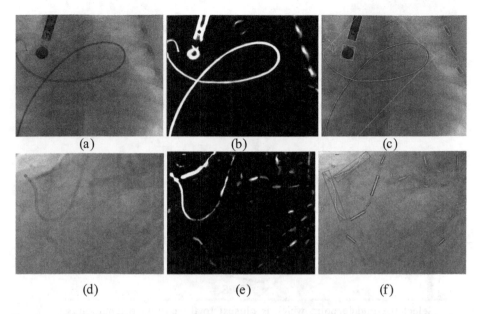

Fig. 4. (a)(d) Origin X-ray images. (b)(e) Images after applying the vessel enhancement filter. (c)(f) Centerlines of contour blocks. Blue rectangles are the line blocks. The green rectangles are the minimum area rectangles of contour blocks. The yellow curves/lines are centerlines. (Color figure online)

To save computational cost, minimum area rectangles of contours are computed. A contour block can be classified as a line block if the width of its minimum area rectangle is less than 8 pixels. 8 pixels is calculated from the average of multiscale $\bar{s} = \dfrac{2+6}{2} = 4$, and $8 = 2\bar{s}$ as the width of minimum area rectangle should be the diameter of wire-like object. If a contour block is a line block, the centerline can be approximated by the middle straight line of the minimum area rectangle (Fig. 4f). Otherwise, the centerline of the contour block can be computed by using the algorithm described in Fig. 5.

In the hierarchy of contours, outer contours are called as parent and inner ones as child. If contours have children contours, it must have loops or intersections. In the final step of the algorithm, if the algorithm finds several middle points, the middle point which is closest to the current orientation vector will be selected. However, In the case of contour self-intersection, in order to increase the robustness, the algorithm will continue to find new middle points along the current orientation vector. The location of self-intersection can be detected by using children contours. The new orientation vectors are calculated from the new middle points. The middle points which create a sharp turning ($<90°$) between the current and new orientation vectors are removed. The final result is a smooth curve or line.

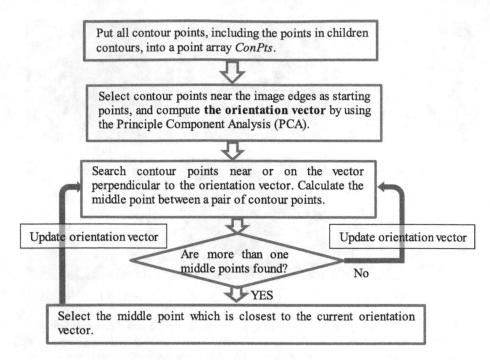

Fig. 5. Flow chart of finding the centerline from a contour block

2.3 Object Classification

Although the vessel enhancement filter can enhance and detect wire-like objects, it could still be affected by some image artifacts, such as calcium deposits and rib bone boundaries (Fig. 1). In order to recognize image artifacts, a k-nearest neighbour (KNN) algorithm is used to separate target objects, such as guiding catheters and guidewire from image artifacts. Instead of applying KNN on the whole image to detect artifacts, a localized KNN image classifier is developed to process only wire-like objects, which are obtained from the previous object extraction step. First, 50 sample images of artifacts (negative data) and 50 sample images of target objects (positive data) are extracted from line blocks or contour blocks. They are manually labelled. The orientation vectors are computed and sample images are organized around the detected main axis. Then, the sample images were flipped along the X and Y axis to create a total of 200 sample images for both positive and negative data. This not only increases the number of training samples, but also solves the asymmetrical problem. The size of sample images is 20×40 pixels. Figure 6 gives some examples. There are some bended wire images in positive data, which gives some flexibilities for KNN to recognize bended wires or catheters. All positive and negative sample images are normalized so that they all have the same value of the average intensity. This step is to prevent the wrong classification which is caused by the difference in the average intensity between two images. To normalize the image intensity, the image is subtracted by the intensity mean and then divided by the standard deviation of the intensity.

(a) Positive data including guiding catheters and guidewires

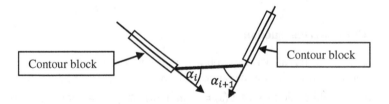

(b) Negative data such as calcium deposit and rib bones

Fig. 6. Training samples for KNN (before the intensity normalization).

In current implementation, 5 nearest neighbours are used for KNN classifications. KNN uses Euclidean distances which are $d(p,q) = \sqrt{\sum_{i-1}^{N}(p_i - q_i)^2}$. Some of failed classifications are illustrated in Fig. 7.

(a)   (b)   (c)

Fig. 7. Failed classifications for negative data (artifacts were classified as wire objects). (a) Intersection with a catheter. (b) X-ray contrast agent injection. (c) Calcium deposit.

Contour block   Contour block

α_i α_{i+1}

Fig. 8. Connecting angles between 2 contour blocks

As in the previous step, centerlines of contour blocks have been already computed, sub-images of 20×40 pixels can be extracted using centerlines and they will be normalized before feeding into the KNN image classifier. An example of classification is shown in Fig. 9d and f.

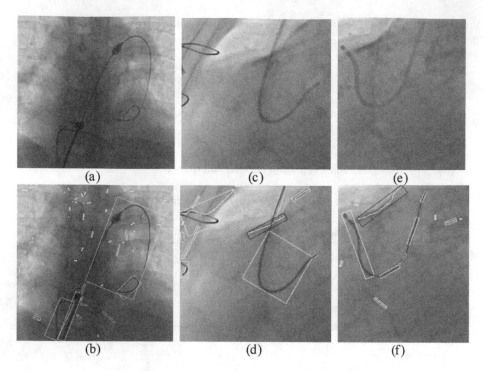

Fig. 9. (a)(c)(e) Origin X-ray images. (b)(d)(f) Purple lines are the detected guidewires. (d)(f) Blue rectangles are the contour blocks which are classified as image artifacts. The figures are best viewed in color. (Color figure online)

2.4 Wire Curve Reconstruction

As there are many different ways to connect neighbour contour blocks, a minimum energy method is used to choose the smooth curve for the completed guiding catheter or guidewire. It could be defined as $E(s) = \int \left| d^2C(s)/ds^2 \right|^2 ds$, where $d^2C(s)/ds^2$ is the second order derivatives of catheter/guidewire curves. When the curve has C^1 continuity, it is a smooth curve. This mean that there are no sharp turning when connecting two contour blocks. Sharp turning is defined as $(\alpha_i + \alpha_{i+1}) > 90°$. A modified discrete energy function $E(s) = n + \sum_{i=0}^{n-1}(-cos\alpha_i)$, where α_i is the connecting angle between the contour block and the connecting line (see Fig. 8).

The distances between neighbour blocks should be also taken into consideration in the energy function. The final energy function is defined as $E(s) = n + \sum_{i=0}^{n-1}(-cos\alpha_i) + n * \dfrac{\sum_{i=0}^{n/2-1} d_i}{L}$ where d_i is the distance between two neighbour blocks and L is the total length of reconstructed wire. The final part of the energy function is to minimize the distances between two neighbour blocks. The curve reconstruction algorithm uses a local optimization strategy as it is designed for real time processing. If the algorithm has to calculate all possible reconstruction paths (permutations), it will not be efficient for real-time processing. Also it will introduce delays in processing some images in the image sequence and will result in inconsistent frame rate of detection, as the number of found blocks varies from one image to another. Instead of calculating all permutations of all found blocks, we start with the longest blocks and search both ends for the next few blocks (in the current implementation, 3 blocks are used). After searching all possible paths, the one with minimum value of energy function is chosen.

3   Results

The localized machine learning method (KNN), was trained using 200 positive and negative sub-images, which are manually extracted from 25 X-ray images. They were acquired during 2 clinical cases. The proposed detection method was tested on 8 image sequences (1102 images), acquired in 3 different clinical cases. The frame size of each sequence is 512×512, with the pixel size between 0.368 mm and 0.433 mm. To establish ground truth for evaluation, guiding catheters or guidewires are manually annotated by a clinical expert. An annotated object starts from the edge of the image and ends at its tip. They are used as the ground truth for accuracy tests. The overall detection precision is 0.87 ± 0.53 mm, which is defined as the average of shortest distances from points on a detected object to the corresponding annotated object. Furthermore, the shortest distances from the points in the annotated object to the detected object are calculated, which carries the penalty for not recovering the total length of the object in the ground truth. The result is 0.91 ± 0.62 mm, which is slight worse than the previous evaluation. The tip tracking precision is 0.54 ± 0.36 mm for catheter tip and guidewire tip. The overall successful rate of detection is 83.4%. For the technique to be acceptable in clinical practice, failed detections are considered to be the ones where any points on the detected object has larger errors than a preset threshold (e.g., a threshold of 4 image pixels [7] is used in this evaluation, which is the average radius of guidewires).

The major computational load of the proposed method is the vessel enhancement filter algorithm. The software of the proposed method was implemented using OpenCV, and currently achieves a frame rate of 15 fps using a single-threaded CPU implementation. The performance was evaluated on an Intel Core i7 2.7 GHz laptop.

4   Conclusions and Discussions

This paper describes a real-time guiding catheter and guidewire detection method. The proposed method does not require any user interaction or prior models. As the method is integrated with a localized machine learning algorithm, it can robustly distinguish between wire-like target objects and image artifacts. Therefore, the proposed method can efficiently and robustly work on low-dose X-ray images with a frame rate of 15 fps. This frame rate is considered as real-time for cardiovascular intervention procedures as the average maximum frame rate for modern intervention X-ray systems is 15 fps. The detected models of guiding catheters or guidewires could be fed into a template-based tracking method, which can produce an even faster guidewire tracking method.

The proposed method is not limited to detect or track guiding catheters or guidewires. With some modifications of the localized machine learning algorithm, it can potentially detect intervention devices such as balloons, stents, and coils. Therefore, based on the proposed method, a general detection and tracking framework could be developed and used in CCI procedures, as well as other cardiac intervention procedures.

5   Disclaimer

Concepts/information are based on research and are not commercially available.

References

1. Yilmaz, A., Javed, O., Shah, M.: Object tracking: a survey. ACM Comput. Surv. **38**(4), 13 (2006)
2. Kass, M., Witkin, A., Terzopoulos, D.: Snakes: active contour models. Int. J. Comput. Vision **1**(4), 321–331 (1987)
3. Zhu, S.C., Yuille, A.L.: Forms: a flexible object recognition and modeling system. Int. J. Comput. Vision **20**, 187–212 (1996)
4. Palti-Wasserman, D., Brukstein, A.M., Beyar, R.: Identifying and tracking a guide wire in the coronary arteries during angioplasty from x-ray images. IEEE Trans. Biomed. Eng. **44**(2), 152–164 (1997)
5. Baert, S.A.M., Viergever, M.A., Niessen, W.J.: Guide wire tracking during endovascular interventions. IEEE Trans. Med. Imaging **22**(8), 965–972 (2003)
6. Barbu, A., Athitsos, V., Georgescu, B., Boehm, S., Durlak, P., Comaniciu, D.: Hierarchical learning of curves application to guidewire localization in fluoroscopy. In: CVPR (2007)
7. Wang, P., Chen, T., Zhu, Y., Zhang, W., Zhou, S.K., Comaniciu, D.: Robust guidewire tracking in fluoroscopy. In: CVPR (2009)
8. Pauly, O., Heibel, H., Navab, N.: A machine learning approach for deformable guide-wire tracking in fluoroscopic sequences. In: Jiang, T., Navab, N., Pluim, J.P.W., Viergever, M.A. (eds.) MICCAI 2010. LNCS, vol. 6363, pp. 343–350. Springer, Heidelberg (2010). doi: 10.1007/978-3-642-15711-0_43
9. Frangi, A.F., Niessen, W.J., Vincken, K.L., Viergever, M.A.: Multiscale vessel enhancement filtering. In: Wells, W.M., Colchester, A., Delp, S. (eds.) MICCAI 1998. LNCS, vol. 1496, pp. 130–137. Springer, Heidelberg (1998). doi:10.1007/BFb0056195
10. Otsu, N.: A threshold selection method from gray-level histograms. IEEE Trans. Syst. Man Cybern. **9**(1), 62–66 (1979)
11. Fazlali, H.R., et al.: Vessel region detection in coronary X-ray angiograms. In: International Conference on Image Processing (2015)
12. Ma, Y.L., Gogin, N., Cathier, P., Housden, R.J., Gijsbers, G., Cooklin, M., O'Neill, M., Gill, J., Rinaldi, C.A., Razavi, R., Rhode, K.S.: Real-time x-ray fluoroscopy-based catheter detection and tracking for cardiac electrophysiology interventions. Med. Phys. **40**(7), 071902 (2013)
13. Hecht, E.: Optics, 4th edn. Addison Wesley, San Francisco (2002)

Feature Tracking Cardiac Magnetic Resonance via Deep Learning and Spline Optimization

Davis M. Vigneault[1,2,3]($\boxtimes$), Weidi Xie[1], David A. Bluemke[2],
and J. Alison Noble[1]

[1] Institute of Biomedical Engineering, Department of Engineering,
University of Oxford, Old Road Campus Research Building,
Roosevelt Dr, Oxford OX3 7DQ, UK
davis.vigneault@gmail.com
[2] Department of Radiology and Imaging Sciences, Clinical Center,
National Institutes of Health, 10 Center Drive, Bethesda, MD 20814, USA
[3] Sackler School of Graduate Biomedical Sciences,
Tufts University School of Medicine, 136 Harrison Ave, Boston, MA 02111, USA

Abstract. Feature tracking Cardiac Magnetic Resonance (CMR) has recently emerged as an area of interest for quantification of regional cardiac function from balanced, steady state free precession (SSFP) cine sequences. However, currently available techniques lack full automation, limiting reproducibility. We propose a fully automated technique whereby a CMR image sequence is first segmented with a deep, fully convolutional neural network (CNN) architecture, and quadratic basis splines are fitted simultaneously across all cardiac frames using least squares optimization. Experiments are performed using data from 42 patients with hypertrophic cardiomyopathy (HCM) and 21 healthy control subjects. In terms of segmentation, we compared state-of-the-art CNN frameworks, U-Net and dilated convolution architectures, with and without temporal context, using cross validation with three folds. Performance relative to expert manual segmentation was similar across all networks: pixel accuracy was $\sim 97\%$, intersection-over-union (IoU) across all classes was $\sim 87\%$, and IoU across foreground classes only was $\sim 85\%$. Endocardial left ventricular circumferential strain calculated from the proposed pipeline was significantly different in control and disease subjects (-25.3% vs -29.1%, $p = 0.006$), in agreement with the current clinical literature.

Keywords: Regional cardiac function · Cardiac magnetic resonance · Deep convolutional neural networks · Quadratic basis splines · Least squares optimization

1 Introduction

Quantification of regional cardiac function is of utmost importance in the characterization of subtle abnormalities which may precede changes in global

© Springer International Publishing AG 2017
M. Pop and G.A. Wright (Eds.): FIMH 2017, LNCS 10263, pp. 183–194, 2017.
DOI: 10.1007/978-3-319-59448-4_18

Match Successive Pairs of Frames

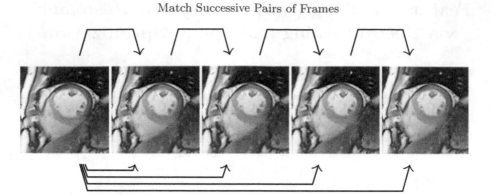

Match Each Frame to End Diastole

Fig. 1. Schematic of traditional feature tracking methods. Each arrow represents a single pairwise registration between a fixed and moving image, and produces a displacement field. Through combinations of resampling, averaging, and regularization, these displacement fields are combined to form a final sequence of fields representing cardiac motion.

metrics [1]. Harmonic phase (HARP) analysis [2] of tagged cardiac magnetic resonance (CMR) images is the gold-standard for regional function, but has not been adopted clinically due to lengthy acquisition and analysis. Moreover, HARP analysis is difficult to apply to chambers other than the left ventricle (LV) due to the thinness of the myocardial wall in the atria and right ventricle (RV). Recently, feature tracking (FT) has emerged as a promising alternative to tagging [3]. Because FT-CMR can be applied to balanced steady state free precession (SSFP) images, acquisition of specialized image sequences is avoided. Moreover, because FT-CMR primarily tracks myocardial borders and trabeculation, the thinness of the atrial and right ventricular myocardium does not hinder tracking.

Despite recent interest in FT-CMR, two important challenges remain. First, all current commercially available implementations (MTT, TomTec, CMR42) require manual contouring of one or more cine frames, preventing full automation and reducing reproducibility. Second, FT generally has been implemented by repeatedly applying methods designed to determine a displacement field between a *single pair* of images (e.g., optical flow, block matching, deformable registration), rather than an image *sequence*; either matching successive pairs, or matching each frame to a single reference, typically end diastole (ED, Fig. 1). Each of these approaches has well-known potential drawbacks [4], which may be overcome by empirically optimizing over all frames simultaneously [5].

Here, we propose a method for FT-CMR analysis which overcomes the first of these challenges by using a deep learning approach in place of human contouring. Deep convolutional neural networks have been used to great effect in image classification [6,7], and semantic segmentation [8,9]. Recently, CNNs have also shown state-of-the-art performance in biomedical image analysis [10,11]. CNN segmentation of short axis CMR has been applied to the LV blood-pool [12], the RV blood-pool [13], and both simultaneously [14]. Here, we perform segmentation of the LV myocardium, LV blood-pool, and RV blood-pool. Moreover, we apply the segmentation to patients with HCM, which increases the complexity of the problem due to the highly variable appearance of the LV in these patients.

The second of these challenges we address by fitting quadratic basis splines to the segmentation data jointly, rather than frame by frame, adapting the technique presented in [5,15] in the context of cardiac ultrasound. In our pipeline, *extraction* is performed using a CNN, *tracking* is performed with simultaneous spline optimization, and cardiac strain is estimated from the registered splines.

2 Methods

Broadly, the automated analysis pipeline involves three steps: feature extraction (segmentation), feature tracking (spline fitting), and calculation of functional parameters (strain estimation). These steps are discussed in detail in the following sub-sections.

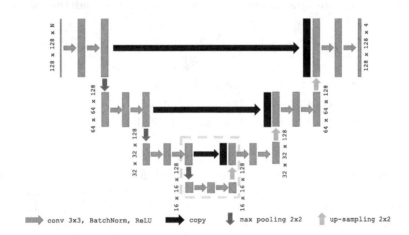

Fig. 2. Architecture of the basic network (Network A). The input image is of size $128 \times 128 \times N$, where N is the number of channels (1 in networks A and B, 2 in networks C and D). Each blue and black box corresponds to a multilingual feature map (black indicates the result of a copy). The dimensions of the feature maps are indicated in the figure as first spatial dimension × second spatial dimension × channels. The number of channels in each feature map is fixed at 128. The dashed yellow box is replaced by dilated convolution in networks B and D. (Color figure online)

2.1 Segmentation

Following the work of [8,10,11], we designed our segmentation architecture as a fully convolutional network. In order to obtain a segmentation map with the same spatial resolution as the input image, up-sampling operators are used to replace the pooling operators in traditional classification networks. This strategy enables our network to segment arbitrarily large images.

As shown in Fig. 2, the architecture consists of a down-sampling path (left) followed by an up-sampling path (right). During the first several layers, the structure resembles the canonical classification CNN [6,7], as a 3×3 convolution, rectified linear unit (ReLU), and 2×2 max pooling are repeatedly applied to the input image and feature maps. In the second half of the architecture, we "undo" the reduction in spatial resolution by performing 2×2 up-sampling, ReLU activation, and 3×3 convolution, eventually mapping the intermediate feature representation back to the original resolution. To provide accurate boundary localization, low-level feature representations from the down-sampling path are concatenated with the feature maps from the up-sampling path. For all layers, we apply 128 trainable kernels. We performed batch normalization [16], which has been shown to increase generalizability, between each pair of convolution and ReLU activation layers.

Table 1. CNN architecture variants considered. Note: ED = End Diastole; DC = Dilated Convolution.

Name	Variant	Input size	Temporal context
Network A	U-Net	$128 \times 128 \times 1$	None
Network B	DC	$128 \times 128 \times 1$	None
Network C	U-Net	$128 \times 128 \times 2$	ED Frame
Network D	DC	$128 \times 128 \times 2$	ED Frame

In addition to this basic architecture (Network A), we varied the amount of temporal context by either inputting the input image alone, or the input image and ED image together. We based this on the intuition that the papillary muscles, which frequently interfere with LV segmentation, are least compacted at ED and may guide the segmentation of the input frame. Additionally, in Networks B and D, the final up-sampling/down-sampling pass was replaced by a dilated convolution (DC). All architectures have $\sim$ 3.1 million trainable parameters. The architectures tested are summarized in Table 1.

2.2 Quadratic Bézier Curve Registration

Following semantic segmentation, contours defining the boundaries of the LV endocardium, LV epicardium, and RV endocardium were extracted through standard morphological operations. In this work, the pixels belonging to these contours are known as "boundary candidates." Unfortunately, these contours cannot be used directly to quantify cardiac function, because they lack anatomical correspondence between frames. It is the aim of this section to describe an optimization procedure for jointly registering a sequence of closed, quadratic Bézier curves to these boundary candidates.

A segment of a closed Bézier curve B of degree d parameterized by $r \in [0, 1]$ is a linear combination of $d + 1$ control points $x_i : 0 \le i \le d$,

$$B_d(r) = \sum_{i=0}^{d} b_{i,d}(r)x_i,$$

where $b_{i,d}(r)$ is the ith Bernstein polynomial of degree d,

$$b_{i,d}(r) = \binom{n}{i}(1-r)^{n-i}r^i,$$

and $\binom{n}{i}$, often read aloud as "n choose i", is the binomial coefficient,

$$\binom{n}{i} = \frac{n!}{i!(n-i)!}.$$

For $d = 2$, $B_2(r)$ expands to

$$B_2(r) = (1-r)^2 x_0 + 2r(1-r)x_1 + r^2 x_2,$$

which may be more conveniently expressed in terms of the monomial basis as

$$B_2(r) = \begin{bmatrix} x_0\ x_1\ x_2 \end{bmatrix} \begin{bmatrix} (1-r)^2 \\ 2r(1-r) \\ r^2 \end{bmatrix} = \begin{bmatrix} x_0\ x_1\ x_2 \end{bmatrix} \begin{bmatrix} 1 & -2 & 1 \\ 0 & 2 & -2 \\ 0 & 0 & 1 \end{bmatrix} \begin{bmatrix} 1 \\ r \\ r^2 \end{bmatrix}.$$

Importantly, the first and second derivatives of $B_2(r)$ with respect to r are trivial to compute.

2.3   Formulating the Optimization

Levenberg-Marquardt least squares optimization [17] is used to register a set of closed, quadratic Bézier curves to the boundary candidates. The parameters $\Delta X \in \mathbb{R}^{2 \times (C \times K)}$ (where C is the number of control points in a single curve and K is the number of cardiac phases) of the optimization are Cartesian displacements to the control points of all template curves across all frames. A fixed number of points $\mathbf{u}_{f,j,r}$ were sampled across each curve. At each step in the optimization, for each of these points, the nearest boundary candidate $\phi(\mathbf{u}_{f,j,r})$ was calculated, where $\phi : \mathbb{R}^2 \to \mathbb{R}^2$. This was computed efficiently by representing the boundary candidate point set at each frame as a K_d tree. The Cartesian components of the distance between the points sampled from the curve and nearest boundary candidate were the residuals of E_{cf}, the first term of the cost function,

$$E_{cf} = \sum_{f,j,r} \| \mathbf{u}_{f,j,r} - \phi \left(\mathbf{u}_{f,j,r} \right) \|^2. \tag{1}$$

Additionally, two regularizers were included to enforce physical constraints of anatomical deformation: control point acceleration and spline curvature.

In our cost function, the control point acceleration regularizer allows information to be shared between frames. At a minimum, regularizing against control point velocity as in [5] is necessary to maintain anatomical consistency (the assumption that, for fixed j and r, $\mathbf{u}_{f,j,r}$ corresponds to the same material point $\forall f$). By regularizing against acceleration rather than velocity, our method additionally encourages smooth, biologically plausible motion. The control point acceleration regularizer E_{ac} was defined as the Cartesian components of the second differences between corresponding vertices $\mathbf{x}_{f,c}$ in three adjacent frames, where $\mathbf{x}_{f,c}$ is the $(f \times C) + c^{\text{th}}$ column of X.

$$E_{ac} = \sum_{f,c} \left\| \begin{bmatrix} 1 & -2 & 1 \end{bmatrix} \begin{bmatrix} \mathbf{x}^{\top}_{(f+2) \bmod K,c} \\ \mathbf{x}^{\top}_{(f+1) \bmod K,c} \\ \mathbf{x}^{\top}_{f,c} \end{bmatrix} \right\|^2 \tag{2}$$

Curvature for segment j in frame f is the second derivative of $B_2(r)$ with respect to r.

$$E_{cv} = \sum_{f,c} \left\| \frac{d^2 B_2(r)}{dr^2} \right\|^2 \tag{3}$$

The overall optimization problem may then be written in terms of Eqs. 1, 2, and 3 and corresponding scaling factors. Scaling factors $\rho_{cf} = 10.0$, $\rho_{ac} = 1.0$, and $\rho_{cv} = 0.1$ were set empirically to prevent any single term from dominating the optimization.

$$E = \min \left(\rho_{cf} E_{cf} + \rho_{ac} E_{ac} + \rho_{cv} E_{cv} \right)$$

Two points relating to computational efficiency are worth noting. First, the Jacobians of Eqs. 1, 2, and 3 can all be calculated analytically. By providing explicit Jacobians, we avoid the need for numeric derivatives, which would slow computation precipitously. Moreover, for a given set of correspondences between surface positions and boundary candidates, the Jacobians of all residuals are linear with respect to the Cartesian displacements of the control points and therefore trivial to calculate. Second, each individual residual depends upon a very small number of parameters. Specifically, each individual residual depends on exactly three control points (six parameters). This sparsity is exploited during the optimization to limit the number of components of the Jacobian which must be evaluated, further reducing computational cost.

Following the initial fit, the spline is subdivided and used to initialize a second optimization, and this process is repeated one further time. This multiresolution approach has benefit over registering the highly subdivided spline directly, which can be sensitive to initialization.

3   Experiments

3.1   Segmentation

The LV myocardium, LV blood-pool, and RV blood-pool were manually segmented in 189 short axis 2D+time volumes (basal, equatorial, and apical cine series from each of 63 subjects). The papillary muscles of the LV were excluded from the myocardium. The subjects were partitioned into three folds of approximately equal size such that the images from any one subject were present in one fold only. The volumes were cropped to 128×128 pixels in the spatial dimensions, and varied from 25 to 50 pixels in the time dimension, totaling 2706, 3000, and 2775 images in the three folds, respectively. For each of the four architectures (U-Net and DC with and without temporal information), three models were trained on two folds and tested on the remaining fold. The images were histogram equalized and normalized to zero mean and unit standard deviation before being input into the CNN. The network weights were initialized with orthogonal weights [18], and were trained with standard stochastic gradient descent (SGD) with momentum (0.9) by optimizing categorical cross entropy. Learning rate was initialized to 0.01 and decayed by 0.1 every 32 epochs. To avoid over-fitting, we used considerable data augmentation (horizontal and vertical flipping, random translations and rotation) and a weight decay of 10^{-4}. Accuracy was measured as pixel accuracy between the prediction and manual segmentations. The model was implemented in the Python programming language using the Keras interface to Tensorflow [19], and trained on one NVIDIA Titan X graphics processing units (GPU) with 12 GB of memory. For all network architectures, it took roughly 200 s to iterate over the entire training set (1 epoch). At test time, the network predicted segmentations at roughly 75 frames per second (real-time).

3.2   Tracking

Short axis (SA) scans from 42 subjects with overt hypertrophic cardiomyopathy (HCM) and 21 control subjects were segmented as described above. For each scan, the LV endocardium, LV epicardium, and RV endocardium were tracked using the spline optimization method. The tracking algorithm was implemented in the C++ programming language using the Insight Toolkit (ITK) for reading, writing, and manipulating images and point sets, and using the Ceres Solver for least squares optimization. The three registration passes took $\sim 2.1s$ per cine sequence.

3.3   Regional Function

Following tracking, registered splines from the 63 subjects were used to calculate global strain. For each structure in each SA plane, global strain was compared between HCM and control subjects using the Student's t test.

4 Results

In terms of segmentation, performance relative to expert manual segmentation was similar across all networks: mean pixel accuracy was ~ 97%, intersection-over-union (IoU) across all classes was ~ 87%, and IoU across foreground classes only was ~ 84%. However, inspection of the images revealed that a single subject with severe, nonuniform illumination was incorrectly segmented by all networks, with a disproportionate effect on mean performance metrics. For this reason, median values are also reported. Broadly, performance improved with the addition of temporal context over the target frame alone, and with the dilated convolution (DC) networks compared with the U-Net networks (Table 2). Therefore, Network D was selected to provide segmentations for the tracking data.

Table 2. Network performance compared with expert manual segmentations.

Name	Description	Metric	Pixel accuracy	IoU (All)	IoU (Foreground)
Network A	U-Net, No context	Mean	0.977	0.874	0.838
		Median	0.981	0.885	0.851
Network B	DC, No context	Mean	0.977	0.876	0.840
		Median	0.981	0.886	0.853
Network C	U-Net, Context	Mean	0.976	0.874	0.838
		Median	0.981	0.887	0.855
Network D	DC, Context	Mean	0.976	0.873	0.837
		Median	0.981	0.888	0.855

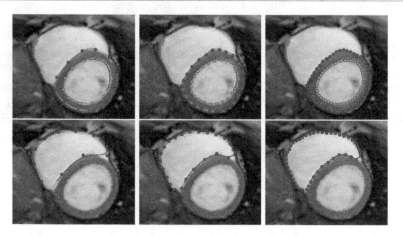

Fig. 3. Spline registration was conducted in successive passes (left to right), where the output of one was subdivided and used to initialize the next. LV and RV tracking results are shown above and below, respectively. Note especially regions of acute curvature, such as the insertion of the RV on the LV, which improves from left to right as the granularity of the spline increases.

Table 3. Circumferential strain results. (†: Significant at the $p < 0.05$ level.)

Plane	Structure	Control (%)	Overt (%)	p
Base	LV endocardium	−26.3	−29.0	0.098
	LV epicardium	−11.8	−11.7	0.952
	RV endocardium	−11.9	−14.3	0.529
Midslice	LV endocardium	−25.3	−29.1	0.006†
	LV epicardium	−8.8	−8.5	0.661
	RV endocardium	−8.0	−12.7	0.152
Apex	LV endocardium	−25.5	−28.6	0.110
	LV epicardium	−10.7	−9.9	0.470
	RV endocardium	−13.8	−12.4	0.566

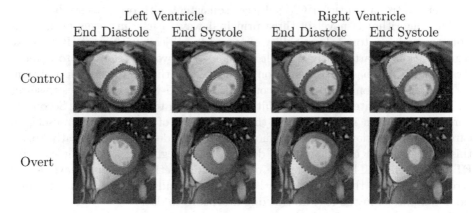

Fig. 4. Representative segmentation and tracking results in a control subject (top) and a patient with overt HCM.

Tracking was performed in three passes (Fig. 3), where the output of one pass was subdivided and passed to the next pass. In each pass, the contours tighten towards the segmentation, allowing for acute structures such as the RV insertion points to be better described. Compared with registering a highly subdivided spline directly, this technique avoids local minima and converges faster.

The relationship between strain values measured in the control and overt groups was consistent with other studies [20]. In particular, circumferential strain measured in the equatorial LV endocardium was higher (more negative) in overt subjects relative to control subjects (−29.1 vs −25.3, $p = 0.006$). Detailed circumferential strain results are given in Table 3.

Representative segmentation and tracking results are shown for control and overt subjects (Fig. 4). The model learned to avoid the papillary muscles of the LV myocardium and performed well even in subjects with severe hypertrophy. Tracking visually followed the contours of the segmentation closely.

5   Conclusions

Measuring cardiac function in a fully automated way from SSFP CMR has the potential to simplify quantification of regional cardiac function, and expedite clinical adoption. We have presented a fully automated pipeline for cardiac segmentation, tracking, and estimation of cardiac strain. We obtained segmentation results with and without temporal context in U-Net and DC architectures, and found improvements with temporal context, as well as in DC architectures. The best-performing architecture (Network D) had a median pixel accuracy of 0.981, all-class IoU of 0.887, and foreground IoU of 0.855. We then presented a feature tracking algorithm taking these segmentations as input and jointly optimizing a set of quadratic splines over all frames simultaneously. We applied this segmentation and tracking to the LV endocardium, LV epicardium, and RV endocardium of healthy and disease subjects, and found statistically significant differences between control and overt HCM subjects consistent with previous studies.

Our algorithm is novel in three principal ways. First, in terms of application, the wide anatomical variability observed in subjects with overt HCM make segmentation a particularly difficult problem; this work is the first to demonstrate that CNN-based segmentation is effective in these subjects. In addition, we have directly compared dilated convolution and U-Net architectures to select an appropriate state-of-the-art architecture solution to this problem. Second, a persistent problem in FT-CMR is the interference of the papillary muscles in cardiac segmentation; we have demonstrated that deep learning neatly solves this problem, and are unaware of deep learning segmentation being used for FT-CMR before. Third, the spline optimization method presented avoids the errors inherent to the various pairwise sequential and reference frame formulations ubiquitous in the feature tracking literature to date.

Notably, all networks failed to segment a single case with severe nonuniform illumination. Augmentation during training to counteract this effect will be the subject of future work. Moreover, because only edge features are tracked, our method suffers from the so-called "aperture-problem," such that anatomical correspondence may not be reliable. In future work, we will incorporate features from our pre-trained CNN into the spline optimization to mitigate this effect. However, this may be a fundamental limitation of FT-CMR where trabeculation is minimal, such as when measuring LV endocardial strain in the basal slice.

In conclusion, we have presented a fully automated pipeline which addresses a number of longstanding challenges to the adoption of FT-CMR, and tested this pipeline successfully in the context of a difficult clinical problem.

Acknowledgements. D. Vigneault is supported by the NIH-Oxford Scholars Program and the NIH Intramural Research Program. W. Xie is supported by the Google DeepMind Scholarship, and the EPSRC Programme Grant Seebibyte EP/M013774/1.

References

1. Tee, M., Noble, J.A., Bluemke, D.A.: Imaging techniques for cardiac strain and deformation: Comparison of echocardiography, cardiac magnetic resonance and cardiac computed tomography. Expert Rev. Cardiovasc. Ther. **11**, 221–231 (2013)
2. Osman, N.F., McVeigh, E.R., Prince, J.L.: Imaging heart motion using harmonic phase MRI. IEEE Trans. Med. Imaging **19**, 186–202 (2000)
3. Pedrizzetti, G., Claus, P., Kilner, P.J., Nagel, E.: Principles of cardiovascular magnetic resonance feature tracking and echocardiographic speckle tracking for informed clinical use. J. Cardiovasc. Magn. Reson. 1–12 (2016)
4. Wong, K.C.L., Tee, M., Chen, M., Bluemke, D.A., Summers, R.M., Yao, J.: Regional infarction identification from cardiac CT images: A computer-aided biomechanical approach. Int. J. Comput. Assist. Radiol. Surg. **11**(9), 1573–1583 (2016)
5. Stebbing, R.: Model-Based Segmentation Methods for Analysis of 2D and 3D Ultrasound Images and Sequences. DPhil. University of Oxford (2014)
6. Krizhevsky, A., Hinton, G.E.: ImageNet classification with deep convolutional neural networks. In: NIPS (2012)
7. Simonyan, K., Zisserman, A.: Very deep convolutional networks for large-scale image recognition. In: ICLR, pp. 1–14 (2014)
8. Long, J., Shelhamer, E., Darrell, T.: Fully convolutional networks for semantic segmentation. In: CVPR, pp. 3431–3440, 7–12 June 2015
9. Yu, F., Koltun, V.: Multi-scale context aggregation by dilated convolutions. In: ICLR, pp. 1–9 (2016)
10. Ronneberger, O., Fischer, P., Brox, T.: U-Net: Convolutional networks for biomedical image segmentation. In: Navab, N., Hornegger, J., Wells, W.M., Frangi, A.F. (eds.) MICCAI 2015. LNCS, vol. 9351, pp. 234–241. Springer, Cham (2015). doi:10.1007/978-3-319-24574-4_28
11. Xie, W., Noble, J.A., Zisserman, A.: Microscopy cell counting with fully convolutional regression networks. In: MICCAI Workshop, pp. 1–10 (2015)
12. Poudel, R.P.K., Lamata, P., Montana, G.: Recurrent Fully Convolutional Neural Networks for Multi-slice MRI Cardiac Segmentation (2016). arXiv.com
13. Luo, G., An, R., Wang, K., Dong, S., Zhang, H.: A deep learning network for right ventricle segmentation in short-axis MRI. Comput. Cardiol. **43**, 485–488 (2016)
14. Tran, P.V.: A Fully Convolutional Neural Network for Cardiac Segmentation in Short-Axis MRI, pp. 1–21 (2016). arXiv.com
15. Stebbing, R.V., Namburete, A.I., Upton, R., Leeson, P., Noble, J.A.: Data-driven shape parameterization for segmentation of the right ventricle from 3D+t echocardiography. Med. Image Anal. **21**, 29–39 (2015)
16. Ioffe, S., Szegedy, C.: Batch normalization: Accelerating deep network training by reducing internal covariate shift. In: ICML, vol. 37, pp. 81–87 (2015)
17. Marquardt, D.W.: An algorithm for least-squares estimation of nonlinear parameters. J. Soc. Ind. Appl. Math. **11**(2), 431–441 (1963)
18. Saxe, A.M., McClelland, J.L., Ganguli, S.: Exact solutions to the nonlinear dynamics of learning in deep linear neural networks. In: ICLR, pp. 1–22 (2014)

19. Barham, P., Chen, J., Chen, Z., Davis, A., Dean, J., Devin, M., Ghemawat, S., Irving, G., Isard, M., Kudlur, M., Levenberg, J., Monga, R., Moore, S., Murray, D.G., Steiner, B., Tucker, P., May, D.C., Brain, G.: TensorFlow: A system for large-scale machine learning (2016)
20. Vigneault, D.M., Yang, E., Chu, C.L., Ho, C.Y., Bluemke, D.A.: Left ventricular strain gradient is abnormal in hypertrophic cardiomyopathy: Assessment by CMR feature tracking. Radiol. Soc. North Am. (2014). (Chicago, IL)

Noise Sensitive Trajectory Planning for MR Guided TAVI

Mustafa Bayraktar[1]($\boxtimes$), Erol Yeniaras[2], Sertan Kaya[3], Seraphim Lawhorn[4], Kamran Iqbal[5], and Nikolaos V. Tsekos[6]

[1] Bioinformatics Department, University of Arkansas at Little Rock, Little Rock, USA
mxbayraktar@ualr.edu
[2] Boeing, Chicago, USA
eyeniaras@gmail.com
[3] Monsanto, St. Louis, USA
sertankaya51@gmail.com
[4] Department of Mathematics and Statistics, University of Arkansas at Little Rock, Little Rock, USA
sxlawhorn@ualr.edu
[5] Systems Engineering, University of Arkansas at Little Rock, Little Rock, USA
kxiqbal@ualr.edu
[6] Computer Science, University of Houston, Houston, USA
nvtsekos@central.uh.edu

Abstract. Image-guided, pre-operative planning is fast becoming the gold standard for navigating real-time robotic cardiac surgeries. Planning helps the surgeon utilize the amended quantitative information of the target area and assess the suitability of the offered intervention technique prior to surgery. In apex access aortic valve replacements, safe zone generation for the penetration of delivery module along the left ventricle (LV) is a crucial step to prevent untoward cases from emerging. To address this problem, we propose a computational core, which is to locate left ventricle borders and specifically papillary muscles (PM), create an obstacle map along the left ventricle (LV), and ultimately extract a dynamic (off-line) trajectory for tool navigation. To this end, we first applied an isotropic diffusion on short-axis (SA) cardiac magnetic resonance (CMR) images. Second, we utilized an active contour model to determine the LV border. Third, we clustered the LV crops to locate the PM. Finally, we computed the centroids of each of the LV segments to determine the safest path for an aortic delivery module.

1 Introduction

Image guidance for cardiac surgeries has been used primarily to provide information for planning the performance of complex interventions into the beating heart [1]. Cardiac magnetic resonance (CMR)-guided imaging—especially for this case—during the pre-surgery stage provides the capability of assessing the

© Springer International Publishing AG 2017
M. Pop and G.A. Wright (Eds.): FIMH 2017, LNCS 10263, pp. 195–203, 2017.
DOI: 10.1007/978-3-319-59448-4_19

adequacy of the proposed intervention technique. There are three critical variables that need to be assessed quantitatively to conduct a promising plan for a transapical-access, aortic-valve implantation, which are the following:

1. The angle between the LV and aorta,
2. The aortic root area, and
3. The dynamic safest path calculation along the LV. These are the most crucial markers during the planning stage [2].

Among these three variables, the safest dynamic corridor calculation is required for orienting the delivery module safely along the determined LV corridor without damaging the heart walls or mitral valve leaflets, as well as avoiding possible untoward cases during transapical access. In this paper, we created a computational pipeline to determine the safest path that takes into account the spatiotemporal movement of the LV and the existence of papillary muscles on any LV slices. Our method is to avoid harm of the heart wall from the delivery tool and to have the tool maintain a safe distance from the papillary muscles.

Up till now, a considerable number of approaches have been proposed to compute the safest corridor for the route of the delivery module in TA-AVI. Yeniaras, et al. [3] offered a method of updating a cylindrical corridor on-the-fly, which leans from the apex to the aortic annulus, by projecting multi-slice dynamic short-axis (SA) MR images onto single-slice, real-time, Long-Axis (LA) MR images. Zhou, et al. [4] used particle filters to trace the landmarks of the heart, such as apices, mediums, valves, and centroids using LA and SA images. Both methods rely on the fact that the safest path can be concatenated on the centroids of SA slices. However, these authors used cine images and did not take the papillary muscles into consideration, which caused the pre-operative planning to be less accurate and unrealistic. Bayraktar, et al. [5–7] adopted similar methods to [3], such as delineating the LV along the heart cycle by using an active-contour model, and computed the centroids of segments to generate a safe path along the LV. However, this method also did not examine the impact of papillary muscles on the computation of centroids. This poses a contrast to the logic of path planning that requires the delivery tool to maintain a considerable distance from heart walls and papillary muscles. To this end, as an enhancement to the existing path-planning methods for aortic delivery, we propose to use density-based spatial clustering of applications with noise (DBSCAN) [8,9] to locate so-called islands (papillary muscles) on the LV crops that are extracted by a hybrid active-contour model.

2   Methodology

We trace the LV boundary in a spatiotemporal fashion. To perform that, we first apply Perona-Malik filtering on the SA slices. This helps us smooth the papillary muscles on the surface of the LV. Then, we apply a region-based, active-contour model to delineate the LV border. Note that we locate a rectangular region of interest (ROI) anywhere on the LV, and this evolves toward the edge of the endo-cardium. The center of the segmented contour is propagated over the slices as a seed of a chosen ROI so that automation is confirmed for LV segmentation throughout MR sequences. Below are technical details about the computational pipeline.

2.1 Filtering

Perona-Malik [10] filtering constrains the smoothing at strong edges to preserve the semantically significant information. The diffusivity function is derived as in the following

$$g(\nabla(I)) = \frac{1}{1 + \sqrt{1 + \frac{\nabla(I^2)}{\gamma^2}}} \tag{1}$$

where $g\nabla(I)$ represents the diffusion coefficient, and $\nabla(I)$ the gradient map of image I. As can be inferred from Eq. 1, $\nabla(I)$ and g are inversely proportional to maintain the philosophy of the Perona-Malik method, which is to keep the strong edges. γ is a constant to control the sensitivity against gradients on the image domain. The diffusion process will decline at the regions where $|\nabla I| \gg \gamma$.

2.2 Segmentation

LV border delineation is performed in this step. We adopted a region-based active contour method proposed by Chan-Vese [11], which is to deform a curve toward object outlines using global image statistics. A known weakness of this method is to be sensitive to intensity inhomogeneity and voluminous noises in the vicinity of the LV edge (i.e. papillary muscles). Notably, we overcame this handicap by the aforementioned filtering method. The rationale behind Chan-Vese method is essentially originated from Mumford-Shah, and it is to approximate the image I based on a piecewise function u as represented in the following equation,

$$\arg \min_{u,C} = \mu Length(C) + \lambda \int_{\Omega} (I(x) - u(x))^2 dx + \int_{\Omega/C} |\nabla u(x)|^2 dx \tag{2}$$

where C is the contour to evolve, and u can be discontinuous. The first term is the regularizer of C; the second term is called data term, which is to ensure that u is close to I; and the last term is the smoothing term to ensure differentiability of u on Ω/C. The key difference between Chan-Vese and Mumford-Shah are additional terms which are designed to penalize the area enclosed by the curve, and u is expressed in a simplified way as in the following function

$$u(x) = \begin{cases} c_1, & \text{where x is inside C} \\ c_2, & \text{where x is inside C} \end{cases} \tag{3}$$

where x represents a pixel of the given image I, C is the border of a closed curve, and c_1 and c_2 are the features inside and outside of C. This method is to approximate the I by minimizing the $u(c_1, c_2)$.

$$\arg\min_{c_1,c_2,C} = \mu Length(C) + \upsilon Area(inside(C))$$

$$+ \lambda_1 \int_{interior(C)} \mid I(x) - c_1 \mid^2 \qquad (4)$$

$$+ \lambda_2 \int_{exterior(C)} \mid I(x) - c_2 \mid^2 dx$$

The first and second terms are the length and area regularizers of the boundary and the region enclosed by C, sequentially. The third and fourth terms are to penalize the difference between the input image I and piecewise constant model u. Segmentation is obtained by finding a local minimizer for this problem, and note that we adopted intensity means for these features (c_1 and c_2) as in the original Chan-Vese paper. Minimization of this functional can be performed by using level sets, and more numerical details can be found in [11].

2.3   Clustering of LV Crops

We project the delineated contour, which is to represent the LV boundary, onto the actual slice and crop it. We feed the extracted crops to DBSCAN, so that we are able to determine the islands on the LV surface which must be considered as obstacles in the path-planning process.

DBSCAN performs clustering over the data points (here pixels) based on their spatial connectivity within a pre-determined radius (ϵ). The user determines the minimum number of connected elements ($minPTS$) that are to construct a cluster. In Algorithm 1 $Data$ represents the LV crop, and $ballpoints$ is used for the clusters, and C represents the number of clusters as the output of DBSCAN.

2.4   Path Planning

Cardiac SA slices provide a dense view of the area that leans from apex to base. Being parallel to each other, and collected from the same field of view imbues SA slices with this property. Motivated from the fact that we trace every single heart beat ($t = 1$ to 25) and spatio-slices ($s = 1$ to 7) to determine the safest path.

In order to create a cylindrical view, we concatenated the segmented 2D contours along the z axis by a certain distance—slice thickness information can be obtained from DICOM header—and that leads us to have 3D+t model of the LV. To determine the safest path along the cylindrically modeled LV, we used the formula which computes centroid of irregular 2D contours [6]. Notably, since we aim to extract a 'papillary muscle sensitive' trajectory, a formula which is to compute centroids of doughnut-like contours is required. To this end, we applied the following formula shown in Eq. 5 especially for the mid-ventricle slices on which PM appear, where A and r respectively represent the area and centroids of circumscribing polygon and holes in them.

$$Centroid = \frac{A_{out} * r_{out} - A_{in} * r_{in}}{A_{out} - A_{in}} \qquad (5)$$

Algorithm 1. DBSCAN Pseudocode

```
1: function DBSCAN(Data, EPS, minPTS)
2:     C=0
3:     for  each unvisited point (P) in Data  do
4:         MarkPvisited
5:         ballpoints = regionQuery(P, epsilon)
6:         if  sizeof(ballpoints) < minPTS then
7:             P is Noise
8:         else
9:         end if
10:    end for
11: end function
12: C = nextcluster
13: expandCluster(P, ballpoints, C, EPS, minPTS)
14: function EXPANDCLUSTER(P, ballpoints, C, EPS, minPTS)
15:     add P to Cluser C
16:     for  eachpointP'inballpoints do
17:         if P'isnotvisited then
18:             markP'asvisited
19:             ball_points' = regionQuery(P', ε)
20:             if  sizeof(ballpoints') >= minPTS then
21:                 ballpoints = ballpointsjoinedwithball_points'
22:             end if
23:             if  P'isnotyetmemberofanycluster then
24:                 addP'toclusterC
25:             end if
26:         end if
27:     end for
28: end function
29: regionQuery(P, EPS): return all points within the n-dimensional sphere centered
    at P with radius epsilon (including P)
```

What we desire is that the found trajectory does not contain any abrupt corners, especially not those that would thwart the suitability of robotic delivery tools. To accomplish that, we applied spline-smoothing on the preliminary trajectory using the 'fit' routine provided by Matlab. This function confines the output to the data term, yet smoothens the output by minimizing the energy using the Levenberg-Marquardt method. This method finds a search direction for the local minima that is interpolated between the gradient descent and Gauss-Newton methods.

3   Data Set and Experiments

The data is multi-slice, non-triggered, free-breathing imaging (that means real-time gated), and collected from a true fast imaging with steady-state precession (TrueFISP), at a $TACQ = 70.50$ ms per slice (Pixel Spacing: 1.25×1.25;

FOV: 299 × 399; TR: 70.50 ms; TE: 1.03 ms; Matrix: 101*101; and Slice Thickness: 8 mm). The computations have been performed on a laptop PC (Intel 2.5 GHz processor with 8 GB RAM). The data set has 165 SA CMR slices.

3.1   LV Segmentation

We evaluated the accuracy of the segmentation method using two metrics, i.e., a Dice matrix and cross correlation between our automatic segmentation and the ground truth. The ground truth is obtained by manual contouring of the LV border performed by an expert MRI interpreter. We selected γ as 20 when performing filtering, which controls the diffusion conduction. Table 1 shows the results of our method. Higher values pose the risk of extreme smoothing, which may cause significant data loss. Correlation coefficients suggest the pixel-wise similarity between manually and automatically extracted crops. The Dice matrix, which measures the overlapping area of two designated crops, is commonly used for segmentation evaluation. We registered the mean Dice and Correlation with their respective standard deviations. As can be seen in Table 1, we obtained very minute standard deviation scores in the segmentation phase.

Table 1. Evaluation of segmentation method with respect to the ground truth

Method	Dice	Correlation
Our method	0.8798 ± 0.0078	0.9426 ± 0.0046
Without perona-malik	0.8436 ± 0.0123	0.7315 ± 0.0178

3.2   Clustering

We qualitatively assessed the suitability of the DBSCAN in terms of capturing the PM on LV crops. Since the intensity difference is very distinct between the crop and the background (of which the intensity value is zero), we complimented it to 255 and selected 0.005 as the minimum radius and 10 for the minimum points to construct a cluster. Note that, as can be seen in Table 2, the constructed clusters are represented by four colors that express the number of clusters (including the noise). We searched the LV crops that have four or more clusters, because the noiseless surface itself, the background, and the noise already comprise three clusters. Consequently, throughout the data set we found 41 slices that meet the criteria. Due to the sheer scale of the tiny structure of the PM, we validated the accuracy of DBSCAN qualitatively. As can be seen in Table 2, clustering worked properly. After clustering, we obtained areas and the centroids of the islands on LV surfaces using a regionprops routine provided by Matlab. Finally, we used Eq. 5 and computed the centroids of the doughnut-like polygons.

Table 2. Qualitative results of DBSCAN on selected two LV crops. Light brown color represents the background, solid amber color represents the LV surface and blue stars represent the PM

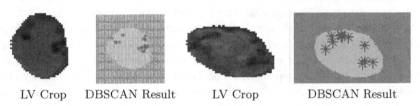

LV Crop   DBSCAN Result   LV Crop   DBSCAN Result

3.3   Path Evaluation

We evaluated accuracy of the trajectory found by our method using its Hausdorff distance, which requires the ground truth to do a comparison. Qualitative and quantitative results are displayed in Tables 3 and 4, respectively. We applied the centroid formula [6] on manually contoured polygons. Note that the centroids of 41 slices with dense PM are separately marked. The interpreter adopted the philosophy of Eq. 5 that computes the centroids by considering the weight (area) of the holes. In addition to this, in order to avoid abruptness in trajectory, we smoothed the array where we store the centroids. We selected the smoothing term λ as 0.4.

Table 3. A represents the segmented mid-ventricle; b,basal; and c, again mid-ventricle. d, e, f represent views of the dynamic safest path

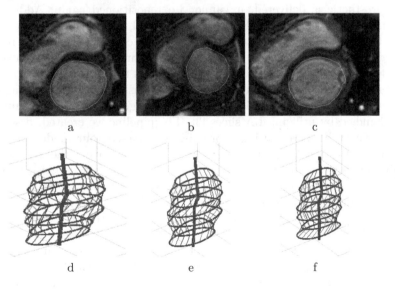

a   b   c

d   e   f

We computed the Hausdorff distance between the automatically and manually found trajectories for each time frame and spatial slice. This provides a Euclidean metric for the error of how the automatically found trajectory deviates from the ground truth. Recall that the ground truth trajectories for each time frame are based on manually segmented contours and the computation of their centroids. The max/mean and mean values are temporal-wise statistical variables to show the accuracy of the dynamic trajectory as displayed in Table 4. These statistical variables represent 7 spatial slices, where each of them yields a composition of 25 time frames. In Table 4, evaluation of [3] shows the registration error of their respective trajectories found using SA slices against the LA slices.

Table 4. Evaluation of segmentation method with respect to the ground truth

Method	Max	Min	Mean
Our method	1.3715	0.4204	0.8135
Without Perona-Malik	3.8367	0.3008	0.9793
Yeniaras et. al	2.1583	1.847	2
Bayraktar et. al	1.4036	0.1925	0.7692

4   Discussion

We introduced an off-line, preoperative planning approach, aiming to be the ground for a real-time Transapical Aortic Valve Replacement in a beating heart. Our work addresses a crucial basis in pre-operative planning, because with TAVI, once a prosthetic is delivered, it cannot be repositioned again. Additionally, proper navigation avoids the obstruction of mitral valve leaflets and coronary arteries that feed the heart muscles [12]. Our work can be expanded to intraoperative guidance, registering the constructed cylindrical path on the linked LA slice of the used SA slices. In that case, the heart's translational and rotational movement along its long axis must be taken into consideration. Additionally, vertical motility with respect to the MR bed emerging, due to free breathing, can also be handled by utilizing LA slices [3]. The pipeline we created, first, delineates the cavity boundary using a promising active-contour model and builds a dynamic 3D + t cylinder along the LV by concatenating the 2D contours. Notably, the bottom-up LV border segmentation deploys a de-noising method based on isotropic diffusion that filters extremely noisy MR slices. As an enhancement of existing path-planning approaches, we took the PM into consideration for dynamic trajectory generation. We also smoothed the trajectory (in each time frame) to avoid the zigzags that cause difficulty in navigating the delivery tool. As an output of this work, the obtained cylindrical structure (3D+t) is potent for offering volumetric depth information for nominally sized aortic-valve delivery tools. Future work contains registering the segmented SA contours onto Long Axis slices that can mimic a real-time procedure, and more assessments about the proximity of trajectory from the papillary muscles.

Acknowledgments. This work is partially supported by Arkansas INBRE Bioinformatics program.

References

1. Karar, M.E., John, M., Holzhey, D., Falk, V., Mohr, F.-W., Burgert, O.: Model-updated image-guided minimally invasive off-pump transcatheter aortic valve implantation. In: Fichtinger, G., Martel, A., Peters, T. (eds.) MICCAI 2011. LNCS, vol. 6891, pp. 275–282. Springer, Heidelberg (2011). doi:10.1007/978-3-642-23623-5_35
2. Navkar, N.V., Yeniaras, E., Shah, D.J., Tsekos, N.V., Deng, Z.: Generation of 4D access corridors from real-time multislice MRI for guiding transapical aortic valvuloplasties. In: Fichtinger, G., Martel, A., Peters, T. (eds.) MICCAI 2011. LNCS, vol. 6891, pp. 251–258. Springer, Heidelberg (2011). doi:10.1007/978-3-642-23623-5_32
3. Yeniaras, E., Navkar, N.V., Sonmez, A.E., Shah, D.J., Deng, Z., Tsekos, N.V.: Mr-based real time path planning for cardiac operations with transapical access, pp. 25–32 (2011)
4. Zhou, Y., Yeniaras, E., Tsiamyrtzis, P., Tsekos, N., Pavlidis, I.: Collaborative tracking for MRI-guided robotic intervention on the beating heart. In: Jiang, T., Navab, N., Pluim, J.P.W., Viergever, M.A. (eds.) MICCAI 2010. LNCS, vol. 6363, pp. 351–358. Springer, Heidelberg (2010). doi:10.1007/978-3-642-15711-0_44
5. Bayraktar, M., Sahin, B., Yeniaras, E., Iqbal, K.: Applying an active contour model for pre-operative planning of transapical aortic valve replacement. In: Linguraru, M.G., Oyarzun Laura, C., Shekhar, R., Wesarg, S., González Ballester, M.Á., Drechsler, K., Sato, Y., Erdt, M. (eds.) CLIP 2014. LNCS, vol. 8680, pp. 151–158. Springer, Cham (2014). doi:10.1007/978-3-319-13909-8_19
6. Bayraktar, M., Kaya, S., Yeniaras, E., Iqbal, K.: Trajectory smoothing for guiding aortic valve delivery with transapical access. In: Shekhar, R., Wesarg, S., González Ballester, M.Á., Drechsler, K., Sato, Y., Erdt, M., Linguraru, M.G., Oyarzun Laura, C. (eds.) CLIP 2016. LNCS, vol. 9958, pp. 44–51. Springer, Cham (2016). doi:10.1007/978-3-319-46472-5_6
7. Bayraktar, M.: Image Guided Preoperative Planning for Aortic Valve Replacement. Master's thesis, University of Houston, Texas, USA (2011)
8. Sander, J., Ester, M., Kriegel, H.-P., Xiaowei, X.: Density-based clustering in spatial databases: The algorithm GDBSCAN and its applications. Data Min. Knowl. Discovery 2(2), 169–194 (1998)
9. Suer, S., Kockara, S., Mete, M.: An improved border detection in dermoscopy images for density based clustering. BMC Bioinform. 12(S-10), S12 (2011)
10. Perona, P., Malik, J.: Scale-space and edge detection using anisotropic diffusion. IEEE Trans. Pattern Anal. Mach. Intell. 12(7), 629–639 (1990)
11. Vese, L.A., Chan, T.F.: A multiphase level set framework for image segmentation using the mumford and shah model. Int. J. Comput. Vis. 50(3), 271–293 (2002)
12. Li, M., Mazilu, D., Wood, B.J., Horvath, K.A., Kapoor, A.: A robotic assistant system for cardiac interventions under MRI guidance (2010)

3D Coronary Vessel Tracking in X-Ray Projections

Emmanuelle Poulain[1,2(✉)], Grégoire Malandain[2], and Régis Vaillant[1]

[1] GE Healthcare, 78530 Buc, France
{emmanuelle.poulain,regis.vaillant}@ge.com
[2] Université Côte d'Azur, Inria, CNRS, I3S, Sophia Antipolis, France
gregoire.malandain@inria.fr

Abstract. Fusing pre-operative CT angiography with per-operative angiographic and fluoroscopic images is considered by physicians as a potentially useful tool for improved guidance. To be adopted, this tool requires the development of tracking methods adapted to the deformations of the arteries caused by the cardiac motion. Here, we propose a 3D/2D temporal tracking of one coronary vessel, based on a spline deformation, using pairings with a controlled 2D stretching or contraction along the paired curves and a preservation of the length of the 3D curve. Experiments were conducted on a database of 10 vessels from 5 distinct patients, with dedicated metrics assessing both the global registration and the local coherency of the position along the vessel. The proposed results demonstrate the efficiency of the proposed method, with an average standard deviation of 2 mm for the localization of landmarks.

Keywords: Deformable registration · Tracking · Coronary arteries · X-ray · Computed tomography angiography · CTA

1 Introduction

Percutaneous Coronary Intervention (PCI) is a minimally procedure which is used to treat coronary artery narrowing. The workflow of the procedure is pretty standard. The physician intervenes on the patient under the guidance of an x-ray imaging system. A guidewire is navigated in the diseased coronary artery. Before this interventional step, x-ray images with injection of a contrast agent are acquired to observe the shape of the vessels according to the projection angles which will be used for the intervention. During the guidewire navigation, the lesion is crossed and in some cases, the physician could benefit from a visual assessment of the coronary wall which may present plaques more or less calcified. The x-ray imaging interventional system used for per-operative guidance is not able to display this information mostly by lack of density resolution. On the contrary, Computed Tomography Angiography (CTA) is a modality which has the capability of capturing both the artery lumen and the characteristics of the vessel wall. To truly help physician to exploit this information during the course of the procedure, registering these two modalities would be useful.

© Springer International Publishing AG 2017
M. Pop and G.A. Wright (Eds.): FIMH 2017, LNCS 10263, pp. 204–215, 2017.
DOI: 10.1007/978-3-319-59448-4_20

[1,3,9] have addressed this problem mostly in the case of images acquired at the same cardiac phase. In this situation and in a first analysis, the geometric difference between the two imaging situations can be described by a rigid transformation combined with a perspective projection since the CTA is a 3D dataset and the angiographic images correspond to a central projection with the x-ray source being the focal point. The following step is then to take into account the cardiac motion which may be captured or not in the CTA. If CTA can deliver multiple volumes corresponding at several phases of the cardiac cycle, it is at the expense of additional ionizing radiation to the patient. [2] has proposed to adapt a generic model of the cardiac motion to end-diastolic CTA dataset in order to register the two modalities along the cardiac cycle. This strategy raises the question of the validity of the generic model. In this work, we propose to explore another direction. The main interest of the physician is in the diseased vessel and he selects a projection angle by rotating the gantry such that the vessel of interest shows up relatively central in the image with limited foreshortening and minimum superimposition. Then to provide effective assistance during the guidewire navigation, the objective is to build a 3D deformation of the 3D vessel extracted from the CTA consistent with the apparent motion in the 2D angiographic sequence. So, any relevant information visible in the CTA volumes such as a plaque can be tracked in the 2D sequence. In [5], the proposed algorithm handles the non-rigid component of the cardiac motion by deformation in the image plane which is a simplification. In this work, we propose a method derived from [3] to track a coronary artery along the cardiac cycle with the objective of maintaining the consistency of the position along the vessels. We propose as in [6] to apply a 3D length preservation constraint as the coronary artery anatomically preserves its length along the cardiac cycle. In the following, we will describe the proposed method and explain the assessment strategy which includes metrics evaluating the registration and a specific metric related to the consistency of the position along the vessel.

2   Method

Before introducing the method, we first describe the data we have at hand. The 3D information is extracted from a Computed Tomography Angiography scan by a fully automated commercial product providing a segmentation of the coronary vessel structure. The coronary vessels are separated between the right and the left coronary and the different branches are represented by their centerlines which are represented by a tree T according to the anatomic structure of the vessels which separate in different branches at the bifurcations. From this structure we extract one vessel of interest V which is tracked along the consecutive images of the x-ray record sequence. Even if the 3D model of the coronary vessels can be depicted by a tree, this may not be the case for the x-ray projection. Indeed, self superimpositions create crossings. The vessel segmentation may also cause over segmentation or miss some vessels. x-ray projections are segmented with an Hessian based vessel enhancement technique, and vessel like structures are extracted forming a set of curves which corresponds to the centerlines of the vessel [4]. The segmented object is organized in a graph by applying standard processing methods to connect neighboring centerlines. Considering the consecutive images obtained in the sequence of N images by performing the acquisition after injection of the contrast agent, we obtain a set of graphs $\mathcal{G} = \{\mathcal{G}_1, \ldots, \mathcal{G}_N\}$.

We initiate the registration by identifying the initial rigid transformation, T° which maps $\mathcal{T}$ to the element $\mathcal{G}_1 \in \mathcal{G}$ corresponding to the same diastolic cardiac phase as the pre-operative CT image [3].

The aim of the proposed tracking method is to track the vessel V in all the consecutive phases of the cardiac motion, which necessitates to deform it. A spline description is a tool suited for this objective and the deformation can be represented by the optimization of its parameters, the control points. The registration itself is based on a two steps mechanism with first the determination of pairings between the projected curve describing the vessel V and the centerlines represented through a graph structure. Second, the parameters are determined by minimizing an energy depending on the distance between the paired points and constraints on the vessel V.

2.1 Problem Modeling

The 3D temporal tracking requires an a priori 3D model of vessels as introduced in [7,11]. They are represented by their centerline which is a 3D curve. The spline functions support a compact and smooth description of curves which can be continuously deformed by changing the position of the control points.

We thus fit an approximating cubic spline curve C as in [8], using a centripetal method such that:

$$\{C(u) \mid u \in [0,1]\} \approx V$$

More precisely the spline is defined as:

$$C(u) = \sum_{i=1}^{n} N_{i,p}(u) P_i \tag{1}$$

where $N_{i,p}$ is the ith B-spline of degree p, P_i the ith control point, u the spline abscissa (between 0 and 1). Thanks to the choice of this model consecutive deformations can be represented by the optimization of the spline parameters. The set of control points to register the 3D vessel with the graph $\mathcal{G}_t$ is determined by solving this optimization problem:

$$\hat{\mathcal{P}}_t = \operatorname{argmin}_{\mathcal{P} \in \mathbb{R}^{3n}} E_d(C_{\mathcal{P}}, \mathcal{G}_t) + \beta E_r(C_{\mathcal{P}}) \tag{2}$$

t denotes the temporal index of the frame, $E_d()$ and $E_r()$ are respectively the data attachment and the regularization energy terms. In the following $\mathcal{P}_t$ denotes the set of control points for frame t while $\mathcal{P}_1^{init}$ denotes the set of control points for the 3D vessel after the pose estimation T° for frame 1. An initial position is used for the 3D vessel to build the data attachment term: it is the 3D vessel/spline $C_{\mathcal{P}_1^{init}}$ issued from the pose estimation for the first frame $t = 1$ or

$C_{\mathcal{P}_{t-1}}$ for frame $t > 1$. For the sake of simplicity, t will be omitted in the following. This 3D curve is projected onto the angiographic frame and is denoted c. A 2D curve v corresponding to the projected 3D curve is extracted from the graph $\mathcal{G}$ as described in [3].

Data Attachment Term. The data attachment term $E_d()$ is a sum of 3D residual distances issued from 3D to 2D pairings. The simplest method to build pairings is to use the closest neighbor scheme (as in the ICP). In [10], a variant of this approach is proposed: the idea is to represent the cardiac motion by covariance matrices on the different parameters describing the coronary tree. For this one, a generative 3D model is employed, i.e. a model including a probabilistic distribution of position for the arterial segment. The concept of distance is then extended from standard Euclidean distance to Mahalanobis distance. This geometrically oriented analysis does not include the constraint of ordered pairing as proposed in [3] where it is shown that a point pairing that respects the order along paired curves yields better results than the closest neighbor scheme. Such an ordered pairing was obtained by the means of the Fréchet distance, that allows *jumps* between paired points. In presence of vessel deformation, we observed that the coherency of the obtained pairings can be discussed. So we propose to constrain the pairing construction with a 2D elongation preservation.

We first recall the Fréchet distance and its induced pairing [3]. Let $c = \{c_1, \ldots, c_{n_c}\}$ and $v = \{v_1, \ldots, v_{n_V}\}$ be the 2D curves to be paired. The points c_i are obtained as projection of points $C_{\mathcal{P}}(\bar{u}_i)$ from the 3D spline which represents the vessel. The points v_i are the discrete points forming the centerline of the vessel extracted from the angiographic images. The point pairings are entirely defined by a single injective function $F : \mathbb{N} \to \mathbb{N}$. The Fréchet distance is defined as:

$$\begin{cases} F(1) = \mathrm{argmin}_{i_v \in I_v} \|v_{i_v} - c_1\| \text{ with } I_v = \{1, \ldots, jump\} \\ F(i_c) = \mathrm{argmin}_{i_v \in I_v} \|v_{i_v} - c_{i_C}\| \text{ with } I_v = \{F(i_c - 1), \ldots, F(i_c - 1) + jump\} \end{cases}$$

with *jump* a parameter controlling the length of allowed jumps in pairings. Looking at the pairing produced by this metric (as in Fig. 1, left), we observed that the simple application of the criteria of minimizing the pairing length may lead to irregular pairings. When computing rigid transformations as in [3], the least squares estimation introduces enough robustness to handle them. However, when dealing with non-linear transformations, the final result may be influenced. Inspired by the Fréchet distance, we present a pairing function which aims to build a pairing function that advances at the same speed along the 2D curves to be paired. Let consider a distance d which will compute the length of 2D curves:

$$d : \mathbb{N}^2 \to \mathbb{R} \quad , \quad d(p_1, p_2) = \sum_{i = p_1 + 1}^{p_2} \|c_i - c_{i-1}\|$$

We will define F as:

$$\begin{cases} F(1) = \operatorname{argmin}_{i_v \in I_v} \|v_{i_v} - c_1\|^2 + \lambda d(v_1, v_{i_v})^2 \text{ with } I_v = \{1, \dots, jump\} \\ F(i_c) = \operatorname{argmin}_{i_v \in I_v} \|v_{i_v} - c_{i_c}\|^2 + \lambda (d(v_{F(i_c-1)}, v_{i_v}) - d(c_{i_c-1}, c_{i_c}))^2 \\ \qquad \text{with } I_v = \{F(i_c - 1), \dots, F(i_c - 1) + jump\} \end{cases} \tag{3}$$

with λ proportional to the local distance between the neighborhood of i_c and i_v. This function favors point pairings between points which are approximately at the same distance from theirs respective neighborhoods.

$F()$ provides 2D point pairings $(v_{F(i)}, c_i)$ between the 2D curves v and c. To compute 3D deformations, we have to define 3D point pairings. $c_i \in c$ is associated to its corresponding 3D point $C_{\mathcal{P}}(\bar{u}_i)$. The 3D point $V'_{F(i)}$ corresponding to $v_{F(i)}$ is the point from the backprojected line issued from $v_{F(i)}$ that is the closest to $C_{\mathcal{P}}(\bar{u}_i)$. The data attachment term is finally:

$$E_d(C_{\mathcal{P}}, \mathcal{G}_t) = \sum_{i=1}^{n_C} \|V'_{F(i)} - C_{\mathcal{P}}(\bar{u}_i)\|^2 \tag{4}$$

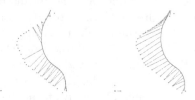

Fig. 1. The figure depicts pairings (blue) between projected 3D vessel c (magenta) and 2D vessels v (red), as observed locally on a case. On the left the parings are made with Fréchet, on the right with weighted Fréchet. Pairings are more regular with weighted Fréchet. (Color figure online)

Regularization Term. The regularization term aims at minimizing the 3D elongation of C:

$$E_r(C_{\mathcal{P}}) = \sum_{j=1}^{J} (\|C_{\mathcal{P}}(e_j) - C_{\mathcal{P}}(e_{j-1})\| - l_j)^2 \text{ with } e_j = \frac{j}{J} \text{ and } l_j$$

$$= \|C_{\mathcal{P}_1^{init}}(e_j) - C_{\mathcal{P}_1^{init}}(e_{j-1})\| \tag{5}$$

J is the number of interval used to enforce the length constraint all along the vessel.

Energy Minimization. This global energy $E_d(C_{\mathcal{P}}, \mathcal{G}_t) + \beta E_r(C_{\mathcal{P}})$ is minimized via a gradient descent. Thanks to the spline description of the 3D curve, the analytic expression of the gradient is used for the gradient descent. The pairings are recomputed along the descent every 1000 iterations. The minimization is stopped when the gradient norm is below a threshold, whose value has been chosen in preliminary experiments.

3   Performance Evaluations

Qualitative evaluation of the performance of the proposed algorithms can first be done by a visual control of the deformation of the projected deformed vessel over the angiographic image along the cardiac cycle. We also propose three quantitative measures. The first two corresponds to methodological expectations on the performance but does not cover directly the intended clinical application. The third one replicates more closely the expectations from a clinical standpoint.

3.1   2D Curve Distance

This is an indirect measure of the quality originally proposed in [11]. Its intent is not to evaluate the correctness of the selected vessel in the angiographic image. This measure indicates only if the deformation of the vessel V has adapted well to the observed projection in the angiographic image. For each registration the distance between the projected curve $\mathcal{P}(C)$ and its corresponding 2D vessel v is computed, with $\mathcal{P}$ the projection matrix. Our measure of 2D curve distance is:

$$cd = \frac{1}{n_C} \sum_{i=1}^{n_C} \|\mathcal{P}(C(i)) - v_{closest(i)}\| \tag{6}$$

with $closest(i)$ the index in v of to the closest point to $\mathcal{P}(C(i))$, n_C the number of retained sample points in C. C is a continuous curve defined by an analytic representation based on spline. For this evaluation measure and also the next one, we select a number of points along the curve. We take them equally spaced.

3.2   Shape Preservation

For this analysis, we start from the idea that the vessel shall return to its initial state if the tracking is performed on a series of consecutive images which start and end by the same image. Let N the number of angiographic images in a sequence which covers a cardiac cycle, the tracking is done from the frame 1 to the frame N, resulting in N 3D curves corresponding to the same vessel of interest temporally tracked, $C = \{C_1, \ldots, C_N\}$. One can then generate the reverse sequence starting from image $N - 1$ down to image 1 and continue to apply the tracking algorithm. The result is an other set of 3D curves $C' = \{C'_{N-1}, \ldots, C'_1\}$. The similarity of the curves C'_k and C_k is an indirect measure of the performance of the tracking algorithm. To measure the similarity, we chose to compute the distance between C_1 and C'_1 which are respectively the first and last curves of the forward and backward tracking. Our measure of shape preservation is:

$$sp = \frac{1}{n_C} \sum_{i=1}^{n_C} \|C_1(i) - C'_1(i)\| \tag{7}$$

with n_C the number of points in C and C'.

3.3 Landmark Tracking

From the point of view of the clinical application, this is the most important measure. The idea is to evaluate if a location defined along the coronary vessel is correctly tracked with the beating heart. A location in the vessel V is defined by its curvilinear abscissa. In the angiographic image, identifying a fixed point is more challenging.

To do so, we first manually point an easily identifiable landmark along the 2D vessel that correspond to the 3D vessel of interest. Vessel bifurcations are natural candidates for such landmarks, and we manually have pointed one bifurcation along the x-ray sequence for each 3D vessel to be registered. To decide whether the same 3D point of the tracked vessel is paired to this ground truth, we use the curvilinear abscissas u (along the spline) of the paired 3D points to the bifurcation. A perfect tracking (along with a perfect manual identification of the bifurcation) should yield the same curvilinear abscissa for all paired 3D points, thus the standard deviation of all curvilinear abscissas is an adequate measure to assess the tracking.

Formally, let $U = \{u_1, \ldots, u_N\}$ be the N abscissas along the x-ray sequence of the paired 3D points, eg. $C(u_t)$ is paired with the bifurcation/landmark in frame t, and $\bar{u}$ be the average value over U, the proposed measure is

$$lt = \sqrt{\frac{1}{N} \sum_{i=1}^{N} (u_i - \bar{u})^2}$$

4 Results

To assess the performance of the proposed approach, we use anonymous data collected after informed patient consent for use in this type of investigation. These data come from five different patients. Both the CT scan and the angiographic images are available. We selected in the angiographic sequences a sub-sequence of 20 images which covers a full cardiac cycle or a bit more depending on the patient case. The CT scans have been pre-processed to extract the coronary vessel trees as described above. Several 3D vessels may have been selected for a given patient, yielding a total of 10 different tracking experiments. Each of them is analyzed separately from the other. Selection is based on the available angiographic views and the vessels are selected as the ones that could be the object of an interventional procedure. In the following, we propose to compare three different settings of the proposed algorithms: the standard Fréchet approach to determine pairings between two curves, the weighted Fréchet as described above with and without the constraint on the length of the vessel V. λ has been set to $\frac{l_d^2}{d_p(c)^2}$, with l_d the local distance between neighborhoods of two points and $d_p(c)$ the average distance between points in c curve. To tune β we observed the length variations of the deforming vessel (see Fig. 2) and the data attachment term value at convergence for different values of β on few cases. First, we observe

for $\beta \geq 100$ a variation smaller than 0.1% despite an apparent 2D length variation on the fluoroscopic plane of 5%. Second, we notice that the data attachment
term value at convergence remain the same for every tested values of β, we thus
conclude that a big β does not alter the closeness of the two curves. Considering
this last point and the fact that the length constraint corresponds to the exact
physical behavior of the coronary during the contraction, we set $\beta = 1000$ such
that the deformed vessel keeps a constant length.

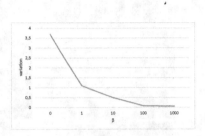

Fig. 2. Variation of the 3D length of a vessel depending on the β parameter, expressed
as a percentage of the initial 3D length.

Figures 3 and 4 are examples of the obtained results for two cases. The
average execution time per image is 30 s on an Intel R Core TM i7-4712HQ
CPU on a virtual machine. The code has not been optimized.

4.1 2D Curve Distance

As explained in Sect. 3.3, we have computed the 2D curve distance for the different considered cases. The obtained results are displayed in Fig. 5 with the
weighted Fréchet approach. The values fall in the range of 0.15 mm with a maximum of 0.25 mm. In this figure, we display for each frame from 1 to 20 the
value of the distance (Eq. 6). Contrary to [11], we do not observe a variation
with the cardiac cycle. So we draw the conclusion that the deformation capability of our 3D vessel is sufficient to follow the deformation of the arteries. In
all the considered cases through observations by an experienced reader, we have
validated that the tracked arteries are the correct ones. The application of the
length constraint does not restrain the capability of the model to deform.

4.2 Shape Preservation

The objective of this measure is to evaluate the correctness of the deformation of
this vessel by looking at a case for which the ground truth is known by design of
the test. Left of Table 1 presents the results obtained on our ten different cases.
When the length constraint is not applied, the results show that the deformation
is not well controlled. In the first case, we are close to about 40 mm of difference
between the two curves. This is to be compared with the typical length of a

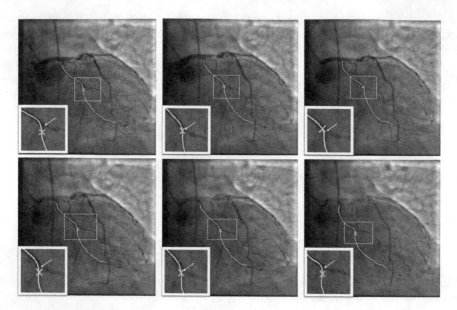

Fig. 3. Tracking results for one patient over one cardiac cycle. The yellow curve represents the projected 3D vessel, the blue cross represents the point tracked as the bifurcation, and the white arrow designs the bifurcation as marked by an experienced reader. Those images come from a 20 frames sequence. This figure shows the frames 1, 5, 9, 13, 17, 20, from left to right, up and down. (Color figure online)

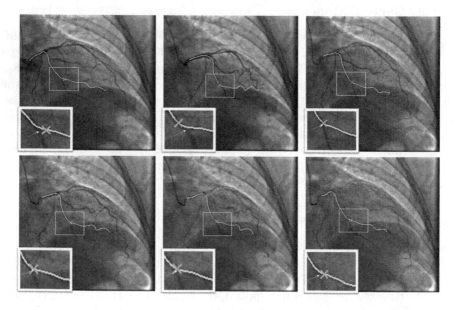

Fig. 4. Tracking results for a second patient. Same conventions than in Fig. 3

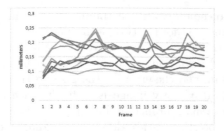

Fig. 5. The 2D curve distance measure (*cd*, see Eq. 6), for the ten cases. The values are plotted for the 20 successive frames of the selected sub-sequence.

coronary vessel that is about 80 mm. The two other techniques which apply the 3D length constraint give similar results. Overall the average 3D distance between the two curves after a tracking over 40 projections is inferior to 7 mm which is pretty encouraging. We observe some variations from case to case, which are very likely to be caused by the complexity of the motion. For some patients, the motion is mostly a translation and a rotation in the image plane plus some large scale contraction. In some other cases, the arteries are more tortuous and along the cardiac cycle they can fold/unfold. These cases are more challenging.

4.3 Landmark Tracking

The results obtained with this method are presented on the right of Table 1. The presented value is the standard deviation of the set of curvilinear abscissa for

Table 1. Results in millimeters of the 3 methods for shape preservation and landmark tracking evaluations on 10 vessels from 5 patients.

Shape preservation			Landmark tracking		
Vessels	Fréchet	Weighted Fréchet	Weighted Fréchet WLC[a]	Frchet	Weighted Fréchet	Weighted Fréchet WLC[a]
1	7.96	7	39.81	1.73	2.21	14.5
2	11.11	8.7	x	2.64	3.2	x
3	2.19	1.4	1.43	1.72	2.1	1.01
4	4.21	3	2.44	2.62	1.3	1.14
5	1.8	1.8	5.02	1.12	0.7	2.75
6	7	3.18	4.9	3.88	4.1	2.79
7	4.94	7.3	6.4	1.9	2.4	1.38
8	2.62	1.7	1.92	1	1.2	1.12
9	11.5	9.6	8.43	2.1	2.1	2.52
10	10.1	10.5	7.92	1.63	1.9	3.01
average	6.33	5.42	8.68	2.02	2.11	3.36

[a]WLC = Without Length Constraint

the point associated to the bifurcation manually marked in each projection. The average value is 2.02 and 2.11 for the two methods with the 3D length constraint. When the length constraint is not applied, we have a larger value as anticipated. This observed standard deviation accounts for two sources of errors: the error created by the algorithm and the error on the ground truth. Marking the exact location of a bifurcation in x-ray image is difficult since the bifurcated vessel may superimpose to the main branch and cannot be distinguished from it. The length of this superimposition varies along the cardiac cycle. This problem is also more or less pronounced depending on the apparent angle at the level of the bifurcation.

5   Discussion and Conclusion

We have presented a method to track a coronary artery from 3D to 2D. Starting from a first registration of the complete tree obtained at the same phase, we then focus our attention on a single vessel in the idea of mimicking the interventional procedure whose objective is to treat a diseased vessel. The key point of the algorithm approach that we have proposed is to establish pairings using the Fréchet algorithm. These pairings are done between the projected 3D vessel and the centerlines that were segmented in the angiographic image. The 3D vessel is then deformed to minimize the total pairing length in the 3D space and under the constraint of length preservation. This constraint is meaningful because this property is respected by the coronary vessel along the cardiac cycle and because it is applied to the 3D curve and not its projection. In the evaluation, we have observed that the absence of this constraint degrades the result even if the tracking remains apparently correct. The individual trajectories of points are not constrained a priori to be smooth. In the evaluation, we have looked a posteriori at the trajectory of specific points placed at the bifurcations and we found that their projections follow pretty well the apparent movement of the bifurcations in the x-ray projections. We have also evaluated the Fréchet pairing algorithm and a variant. Main difference is the introduction of an additional contributor in the optimized criteria. The criteria is based on the variation of the apparent length in the successive pairings. As observed in Fig. 1, the obtained pairings are more meaningful than with the standard Fréchet method. Looking at the results on the ten test cases and with the different evaluation strategies implemented, this difference translate in an improvement for the shape preservation criteria. Interestingly the average distance is about 5.5 mm after a tracking performed over 40 frames. Typical length of the selected coronary arteries is about 80 mm. For the landmark tracking, the average value is about 2 mm in both variants of the Fréchet method which also accounts for the imprecision in the definition of the bifurcation location in the angiographic image. This last result appears very encouraging since it is in the range of the expectation of the physician who expects to get information on the vessel wall at the place where the tip of the guidewire is positioned. The exact position in the artery of the guidewire tip is also changing slightly with the cardiac motion. In a further step, methods to register the position of the guidewire tip observed in the subsequent fluoroscopic sequence with the angiographic projection will be developed. A more complete evaluation of the accuracy could also be done by performing in parallel to the angiographic acquisition some intravascular images which are able to display the vessel wall properties. After proper co-registration of these intravascular images with the angiographic images, it would be then possible to compare the two modalities intravascular and CT views of the vessel wall and to quantify the geometrical differences from an algorithmic standpoint.

References

1. Aksoy, T., Unal, G., Demirci, S., Navab, N., Degertekin, M.: Template-based CTA to x-ray angio rigid registration of coronary arteries in frequency domain with automatic x-ray segmentation. Med. Phys. **40**(10), 1903–1918 (2013)
2. Baka, N., Metz, C.T., Schultz, C., Neefjes, L., van Geuns, R.J., Lelieveldt, B.P.F., Niessen, W.J., van Walsum, T., de Bruijne, M.: Statistical coronary motion models for 2D+ t/3D registration of x-ray coronary angiography and CTA. Med. Image Anal. **17**(6), 698–709 (2013)
3. Benseghir, T., Malandain, G., Vaillant, R.: A tree-topology preserving pairing for 3D/2D registration. Int. J. Comput. Assist. Radiol. Surg. **10**(6), 913–923 (2015)
4. Frangi, A.F., Niessen, W.J., Vincken, K.L., Viergever, M.A.: Multiscale vessel enhancement filtering. In: Wells, W.M., Colchester, A., Delp, S. (eds.) MICCAI 1998. LNCS, vol. 1496, pp. 130–137. Springer, Heidelberg (1998). doi:10.1007/BFb0056195
5. Gatta, C., Balocco, S., Martin-Yuste, V., Leta, R., Radeva, P.: Non-rigid multi-modal registration of coronary arteries using SIFTflow. In: Vitrià, J., Sanches, J.M., Hernández, M. (eds.) IbPRIA 2011. LNCS, vol. 6669, pp. 159–166. Springer, Heidelberg (2011). doi:10.1007/978-3-642-21257-4_20
6. Groher, M., Zikic, D., Navab, N.: Deformable 2D–3D registration of vascular structures in a one view scenario. IEEE Trans. Med. Imaging **28**(6), 847–860 (2009)
7. Heibel, T.H., Glocker, B., Groher, M., Paragios, N., Komodakis, N., Navab, N.: Discrete tracking of parametrized curves. In: IEEE Conference on Computer Vision and Pattern Recognition, CVPR, pp. 1754–1761. IEEE (2009)
8. Piegl, L., Tiller, W.: The NURBS Book. Springer, Heidelberg (2012)
9. Ruijters, D., ter Haar Romeny, B.M., Suetens, P.: Vesselness-based 2D–3D registration of the coronary arteries. Int. J. Comput. Assist. Radiol. Surg. **4**(4), 391–397 (2009)
10. Serradell, E., Romero, A., Leta, R., Gatta, C., Moreno-Noguer, F.: Simultaneous correspondence and non-rigid 3D reconstruction of the coronary tree from single x-ray images. In: 2011 IEEE International Conference on Computer Vision (ICCV), pp. 850–857. IEEE (2011)
11. Shechter, G., Devernay, F., Coste-Maniere, E., McVeigh, E.R.: Temporal tracking of 3D coronary arteries in projection angiograms. In: Medical Imaging 2002, pp. 612–623. International Society for Optics and Photonics (2002)

Electrophysiology: Mapping and Biophysical Modelling

A Parameter Optimization to Solve the Inverse Problem in Electrocardiography

Gwladys Ravon[1,2,3]([✉]), Rémi Dubois[1,2,3], Yves Coudière[1,4,5], and Mark Potse[1,4,5] [iD]

[1] IHU Liryc, Electrophysiology and Heart Modeling Institute, 33000 Pessac, France
gwladys.ravon@ihu-liryc.fr
[2] Univ. Bordeaux, CRCTB, U1045, Bordeaux, France
[3] INSERM, CRCTB, U1045, Bordeaux, France
[4] CARMEN Research Team, Inria, Bordeaux, France
[5] Bordeaux University, IMB UMR 5251, 33400 Talence, France

Abstract. The main challenge of electrocardiography is to retrieve the best possible electrical information from body surface electrical potential maps. The most common methods reconstruct epicardial potentials. Here we propose a method based on a parameter identification problem to reconstruct both activation and repolarization times. The shape of an action potential (AP) is well known and can be described as a parameterized function. From the parameterized APs we compute the electrical potentials on the torso. The inverse problem is reduced to the identification of all the parameters. The method was tested on *in silico* and experimental data, for single ventricular pacing. We reconstructed activation and repolarization times with good accuracy accurate (CC between 0.71 and 0.9).

1 Introduction

The main challenge of electrocardiography is to retrieve the best possible electrical information from body surface electrical potential maps (BSPM). The most common approach relies on the inverse solution of the Laplace equation in the torso. It reconstructs epicardial potential maps from the BSPM. This technique requires a regularization strategy to deal with the ill-posedness of the problem, and a discretization method to approximate the Laplace equation. A Tikhonov regularization and the Method of Fundamental Solutions (MFS) are commonly used, as proposed by Wang and Rudy [1]. Relevant activation maps can be retrieved from this inverse solution, though it provides signals with a lower amplitude. Nevertheless the reconstruction of accurate activation maps and repolarization maps remains a very challenging problem.

Alternative formulations have been proposed Liu *et al.* [2] and by Van Oosterom *et al.* [3], in order to reconstruct directly the activation times (ATs). The method proposed by Liu *et al.* looks for the three-dimensional activation sequence in the ventricular muscle. The method of Van Oosternom *et al.* considers both epicardium and endocardium. These approaches still rely on a regularization technique and are not designed to obtain repolarization maps.

© Springer International Publishing AG 2017
M. Pop and G.A. Wright (Eds.): FIMH 2017, LNCS 10263, pp. 219–229, 2017.
DOI: 10.1007/978-3-319-59448-4_21

We introduce a new technique that aims at recovering both the activation and repolarization maps on the epicardium. We first focus on single ventricular pacing cases. We propose an approach based on a parameter identification (PI) problem. As epicardial potentials are difficult to parameterize, we rather represent the action potential (AP) as a function of 4 parameters; namely the amplitude, activation time, plateau phase duration and repolarization slope. The final parameter identification problem consists of identifying these 4 space-dependent parameters from the complete BSPM sequence. This method solves the whole electrical sequence: depolarization and repolarization. Since it introduces *a priori* the shape of the AP, no regularization is needed. The nonlinear least squares parameter identification problem is solved by a gradient method.

The method was tested on both *in silico* and *ex vivo* experimental data. We found that activation maps from PI were at least as good as those from the MFS. Accuracy of reconstructed repolarization maps and torso potentials were also discussed.

2   Methods

2.1   Parameterization of the Action Potential

Following the work of Van Oosterom [4] we define the transmembrane potential (TMP) as the function:

$$V_{\mathrm{m}}(t, x) = \mathcal{A}\mathcal{F}(\alpha, t - \tau)\mathcal{F}(-\alpha_{\mathrm{R}}, t - (\tau + \tau_{\mathrm{R}})), \qquad (1)$$

where $\mathcal{F}(\alpha, \zeta) = \dfrac{1}{1 + \exp^{-\alpha\zeta}}$. α is the constant slope of the depolarization. Its value $(3.3\,\mathrm{ms}^{-1})$ is taken from the same study [4]. $\mathcal{A}$ is the amplitude, τ the activation time, τ_{R} the plateau phase duration and α_{R} the slope of the repolarization (see Fig. 1). Each of these 4 parameters may be space-dependent. Note that in our study the amplitude does not have physiological value. It is a qualitative parameter made to fit the amplitude of the given BSPM.

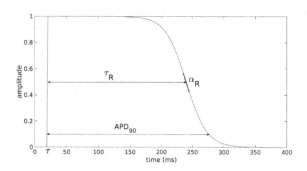

Fig. 1. Parameterized action potential with $\tau = 20\,\mathrm{ms}$ and $\tau_{\mathrm{R}} = 220\,\mathrm{ms}$.

2.2   Mapping the TMP to the BSPM

Given the TMP V_m on each point x_j of the epicardium, we compute the extra-cellular potentials:

$$\phi_\mathrm{e}(x_j, t) = \overline{V_\mathrm{m}}(t) - V_\mathrm{m}(x_j, t) \quad j = 1, \ldots, N_\mathrm{H}, \tag{2}$$

where N_H is the number of points on the heart surface. $\overline{V_\mathrm{m}}(t)$ is the spatial mean of V_m at each time. This formulation is derived from the monodomain model [5].

Finally, we approximate the solution of the Laplace equation far from the heart surface by [6, 7]:

$$\phi_\mathrm{T}(y, t) = \sum_{j=1}^{N_\mathrm{H}} \frac{1}{4\pi \|x_j - y\|} \phi_\mathrm{e}(x_j, t), \tag{3}$$

where y is any point on the body surface. These ϕ_T will be compared to the BSPM.

2.3   The Parameter Identification Problem

As a consequence we look for the parameter set $\mathcal{P} = (\mathcal{A}, \tau, \tau_\mathrm{R}, \alpha_\mathrm{R})$ that minimizes the least squares error

$$J(\mathcal{P}) = \frac{1}{2} \sum_{k=k_1}^{k_2} \sum_{i=1}^{N_\mathrm{T}} (\phi_\mathrm{T}(y_i, t_k) - \phi^\star(y_i, t_k))^2, \tag{4}$$

where $(y_i)_{i=1\ldots N_\mathrm{T}}$ are the N_T electrode locations on the body surface, $(t_k)_{k=k_1\ldots k_2}$ is the time sequence of interest, and $(\phi^\star(y_i, t_k))$ are the measured BSPM.

However, in order to improve the convergence of the method we make the following choices:

- the amplitude and the repolarization slope are set to a constant over the whole epicardium
- τ and τ_R are space-dependent. $(\tau_j)_j$ and $(\tau_{\mathrm{R},j})_j$ are taken at the same location on the epicardium as $(x_j)_j$
- the parameter set is split into the depolarization subset $\mathcal{P} = (\mathcal{A}, (\tau_j)_j)$ and the repolarization subset $\mathcal{P} = ((\tau_\mathrm{R})_j, \alpha_\mathrm{R})$.

During the depolarization phase of a paced or normal beat we can assume that $t \ll \tau + \tau_\mathrm{R}$, so that $V_\mathrm{m}(t, x) \simeq \mathcal{A}\mathcal{F}(\alpha, t - \tau)$. Hence we solve for $\mathcal{P} = (\mathcal{A}, (\tau_j)_j)$ in (4) on a time interval $[t_{k_1}, t_{k_2}]$ that covers the total QRS interval, but contains no T wave (see Fig. 2). Still in order to improve the convergence, we split the identification in two steps: we first identify the constant amplitude $\mathcal{A}$, and then the ATs $(\tau_j)_j$. We apply a standard MFS with a regularization method [1]. This gives us electrical potentials on the heart surface. We then compute the

ATs as the time with the highest negative slope. When we identify the amplitude the parameter set $\mathcal{P}$ is simply the singleton $\{\mathcal{A}\}$, and ATs obtained from the MFS are an input in our PI problem. Once we have this optimized amplitude $\mathcal{A}^\star$ it becomes an input and the cost function J is minimized for $\mathcal{P} = ((\tau_j)_j)$.

During the repolarization phase, we solve for $\mathcal{P} = ((\tau_{R,j})_j, \alpha_R)$ in (4) on a time interval $[t_{k_1}, t_{k_2}]$ that covers the total extent of the T waves (see Fig. 2). We first identify the constant slope (e.g. $\mathcal{P} = (\alpha_R)$). The input plateau phase durations are coarsely determined from the given BSPM. Once we have the optimum $\alpha_R^\star$ we minimize J for $\mathcal{P} = ((\tau_{R,j})_j)$. Optimum $\mathcal{A}^\star$, $(\tau_j^\star)_j$ and $\alpha_R^\star$ are an input in this PI problem.

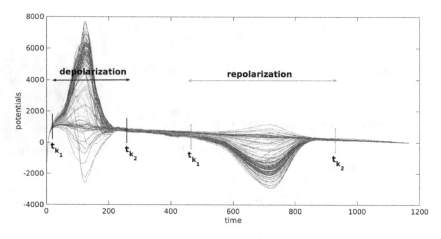

Fig. 2. Example of choice for k_1 and k_2 for experimental signals. Each curve represents the signal of one of the torso electrodes.

These four nonlinear least squares problems are solved by the gradient descent method. The explicit gradient of the cost function J with respect to the unknown parameters $\mathcal{P}$ is calculated analytically. If we wait for the method to converge overfitting occurs. It means that there are small changes in the cost function but the quality of the activation map decreases. To avoid overfitting, we couple the gradient descent method with an early stopping criterion based on the shape of the learning curve. For each gradient descent method, an initial guess is required. For the amplitude, we obtain this guess manually with a dichotomy approach. For the ATs, we use the ATs computed from the MFS. The initial guess for the slope is arbitrarily set to 1. The overall algorithm is summarized below.

Algorithm 1.1

Depolarization phase
1: Standard MFS on interval $[t_{k_1}, t_{k_2}]$
2: Compute ATs from MFS solution
3: Gradient descent to optimize the amplitude (MFS ATs as input)
4: Gradient descent to optimize the ATs (optimum $\mathcal{A}^*$ as input)
5: Gradient descent to optimize the amplitude (optimum τ^* as input)

Repolarization phase
1: Input: $\mathcal{A}^*$ and τ^*, from torso signal for τ_R
2: Gradient descent to optimize the constant slope α_R
3: Gradient descent to optimize the $(\tau_{R,j})_j$ (optimum α_R^* as input)
4: Gradient descent to optimize the slope (optimum τ_R^* as input)

---

3 Results

3.1 *in silico* data

In order to create *in silico* testing data, a simulation was run on an anatomically realistic 3D geometry of the torso, including heart, blood vessels, lungs and skeletal muscle. Each organ had its own conductivity. Propagating action potentials were generated using a monodomain reaction-diffusion model with a Ten Tusscher membrane model [8]. An anisotropic human heart model at 0.2 mm resolution was used for this purpose. An anisotropic Laplace equation was solved in the torso volume with a finite difference method [9]. We had access to the transmembrane potentials on the subepicardium and extra-cellular potentials on the epicardium. ATs were calculated from the epicardial potentials with the same method as the ATs from the MFS solution. Repolarization times were computed from epicardial potentials as the time with the highest positive slope during the repolarization phase.

On the same model anatomy, two different simulations were run: a left ventricular (LV) pacing, and a right ventricular (RV) one. For both cases, we identified our parameters following the algorithm detailed in Algorithm 1.1.

We first looked at the ATs. For the LV pacing the method converged to satisfactory ATs on the whole heart (Fig. 3, left). Our method gave a better range of ATs than the MFS. However for both methods the pacing site was not well localized. For the RV pacing, late activations were better reconstructed than the early ones. The correlation coefficient (CC) was the same for both methods but the distribution of points in the scatter plots was different. Indeed for the MFS we observed clustering of points along horizontal lines. This means that there were discontinuities in the distribution of ATs. These discontinuities were not consistent with the reference ATs (Figs. 3, right and 4). On the map on the right, discontinuities are clearly visible between the dark blue and light green parts and between the dark green and orange parts.

In Fig. 5 we compare APD90 calculated from the reference APs and from the reconstructed ones, for both LV and RV pacing. First of all, we had a larger

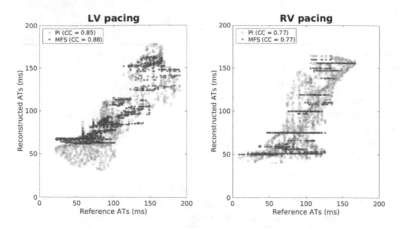

Fig. 3. Scatter plot of the ATs for *in silico* data. For each point, the x coordinate is the reference AT and the y coordinate is the corresponding reconstructed AT (PI in red, MFS in blue). (Color figure online)

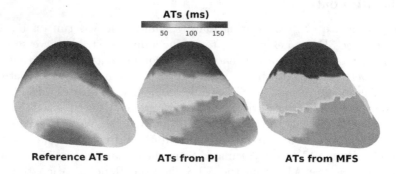

Fig. 4. Activation maps of the RV pacing *in silico* case.

range of APD90 from the PI, especially for the RV pacing. Moreover, there was a clear difference in the repartition of APD90 between the LV and the RV. This difference was not correctly reconstructed with the PI approach.

We compare the repolarization times in Fig. 6. For the PI, the CC were 0.9 and 0.71 for the LV and RV pacing, respectively. These values were close to those for the MFS but we observed discontinuities in the distribution of points, as for the ATs (see Fig. 7). Note that the quality of the reconstruction from the PI method was better for the repolarization times than for the APD90. This can be explained by the fact that if we had an error in the parameter τ, it would imply another error in the parameter τ_R to fit the repolarization phase correctly.

Finally we compare signals on the torso. Reconstructed potentials were computed from Eqs. (1), (2) and (3) with the optimized parameters. In both cases the amplitude was optimized to fit the given BSPM. Figure 8 shows given and reconstructed potentials on two torso electrodes. For the LV pacing the amplitude of the signal was better fitted on the first electrode than on the second.

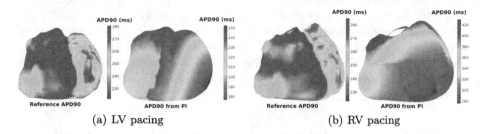

(a) LV pacing   (b) RV pacing

Fig. 5. Comparison of APD90. The reference APD90 are obtained on the subepicardium mesh and the reconstructed APD90 on the epicardium mesh. Anteroposterior view.

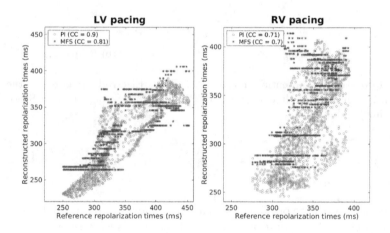

Fig. 6. Scatter plot of the repolarization times for *in silico* data.

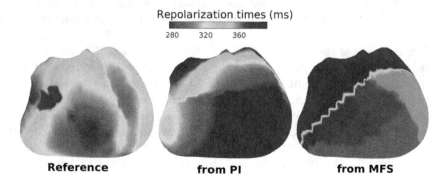

Fig. 7. Repolarization maps of the RV pacing *in silico* case.

Both depolarization and repolarization phases were quite well identified. These two electrodes were representative for the 252 torso electrodes. For the RV pacing, on the same electrodes, the reconstruction was less accurate. Specifically, the repolarization was inverted on the first electrode, while on the second the depolarization was. This was due to the incorrect activation times on the right ventricle. In this case, the chosen electrodes exhibit some of the worst reconstructed potentials.

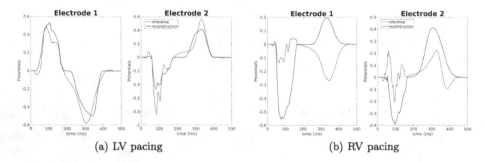

(a) LV pacing   (b) RV pacing

Fig. 8. Reconstructed potentials for *in silico* data. Electrode 1: close to the heart, Electrode 2: right hip. Red line: reference BSPM; blue line: reconstructed BSPM. (Color figure online)

3.2   Experimental Data

Experimental data were obtained from a Langendorff-perfused pig heart with an oxygenated electrolytic solution. Epicardial potentials were recorded with a flexible electrode sock (108 electrodes) placed over the epicardium. The heart was placed inside a human-shaped tank. BSPM were recorded from 128 electrodes, simultaneously with epicardial potentials. LV and RV pacing were performed with 1 Hz frequency. We removed signals from bad leads (e.g. not well connected) and the baseline. Finally we made a signal averaging over the whole recording to obtain a single beat.

We first looked at the ATs. As for the *in silico* data, it was difficult to localize the pacing sites precisely. Nevertheless the reconstructed ATs were satisfactory (Fig. 9). We noticed a slight improvement of our method compared to MFS for the LV pacing (CC: 0.85 vs 0.82). On the RV pacing (CC: 0.92 vs 0.89) improvement was more obvious when we compared the distribution of points.

Scatter plots of the repolarization times are presented in Fig. 10. For both cases, the reconstructed repolarization phase was shorter than the real. The CC were very close for the PI and the MFS but with the PI the distribution was smoother.

Reconstructed potentials are presented in Fig. 11. With our method we were able to identify both depolarization and repolarization phases. The amplitude of

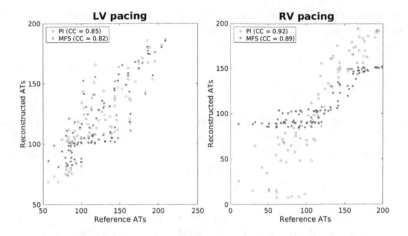

Fig. 9. Scatter plot of the ATs for experimental data

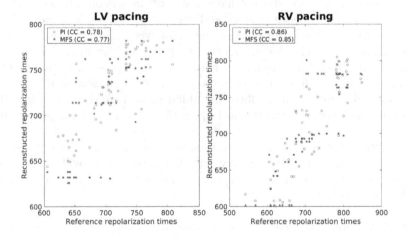

Fig. 10. Scatter plot of the repolarization times for experimental data.

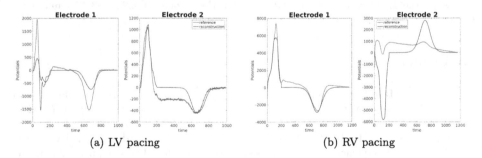

Fig. 11. Reconstructed potentials for experimental data. Electrode 1: close to the heart, Electrode 2: right hip. Red line: reference BSPM; blue line: reconstructed BSPM (Color figure online)

the signals was well reproduced thanks to the optimized amplitude $\mathcal{A}^\star$. We did not have any inversions of the T wave. The chosen electrodes were representative for all the electrodes.

4　Conclusion

We presented a parameter optimization method to solve the inverse problem of electrocardiography. Our method relies on a parameterization of the AP. Our main objective was to develop a method that gives precise information on the repolarization phase, in addition to information on the depolarization phase. Moreover, this approach gives access to all the properties of a local AP, like APD90.

We had to make some choices to ensure a good convergence and avoid over-fitting: constant amplitude and slopes; split the identification process in two steps. Compared to the MFS, our PI gave better activation maps: we obtained a better range of ATs and did not have any artificial discontinuities in the distribution of the ATs on both *in silico* and experimental data. The method fitted the repolarization phases quite accurately. However, having a good fit on the repolarization time could hide an error in the plateau phase duration. Indeed an error in the AT would lead to an error in the plateau phase duration to fit the repolarization phase. Reconstructed torso potentials were close to the measured ones. Especially, optimized amplitude enabled to fit BSPM amplitudes. Moreover, both depolarization and repolarization phases were well caught on all the torso.

To improve the quality of the results, we plan to add either the septum or the endocardium into the geometry. Another development will be to consider piecewise constant amplitudes and repolarization slopes, instead of constant over the whole epicardium.

Acknowledgments. This study received financial support from the French Government as part of the "Investments of the Future" program managed by the National Research Agency (ANR), Grant reference ANR-10-IAHU-04. This work was granted access to the HPC resources of TGCC under the allocation x2016037379 made by GENCI.

References

1. Wang, Y., Rudy, Y.: Application of the method of fundamental solutions to potential-based inverse electrocardiography. Ann. Biomed. Eng. **34**, 1272–1288 (2006)
2. Liu, Z., Liu, C., He, B.: Noninvasive reconstruction of three-dimensional ventricular activation sequence from the inverse solution of distributed equivalent current Density. IEEE Trans. Med. Imag. **25**, 1307–1318 (2006)
3. van Oosterom, A., Oostendorp, T.: On computing pericardial potentials and current densities in inverse electrocardiography. J. Electrocardiol. **25S**, 102–106 (1993)

4. van Oosterom, A., Jacquemet, V.: A parameterized description of transmembrane potentials used in forward and inverse procedures. Int. Conf. Electrocardiol. **6**, 5–8 (2005)
5. Labarthe, S.: Modélisation de l'activé électrique des oreillettes et des veines pulmonaires. Thesis, University of Bordeaux (2013)
6. Malmivuo, J., Plonsey, R.: Bioelectromagnetism: Principles and Applications of Bioelectric and Biomagnetic Fields. Oxford University Press, USA (1995)
7. Macfarlane, P.W., van Oosterom, A., Pahlm, O., Kligfield, P., Janse, M., Camm, J.: Comprehensive Electrocardiology. Springer, London (2010)
8. ten Tusscher, K.H.W., Noble, D., Noble, P.J., Panfilov, A.V.: A model for human ventricular tissue. Am. J. Physiol. H. **286**, H1573–H1589 (2004)
9. Potse, M., Dubé, B., Vinet, A.: Cardiac anisotropy in boundary-element models for the electrocardiogram. Med. Biol. Eng. Comput. **47**(7), 719–729 (2009)

Sparse Bayesian Non-linear Regression for Multiple Onsets Estimation in Non-invasive Cardiac Electrophysiology

Sophie Giffard-Roisin[1(✉)], Hervé Delingette[1], Thomas Jackson[2],
Lauren Fovargue[2], Jack Lee[2], Aldo Rinaldi[2], Nicholas Ayache[1], Reza Razavi[2],
and Maxime Sermesant[1(✉)]

[1] Université Côte d'Azur, Inria, Nice, France
{sophie.giffard-roisin,maxime.sermesant}@inria.fr
[2] Department of Biomedical Engineering, King's College London, London, UK

Abstract. In the scope of modelling cardiac electrophysiology (EP) for understanding pathologies and predicting the response to therapies, patient-specific model parameters need to be estimated. Although personalisation from non-invasive data (body surface potential mapping, BSPM) has been investigated on simple cases mostly with a single pacing site, there is a need for a method able to handle more complex situations such as sinus rhythm with several onsets. In the scope of estimating cardiac activation maps, we propose a sparse Bayesian kernel-based regression (relevance vector machine, RVM) from a large patient-specific simulated database. RVM additionally provides a confidence on the result and an automatic selection of relevant features. With the use of specific BSPM descriptors and a reduced space for the myocardial geometry, we detail this framework on a real case of simultaneous biventricular pacing where both onsets were precisely localised. The obtained results (mean distance to the two ground truth pacing leads is 18.4 mm) demonstrate the usefulness of this non-linear approach.

Keywords: ECG imaging · Personalisation · Relevance Vector Machine · Cardiac electrophysiology

1 Introduction

Modelling cardiac electrophysiology (EP) can help in understanding pathologies and predicting the response to therapies such as cardiac resynchronization therapy (CRT). However estimating accurately patient-specific model parameters is then crucial, and it often involves invasive measurements [1]. In order to replace these invasive measurements -risky for the patient-, Giffard-Roisin et al. [2] personalised the cardiac EP model from body surface potential mappings (BSPM). The onset activation location and the global conduction velocity were estimated in different pacing locations from several patients. However, personalisation may often be needed in more complex situations, such as multiple

© Springer International Publishing AG 2017
M. Pop and G.A. Wright (Eds.): FIMH 2017, LNCS 10263, pp. 230–238, 2017.
DOI: 10.1007/978-3-319-59448-4_22

activation onsets, heterogeneous myocardial tissue (scar) or a particular pathology. Estimation of heterogeneous myocardial conduction using a Bayesian framework has been recently explored by Dhamala et al. [3], but the other parameters such as the onset are supposed to be known, and the uncertainty on the result is not estimated. The contributions of this work are the extension of a cardiac EP personalisation in order to handle non-linear situations (from single to multiple onsets) and to acquire information relative to the confidence on the results. The methodology, based on a relevance vector regression and on a myocardial shape dimension reduction, is tested and compared with the method from [2] on a real biventricular pacing dataset. The methodology is summarised in Fig. 1.

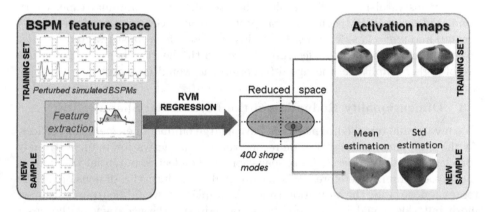

Fig. 1. Pipeline: The RVM (relevance vector machine) regression is performed between BSPM features and a reduced shape space where the patient-specific training activation maps were projected. For a new sample (BSPM data) we estimated its activation map and the regression error. In this paper, the onset locations were further extracted from this estimated activation map.

2   Materials and Methods

2.1   Clinical Data

In this study, we considered a patient dataset composed of BSPM signals, ventricular myocardial geometry (since only the QRS complex is studied, the atria were not included in the model), torso leads and pacing leads locations. The BSPM potentials (from a CardioInsight jacket) were acquired during the procedure at a sampling rate of 1 kHz, with a number of torso sensors of 205. The protocol of this study was approved by the local research ethics committee. The approximated myocardial surface as well as the location of the torso sensors and the pacing leads were extracted from 3D imaging (CT scanner). On a rigidly registered generic and volumetric mesh, cardiac fiber orientations were estimated with a rule-based method (elevation angle between $-70°$ to $70°$).

2.2   Non-invasive Personalisation of a Cardiac EP Model

We used a fast forward EP model derived from the anisotropic Mitchell-Schaeffer cardiac model and a current dipole formulation for computing simultaneously the cardiac electrical sources and body surface potentials [2]. From the generation of a large database of simulated transmural myocardial potentials and torso signals, we learned patient-specific parameters of the EP cardiac model. In order to retain the important aspects of the QRS complex from the BSPM signals, specific descriptors on the normalized BSPM signals were extracted (such as timings, area under the curve, sign of extremum). Specifically, the location of the onset activation was estimated by firstly regressing the activation times of the transmural myocardial mesh, before localizing the minimal time. Every simulated set of BSPM signals v_i was perturbed by a Gaussian random noise of mean 0 and $std = 2e - 3 \times norm(v_i)$. This was done for robustness and in order to give more confidence to electrodes closer to the heart: where the potentials are of higher amplitude, the signal-to-noise ratio would be larger.

2.3   Dimensionality Reduction of the Myocardial Shape

The myocardial tetrahedral mesh can have a large number of elements or vertices. At the same time, the signal to be reconstructed, the activation map, is strongly correlated spatially due to the propagation of the electric potential throughout the myocardium. Therefore, it is meaningful to reduce the dimension of the regression variable, the activation times. A simple way would be to use a coarser mesh but this would be at the expense of reducing the accuracy of the onset locations. Instead, we propose to use a hierarchical decomposition of the mesh, naturally provided by the eigenmodes of a structural matrix. To this end we chose the eigen-decomposition of the stiffness matrix associated with the Laplacian operator. This decomposition has been widely used in various spectral shape analysis [4,5] and is closely related to the modes of vibration of the myocardium. The extracted eigenvectors are naturally sorted by ascending order to spatial frequency. By selecting the first few eigenmodes, we only keep the large spatial variations. If we call $\mathbf{t}$ the vector of N activation times at each vertex of the myocardial mesh, we get the following reduction and reconstruction formulas:

$$\mathbf{t}_{red} = \mathbf{V}_M \mathbf{t} \; ; \; \mathbf{t}_{rec} = \mathbf{V}_M{}^T \mathbf{t}_{red}$$

with $\mathbf{t}_{red}$ the coordinates of $\mathbf{t}$ in the reduced space, $\mathbf{V}_M$ the $N \times M$ matrix of the first M eigenvectors of the stiffness matrix, and $\mathbf{t}_{rec}$ the reconstructed activation times. The matrix V_M is independent of $\mathbf{t}$ and is thus computed only once. An example of reconstructed activation map (on $14\,\mathrm{K}$ vertices) using 400 modes is shown in Fig. 2(c). From Fig. 2(a), we can see that the mean reconstruction error is less than $2\,\mathrm{ms}$ (max: $7\,\mathrm{ms}$) for 400 modes.

2.4   Parameter Estimation Using Relevance Vector Regression

In order to regress the myocardial activation times from the BSPM features, we use the Relevance Vector Machine (RVM) regression method [6]. This approach

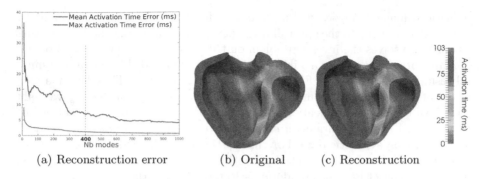

(a) Reconstruction error   (b) Original   (c) Reconstruction

Fig. 2. Example of reconstruction of an activation map (on 14430 vertices) from the eigenvectors of the stiffness matrix: (a) Reconstruction error wrt. the number of modes (b) original activation map (c) reconstructed activation map from 400 modes.

performs sparse kernel regression based on a sparsity inducing prior on the weight parameters within a Bayesian framework. Unlike the commonly used Elastic-Net or Lasso approaches (based on L1 Norm a.k.a Laplacian prior), the RVM method does not require to set any regularization parameters through cross-validation. Instead, it automatically estimates the noise level in the input data and performs a trade-off between the number of basis (complexity of the representation) and the ability to represent the signal. Furthermore, unlike SVM regression or Elastic-Net, it provides a posterior probability of each estimated quantity which is reasonably meaningful if that quantity is similar to the training set. In our setting, we used Gaussian kernels for the non-linear regression whose variance of the kernel parameters need to be defined. The RVM regression only selects the input BSPM feature set that can best explain the activation map in the training set, thus limiting the risk of overfitting. RVM is a multivariate but single-valued approach and therefore the regression was directly performed on the reduced space of Sect. 2.3: only 400 regressions are needed to perform an estimation of the 14 K activation times. We used a Gaussian radial basis function with a kernel bandwidth of 1e4 (from cross-validation). On an EliteBook Intel Core i7, a regression of 1000 training samples and 1235 features runs in 40 s.

3 Application to the Personalisation of a Simultaneous Biventricular Pacing

3.1 Simultaneous Biventricular Pacing Personalisation

We consider here a non-ischaemic implanted CRT patient that underwent a pacing lead optimization procedure with a BSPM device. In particular, the 205 working torso electrodes recorded a biventricular simultaneous pacing. The anatomy as well as the location of the torso sensors and the pacing leads were extracted from CT images. Because of important artifacts coming from the pacemaker, only a coarse epicardial geometry of the myocardium is visible. That is why a generic volumetric myocardial mesh of roughly 65 K tetrahedra was manually registered and scaled to the epicardial geometry. The CardioInsight Technologies software also solves the inverse problem on this epicardial surface (based on the standard formulation using a Tikhonov regularization and the generalized minimal residual algorithm), see Fig. 3(a). In this work, we used this activation map as part of the evaluation of our method. From this simultaneous pacing, we want to retrieve the two onset locations. Locating both activation onset locations by means of an EP model parameter estimation has first been studied by He et al. [7], but only on synthetic data. For this particular goal, our simulated training set was composed of 1000 simulations with fixed conduction velocity (0.5 m/s) and with two onset locations randomly selected on the surface of the myocardial mesh (endocardium and epicardium).

3.2   Results

We compare here the results of the activation map regression of (i) the Kernel Ridge Regression provided by [2], (ii) the RVM regression independently on each point of the cardiac mesh and (iii) the RVM regression using the reduced shape space (Fig. 3). The two ground truth onset locations (pacing sites) are the red dots. On top is the flattened representation of the left ventricular endocardium, where the apex is at the center [8]. The activation map provided by the

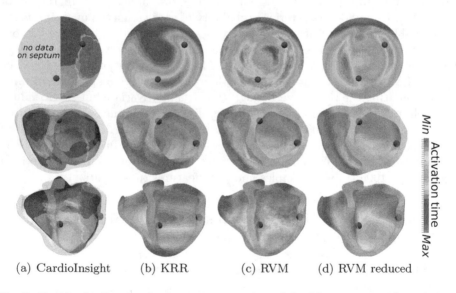

    (a) CardioInsight   (b) KRR   (c) RVM   (d) RVM reduced

Fig. 3. Results for locating both onsets: ground truth lead locations are the red dots. (a) CardioInsight inverse solution (b) kernel ridge regression activation map result (c) relevance vector regression separately for the 14 K vertices (d) relevance vector regression on 400 modes of the reduced space. The color maps are showing short activation times (red) to large ones (purple). On top: flattened representation of the left endocardium in the conventional orientation (i.e. anterior on top). (Color figure online)

CardioInsight system on the coarse surface mesh is shown in Fig. 3(a). Although this map is not precise and cannot be used as a ground truth, it gives an idea of the wave shape. We can see that the kernel ridge regression acts as a blurry mixture of both initializations, not able to separate them. The RVM regression performed on each vertex captures the two onset zones, but the resulting activation maps are noisy and the two minimal values are not easy to capture. The error distances to the true pacing leads are 31.7 mm (left lead) and 41.2 mm (right lead). On the other hand, the RVM performed on the reduced space captures two onset locations with an error distance of less than 23 mm for both leads, while having a smooth solution. The mean number of relevance vectors retained was 213 (out of 1000 samples).

The RVM provides the result as a Gaussian probability distribution, where the retained solution of each regression was the mean. By looking at the estimated variance across each mode, we can estimate the confidence on our result. On Fig. 4 are plotted the projections on the first spectral modes of the simulated activation maps (training set) as well as the estimated solution. We have also represented the confidence interval (± standard deviation, std). First, we can see that the retained sample lies inside the training set point cloud, which is important for RVM to perform well. We have projected the point-wise estimated std onto the myocardial mesh (Fig. 5(a)) where we can see that all the vertices from the left ventricle have an estimated std below 17 ms (and below 14 ms for the septum). Although higher values (max: 28.9 ms) are found in the right ventricle, we can notice that the estimated activation times from this region were greater than 100 ms.

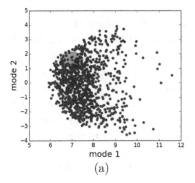 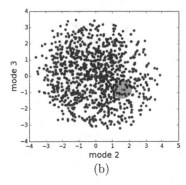

(a)	(b)

Fig. 4. Confidence interval in the reduced space (± std): (a) projection on modes 1 and 2 (b) projection on modes 2 and 3. blue dots: projection of the 1000 simulated activation maps used for training. Red dot: estimated activation map. (Color figure online)

However, the confidence on the activation times is not sufficient to quantify the onset location error. This is why we additionally randomly sampled 10 K coordinates from the estimated Gaussian distribution in the reduced space (see red ellipsoid from Fig. 4). We then reconstructed the activation maps for each

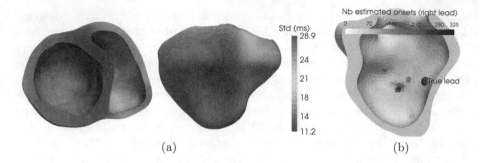

Fig. 5. (a) Standard deviation (std) of the estimated activation time, in milliseconds. All the points in the left ventricle have an estimated std below 17 ms. (b) Locations of the right onset when randomly sampling the estimated Gaussian distribution of the reduced space (10 K samples).

one of them and retrieved their minima as estimations of left and the right onsets (see Fig. 5(b) for the right onset). For the most probable location, the distance error to the ground truth location was 12.3 mm for the left lead and 24.5 mm for the right lead. The standard deviation of the error was 18 mm for the left onset and 11.4 mm for the right onset. We can notice that the left onset was located more accurately, however with less confidence.

4   Discussion

The presented method holds several limitations. One current limitation of the RVM regression is the fact that it only performs multivariate regression of a single output scalar value but not of a vector value (unlike KRR for instance) although multivalued versions have been recently proposed [9]. Concerning the reconstruction from the reduced space, we also tested it with sharp edges in the activation map, like scar tissue blocking the activation wave. Despite higher errors of activation times in the scar, the boundaries of the scar were well captured so we hope we could extend our method to ischemic patients.

The estimations of the covariance matrix of the activation map are provided as a by-product of the RVM regression, giving an estimation of the regression error. However, other sources of error may be taken into account in order to have a full apprehension on the confidence of the result. We think it is the reason why the cloud of possible onsets of Fig. 5(b) did not exactly include the true lead. If the torso sensors were well localized by the CT scan, we had to use a generic mesh because no precise myocardial shape was available. We would like to test our method on patient-specific myocardial meshes which will also help in modelling the error source when using generic meshes. Regarding the errors coming from the BSPM data, they were considerably reduced by the fact that the disconnected or bad leads were automatically excluded, and that we added some Gaussian noise to our simulated BSPM for robustness. Finally, we should also evaluate the EP modelling errors, for instance by analyzing the discrepancy between the true and simulated BSPM signals.

This new method has been tested on only one patient so more experiments are needed for its validation. As a future work, we want to exploit the generic properties of this approach to robustly estimate additional cardiac parameters, like the (local) conduction velocity for instance.

5 Conclusion

We presented a new methodology for multiple onset estimation from BSPM and applied it to one clinical dataset. This was integrated in a novel framework for the personalisation of cardiac EP parameters from BSPM data. This method relies on the generation of a simulated patient-specific database on which a relevance vector regression estimates the activation map from a new set of BSPM signals. As input, specific shape-related features were extracted from the BSPM. As output, the activation map was projected onto a reduced space defined from myocardial shape oscillations. This pipeline enables the estimation of parameters in more complex situations, as the example presented here with the location of two onsets from a real biventricular pacing BSPM sequence. We think this method could be useful for a generalization of cardiac EP personalisation, one of the advantages being the confidence on the regression provided by the RVM.

Acknowledgments. The research leading to these results has received funding from the Seventh Framework Programme (FP7/2007-2013) under grant agreement VP2HF n°611823.

References

1. Sermesant, M., Chabiniok, R., Chinchapatnam, P., et al.: Patient-specific electro-mechanical models of the heart for the prediction of pacing acute effects in crt: A preliminary clinical validation. Med. Image Anal. **16**(1), 201–215 (2012)
2. Giffard-Roisin, S., Jackson, T., Fovargue, L., Lee, J., Delingette, H., Razavi, R., Ayache, N., Sermesant, M.: Non-invasive personalisation of a cardiac electrophysiology model from body surface potential mapping. IEEE Trans. Biomed. Eng. (2016). doi:10.1109/TBME.2016.2629849
3. Dhamala, J., Sapp, J.L., Horacek, M., Wang, L.: Spatially-adaptive multi-scale optimization for local parameter estimation: application in cardiac electrophysiological models. In: Ourselin, S., Joskowicz, L., Sabuncu, M.R., Unal, G., Wells, W. (eds.) MICCAI 2016. LNCS, vol. 9902, pp. 282–290. Springer, Cham (2016). doi:10.1007/978-3-319-46726-9_33
4. Reuter, M., Wolter, F., Peinecke, N.: Laplace-spectra as fingerprints for shape matching. In: ACM Symposium on Solid and Physical Modeling (2005)
5. Umeyama, S.: An eigendecomposition approach to weighted graph matching problems. IEEE Trans. Pattern Anal. Mach. Intell. **10**(5), 695–703 (1988)
6. Tipping, M.E., Faul, A.C.: Fast marginal likelihood maximisation for sparse Bayesian models. In: AISTATS (2003)

7. He, B., Li, G., Zhang, X.: Noninvasive three-dimensional activation time imaging of ventricular excitation by means of a heart-excitation model. Phys. Med. Biol. **47**(22), 4063 (2002)
8. Soto-Iglesias, D., Butakoff, C., Andreu, D., Fernández-Armenta, J., Berruezo, A., Camara, O.: Integration of electro-anatomical and imaging data of the left ventricle: an evaluation framework. Med. Image Anal. **32**, 131–144 (2016)
9. Le Folgoc, L., Delingette, H., Criminisi, A., Ayache, N.: Sparse Bayesian registration of medical images for self-tuning of parameters and spatially adaptive parametrization of displacements. Med. Image Anal. **36**, 79–97 (2017)

Estimation of Local Conduction Velocity from Myocardium Activation Time: Application to Cardiac Resynchronization Therapy

Thomas Pheiffer[1], David Soto-Iglesias[2,3], Yaroslav Nikulin[1],
Tiziano Passerini[1], Julian Krebs[1], Marta Sitges[3], Antonio Berruezo[3],
Oscar Camara[2], and Tommaso Mansi[1(✉)]

[1] Siemens Medical Solutions, Medical Imaging Technologies, Princeton, NJ, USA
tommaso.mansi@siemens.com
[2] Universitat Pompeu Fabra, Barcelona, Spain
[3] Cardiology Department, Thorax Institute, Hospital Clinic, Barcelona, Spain

Abstract. As models of cardiac electrophysiology (EP) are maturing, an increasing effort is being put in their translation to the bed side, in particular for abnormal cardiac rhythm diagnosis and therapy planning. However, the parameters that govern these models need to be estimated from noisy and sparse clinical data in an efficient and precise way, which is still an unsolved challenge. Invasive cardiac mapping provides the richest EP information available today. This paper proposes a new method to estimate a local map of electrical conductivities of the bi-ventricular heart by applying the back-propagation error concept, widely used in neural networks. The method works when either endocardial or epicardial activation time maps are available, and can cope with heterogeneous cardiac tissue. The method was evaluated on synthetic data, showing significantly increased performance in goodness of fit compared to a global parameter estimation approach. The resulting predictive power of the personalized model for cardiac resynchronization therapy was then assessed on 16 swine models of left bundle branch block with rich imaging and EP data before and after CRT. With the proposed personalization, the average error in activation time post CRT was 10 ± 4.5 ms, lower than the observed pre/post-CRT difference of 26.3 ± 16.8 ms.

1 Introduction

Computational models of cardiac electrophysiology (EP) are reaching a level of maturity that enables the development of new tools to support clinical management of cardiac rhythm diseases. Application to atrial or ventricular arrhythmias [1], bundle branch blocks [12], ablation therapy [2] and cardiac resynchronization therapy (CRT) [10] have been explored. Yet, one crucial challenge that still needs to be addressed is the efficient, robust and precise estimation of the parameters that govern the equations associated to EP models, so as to individualize them and capture the specific patho-physiology of the patient under

© Springer International Publishing AG 2017
M. Pop and G.A. Wright (Eds.): FIMH 2017, LNCS 10263, pp. 239–248, 2017.
DOI: 10.1007/978-3-319-59448-4_23

consideration [3]. Due to the sparsity and noisy nature of clinical data (imaging, ECG or, in the best case, electro-anatomical mappings (EAM)), not all the parameters can be observed and assumptions are required. Also, due to the computational burden, only a few parameters are estimated and uniformity assumptions, or even default values determined from animal experiments, are often used [5,6,11].

This paper focuses on the estimation of local, point-wise conduction velocities within the bi-ventricular myocardium from activation maps. Several approaches have been proposed in the past, including coarse-to-fine optimization [8] or with uncertainty quantification [15]. A comprehensive review can be found in [3]. Yet, the local estimation of the conduction velocity remains a challenge due to the high number of unknown variables and computational demand.

Inspired by neural network theories [4], we propose a novel, back-propagation technique to estimate the electrical conductivity along the edges of a volumetric mesh representing the bi-ventricular myocardium (Sect. 2). The approach assumes a front propagation without re-entry and the availability of at least one of endo- and epi-cardial maps of local activation times (LAT). Given measured LAT and a graph-based model of cardiac EP, the errors between simulated and measured LATs are iteratively propagated back with respect to the front-wave to adjust the conduction velocities of every edge of the mesh. The approach copes with different tissue types and anisotropy. The method was verified on synthetic data (Sect. 3). The algorithm performance in predicting the electrical response of CRT was then evaluated on 16 comprehensive swine datasets, showing promising generalization performance. Section 4 concludes the paper.

2   Methods

2.1   Forward Model of Cardiac Electrophysiology

Anatomical Model. The anatomical model is estimated following the framework described in [6]. In brief, machine learning algorithms are employed to efficiently segment the left (LV) and right (RV) ventricle endocardia and LV epicardium from cine MRI. The surfaces are then fused together and a volumetric tetrahedral mesh created. By leveraging the point-correspondences of the segmented meshes, myocardium fibers and mesh tags (LV/RV septum, LV/RV endocardium) are defined automatically (Fig. 1).

EP Model. The resulting volumetric mesh defines the computational domain for solving the EP equations to get point-wise LATs. This work relies on a graph-based EP model, termed GraphEP, in which the LAT at each point of the mesh is calculated given LV, RV septal and device activation points using a shortest-path algorithm adapted to the EP use case [6,14,15]. A generalized edge weight is calculated such that the conduction velocity along each edge takes into account the different tissues it traverses and the local anisotropy. Let $\mathbf{p}_i$ and $\mathbf{p}_j$ be two connected mesh points. The generalized edge weight w_{ij}, measured in seconds, corresponds to the time needed for the action-potential to travel from $\mathbf{p}_i$ to $\mathbf{p}_j$:

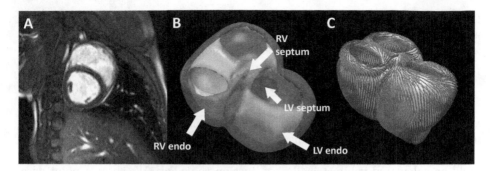

Fig. 1. Left: MR image volume from which the cardiac structures are segmented. Middle: segmentation of the heart structures as a transparent mesh. Right: anatomical model used for the computation, with color lines representing myocardium fibers.

$w_{ij} = l_{ij}/c_{ij}$, where $l_{ij} = \sqrt{(\mathbf{e}_{ij}^T D \mathbf{e}_{ij})}$, $\mathbf{e}_{ij} = \mathbf{p}_i - \mathbf{p}_j$, D is the anisotropy tensor defined as $D = (1 - r)\mathbf{f}_{ij}\mathbf{f}_{ij}^T + rI$, $\mathbf{f}_{ij}$ is the fiber direction along the edge and r the anisotropy ratio ($r = 0.3$). c_{ij} is the apparent conduction velocity in m/s along the edge approximated linearly from the conduction velocity c_i and c_j and the different tissue types the edge traverses. In other words, the EP model is essentially a Dijkstra shortest path propagation of activation time along the mesh edges, in which the cost is controlled by the parameters described here.

EP Activation Model. Intrinsic cardiac stimulation is modeled by an instantaneous activation of the LV and RV septum, to mimic the effects of the His bundle. In terms of the Dijkstra graph model, this means setting the point-wise activation time to zero on these mesh regions. Fast activation from the Purkinje network is modeled assigning different conduction velocity to a smooth, thin layer of nodes distributed all over the endocardial surfaces of the LV and RV (c_{LV} and c_{RV}). The thickness of this fast conducting layer is set to 3 mm to model swine Purkinje system, which goes deeper within the myocardium than in humans. With these conditions and material properties set, the shortest-path propagation is calculated to obtain the full EP activation across the heart.

2.2 Local Estimation of Conduction Velocities

Electro-anatomical mapping data (EAM) is integrated with the MRI surface using a quasi-conformal mapping technique (QCM). QCM takes advantage of the existence of a homeomorphism between the LV endocardial surface and a 2D disk. By mapping both MRI and EAM surfaces to the same disk we can easily establish a piecewise linear homeomorphism between the two surfaces, as in [13].

The personalization algorithm is initialized by estimating three conduction velocities (c_{LV}, c_{RV} and c_{Myo}) that minimize the sum of squared distances (SSD) between measured and simulated LATs at the points where measurement is available. These three values are the velocity in the Purkinje layer of the left ventricle, in the Purkinje layer of the right ventricle, and in the myocardium, respectively.

This initialization step is hereafter referred to as the global personalization step, as these conduction velocities are merely estimated as a single value for each particular tissue region. Formally, the SSD objective is defined as:

$$D = \sum_{i=0}^{N}(m_i - c_i)^2 \tag{1}$$

where m_i is the measured activation time at point i, and c_i is the calculated activation time at point i. The distance is calculated over all the N points where measured activation time data is available. Note that the conduction velocities could be calculated volumetrically, e.g. just on the edges. We chose a point-wise implementation primarily for ease of integration in the rest of our data processing workflow, particularly for visualization purposes. Point-wise conductivities also allow modeling an apparent conductivity when the edge crosses different tissue types. A standard, trust-region technique is employed [7]. Next, the generalized edge weights are estimated w_{ij}. Let $\mathcal{L}$ be the loss between measured and computed LATs, $\mathcal{L} = \sum_{i=0}^{N}(t_{\mathrm{m},i} - t_{\mathrm{c},i})^2$, where N is the total number of points where measurement is available, $t_{\mathrm{m},i}$ and $t_{\mathrm{c},i}$ are the measured and computed activation times at point i, respectively. If the EP wave propagates without re-entries, one can see the mesh nodes arranged in layers approximatively parallel to the EP iso-chrones, similar to a neural network layer. The input layer is thus the set of activation points, and the output layer can be defined as the set of nodes where LAT measurement is available, to enable back-propagation parameter estimation. Thus after the initial global personalization, we seek to personalize the point-wise conductivities based on back-propagation along the edges until the error at the data points reaches a convergence tolerance. The edges which are not reached by the backpropagation algorithm keep their conductivity values assigned during the global personalization. A variant of gradient descent with step α is used to adjust the generalized edge weights w_{ij} so as to minimize $\mathcal{L}$: $w'_{ij} = w_{ij} - \alpha \frac{\partial \mathcal{L}}{\partial w_{ij}}$. Assuming that the LAT at point j is known, the partial derivative can be written as:

$$\frac{\partial \mathcal{L}}{\partial w_{ij}} = \frac{\partial \mathcal{L}}{\partial t_{\mathrm{c},j}} \frac{\partial t_{\mathrm{c},j}}{\partial w_{ij}} \tag{2}$$

As a property of the backpropagation framework implemented on the mesh graph, the edge weight w_{ij} only effects the overall loss function $\mathcal{L}$ based on its affect on the output at point j. Note that this statement follows from the perspective of the personalization step, in which the errors are propagated backwards rather than forward. Thus, Eq. 2 writes:

$$\frac{\partial \mathcal{L}}{\partial w_{ij}} = -2(t_{\mathrm{m},j} - t_{\mathrm{c},j})\frac{\partial t_{\mathrm{c},j}}{\partial w_{ij}}$$

Exploiting the fact that the solution is a shortest path solution, the second term becomes:

$$\frac{\partial t_{\mathrm{c},j}}{\partial w_{ij}} = \begin{cases} 1 & \text{if } \arg\min_{k \in \text{neighbors}(j)}(t_{\mathrm{c},k} + w_{kj}) = i \\ 0 & \text{otherwise} \end{cases}$$

If the LAT at point j is unknown, Eq. 2 is unrolled to the previous layers through the chain rule:

$$\frac{\partial \mathcal{L}}{\partial t_{c,j}} = \sum_{k \in \text{neighbors}(j)} \frac{\partial \mathcal{L}}{\partial t_{c,k}} \frac{\partial t_{c,k}}{\partial t_{c,j}}$$

with

$$\frac{\partial t_{c,k}}{\partial t_{c,j}} = \begin{cases} 1 & \text{if arg min}_{i \in \text{neighbors}(k)} (t_{c,i} + w_{ki}) = j \\ 0 & \text{otherwise} \end{cases}$$

and so on until the activation points are reached.

Once the edge weights are estimated, the point-wise conduction velocities c_i are derived. When both edge extremities belong to the same tissue type, c_i and c_j are trivially obtained. In the case they belong to different tissue types, the problem becomes ill-posed. To address this challenge, we assume the ratio of conduction velocities between different tissue types stays constant throughout the heart. The ratios are obtained directly from the initial personalization (c_{LV}/c_{RV}, c_{LV}/c_{Myo} and other combinations, including scar and border zone when available). As an example, let point i belongs to LV endocardium and point j belongs to myocardium, L the distance between i and j, l the distance between i and the boundary between tissue types, c the estimated apparent conductivity along this edge and ρ the ratio c_j/c_i. In this case, we have:

$$c_i = \frac{\rho l + (L - l)}{\rho L} c, \ c_j = \rho c_i \tag{3}$$

Because each edge is processed independently, one vertex may have several estimated conduction velocities. In this case, the average value is taken and clamped within a physiological range for stability.

3 Experiments and Results

Data Acquisition Protocol. 16 pigs (average weight $= 34$ $(30/35)$ kg) were studied, a sub-set of a larger database [9]. A left bundle branch block (LBBB) was induced in all animals using radio-frequency ablation. A cardiac resynchronization therapy (CRT) device was then positioned in the animal model. The RV lead was placed at the apex while the LV lead was positioned through a lateral or antero-lateral position of the LV epicardial surface. All pigs received an MRI a week before the experiment. LV endocardial and epicardial EAMs were acquired on the day of the experiment after LBBB (hereafter referred to as baseline), and with CRT pacing. An example of the data from these pig experiments is shown in Fig. 3. In all our experiments, both endocardial and epicardial EAMs were used. With respect to the gradient method, in all cases we used a stopping criteria of 0.01 ms in the activation time fitting. The maximum number of iterations allowed was 30 for the dense GraphEP personalization. All cases were observed to converge within this number of iterations.

Verification on Synthetic Data. The proposed GraphEP personalization method was first evaluated on synthetic data with known conductivity. The forward model was used to calculate "ground-truth" point-wise LAT on the endocardial and epicardial surface. Nine different distribution patterns of conductivity and LAT were generated (Fig. 2). First, conduction velocities were set within the endocardium layer prescribing the Purkinje network by (1) linearly varying conductivity values along the X, Y and Z axis of the anatomy within the interval [1.5, 4.0] m/s (3 models), (2) random velocities per American Heart Association (AHA) regions, within the same range (3 models), and (3) 10 random regions defined using region growing around random seeds (3 models). The endocardial conduction velocity was then propagated throughout the myocardium by assuming a constant Purkinje/myocardial cells ratio of 3. This procedure was repeated on two heart meshes generated from two different pigs, resulting in a total of 18 synthetic datasets. Finally, a virtual CRT was performed by placing an RV and LV electrode at standard location to test the predictive power of the personalized model.

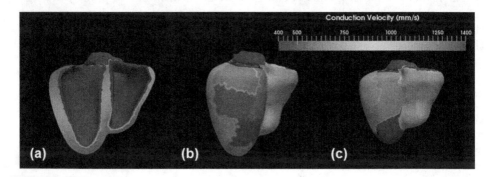

Fig. 2. Conductivity distribution of the synthetic datasets with (a) linearly varying distribution along the Y axis, (b) distribution according to AHA regions, and (c) region-growing regions based on random seeds.

The computed activation times normalized with respect to the simulated QRS duration is reported for both global and local personalization in Table 1. The proposed estimation method effectively estimated the ground truth conductivities, yielding an LAT error of less than 1 ms in average, compared with 4 ms for a global personalization. In all experiments, over-fitting was not severe since the CRT prediction was also significantly more accurate with locally estimated conductivities (1.3 ± 0.4 vs. 6.2 ± 2.6 normalized AT error respectively). Interestingly, the performance of the proposed method was relatively insensitive to the distribution of conduction velocities, as highlighted by the min and max normalized LAT errors calculated as in Eq. 1. (Table 1).

Table 1. LAT errors between computed and ground truth on synthetic datasets, normalized with respect to QRS duration. For both baseline and CRT, the proposed local personalization yielded significant improvements in terms of accuracy (paired t-test, $p < 0.5$, values reported as mean $\pm$ SD (min, max)).

	Global	Local
Baseline	$4.4 \pm 1.4(2.4, 7.2)$	$0.6 \pm 0.2(0.2, 1.2)$
CRT	$6.2 \pm 2.6(1.9, 11)$	$1.3 \pm 0.4(0.8, 2)$

Evaluation on Animal Data for Cardiac Resynchronization Therapy.
The result of the personalization after the global initialization and after the local refinement are compared. Both endo- and epi-cardial measured EAMs were used. The baseline and CRT simulation errors compared to the measurements for all pigs are presented in Table 2, with example visualizations of measured and simulated activation maps shown in Fig. 3. The results confirmed the trend observed in synthetic data. Table 2 shows that the proposed local estimation method consistently resulted in lower mean LAT error across the whole heart and also when considering each ventricle individually, in the case of both baseline rhythm and after CRT. It is also interesting to note the lower standard deviation of LAT errors for the proposed method, which points toward the improved ability to accurately personalize local variations in conductivity across the heart. In both the global and local personalization methods, the residual error was consistently higher for CRT than the baseline. This reflects the complexity of accurately modeling all of the parameters involved in CRT, such as lead placement, which could only be approximated from the EAMs as the center of the early activation areas. It is especially important to note that the QRS length at baseline LBBB was measured to be on average 46.6 ± 18.3 ms, and after CRT was found to be 70.3 ± 14.0 ms (in some animals, the electrodes were not placed optimally). As a result, the simulated LAT errors were below the activation difference in pre/post-CRT.

Computation time. In this experiment, a personalization took less than one minute without code optimization on a tetrahedral mesh of approximately 14,000 vertices using a standard laptop (Intel Core i7-4800MQ 2.70 GHz, 8 GB of RAM), owing to the fast graph-based algorithms.

Table 2. Absolute LAT errors (in ms) for baseline and CRT simulations in the swine dataset. Values reported as mean $\pm$ SD, per ventricle and over the entire bi-ventricular myocardium. Goodness of fit and predictive power significantly improved after local personalization.

	Global			Local		
	WholeHeart	LV	RV	WholeHeart	LV	RV
Baseline	12.2 ± 5.2	8.8 ± 4.0	11.8 ± 5.3	3.7 ± 1.7	4.8 ± 2.0	2.0 ± 0.9
CRT	14.8 ± 6.6	24.4 ± 11.7	25.8 ± 12.0	10.0 ± 4.5	15.4 ± 6.7	17.8 ± 8.1

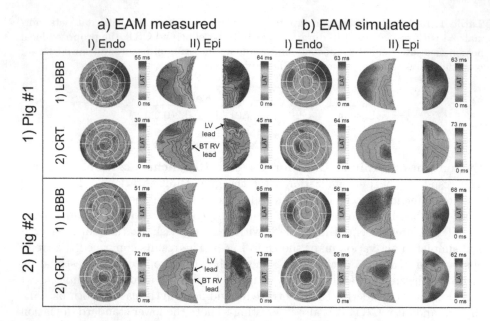

Fig. 3. Example data from two pigs showing the measured EAMs during baseline LBBB rhythm and CRT activation, with the corresponding simulated EAMs using the proposed local personalization of conductivity. The approach yielded qualitatively similar maps at baseline and during CRT, suggesting a good fit to the data.

4   Discussion and Conclusion

This paper presented a novel method to estimate the electrical conduction velocity at every point of the myocardium. The proposed method has a number of advantages. It allows personalization of heart activation patterns to the resolution of the computational domain mesh, while being extremely computationally efficient. The method is also adaptable to varying levels of sparsity in the measured LAT input values. However, one disadvantage of using a variant of back-propagation is the potential over-fitting of the model. Yet, the obtained CRT predictions suggest predictions are still much more precise with the proposed method than with global estimation. Further investigation is needed to evaluate the sensitivity of the method to noise and missing data. Possible future directions are to incorporate prior knowledge of the heart conduction pathways to regularize the personalization, add measurement uncertainty to the estimation process, and further validate the approach.

Disclaimer: This feature is based on research, and is not commercially available. Due to regulatory reasons its future availability cannot be guaranteed.

References

1. Arevalo, H.J., Vadakkumpadan, F., Guallar, E., Jebb, A., Malamas, P., Wu, K.C., Trayanova, N.A.: Arrhythmia risk stratification of patients after myocardial infarction using personalized heart models. Nat. Commun. **7** (2016)
2. Bayer, J.D., Roney, C.H., Pashaei, A., Jaïs, P., Vigmond, E.J.: Novel radiofrequency ablation strategies for terminating atrial fibrillation in the left atrium: a simulation study. Front. Physiol. **7**, 108 (2016)
3. Chabiniok, R., Wang, V.Y., Hadjicharalambous, M., Asner, L., Lee, J., Sermesant, M., Kuhl, E., Young, A.A., Moireau, P., Nash, M.P., et al.: Multiphysics and multiscale modelling, data-model fusion and integration of organ physiology in the clinic: ventricular cardiac mechanics. Interface Focus **6**(2), 20150083 (2016)
4. LeCun, Y., Bengio, Y., Hinton, G.: Deep learning. Nature **521**(7553), 436–444 (2015)
5. Marchesseau, S., Delingette, H., Sermesant, M., Cabrera-Lozoya, R., Tobon-Gomez, C., Moireau, P., I Ventura, R.F., Lekadir, K., Hernandez, A., Garreau, M., et al.: Personalization of a cardiac electromechanical model using reduced order unscented kalman filtering from regional volumes. MedIA **17**(7), 816–829 (2013)
6. Neumann, D., Mansi, T., Itu, L., Georgescu, B., Kayvanpour, E., Sedaghat-Hamedani, F., Amr, A., Haas, J., Katus, H., Meder, B., et al.: A self-taught artificial agent for multi-physics computational model personalization. MedIA **34**, 52–64 (2016)
7. Powell, M.J.: The bobyqa algorithm for bound constrained optimization without derivatives. Cambridge NA Report NA2009/06, University of Cambridge, Cambridge (2009)
8. Relan, J., Chinchapatnam, P., Sermesant, M., Rhode, K., Ginks, M., Delingette, H., Rinaldi, C.A., Razavi, R., Ayache, N.: Coupled personalization of cardiac electrophysiology models for prediction of ischaemic ventricular tachycardia. Interface Focus **1**, 396–407 (2011)
9. Rigol, M., Solanes, N., Fernandez-Armenta, J., Silva, E., Doltra, A., Duchateau, N., Barcelo, A., Gabrielli, L., Bijnens, B., Berruezo, A., et al.: Development of a swine model of left bundle branch block for experimental studies of cardiac resynchronization therapy. JCTR **6**(4), 616–622 (2013)
10. Sermesant, M., Chabiniok, R., Chinchapatnam, P., Mansi, T., Billet, F., Moireau, P., Peyrat, J.M., Wong, K., Relan, J., Rhode, K., et al.: Patient-specific electromechanical models of the heart for the prediction of pacing acute effects in crt: a preliminary clinical validation. MedIA **16**(1), 201–215 (2012)
11. Sermesant, M., Moireau, P., Camara, O., Sainte-Marie, J., Andriantsimiavona, R., Cimrman, R., Hill, D.L., Chapelle, D., Razavi, R.: Cardiac function estimation from mri using a heart model and data assimilation: advances and difficulties. MedIA **10**(4), 642–656 (2006)
12. Sohal, M., Shetty, A., Niederer, S., Lee, A., Chen, Z., Jackson, T., Behar, J.M., Claridge, S., Bostock, J., Hyde, E., et al.: Mechanistic insights into the benefits of multisite pacing in cardiac resynchronization therapy: the importance of electrical substrate and rate of left ventricular activation. Heart Rhythm **12**(12), 2449–2457 (2015)
13. Soto-Iglesias, D., Butakoff, C., Andreu, D., Fernández-Armenta, J., Berruezo, A., Camara, O.: Integration of electro-anatomical and imaging data of the left ventricle: an evaluation framework. MedIA **32**, 131–144 (2016)

14. Wallman, M., Smith, N.P., Rodriguez, B.: A comparative study of graph-based, eikonal, and monodomain simulations for the estimation of cardiac activation times. IEEE Trans. Biomed. Eng. **59**(6), 1739–1748 (2012). http://ieeexplore.ieee.org/document/6178774/
15. Wallman, M., Smith, N.P., Rodriguez, B.: Computational methods to reduce uncertainty in the estimation of cardiac conduction properties from electroanatomical recordings. MedIA **18**(1), 228–240 (2014)

Variance Based Sensitivity Analysis of I_{Kr} in a Model of the Human Atrial Action Potential Using Gaussian Process Emulators

Eugene T.Y. Chang[1,2], Sam Coveney[3], and Richard H. Clayton[1,2(✉)]

[1] Insigneo Institute of In-Silico Medicine, University of Sheffield, Sheffield, UK
r.h.clayton@sheffield.ac.uk
[2] Department of Computer Science, University of Sheffield, Sheffield, UK
[3] Department of Physics and Astronomy, University of Sheffield, Sheffield, UK

Abstract. Cardiac cell models have become valuable research tools, but biophysically detailed models embed large numbers of parameters, which must be fitted from experimental data. The provenance of these parameters can be difficult to establish, and so it is important to understand how parameter values influence model behaviour. In this study we examined how model parameters influence the repolarising current I_{Kr} in the Courtemanche-Ramirez-Nattel model of the human atrial action potential. We used a statistical approach in which Gaussian processes (GP) are used to emulate the model outputs. A GP emulator can treat model inputs and outputs as uncertain, and so can be used to directly calculate sensitivity indices. We found that 3 of the 10 parameters influencing I_{Kr} had a strong influence on APD_{70}, APD_{90}, and Dome V_m. These three parameters scale the magnitude of the I_{Kr} gating variable time constant and the voltage dependence of the steady state activation curve, and these mechanisms act to modify the amplitude of I_{Kr} during repolarisation. This study highlights the potential value of statistical approaches for investigating cardiac models, and that uncertainties or errors in parameters resulting from attempts to fit experimental data during model development can ultimately affect model behaviour.

1 Introduction

Since the introduction of the first model of cardiac cellular electrophysiology over 50 years ago [12], models of electrical activation and recovery in cardiac cells have become important research tools. The present generation of cardiac cell models represent not only changes in transmembrane potential resulting from movement of ions through the cell membrane, but also the diffusion, storage, release, and uptake of Ca^{2+} within the cell [5]. Typically, cardiac cell models are a stiff, nonlinear system of coupled ordinary differential equations, and are solved using a numerical scheme to obtain time series of membrane voltage, intracellular Ca^{2+} concentration, and other quantities of interest.

Each of the equations involve parameters; for equations describing transmembrane current flow through ion channels, pumps, and exchangers these parameters typically include a maximum current density per unit membrane area,

© Springer International Publishing AG 2017
M. Pop and G.A. Wright (Eds.): FIMH 2017, LNCS 10263, pp. 249–259, 2017.
DOI: 10.1007/978-3-319-59448-4_24

and other parameters that regulate the dynamic behaviour of the current. The parameters are fitted from experimental data often following the approach pioneered by Hodgkin and Huxley [6]. Many models take a modular approach to building the full suite of equations representing transmembrane current flow, with re-use of parameters from older models and experiments. The provenance of these parameters is not always easy to establish [11], yet the influence of uncertain parameters on model behaviour is difficult to assess because of model complexity.

Recent studies have begun to address this problem by examining the sensitivity of model outputs such as action potential duration (APD) to variable model parameters [2,16]. These studies have concentrated on maximum conductances of ion channels, but even with this subset the potential parameter space to explore is vast. Although cardiac cell models are relatively cheap to compute, a comprehensive exploration of very high dimensional parameter space remains computationally demanding.

An alternative approach is to build a statistical model (an *emulator* or metamodel), which acts as a fast running surrogate for the original model or *simulator*. This approach has been used to examine models of systems including atmospheric pollution [8] and galaxy formation [17], where the emulator is a Gaussian process (GP) [13]. A particular advantage of a GP emulator is that the simulator parameters, or *inputs*, can be treated as uncertain so that they are represented by a distribution rather than a fixed value. Using Gaussian (normal) distributions allows direct calculation of expected mean and variance of an output given uncertainty in the inputs. The proportion of output variance that is accounted for by variance in each input is then a first order sensitivity index [13]. Furthermore, the main effect of model inputs can be directly calculated, showing how a single input affects an output given specified distributions on the other inputs. This approach has been used to examine cardiac cell models, where inputs were ion channel maximum conductances and outputs were features of the action potential [3,7].

The rapidly inactivating K^+ current I_{Kr} regulates repolarisation in cardiac myocytes, and is an important pharmaceutical target. The aim of the present study was therfore to undertake sensitivity analysis of the I_{Kr} channel in the Courtemanche-Ramirez-Nattel (CRN) model of the human atrial action potential [4].

2   Methods

2.1   CRN Model Inputs and Outputs

The equations describing the I_{Kr} current in the CRN model [4] are given below. The current density I_{Kr} is given by

$$I_{Kr} = \frac{g_{Kr} x_r \left(V_m - E_K \right)}{1.0 + \exp \left[\dfrac{V_m + Kr_1}{Kr_2} \right]}. \tag{1}$$

where the gating variable x_r varies between 0 and 1, and is given by

$$\frac{\mathrm{d}x_r}{\mathrm{d}t} = \frac{x_{r\infty} - x_r}{\tau_{xr}};$$

(2)

where the gating variable time constant τ_{xr} and steady state activation $x_{r\infty}$ depend on transition rates α_{xr} and β_{xr};

$$\alpha_{xr} = Kr_3 \frac{V_m + Kr_4}{1.0 - \exp\left[-\dfrac{V_m + Kr_4}{Kr_5}\right]},$$

(3)

$$\beta_{xr} = Kr_6 \frac{V_m - Kr_7}{\exp\left[\dfrac{V_m - Kr_7}{Kr_8}\right] - 1.0},$$

(4)

$$\tau_{xr} = \frac{1.0}{\alpha_{xr} + \beta_{xr}},$$

(5)

$$x_{r\infty} = \left[1.0 + \exp\left(-\frac{V_m + Kr_9}{Kr_{10}}\right)\right]^{-1}.$$

(6)

Each of the 10 parameters labelled Kr_1 to Kr_{10} appear as numbers without units in the original formulation, and in this study we examined variation in the range $0.5\times$ to $1.5\times$ these values as shown in shown in Table 1. The maximum conductance g_{Kr} was set to a baseline value of $0.0294 \; nS/pF$ and E_K was set to $-86.7653 \; mV$ for fixed intracellular K^+ concentration, as described below.

Table 1. Baseline values for each input, and range over which each input was varied for design data.

Parameter	Baseline	Range
Kr_1	15.0	(7.5–22.5)
Kr_2	22.4	(11.2–33.6)
Kr_3	3.0×10^{-4}	$(1.5-4.5) \times 10^{-4}$
Kr_4	14.1	(7.05–21.15)
Kr_5	5.0	(2.5–3.75)
Kr_6	7.3898×10^{-5}	$(3.6949-11.0847) \times 10^{-5}$
Kr_7	3.3328	(1.6664 –4.9992)
Kr_8	5.1237	(2.5619–7.6856)
Kr_9	14.1	(7.05–21.15)
Kr_{10}	6.5	(3.25– 9.75)

We identified seven outputs that describe features or biomarkers of the action potential, based on previous work [2,3], and these are illustrated in Fig. 1.

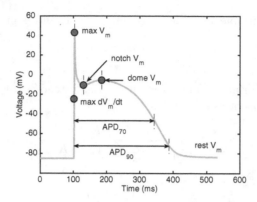

Fig. 1. Action potential features

2.2   Implementation of CRN Model

The CRN model equations were implemented in Matlab (Mathworks, CA), using code automatically generated from the CellML repository (http://cellml.org). The cell models were solved using the Matlab `ode15s` time adaptive solver for stiff systems of ODEs, with tolerances set to 10^{-6}. The CRN model does not properly balance intracellular ion concentrations [18], so we fixed $[Na^+]_i$ and $[K^+]_i$ at baseline values of 11.1 mM and 139.0 mM respectively.

2.3   Emulator Design Data

For fitting and evaluating the GP emulators, we obtained design data comprising a set of 500 simulator runs. For each run, a value for each input was selected from the range shown in Table 1, using Latin hypercube sampling to ensure an even distribution of points in the input space. To achieve a stable action potential duration, each simulator run included 40 stimuli of strength $-2.0\,nA$ and duration 2 ms delivered at a 1000 ms cycle length to represent a resting human heart rate. The final action potential in this sequence was used to obtain the outputs. The distribution of action potentials and I_{Kr} gating dynamics are shown in Fig. 2, where the influence of uncertain inputs on repolarisation is clearly visible.

2.4   Sensitivity Analysis

First order sensitivity indices produced from a GP emulator represent the proportion of total output variance that is accounted for by variance in each input [13]. We assumed that inputs and outputs could be described by a normal distribution, with the mean of each input set to the baseline value and variance set to 0.01 of the range shown in Table 1.

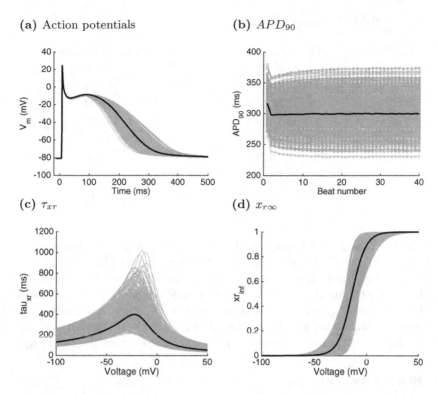

(a) Action potentials

(b) APD_{90}

(c) τ_{xr}

(d) $x_{r\infty}$

Fig. 2. Design data showing (a) final action potential out of a sequence of 40; (b) stabilisation of APD_{90} during the 40 beat sequence; (c) and (d) voltage dependence of steady state activation $x_{r\infty}$ and gating variable time constant τ_{xr} for each set of inputs. Bold lines indicate the model behaviour for baseline values of the inputs, grey lines show design data.

3   Results

3.1   Emulator Fitting and Evaluation

We fitted and evaluated a separate GP emulator for each of the 7 outputs, using an open source Python implementation; details including the URL have been withheld to ensure anonymity of the authors.GP_emu_UQSA (available from http://doi.org/10.5281/zenodo.215521). For each output, the 500 design data were separated into a training set of 450 simulator runs and a test set of 50 simulator runs. Emulator hyperparameters were fitted to the training set with a maximum likelihood approach described previously [3], using the fmin_l_bfgs_b optimisation function available in the Python SciPy package. Fitting was repeated five times to ensure that local maxima were avoided, and the fit with greatest likelihood was selected. The test set was then used to compare outputs predicted by the emulator against those produced by the simulator for the same set of inputs. The difference between emulator and simulator outputs was summarised

Table 2. Fit and evaluation of emulator for each output.

Emulator	Expectation of mean	Expectation of variance	Mahanalobis distance
$dV_m/dtMax.(mV/ms)$	218.13	0.007	63.85
$V_mMax.(mV)$	24.47	5.62×10^{-6}	9.43
$NotchV_m(mV)$	−12.64	0.02	41.92
$DomeV_m(mV)$	−8.62	0.015	43.29
$APD_{70}(ms)$	233.04	59.26	49.99
$APD_{90}(ms)$	300.14	89.69	51.19
$RestingV_m(mV)$	−80.81	0.001	38.12

using the Mahanalobis distance, which for a test set of 50 runs has a reference distribution with mean 50 and standard deviation 10.5 [1] (Table 2).

We considered a Mahanalobis distance between 30 and 70 (*i.e.* ± 2 SD) to indicate a well fitted emulator, and the only emulator that did not meet this criterion was $V_mMax.$. The was not surprising, since I_{Kr} predominantly influences repolarisation, and inspection of the design data showed a change of only ±0.06 mV in $V_mMax.$ arising from inputs varied across the full range.

3.2 Sensitivity Indices

The first order sensitivity indices were obtained using GP_emu_UQSA and are shown in Fig. 3. Each row shows the relative contribution of each input, and

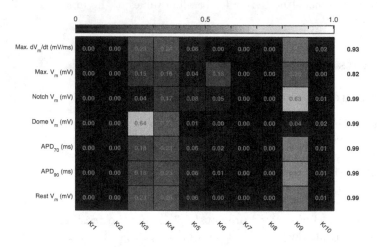

Fig. 3. Sensitivity indices for each combination of input and output. The column to the right is the sum of sensitivity indices for each output. Inputs Kr3 and Kr4 scale the magnitude and voltage dependence of gate activation (Eq. 3), and input Kr9 scales the voltage dependence of steady state activation (Eq. 6).

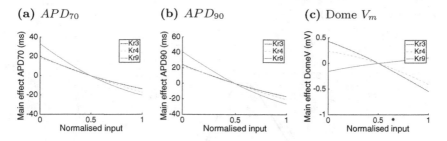

Fig. 4. Main effects of inputs $Kr3$, $Kr4$, and $Kr9$ on (a) APD_{70}, (b) APD_{90}, and (c) dome V_m. Each input is plotted on a normalised scale corresponding to the range in Table 1.

the sum of these contributions is shown in the column to the right of the main figure. For six of the emulators, the sum of sensitivity indices was close to 1.0, indicating that interaction effects are negligible. The sum of sensitivity indices for the $V_m Max.$ emulator was lower, indicating additional variance arising from either the relatively poor fit or possible interactions.

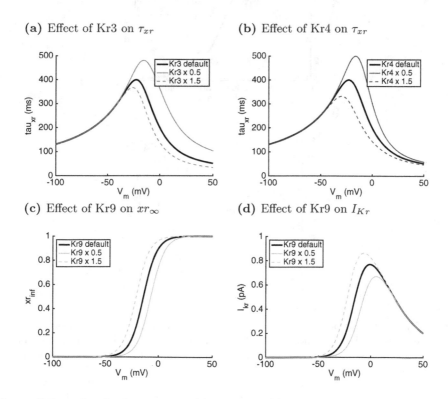

Fig. 5. Effect of multiplying inputs (a) $Kr3$ and (b) $Kr4$ by 0.5 and 1.5 on voltage dependence of τ_{xr}, and effect of multiplying $Kr9$ by 0.5 and 1.5 on (c) xr_∞, and (d) I_{Kr} calculated using steady state values of x_r, xr_∞. Bold lines show baseline model output.

(a)

(b)

(c)

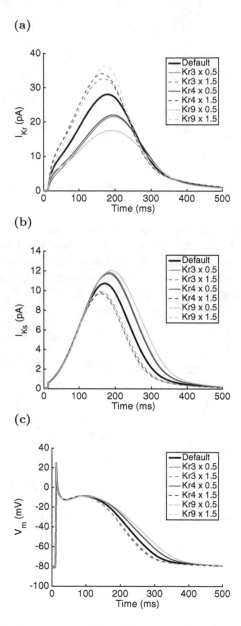

Fig. 6. Effect of multiplying inputs $Kr3$, $Kr4$, and $Kr9$ by 0.5 and 1.5 on the time course of (a) I_{Kr}, (b) I_{Ks}, and (c) V_m. Each plot shows the final action potential of a 40 beat sequence with a cycle length of 1000 ms, and the bold lines show model output for baseline parameter settings.

Three of the inputs had the greatest effect on the outputs: $Kr9$ influenced all outputs except for Dome V_m, $Kr3$ had a strong effect on Dome V_m, and $Kr4$ had an intermediate effect on all outputs. In Fig. 4 we have plotted the main effect (obtained using GP_emu_UQSA) of these three inputs on APD_{70}, APD_{90}, and Dome V_m. The main effect is the change in the expected mean of the emulator output as one input changes while all others are assigned a fixed distribution with mean 0.5 and variance 0.01 (in normalised units). These plots show that increasing all three inputs acts to decrease APD.

3.3   Insight into Model Mechanism

The mechanisms by which $Kr3$, $Kr4$, and $Kr9$ act to modify the action potential shape and duration are shown in Figs. 5 and 6. In Fig. 5 the effect of multiplying $Kr3$, $Kr4$, and $Kr9$ by 0.5 and 1.5 on the voltage dependence of τ_{xr}, xr_∞, and steady state I_{Kr} is shown. An increased $Kr3$ and $Kr4$ resulted in a shorter activation gate time constant, while an increased $Kr9$ resulted in a leftward shift in the voltage dependence of xr_∞, and consequently an increased I_{Kr} at voltages close to 0 mV.

The effects of changes in the gating parameters on I_{Kr}, I_{Ks}, and the action potential are shown in Fig. 6. Increased $Kr3$, $Kr4$, and $Kr9$ all acted to increase the magnitude of I_{Kr}, resulting in more outward current during the plateau phase of the action potential and a shorter APD. The increased outward current led to a more rapid repolarisation, which in turn acted to reduce magnitude of I_{Ks}. The change in I_{Ks} compensated for increased I_{Kr}, but not enough to offset the increase in outward current. The magnitude of other currents was not changed (data not shown).

4   Discussion and Conclusions

In this study we have focussed on how the dynamical behaviour of a single ion channel depends on parameters (or inputs) that are fitted from experimental data. We have used GP emulators to calculate sensitivity indices, and have identified three inputs that have the greatest influence on model outputs. This study builds on previous work that has studied how maximum ion channel conductance influence model outputs [2,3,15,16], and another study that has investigated how the dynamics of I_{Na} influence model behaviour [14]. Taken together, these studies show that tools developed for other modelling communities can be valuable for examining computationally intensive cardiac models, and that uncertainty or errors in fitting cardiac cell model parameters may have an important influence on model behaviour. The present study highlights a number of directions for future research, and these are discussed below.

Several different approaches have been adopted for sensitivity analysis of cardiac models, and these include partial least squares regression [9,16] and a population of models [15]. In this study we chose to construct GP emulators to examine the properties of the I_{Kr} formulation in the Courtemanche model because this approach has already shown promise for analysis of complete cardiac cell models [3,7]. One advantage of GP emulators over other techniques is that the emulator can treat model inputs and outputs as uncertain quantities. Under the assumption that inputs and outputs have a normal distribution then is it possible to directly calculate an output distribution given distributions on model inputs, and this approach is computationally very efficient compared to a more standard Monte Carlo method [3]. Another benefit from a fast-running surrogate of a computationally demanding model is that a large number of model runs can be used to identify sets of model parameters that are consistent with experimental observations, a technique called history matching [17].

To fit the GP emulators, we generated design data by varying each of the inputs $Kr1$ to $Kr10$ in the range $0.5\times$ to $1.5\times$ their baseline value. The baseline values of these inputs are subject to constraints; for example $Kr2$, $Kr3$, $Kr5$, $Kr6$, and $Kr10$ should be positive and non-zero. The range of model inputs over which we trained the emulators was selected so that we could undertake sensitivity analysis without breaking the model. However, it is possible that sensitivities are different outside this range. We also note that the effect of varying input $Kr9$ on the voltage dependence of steady state I_{Kr} shown in Fig. 5(d) results in a curve that no longer fits the experimental data shown in Fig. 3 of [4]. History matching of the I_{Kr} formulation to new experimental data, given our finding that $Kr3$, $Kr4$, and $Kr9$ have a strong influence on model behaviour, would be an interesting future direction and may be more computationally efficient than other approaches [10].

We have only examined the dynamics of I_{Kr} at a single cycle length corresponding to a resting human heart rate, and it is possible that different inputs begin exert a stronger influence at shorter cycle lengths. Recent studies of cycle length dependent sensitivity analysis have concentrated on the effect of inputs that control the maximum flow of current through ion channels, pumps and exchangers [9]. A useful extension of this approach and the present study would be to combine analysis of maximum conductances with model inputs that control ion channel dynamics.

Acknowledgements. This work was funded by the UK EPSRC through grant number EP/K037145/1.

References

1. Bastos, L.S., O'Hagan, A.: Diagnostics for Gaussian process emulators. Technometrics **51**(4), 425–438 (2009)
2. Britton, O.J., Bueno-Orovio, A., Van Ammel, K., Lu, H.R., Towart, R., Gallacher, D.J., Rodríguez, B.: Experimentally calibrated population of models predicts and explains intersubject variability in cardiac cellular electrophysiology. Proc. Nat. Acad. Sci. U.S.A. **110**(23), E2098–E2105 (2013)
3. Chang, E.T.Y., Strong, M., Clayton, R.H.: Bayesian sensitivity analysis of a cardiac cell model using a Gaussian process emulator. PLoS ONE **10**(6), e0130252 (2015)

4. Courtemanche, M., Ramirez, R.J., Nattel, S.: Ionic mechanisms underlying human atrial action potential properties: insights from a mathematical model. Am. J. Physiol. **275**, H301–H321 (1998)
5. Fink, M., Niederer, S.A., Cherry, E.M., Fenton, F.H., Koivumaki, J.T., Seemann, G., Thul, R., Zhang, H., Sachse, F.B., Crampin, E.J., Smith, N.P.: Cardiac cell modelling: observations from the heart of the cardiac physiome project. Prog. Biophys. Mol. Biol. **104**, 2–21 (2011)
6. Hodgkin, A., Huxley, A.: A quantitative description of membrane current and its application to conduction and excitation in nerve. J. Physiol. (London) **117**, 500–544 (1952)
7. Johnstone, R.H., Chang, E.T.Y., Bardenet, R., de Boer, T.P., Gavaghan, D.J., Pathmanathan, P., Clayton, R.H., Mirams, G.R.: Uncertainty and variability in models of the cardiac action potential: can we build trustworthy models? J. Mol. Cell. Cardiol. **96**, 49–62 (2015)
8. Lee, L.A., Carslaw, K.S., Pringle, K.J., Mann, G.W., Spracklen, D.V.: Emulation of a complex global aerosol model to quantify sensitivity to uncertain parameters. Atmos. Chem. Phys. **11**(23), 12253–12273 (2011)
9. Lee, Y.S., Hwang, M., Song, J.S., Li, C., Joung, B., Sobie, E.A., Pak, H.N.: The contribution of ionic currents to rate-dependent action potential duration and pattern of reentry in a mathematical model of human atrial fibrillation. PLoS ONE **11**(3), 1–17 (2016)
10. Loewe, A., Wilhelms, M., Schmid, J., Krause, M.J., Fischer, F., Thomas, D., Scholz, E.P., Dössel, O., Seemann, G.: Parameter estimation of ion current formulations requires hybrid optimization approach to be both accurate and reliable. Front. Bioeng. Biotechnol. **3**, 209 (2015)
11. Niederer, S.A., Fink, M., Noble, D., Smith, N.P.: A meta-analysis of cardiac electrophysiology computational models. Exp. Physiol. **94**(5), 486–495 (2009)
12. Noble, D.: A modification of the Hodgkin-Huxley equations applicable to Purkinje fibre action and pacemaker potentials. J. Physiol. (London) **160**(2), 317–352 (1962)
13. Oakley, J.E., O'Hagan, A.: Probabilistic sensitivity analysis of complex models: a Bayesian approach. J. Roy. Stat. Soc. Ser. B (Stat. Method.) **66**(3), 751–769 (2004)
14. Pathmanathan, P., Shotwell, M.S., Gavaghan, D.J., Cordeiro, J.M., Gray, R.A.: Uncertainty quantification of fast sodium current steady-state inactivation for multi-scale models of cardiac electrophysiology. Prog. Biophys. Mol. Biol. **117**(1), 1–15 (2015)
15. Sánchez, C., Bueno-Orovio, A., Wettwer, E., Loose, S., Simon, J., Ravens, U., Pueyo, E., Rodriguez, B.: Inter-subject variability in human atrial action potential in sinus rhythm versus chronic atrial fibrillation. PLoS ONE **9**(8), e105897 (2014)
16. Sarkar, A.X., Christini, D.J., Sobie, E.A.: Exploiting mathematical models to illuminate electrophysiological variability between individuals. J. Physiol. **590**(Pt 11), 2555–2567 (2012)
17. Vernon, I., Goldstein, M., Bower, R.G.: Galaxy formation: a Bayesian uncertainty analysis. Bayesian Anal. **5**(4), 619–669 (2010)
18. Wilhelms, M., Hettmann, H., Maleckar, M.M., Koivumäki, J.T., Dössel, O., Seemann, G.: Benchmarking electrophysiological models of human atrial myocytes. Front. Physiol. **3**, 487 (2012)

Image-Based Modeling of the Heterogeneity of Propagation of the Cardiac Action Potential. Example of Rat Heart High Resolution MRI

Anđela Davidović[1,2,3]([✉]), Yves Coudière[1,2,3,4], and Yves Bourgault[5]

[1] Inria Bordeaux Sud-Ouest, Talence, France
andjela.davidovic@inria.fr
[2] IHU Liryc, Electrophysiology and Heart Modeling Institute, Pessac, France
[3] Institute of Mathematics of Bordeaux, Talence, France
[4] University of Bordeaux, Talence, France
[5] University of Ottawa, Ontario, Canada

Abstract. In this paper we present a modified bidomain model, derived with homogenization technique from assumption of existence of diffusive inclusions in the cardiac tissue. The diffusive inclusions represent regions without electrically active myocytes, *e.g.* fat, fibrosis etc. We present the application of this model to a rat heart. Starting from high resolution (HR) MRI, geometry is built and meshed using image processing techniques. We perform a study on the effects of tissue heterogeneities induced with diffusive inclusions on the velocity and shape of the depolarization wavefront. We study several test cases with different geometries for diffusive inclusions, and we find that the velocity might be affected by 5% and up to 37% in some cases. Additionally, the shape of the wavefront is affected.

Keywords: Bidomain model · Heterogeneous conductivities · Fibrosis · Multiscale modelling · Image-based modelling

1 Introduction

The standard macroscopic model for the electrophysiology of the heart is the bidomain model [10]. This model is an anisotropic three-dimensional cable equation, that represents the averaged electric behavior of the myocardium. The electrical conductivity tensors for the intracellular and extracellular spaces are anisotropic, with the electrical conduction being the fastest in the fiber direction. In the standard bidomain model the anisotropy ratio in the intracellular space is about 10:1, while in the extracellular space about 2:1.

The bidomain model assumes the existence of uniformly spread myocytes, organized into a dense network [11,14]. It is a reasonable assumption for describing the propagation of the action potential in healthy tissues. In pathological cases this assumption does not hold, as there are regions with large patches of collagen or fibrosis. This is observed in ischemic and rheumatic heart disease, inflammation, hypertrophy, and infarction [3]. In the current modeling of such

© Springer International Publishing AG 2017
M. Pop and G.A. Wright (Eds.): FIMH 2017, LNCS 10263, pp. 260–270, 2017.
DOI: 10.1007/978-3-319-59448-4_25

defects, usually the standard models for the healthy tissues are used, with the model parameters tuned in an ad hoc way.

An explanatory model was proposed for more rigorous tuning of the parameters in such situations, in [5,6]. The main extension of this model w.r.t. the standard bidomain model is the existence of relatively large regions of tissue where there are no myocytes, nor other kind of cells, and these regions are assumed to be passive electrical conductors. They are called **diffusive inclusions**. The diffusive inclusions may represent electrically passive infiltrations in the cardiac tissue like fat, collagen, fibrosis etc.

In this paper we apply the proposed model to a slab of rat heart. We start from HR-MRI data of a rat heart. We perform the image analysis to obtain a computational domain, to define the diffusive inclusions and to find the local volume fractions of the diffusive inclusions. These are then used in the simulations of the modified bidomain model. Finally, we make a study of how different diffusive inclusions might affect the shape and the velocity of depolarization waves.

2   Modified Bidomain Model for Rat Heart

In this section we give an overview of the model proposed in [5,6]. Full derivation of the model is out of the scope of this paper, since we want to focus on the application of this model on a real rat heart.

On the microscopic scale, periodic diffusive inclusions have been embedded into the healthy cardiac tissue, *i.e.* the bidomain model. In order to observe the effects of these inclusions on the macroscopic scale a homogenisation technique has been applied. This approach gives rise to the following two sets of problems:

- homogenised problem - that is in fact the macroscopic model that we were looking for. In our case it is a modified bidomain model, where the effects of inclusions are mainly contained in the conductivity tensors, as expected.
- cell problems - these are correction equations, used to calculate the modified conductivity tensors. They are set on the unit cube of $\mathbb{R}^3$ and represent a rescaled typical volume of interest of cardiac tissue.

The general homogenisation approach is presented in [1,2]. It is a well known technique in mathematical modelling, and has been used for example in the derivation of the bidomain model [11,14].

2.1   Equations

Homogenized Problem. On the microscopic scale it is assumed that diffusive inclusions are periodic with a period ϵ, such that $l \ll \epsilon \ll L$, where L is the tissue scale and l is the cardiac cell scale. Applying the mathematical homogenization technique, assuming that ϵ goes to zero, an averaged non-dimensional model has been derived, and is given as follows

$$(\partial_t v + I_{ion}(v, h)) \xi_B = N \nabla \cdot (\sigma^{i*} \nabla u_i), \qquad \text{in } [0, T] \times \Omega, \qquad (1)$$
$$(\partial_t v + I_{ion}(v, h)) \xi_B = -N \nabla \cdot (\sigma^{e*} \nabla u_e), \qquad \text{in } [0, T] \times \Omega, \qquad (2)$$
$$\partial_t h + g(v, h) = 0, \qquad \text{in } [0, T] \times \Omega, \qquad (3)$$

where u_i and u_e are the intracellular and extracellular potentials, $v = u_i - u_e$ is the transmembrane voltage, I_{ion} is the transmembrane current that is a function of v and h, a state variable, ξ_B is the local volume fraction of the healthy tissue, N is the adimensionalisation parameter that depends on time and space scales and physical properties of the cell as given in [15], and σ^{i*} and σ^{e*} are the modified conductivity tensors.

Cell Problems. The effective intracellular and extracellular conductivity tensors σ_i^* and σ_e^* are obtained by solving the set of so-called cell problems that are defined on the unit cell space, $Y = [0, 1] \times [0, 1] \times [0, 1]$. The unit cell is in fact the rescaled periodic cell of tissue, where one can identify the healthy tissue, Y_B, and the diffusive inclusion, Y_D, see Fig. 1. Note that Y represents a microscopic piece of tissue (of size ϵ), and the tissue is considered to be locally periodic. On a large scale the shape and size of the diffusive inclusions may vary, as illustrated in the following study.

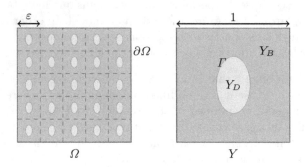

Fig. 1. Left: the idealised full domain, Ω. Right: the unit cell, Y.

On the unit cell we define the Y-periodic functions w_j and w_j^i, $j = 1, 2, 3$, as solutions to the cell problems

$$\nabla \cdot (\sigma_i \nabla w_j^i) = 0, \text{ in } Y_B, \quad \sigma_i (\nabla w_j^i + e_j) \cdot n = 0, \text{ on } \Gamma, \qquad (4)$$
$$\nabla \cdot (\sigma \nabla w_j) = 0, \text{ in } Y, \quad (\sigma_e - \sigma_d)(\nabla w_j + e_j) \cdot n = 0, \text{ on } \Gamma, \qquad (5)$$

where σ_i and σ_e are the intracellular and extracellular conductivity tensors used in the healthy tissue, *i.e.* in the standard bidomain model, and σ_d is the conductivity assumed inside of the diffusive inclusion, n is an outer normal to the boundary Γ and $e_j, j = 1, 2, 3$, are the unit vectors of the standard basis. The conductivity tensor σ in (5) is defined as $\sigma = \sigma_e$, in Y_B, and $\sigma = \sigma_d$, in Y_D.

The periodic functions w_j^i and w_j are then used to define the effective conductivity tensors as:

$$\sigma_{kj}^{i*} = \sigma_{i,kj} \xi_B$$
$$+ \left(\sigma_{i,k1} \int_{Y_B} \partial_{y_1} w_j^i dy + \sigma_{i,k2} \int_{Y_B} \partial_{y_2} w_j^i dy + \sigma_{i,k3} \int_{Y_B} \partial_{y_3} w_j^i dy \right), \quad (6)$$

$$\sigma_{kj}^{e*} = \int_{Y} \sigma_{kj} dy + \left(\int_{Y} \sigma_{k1} \partial_{y_1} w_j dy + \int_{Y} \sigma_{k2} \partial_{y_2} w_j dy + \int_{Y} \sigma_{k3} \partial_{y_3} w_j dy \right), \quad (7)$$

for $j, k = 1, 2, 3$, and $\xi_B = |Y_B|$.

As one may notice, σ^{i*} and σ^{e*} depend not only on the volume fraction of the diffusive inclusion, but on their shape as well.

Ionic Model for the Rat Heart. A reference ionic model that corresponds to the action potential (AP) of the rat heart is given in [13]. In our simulations we use the Mitchell Schaeffer (MS) model [9], fitted to the reference ionic model for the rat heart,

$$I_{ion}(v, h) = \frac{1}{\tau_{in}} h v^2 (v - 1) + \frac{1}{\tau_{out}} v, \quad (8)$$

$$g(v, h) = \begin{cases} \frac{1}{\tau_{open}} (1 - v), & \text{for } v < v_{gate}, \\ -\frac{1}{\tau_{close}} v, & \text{for } v \geq v_{gate}, \end{cases} \quad (9)$$

where I_{ion}, g, v and h have the same meaning as before, and $\tau_{in}, \tau_{out}, \tau_{open}, \tau_{close}$, and v_{gate} are the model parameters. The former four are normally given in units of time and are related to the duration of AP phases: depolarisation, repolarisation, plateau phase and the total AP duration. The parameter v_{gate} is related to the threshold value of transmembrane voltage when the AP in cardiac cell is triggered.

The original MS model is given so that the transmembrane voltage, v, is non-dimensional and is scaled to $[0, 1]$. Here we went a step further and used the approach given in [15] to work with a fully non-dimensional model, using the time scale $T = 10^{-3}$ s and the length scale $L = 10^{-3}$ m. We have used the algorithm given in [12], that enables us to automatically determine new parameters of the MS model, given in Table 1. Note that in the non-dimensional model the conductivities are also scaled with the parameter $\bar{\sigma} = 0.1$ Sm^{-1}, and the non-dimensional values for the anisotropic tensors σ_i and σ_e are given in Table 1.

The MS model is chosen for its mathematical simplicity and easy numerical implementation. It has been shown that it is convenient as well for patient-specific modeling in [16].

Table 1. Non-dimensional bidomain model parameters of the rat heart, as in [15].

τ_{in}	τ_{out}	τ_{open}	τ_{close}	v_{gate}	$\sigma_{i,l}$	$\sigma_{i,t}$	$\sigma_{e,l}$	$\sigma_{e,t}$	N
0.073	8.369	25.743	15.438	0.02	1.741	0.1934	3.906	1.97	0.0125

2.2   Computing Effective Conductivities

To compute the effective conductivities σ^{i*} and σ^{e*}, first we have to solve the Eqs. (4)–(5) on the unit cell Y. Hence, we need values for the conductivities σ_i, σ_e and σ_d. The values for the anisotropic tensors σ_i and σ_e corresponding to the non-dimensional model are given in Table 1. We assume that the fibers are aligned with the x-axis, so σ_i and σ_e are diagonal matrices. In that sense, we have $\sigma_{i,11} = \sigma_{i,l}$ and $\sigma_{i,22} = \sigma_{i,33} = \sigma_{i,t}$, with non-diagonal terms being equal to zero. Similarly, $\sigma_{e,11} = \sigma_{e,l}$ and $\sigma_{e,22} = \sigma_{e,33} = \sigma_{e,t}$, with non-diagonal terms being equal to zero.

The conductivity tensor σ_d is scaled with the same parameter $\overline{\sigma} = 0.1\,\mathrm{Sm}^{-1}$. We chose to test isotropic values $\sigma_d = 0.2, 1.5$ and 3.0, where the former one corresponds to the conductivity of fatty tissue and the latter to isotropic extra-cellular space. Several shapes of the diffusive inclusions have been tested, as shown in the Fig. 2.

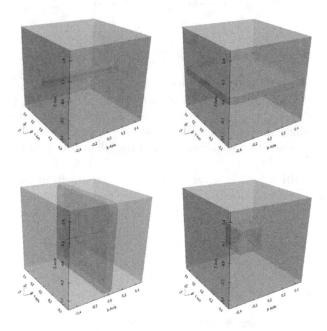

Fig. 2. Test cases for the unit cells in 3D. The shapes of the inclusions are: sticks aligned with fiber direction (here x-axis), plates parallel to fibers, plates perpendicular to fibers, and nearly cubic superellipsoids. For each case we used superellipsoids, with different semi-axes.

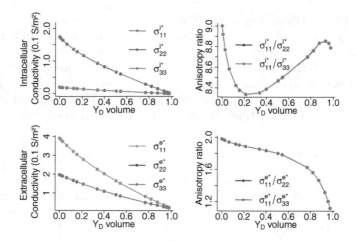

Fig. 3. Effective conductivities and anisotropy ratios for inclusions in the shape of cubes (see Fig. 2 at the bottom right), ranging in volume fraction from 0 to 98%, with fibers aligned with the x-axis.

The cell problems were solved using the finite element approach and *FreeFem++*[1]. To illustrate results we plot the conductivity tensors and the anisotropy ratios for one case (see Fig. 3). In this case we obtain diagonal conductivity tensors with $\sigma_{22}^{i*} = \sigma_{33}^{i*}$ and $\sigma_{22}^{e*} = \sigma_{33}^{e*}$. We observe changes in the anisotropy ratios for both intra and extra-cellular conductivities. It is consistent with [6] where the shape of diffusive inclusions induces significant changes in anisotropy ratios.

Additionally to that, the main direction of propagation might be changed as well. This implies that in certain situations, depending on the tissue structure, one needs to know more than just the fiber direction to recover the correct wavefront of depolarization.

3 Application to a Slab of Rat Heart

3.1 Data on the Rat Heart

HR-MRI. Data were provided by the IHU-Liryc, Bordeaux. The MR Imaging has been performed on the heart of a male Wistar rat. After the proper preparation of the heart, the heart was perfused with MRI contrast agent and fixative, and then stored in contrast/fixative solution until imaging. The heart was imaged using a T1 weighted FLASH (Fast Low Angle SHot) MRI sequence in a Bruker (Ettlingen, Germany) 9.4 T spectroscope with 20 averages and echo time (TE) = 7.9 ms, repetition time (TR) = 50 ms, and flip angle 40°, at a resolution of $50 \times 50 \times 50\mu$m, a matrix size of $256 \times 256 \times 512$ for a field of view of $12.8 \times 12.8 \times 25.6$ mm. For further details see [8].

---

[1] http://www.freefem.org/.

DT-MRI Fiber Directions. The fiber structure in the cardiac tissue leads to anisotropy in the bidomain model. For the simulation of both bidomain and modified model it is important to assess the fiber orientation in the tissue. The diffusion tensor (DT) MRI technique is used for this purpose. The data from DT-MRI are given on an image that is four times coarser than the original HR-MRI data, *i.e.* $64 \times 64 \times 128$ for the same view field. This results in a spacing between voxels that is four times larger, *i.e.* $0.2 \times 0.2 \times 0.2$ mm.

3.2   Image Processing

Segmentation. The semi-manual segmentation of the images has been performed with the software *Seg3D*[2]. The median filter and thresholding on the gray scale were used to define roughly the boundaries of cardiac tissue. Following thresholding, bad pixels have been fixed manually, layer by layer. Cropping tools have been used to define the computational domain, *i.e.* a part of the left-ventricular wall. This is done solely for performance reasons, in order to reduce the computing cost of the simulations.

Diffusive Inclusion Detection and Local Volume Fraction. Using the software *Seg3D* we were able to define the diffusive inclusions in the computational domain. For this we used only a threshold on gray scale, without additional processing, see Fig. 4. The mask of computational domain with flagged subdomain of diffusive inclusions has been exported in .mat file, and Matlab was used for the computation of the local volume fraction of the inclusions. Around each voxel, X, we define a $5 \times 5 \times 5$ window, and count the number n of voxels inside of this window that belong to the diffusive inclusions, $n \in [0, 125]$. The local volume fraction of voxel X is then given as $\xi_D(X) = \frac{n}{125}$.

Mesh Generation. For the mesh generation we used the software *SCIRun*[3], which integrates the call function to tetgen. We have set a minimum radius-edge ratio and imposed a maximum volume constraint on tetrahedra. The result is a fine mesh, that has 351706 nodes and 1924747 tetrahedra, see Fig. 4.

Mapping Data on the Mesh. To be able to use the imaging data in our simulations we have to map them on the mesh nodes. For this purpose, we use the software *SCIRun*. Both, local volume fraction ξ_D and the fiber orientations were mapped on nodes of the mesh.

3.3   Simulations

Settings. Here we set the parameters for the bidomain and modified model (1)–(3). The computational mesh and $\xi_B = 1 - \xi_D$ are set. The non-dimensional

---

[2] http://www.sci.utah.edu/cibc-software/seg3d.html.
[3] http://www.sci.utah.edu/cibc-software/scirun.html.

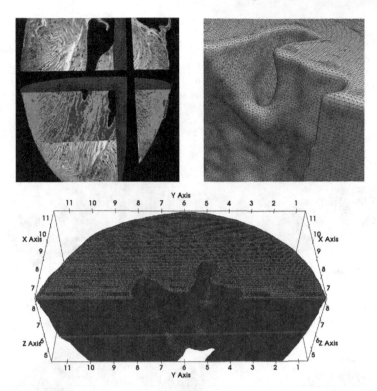

Fig. 4. Up left: segmentation from the HR-MRI data and diffusive inclusions detection. Bottom: fine mesh of the selected domain. Up right: details of the mesh.

parameter N is set to 0.0125, as in the Table 1. The effective conductivity tensors, σ_i^* and σ_e^*, have been computed for several test cases, as in Fig. 2, with the fiber direction assumed to be aligned with the x-axis and for the range of diffusive inclusions volume fractions $\xi_D \in (0, 1)$. Now, for each node in the mesh we assign one of these values depending on the corresponding local value of ξ_D, obtained from the image processing. Finally, we recompute the effective conductivities for each node based on the fiber direction, obtained from DT-MRI images.

We use linear finite elements as implemented in *FreeFem++* to solve the bidomain and modified model (1)–(3). For the time discretization we use the semi-explicit SBDF2 numerical scheme, as proposed in [7], with the time step $dt = 0.05$ ms. The resulting linear system was solved using the conjugate gradient method, to avoid excessive memory usage.

Results. We run simulations for several test cases as given in Table 2. The reference case consists in solving the standard bidomain model without any diffusive inclusions. In the table we report the total depolarization duration of the computational domain, T_D, calculated as the first time $t > 0$ for which all nodes have a value $v > v_{gate}$. From this one we conclude that the velocity is

Table 2. Various test cases and total depolarization time, T_D.

	Ref	Volume fraction from HR-MRI									Artificial scars	
Shapes	–	Sticks (∥)		Plates (∥)		Plates (⊥)		Cubes			Cubes	
$\sigma_d\,[Sm^{-1}]$	–	0.02	0.3	0.02	0.3	0.02	0.3	0.02	0.15	0.3	0.02	0.3
$T_D\,[ms]$	32.66	33.02	31.75	32.26	30.99	44.89	44.07	34.31	32.49	33.3	34.91	34

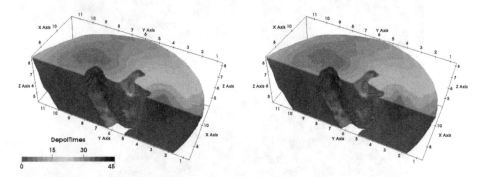

Fig. 5. Depolarization isochrones. From left to right, reference case and parallel plates, $\sigma_d = 0.02\,\mathrm{Sm}^{-1}$. Time between consecutive isochrones is 2.5 ms.

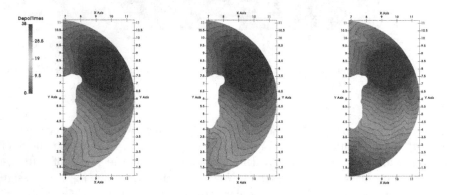

Fig. 6. Depolarization isochrones. From left to right: reference case, parallel plates and perpendicular plates, $\sigma_d = 0.02\,\mathrm{Sm}^{-1}$. Time between consecutive isochrones is 2 ms.

affected by 5 – 7% in most of the cases, and 35% and 37% in the case of plates perpendicular to the fiber direction.

On Fig. 5 we plot the isochrones of depolarization on the boundary of the computational domain. We compare the reference case to the case of inclusions with the shapes of plates parallel to the fiber directions. We can observe a change in the shape of the wavefront. The same isochrones plotted on a cut through the domain are shown in Fig. 6, where contours are separated by 2 ms. As can be seen the shape of the wavefront far from the boundary is more affected by the diffusive inclusions than on the boundary of the domain.

4   Conclusions and Discussion

In some pathological cases the microscopic structure of the cardiac tissue is affected and there is an increase in collagen, fatty or fibrous tissue. These microscopic changes affect the propagation of electrical signals through the heart walls. In models, these changes are usually accounted for through the tuning of model parameters in an ad hoc way.

We have presented a modified bidomain model that has been derived in a rigorous way from the microscopic model of heterogeneous tissue. The modelling assumption is that the fibrous infiltrations are electrically passive, and are organized in a locally periodic way. Then, using the homogenisation technique the modified bidomain model has been derived, where the diffusive inclusions give rise to modified conductivity tensors in the bidomain model. We obtain a direct relation between the modified conductivity tensors and the local size and shape of the diffusive inclusions.

Further, we described a framework to obtain image based distribution of the parameters for the modified bidomain model. We worked on high resolution MRI of the rat heart. Using thresholding we detected the diffusive inclusions in the images and determined their local volume fractions. We used several test cases for the shapes of the inclusions, and computed the modified conductivity tensors for each test case, based on the local volume fractions of the diffusive inclusions. Finally, we ran simulations for all test cases and compared the results to the reference case, without diffusive inclusions.

The results are interesting as we could observe changes in the velocity of propagation, from 5% in some test cases to 37% in others. The largest impact on the velocity and shape of the depolarization wavefront occurred for diffusive inclusions perpendicular to the fiber directions. This is in agreement with previous 2D test cases in [6], where it has been observed that the principal direction of propagation might change depending on the shape and the large volume fraction of diffusive inclusions.

Since we do not have an a priori knowledge on the actual shapes of the diffusive inclusions, we have tested several simple geometries. To our knowledge, there has not been any studies performed that would give us a more precise idea on the shape of passive inclusions in the cardiac tissue.

In this paper we aimed to demonstrate the possibility to rigorously determine model parameters form HR-MRI of the rat heart with a pathological heterogeneous structure. It has a great deal of possible applications to fibrotic disease, ischemic heart disease, infarct scars etc.

Acknowledgments. The authors would like to thank Stephen Gilbert for providing HR-MRI and DT-MRI data of the rat heart, and for the useful discussions about the imaging process. This work was funded by the ANR project HR-CEM reference ANR-13-MONU-0004. This work was partially supported by an ANR grant part of "Investments d'Avenir" program reference ANR-10-IAHU-04.

References

1. Allaire, G.: Homogenization and two-scale convergence. SIAM J. Math. Anal. **23**(6), 1482–1518 (1992)
2. Bensoussan, A., Lions, J.L., Papanicolaou, G.: Asymptotic Analysis for Periodic Structures. North-Holland Publishing Company, Amsterdam (1978)
3. Camelliti, P., Borg, T.K., Kohl, P.: Structural and functional characterisation of cardiac fibroblasts. Cardiovasc. Res. **65**(1), 40–51 (2005)
4. Clayton, R.H., Bernus, O., Cherry, E.M., Dierckx, H., Fenton, F.H., Mirabella, L., Panfilov, A.V., Sachse, F.B., Seemann, G., Zhang, H.: Models of cardiac tissue electrophysiology: progress, challenges and open questions. Prog. Biophys. Mol. Biol. **104**(1), 22–48 (2011)
5. Coudiere, Y., Davidovic, A., Poignard, C.: The modified bidomain model with periodic diffusive inclusions. In: Computing in Cardiology Conference (CinC), pp. 1033–1036. IEEE (2014)
6. Davidovic, A.: Multiscale mathematical modelling of structural heterogeneities in cardiac electrophysiology. Ph.D. thesis, University of Bordeaux (2016). https://hal.archives-ouvertes.fr/tel-01478145
7. Ethier, M., Bourgault, Y.: Semi-implicit time-discretization schemes for the bidomain model. SIAM J. Numer. Anal. **46**(5), 2443–2468 (2008)
8. Gilbert, S.H., Benoist, D., Benson, A.P., White, E., Tanner, S.F., Holden, A.V., Dobrzynski, H., Bernus, O., Radjenovic, A.: Visualization and quantification of whole rat heart laminar structure using high-spatial resolution contrast-enhanced MRI. Am. J. Physiol. Heart Circulatory Physiol. **302**(1), H287–H298 (2012)
9. Mitchell, C.C., Schaeffer, D.G.: A two-current model for the dynamics of cardiac membrane. Bull. Math. Biol. **65**(5), 767–793 (2003)
10. Keener, J.P., Sneyd, J.: Mathematical Physiology. Springer, New York (2009)
11. Neu, J.C., Krassowska, W.: Homogenization of syncytial tissues. Crit. Rev. Biomed. Eng. **21**(2), 137–199 (1992)
12. Ngoma, D.V., Bourgault, Y., Nkounkou, H.: Parameter identification for a non-differentiable ionic model used in cardiac electrophysiology. Appl. Math. Sci. **9**(150), 7483–7507 (2015)
13. Pandit, S.V., Clark, R.B., Giles, W.R., Demir, S.S.: A mathematical model of action potential heterogeneity in adult rat left ventricular myocytes. Biophys. J. **81**(6), 3029–3051 (2001)
14. Pennacchio, M., Savare, G., Franzone, P.C.: Multiscale modeling for the bioelectric activity of the heart. SIAM J. Math. Anal. **37**(4), 1333–1370 (2005)
15. Rioux, M., Bourgault, Y.: A predictive method allowing the use of a single ionic model in numerical cardiac electrophysiology. ESAIM Math. Model. Numer. Anal. **47**(4), 987–1016 (2013)
16. Relan, J., Pop, M., Delingette, H., Wright, G.A., Ayache, N., Sermesant, M.: Personalization of a cardiac electrophysiology model using optical mapping and MRI for prediction of changes with pacing. IEEE Trans. Biomed. Eng. **58**(12), 3339–3349 (2011)

VT Scan: Towards an Efficient Pipeline from Computed Tomography Images to Ventricular Tachycardia Ablation

Nicolas Cedilnik[1(✉)], Josselin Duchateau[2], Rémi Dubois[2], Pierre Jaïs[2], Hubert Cochet[2], and Maxime Sermesant[1]

[1] Université Côte d'Azur, Inria, Sophia Antipolis, France
nicolas.cedilnik@inria.fr
[2] Liryc Institute, Bordeaux, France

Abstract. Non-invasive prediction of optimal targets for efficient radio-frequency ablation is a major challenge in the treatment of ventricular tachycardia. Most of the related modelling work relies on magnetic resonance imaging of the heart for patient-specific personalized electrophysiology simulations.

In this study, we used high-resolution computed tomography images to personalize an Eikonal model of cardiac electrophysiology in seven patients, addressed to us for catheter ablation in the context of post-infarction arrhythmia. We took advantage of the detailed geometry offered by such images, which are also more easily available in clinical practice, to estimate a conduction speed parameter based on myocardial wall thickness. We used this model to simulate a propagation directly on voxel data, in similar conditions to the ones invasively observed during the ablation procedure.

We then compared the results of our simulations to dense activation maps that recorded ventricular tachycardias during the procedures. We showed as a proof of concept that realistic re-entrant pathways responsible for ventricular tachycardia can be reproduced using our framework, directly from imaging data.

Keywords: Electrophysiological modelling · Heart imaging · Ventricular tachycardia · Catheter ablation

1 Introduction

Sudden cardiac death (SCD) due to ventricular arrhythmia is responsible for hundreds of thousands of deaths each year [1]. A high proportion of these arrhythmias are related to ischemic cardiomyopathy, which promotes both ventricular fibrillation and ventricular tachycardia (VT). The cornerstone of SCD prevention in an individual at risk is the use of an implantable cardioverter defibrillator (ICD). ICDs reduce mortality, but recurrent VT in recipients are a source of important morbidity. VT causes syncope, heart failure, painful shocks

© Springer International Publishing AG 2017
M. Pop and G.A. Wright (Eds.): FIMH 2017, LNCS 10263, pp. 271–279, 2017.
DOI: 10.1007/978-3-319-59448-4_26

Fig. 1. Overall pipeline of the approach. Ridge and circuit detection are not presented in this article.

and repetitive ICD interventions reducing device lifespan. In a small number of cases called arrhythmic storms, the arrhythmia burden is such that the ICD can be insufficient to avoid arrhythmic death.

In this context, radio-frequency ablation aiming at eliminating re-entry circuits responsible for VT has emerged as an interesting option. VT ablation has already demonstrated its benefits [2] but still lacks clinical consensus on optimal ablation strategy [3]. Two classical strategies have been developped by electrophysiologists. The first strategy focuses on mapping the arrhythmia circuit by inducing VT before ablating the critical isthmus. The second strategy focuses on the substrate, eliminating all abnormal potentials that may contain such isthmuses.

Both strategies are very time-consuming, with long mapping and ablation phases respectively, and procedures are therefore often incomplete in patients with poor general condition. Many authors have thus shown an interest in developing methods coupling non-invasive exploration and modelling to predict both VT risk and optimal ablation targets [4–8]. Most of the published work in this area relies on cardiac magnetic resonance imaging (CMR), at the moment considered as the gold standard to assess myocardial scar, in particular using late gadolinium enhancement sequences. However, despite being the reference method to detect fibrosis infiltration in healthy tissue, CMR methods clinically available still lack spatial resolution under 2 mm to accurately assess scar heterogeneity in chronic healed myocardial infarction, because the latter is associated with severe wall thinning (down to 1 mm). Moreover, most patients recruited for VT ablation cannot undergo CMR due to an already implanted ICD.

In contrast, recent advances in cardiac computed tomography (CT) technology now enable the assessment of cardiac anatomy with extremely high spatial resolution. We hypothesized that such resolution could be of value in assessing the heterogeneity of myocardial thickness in chronic healed infarcts. Based on our clinical practice, we indeed believe that detecting VT isthmuses relies on detecting thin residual layers of muscle cells-rich tissue inside infarct scars that CMR fails to identify due to partial volume effect. CT image acquisistion is less operator-dependent than CMR, making it a first class imaging modality for automated and reproductible processing pipelines. CT presents fewer contraindication than CMR (it is notably feasible in patients with ICD), it costs less, and its availability is superior.

The aim of this study was to assess the relationship between wall thickness heterogeneity, as assessed by CT, and ventricular tachycardia mechanisms in chronic myocardial infarction, using a computational approach (see the overall pipeline in Fig. 1). This manuscript presents the first steps of this pipeline, in order to evaluate the relationship between simulations based on wall thickness and electro-anatomical mapping data.

2   Image Acquisition and Processing

2.1   Population

The data we used come from 7 patients (age 58 ± 7 years, 1 woman) referred for catheter ablation therapy in the context of post-infarction ventricular tachy-cardia. The protocol of this study was approved by the local research ethics committee.

2.2   Acquisition

All the patients underwent contrast-enhanced ECG-gated cardiac multi-detector CT (MDCT) using a 64-slice clinical scanner (*SOMATOM Definition, Siemens Medical Systems*, Forchheim, Germany) between 1 to 3 days prior to the electro-physiological study. This imaging study was performed as part of standard care as there is an indication to undergo cardiac MDCT before electrophysiological procedures to rule out intra cardiac thrombi. Coronary angiographic images were acquired during the injection of a 120 ml bolus of iomeprol 400 mg/ml (*Bracco*, Milan, Italy) at a rate of 4 ml/s. Radiation exposure was typically between 2 and 4 mSv. Images were acquired in supine position, with tube current modulation set to end-diastole. The resulting voxels have a dimension of $0.4 \times 0.4 \times 1\,\mathrm{mm}^3$.

2.3   Segmentation

The left ventricular endocardium was automatically segmented using a region-growing algorithm, the thresholds to discriminate between the blood pool and the wall being optimized from a prior analysis of blood and wall densities. The

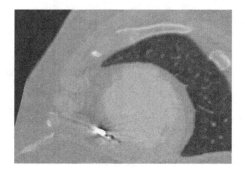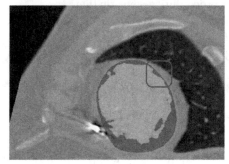

Fig. 2. (Left) original CT-scan image. (Right) wall mask (green). Note the thickness heterogeneity (red box). (Color figure online)

epicardium was segmented using a semi-automated tool based on the interpolation of manually-drawn polygons.

All analyses were performed using the MUSIC software[1] (IHU Liryc Bordeaux, Inria Sophia Antipolis, France). An example of the result of such segmentation can be seen in Fig. 2.

2.4   Wall Thickness Computation

In CT-scan images, only the healthy part of the myocardium is visible, hence wall thinning is a good marker of scar localization and abnormal electro-physiological parameters [9]. To accurately compute wall thickness, and to overcome difficulties related to its definition on tri-dimensional images, we chose the Yezzi et al. method [10]. It is based on solving the Laplace equation with Dirichlet boundary conditions to determine the trajectories along which the thickness will be computed. One advantage of such thickness definition is that it is then defined for every voxel, which is later useful in our model (see Sect. 3.2). An example of the results can be observed in Fig. 3 (left and middle), along with a comparison with the much lower level of detail in scar morphology assessment that is obtained with a voltage map from a sinus rhythm recording.

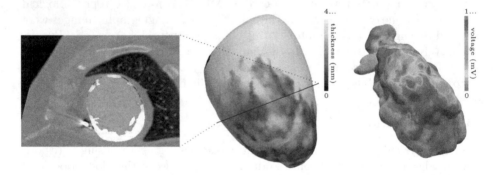

Fig. 3. (Left-Middle) wall thickness as defined in Sect. 2.4. (Right) voltage map recorded during sinus rhythm.

3   Cardiac Electrophysiology Modelling

3.1   Eikonal Model

We chose the Eikonal model for simulating the wave front propagation because of the following reasons:

- It requires very few parameters compared with biophysical models, making it more suitable for patient-specific personalization in a clinical setting.

---

[1] https://team.inria.fr/asclepios/software/music/.

- Its output is an activation map directly comparable with the clinical data.
- It allows for very fast solving thanks to the fast marching algorithm.

We used the standard Eikonal formulation in this study:

$$v(X)||\nabla T(X)|| = 1 \tag{1}$$

The sole parameter required besides the myocardial geometry is the wave front propagation speed $v(X)$ (see Sect. 3.2). More sophisticated biophysical models allow for more precise results; however, this precision relies on an accurate estimate of the parameters and the necessary data to such ends are not available in clinical practice.

This original approach enabled us to parametrize directly the model on the basis of the thickness computed from the images. Additionally, we generated a unidirectional onset by creating an artificial conduction block. This block was needed to represent the refractoriness of the previously activated tissue that the Eikonal model does not include. An illustration of this phenomenon can be observed in the green box in Fig. 4.

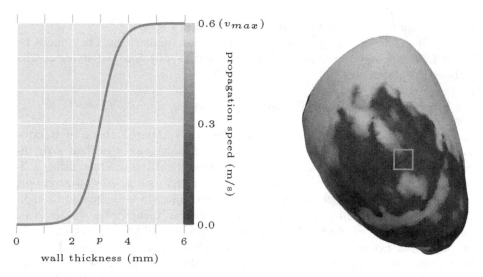

Fig. 4. (Left) Transfer function used to estimate wave front propagation speed from wall thickness. (Right) example resulting speed map. Note the artificial refractory block (dark straight line in green box). (Color figure online)

3.2 Wave Front Propagation Speed Estimation

As there is a link between myocardial wall thickness and its viability, it was possible to parameterize our model from the previous wall thickness computation (see Sect. 2.4). There are basically three different cases:

1. Healthy myocardium where the wave front propagation speed is normal.

2. Dense fibrotic scar where the wave front propagation is extremely slow.
3. Gray zone area where it is somewhere in between.

Instead of arbitrarily choosing a specific speed for the "gray zone", we exploited the resolution offered by our images to come up with a smooth, continuous estimate of the speed between healthy and scar myocardium. This resulted in the following logistic transfer function:

$$v(X) = \frac{v_{max}}{1 + e^{r(p-t(X))}} \qquad (2)$$

where v_{max} is the maximum wave front propagation speed, $t(X)$ the thickness at voxel X, p the inflection point of the sigmoidal function, i.e., the thickness at which we reach $\frac{1}{2}v_{max}$ and r a dimensionless parameter defining the steepness of the transfer function.

More specifically, we chose the following parameters:

- $v_{max} = 0.6$ m/s, as it is the conduction speed of healthy myocardium,
- $p = 3$ mm, as it was considered the gray zone "center",
- $r = 2$, in order to obtain virtually null speed in areas where thickness is below 2 mm.

The resulting transfer function can be visualized in Fig. 4.

The fiber orientations were not included in our simulations. They might be in our final pipeline, but it is worth noting that we obtained satisfying results without this parameter.

4   Electrophysiological Data

In order to evaluate the simulation results, we compared them to electro-anatomical mapping data. As part of the clinical management of their arrhythmia, all the patients underwent an electrophysiological procedure using a 3-dimensional electro-anatomical mapping system (*Rhythmia, Boston scientific*, USA) and a basket catheter (*Orion, Boston scientific*, USA) dedicated to high-density mapping. During the procedure, ventricular tachycardia was induced using a dedicated programmed stimulation protocol, and the arrhythmia could be mapped at extremely high density (about 10 000 points per map). Patients were then treated by catheter ablation targeting the critical isthmus of the recorded tachycardia, as well as potential other targets identified either by pace mapping or sinus rhythm substrate mapping. These maps were manually registered to the CT images.

5   Results

Examples results of the simulations and their comparisons to mapping data are presented in Fig. 5 In each case, a qualitatively similar re-entry circuit is predicted.

Similar visual results were reached for all the VTs, without any further tuning of the model. However, to reach such results we needed to pick the stimulations points and directions very carefully.

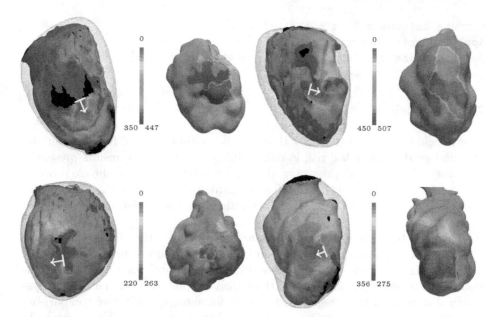

Fig. 5. Comparison of predicted (left) activation maps (in ms) to "ground truth" data (right) obtained from VT recording during RF catheter ablation in 4 different patients. White arrows: starting point and direction of simulated electrical impulse. Both recorded and simulated VT represent one cycle only. Animations of these activation maps are available online. (https://team.inria.fr/asclepios/vt-scan-fimh-2017/.)

6 Implementation

All the activation maps acquired during ventricular tachycardia (10 maps of 10 different circuits acquired in 7 patients) were exported to *Matlab* software (*Mathworks*, USA).

The simulation was computed directly on the voxel data as the resolution of our images represents highly detailed anatomy, and to avoid arbitrary choices inherent to mesh construction. We then mapped the results to a mesh but for visualization purposes only.

We used a custom *Python* package[2] for the thickness computation and the *SimpleITK* fast marching implementation to solve the Eikonal equation.

The artificial conduction block was created by setting the wave front propagation speed to zero in a parametrically defined disk of 15 mm radius, 2 mm behind the stimulation points, orthogonal to the desired stimulation direction.

From the results of the semi-automatic CT image segmentation, the complete simulation pipeline, with an informal benchmark realized on a i7-5500U CPU at 2.40 GHz (using only 1 core), looks as follows:

---

[2] https://pypi.python.org/pypi/pyezzi.

```
Compute thickness from masks [43s]
Apply transfer function [0.5s]
Select pacing site and create refractory block [7s]
Solve the Eikonal equation [0.5s]
```

7   Discussion

This article presents a framework that may be suitable for VT RF ablation targets and prediction of VT risk in daily clinical practice. The results presented here are preliminary but promising, due to the robustness of the image processing, the fast simulation of the model, and the high-resolution of CT scan images. This resolution was crucial in the characterization of the VT isthmuses presented in this study.

The main limitations of the results presented here are the limited sample size and the lack of quantitative evaluation of the simulation results. It is however worth noting that we were able to obtain such results without advanced calibration of the model's parameters. For the latter, we plan to map the electrophysiological data onto the image-based meshes for more quantitative comparison. However it remains challenging due to the shape differences. Heterogeneity in CT images also induce a bias in thickness computation that may require patient-specific personalization of the transfer function parameters.

We believe that in order to fully automatize the pipeline, we need not only to determine a ridge detection strategy, but also to overcome some of the simplifications induced by the Eikonal modelling choice and enhance the circuit characterization.

References

1. Mozaffarian, D., Benjamin, E.J., Go, A.S., Arnett, D.K., Blaha, M.J., Cushman, M., Das, S.R., de Ferranti, S., Després, J.-P., Fullerton, H.J., et al.: Executive summary: heart disease and stroke statistics-2016 update: a report from the american heart association. Circulation **133**(4), 447 (2016)
2. Ghanbari, H., Baser, K., Yokokawa, M., Stevenson, W., Bella, P.D., Vergara, P., Deneke, T., Kuck, K.-H., Kottkamp, H., Fei, S., et al.: Noninducibility in postinfarction ventricular tachycardia as an end point for ventricular tachycardia ablation and its effects on outcomes. Circ. Arrhythmia Electrophysiol. **7**(4), 677–683 (2014)
3. Aliot, E.M., Stevenson, W.G., Almendral-Garrote, J.M., Bogun, F., Calkins, C.H., Delacretaz, E., Bella, P.D., Hindricks, G., Jaïs, P., Josephson, M.E., et al.: EHRA/HRS expert consensus on catheter ablation of ventricular arrhythmias. Europace **11**(6), 771–817 (2009)
4. Ashikaga, H., Arevalo, H., Vadakkumpadan, F., Blake, R.C., Bayer, J.D., Nazarian, S., Zviman, M.M., Tandri, H., Berger, R.D., Calkins, H., et al.: Feasibility of image-based simulation to estimate ablation target in human ventricular arrhythmia. Heart Rhythm **10**(8), 1109–1116 (2013)
5. Arevalo, H., Plank, G., Helm, P., Halperin, H., Trayanova, N.: Tachycardia in postinfarction hearts insights from 3D image-based ventricular models. PloS one **8**(7) (2013). e68872

6. Cabrera-Lozoya, R., et al.: Confidence-based training for clinical data uncertainty in image-based prediction of cardiac ablation targets. In: Menze, B., et al. (eds.) MCV 2014. LNCS, vol. 8848, pp. 148–159. Springer, Cham (2014). doi:10.1007/978-3-319-13972-2_14
7. Chen, Z., Cabrera-Lozoya, R., Relan, J., Sohal, M., Shetty, A., Karim, R., Delingette, H., Gill, J., Rhode, K., Ayache, N., et al.: Biophysical modeling predicts ventricular tachycardia inducibility and circuit morphology: a combined clinical validation and computer modeling approach. J. Cardiovasc. Electrophysiol. 27(7), 851–860 (2016)
8. Wang, L., Gharbia, O.A., Horáček, B.M., Sapp, J.L.: Noninvasive epicardial and endocardial electrocardiographic imaging of scar-related ventricular tachycardia. J. Electrocardiol. 49(6), 887–893 (2016)
9. Komatsu, Y., Cochet, H., Jadidi, A., Sacher, F., Shah, A., Derval, N., Scherr, D., Pascale, P., Roten, L., Denis, A., et al.: Regional myocardial wall thinning at multi-detector computed tomography correlates to arrhythmogenic substrate in post-infarction ventricular tachycardia: assessment of structural and electrical substrate. Circ. Arrhythmia Electrophysiol. 6(2), 342–350 (2013)
10. Yezzi, A.J., Prince, J.L.: An eulerian PDE approach for computing tissue thickness. IEEE Trans. Med. Imaging 22(10), 1332–1339 (2003)

Analysis of Activation-Recovery Intervals from Intra-cardiac Electrograms in a Pre-clinical Chronic Model of Myocardial Infarction

Danielle Denisko[1], Samuel Oduneye[1,2], Philippa Krahn[1,2], Sudip Ghate[2], Ilan Lashevsky[3], Graham Wright[1,2], and Mihaela Pop[1,2(✉)]

[1] Department of Medical Biophysics, University of Toronto, Toronto, Canada
mihaela.pop@utoronto.ca
[2] Sunnybrook Research Institute, Toronto, Canada
[3] Arrhythmia Services Sunnybrook, Toronto, Canada

Abstract. Mapping of intracardiac electrical signals is a well-established clinical method used to identify the foci of abnormal heart rhythms associated with chronic myocardial infarct (a major cause of death). These foci reside in the 'border zone' (BZ) between healthy tissue and dense collagenous scar, and are the targets of ablation therapy. In this work we analyzed detailed features of the electrical signals recorded in a translational animal model of chronic infarct. Specifically, activation maps and bipolar voltages were recorded in vivo from 6 pigs at ~5 weeks following infarct creation, as well as 6 control (normal) pigs. Endocardial and epicardial maps were obtained during normal sinus rhythm and/ or pacing conditions via X-ray guided catheter-based mapping using an electro-anatomical CARTO system. The depolarization and repolarization maps were derived through manual annotation of electro-cardiogram waves, where the peak of the QRS wave marked the time of depolarization and the peak of the T wave marked the recovery time. Then, at each recording point, activation-recovery intervals ARIs (clinical surrogates of action potential duration) were found by subtracting activation times from repolarization times. Overall, we observed that ARI values in the BZ have recovered from the acute stage and were close to values in healthy tissue. In general we observed a weak negative correlation between the activation times and ARI values, also not a significant variation ($p < 0.5$) between mean ARI values in the BZ area and those in the healthy areas.

Keywords: CARTO electro-anatomical mapping · Bipolar voltages · Chronic infarct · Fibrosis · Border zone · Activation-recovery intervals

1 Introduction

Abnormal propagation of the electrical impulse in patients with prior myocardial infarction is a major cause of sudden cardiac death (SCD) in the industrialized world [1]. Structurally, a chronic infarct is comprised of a dense scar (collagenous fibrosis) and a border zone, BZ (also known as the peri-infarct). The latter is a 'heterogeneous' tissue, being comprised of a mixture of viable myocytes and collagen fibers [2]. Electrically,

© Springer International Publishing AG 2017
M. Pop and G.A. Wright (Eds.): FIMH 2017, LNCS 10263, pp. 280–288, 2017.
DOI: 10.1007/978-3-319-59448-4_27

the scar is non-conductive, whereas the BZ has reduced tissue conductivity (i.e., the action potential wave propagates slower than in the normal myocardium) [3]. These structural and electrical heterogeneities can trigger lethal ventricular arrhythmias such as ventricular tachycardia (VT), and ventricular fibrillation (VF) [4]. Thus, an important clinical task is the evaluation of disease severity and localization of BZ (the substrate of VT/VF) in patients with scar-related arrhythmia, along with the assessment of changes in tissue structure and electrical properties during infarct healing.

In the clinical laboratory, non-invasive imaging methods and various (more or less minimally) invasive electrophysiology (EP) tools have been intensively employed in the past decade. To date, several groups have demonstrated that the VT foci reside in the 'border zone' (BZ) between healthy tissue and collagenous scar, and can be structurally identified using MR imaging [5, 6]. Other studies have shown that infarct and BZ regions identified in *ex vivo* and *in vivo* MR images correlated well the location of these areas identified in electrical maps (by optical fluorescence imaging or unipolar/bipolar voltage maps) [7–9]. These studies provided a better tissue characterization in the dense scar and BZ, the latter being the target of a potential curative therapy known as RF thermal ablation [10]. However, there is a lack of thorough characterization of electrical signals in these critical tissue areas, and for this reason conventional unipolar/bipolar EP mapping often fails to localize the BZ in the clinical EP lab [11]. Furthermore, the action potential wave cannot be recorded directly *in vivo* via conventional EP catheters; thus, other methods are sought to identify such electrical characteristics of BZ.

The purpose of this work was to better understand the subtle characteristics of electrical signals in the peri-infarct, using a reproducible pre-clinical animal model of chronic infarction that mimics the post-infarct pathophysiology in humans. Specifically, here we focused on analyzing the features of intra-cardiac electrical signals in the BZ area by studying the *activation-recovery interval* (ARI), a clinical surrogate of action potential duration, APD [12]. We hypothesized that ARIs derived from locally-recorded activation maps (i.e., depolarization and repolarization times) can help identify distinctive diagnostic features in the BZ.

2   Materials and Methods

All *in vivo* experiments included in this study received ethics approval by our research institute. In this work, we performed X-ray guided electro-anatomical mapping studies in a swine model. This particular animal model is advantageous to use in translational frameworks, because the heart size and the infarct pathophysiology are close to those of human heart. Specifically, we performed catheter-based EP mapping experiments in 6 normal and 6 chronically infarcted swine, followed by data analysis to characterize ARIs in healthy tissue, dense scar and BZ.

2.1   Animal Model of Infarction

Myocardial infarction was created in juvenile swine (~30 kg) under X-ray guidance via a balloon catheter (Fig. 1). This catheter (indicated by the white arrow) was inserted and inflated into a major coronary artery (e.g. either the left anterior descendant LAD or the left circumflex artery, LCX), using a 90 min occlusion-reperfusion method. The lesions were then allowed to heal for ~5 weeks as in our previous studies [13]. At this time point post-infarction, edema had resorbed and the core of the chronic lesion is replaced by mature collagen (evident in the Masson Trichrome stain), while the BZ was character-ized by a patchy fibrosis, with altered electrical properties as observed in Fig. 1 from the decreased density in gap junctions, Cx43 (in agreement with [14]). The Cx43 protein is responsible for the cell-to-cell electrical conduction through the flow of ionic current [15]; its inhibition results in a slower conduction of the electrical impulse and triggers VT.

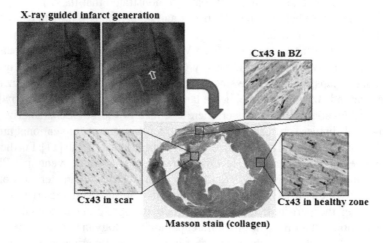

Fig. 1. X-ray guided infarct creation in a swine model (red arrow indicates the cessation of flow during the 90 min occlusion of LAD), along with histological images demonstrating deposition of dense fibrosis in the scar, a mixture of viable and collagen in BZ, and a decreased density in gap junctions (Cx43) in BZ. (Color figure online)

2.2   *In Vivo* Electro-Anatomical Mapping Study and ARI Analysis

Intra-cardiac waves, along with unipolar and bipolar voltage maps, were recorded by a catheter-based invasive method using a contact electro-anatomical mapping system (i.e., CARTO-XP, Biosense, USA). Some maps were recorded from the left ventricle (LV) endocardium and others from the epicardium, and some from both the LV-endo and epicardium. The electrical signals were acquired under sinus rhythm or pacing condi-tions (via a secondary catheter inserted at the apex of right ventricle, RV) from normal and infarcted pigs. During acquisition, cut-off filters were applied to measure only the signals within 30–400 Hz.

Figure 2 shows typical surfacic voltage maps recorded via a QuickStar catheter (2a) from the endocardial LV of an infarcted swine (2b) and a normal swine (2c), along with examples of intracardiac ecg-s waves from healthy tissue (pink), dense scar (red) and BZ (blue/green). These categories were defined based on clinical accepted thresholds: healthy tissue >1.5 mV; dense scar <0.5 mV; and, 0.5 mV < BZ < 1.5 mV [16].

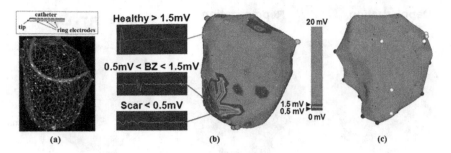

Fig. 2. Example of catheter-based electrical recordings from infarcted and normal pigs. (Color figure online)

For analysis, the activation-recovery intervals (ARIs) were determined for all intra-cardiac unipolar waves using the Wyatt method using the QRS and T waves [12]. Briefly, for each intra-cardiac wave, a local repolarization time (LRT) was found for biphasic and negative deflections using dV/dt_{max} (maximum rate of rise of voltage), whereas for positive deflections we used dV/dt_{min} on the descending limb of the T-wave. Finally, ARI was calculated as the difference between LRT and LAT (Fig. 3), as per the method proposed in [17]. All calculations and annotations were manually performed off-line using the CARTO analysis software.

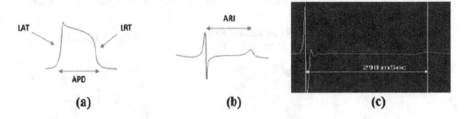

Fig. 3. Relation between (a) APD vs (b) ARI; and (c) an annotated sample of intracardiac ecg signal from a healthy myocardial tissue.

3   Results

Figure 4 shows an example of endocardial activation maps from the LV of a normal pig during normal sinus rhythm, NSR. Note that the earliest time point is in red and the latest activation time is depicted in purple. The propagation of depolarization wave starts in the anterior mid-septum and ends close to the mitral valve (see depolarization map in Fig. 4a). The repolarization follows a similar sequence (see repolarization map in Fig. 4b). The ARI map shows an opposite sequence, with shortest values near mitral valve, and longest near the septum (as clearly depicted in Fig. 4c).

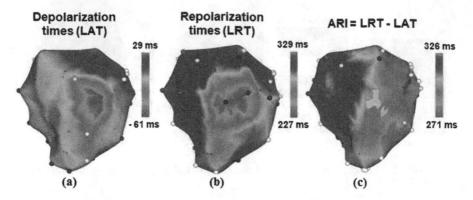

Fig. 4. Endocardial activation times reconstructed from CARTO recordings in the LV of a normal pig: (a) depolarization map (LAT); (b) repolarization (LRT) map; and (c) ARI map.

Figure 5 shows an example of endocardial LV maps from an infarcted heart, obtained under paced conditions (i.e., pacing catheter placed at the apex of RV). The earliest activation in the depolarization (LAT) map at site of the action potential wave break-through the septum, is in red. Notable here is a dispersion of repolarization near the dense scar region as evident in the associated LRT map. In this case, the repolarization sequence is opposite to the activation sequence (as compared to the normal case described in Fig. 4). The spatial pattern of the ARI closely follows the LRT pattern.

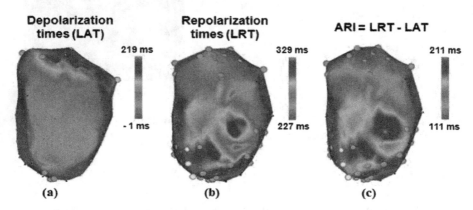

Fig. 5. Endocardial activation times reconstructed from CARTO recordings in the LV of an infarcted pig (note: the LAT map was recorded during pacing from RV): (a) depolarization map; (b) repolarization map; and (c) ARI map.

Figure 6 shows quantitative results of ARI calculations as a function of depolarization time recorded in LV. Notable is the clear separation of NSR data vs paced data, where the pacing cycle length (CL) was 400 ms in the LV of a normal heart (Fig. 6a). In contrast, we observed a significant overlap of the data points in the infarcted hearts in both healthy tissue areas (Fig. 6b) and BZ area (Fig. 6c), where the pacing CL was ~550–600 ms.

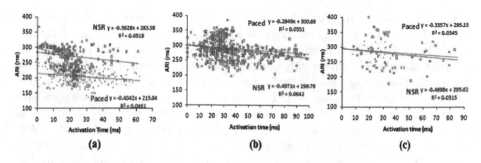

Fig. 6. Plots of ARI as a function of depolarization times (see text for details)

The number of recording points in the CARTO maps varied between 65 to ~200 points (with the less dense maps corresponding to the normal hearts).

Figure 7 shows a comparison between the mean ARI recorded in NSR from the LV-endo vs. epicardium of normal pigs and infarcted pigs (both areas: healthy and BZ). In the infarcted pigs, we obtained significant differences between the mean ARI in LV-endo and in epicardium, in the recordings from both healthy (remote) tissue ($p = 0.020$) and BZ ($p = 0.024$) tissue. Lastly, the differences in mean ARI between LV-endo and epicardium of normal pigs was found not significant ($p = 0.055$). Statistically significant differences were defined by p values <0.05 and were calculated using [18].

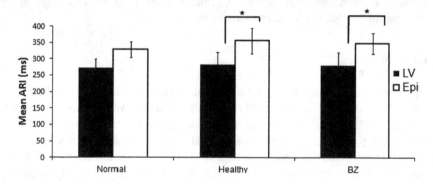

Fig. 7. Comparison between mean ARI in LV-endo vs. mean ARI in epicardium from different zones in normal and infarcted pigs (note: all recordings in NSR). At least 3 datasets were included in each category for averaging and statistical analysis. The error bars represent standard deviation (SD) (* $p < 0.05$).

4 Discussion

Accurate structural and electrical characterization of myocardial tissue in post-infarction remodelling is an important task in order to identify potential arrhythmogenic foci. Previous electrophysiology studies carried out in preclinical porcine models demonstrated that the amplitude of unipolar/bipolar waves can help detect the infarct location in electro-anatomical maps [19, 20]; however, these studies lacked the analysis of intracardiac ecgs in the BZ. To the best of our knowledge, this study is the first to analyze in detail the ARIs (as an APD surrogate) in a pre-clinical model of chronic infarction.

Our mean ARIs and activation time ranges obtained in the normal porcine hearts (controls) agreed well with the only other existing study in normal pigs [17]. We also observed that mean ARI values in the BZ have recovered from the acute stage of infarction and were somewhat close to ARI values in normal tissue. This can be somewhat explained by a recovery of APD in the BZ about one month following infarction, as previously suggested by other studies [3]. Interestingly however, although the mean ARI values averaged over all pigs were not significant ($p < 0.5$), the ARI values were different between BZ and healthy areas per each animal (up to ~40 ms). This result suggests that, to ensure accurate predictions, the parameterization of computer models from ARI maps (an important step in our future work) should be performed on an individual heart basis rather than using mean values or literature values.

The negative correlation between ARI vs activation time, suggests coupling between both variables, which leads to a more simultaneous repolarization process. The AP wave exhibited a time delay between LV and RV propagation, and from endocardium to epicardium breakthrough. The effects of pacing followed the typical APD restitution curve, with an ARI shortening in response to faster pacing (although we acknowledge that the pacing was performed only at CLs of 400 ms, 550 ms and 600 ms). Shorter pacing CL (Fig. 7a) resulted in lower ARI while pacing at a CL similar to NSR CL had little effect (Fig. 7b, c). Another limitation of this study is the relatively small sample size included in each statistical analysis (i.e., the number of datasets included in each category: pacing vs. NSR, LV vs. epicardial mapping, etc.).

5   Conclusion

To sum up, we successfully developed a preclinical framework to analyze the features of electrical signals in an animal model of chronic infarction. We acknowledge that the current data was limited to intracardiac ecg recordings in either sinus rhythm, or at one pacing frequency, or both sinus and (per animal). Thus, our next steps will focus on performing more controlled pre-clinical experiments, with a particular focus on obtaining more pacing frequencies data in the same subject (e.g. 4–5 pacing frequencies besides the sinus rhythm), to enable a thorough analysis of the ARI characteristics during pacing. This will also allow us to perform a parameter fitting for restitution curves. Furthermore, we will focus on personalizing 3D MRI-based computer models per individual heart from *in vivo* electrical maps, enabling more accurate predictions of cardiac wave propagation and simulation of scar-related arrhythmia.

Acknowledgement. This work was financially supported in part by a grant from CIHR (MOP # 93531) and a summer student award D&H (Sunnybrook).

References

1. Stevenson, W.G.: Ventricular scars and VT tachycardia. Trans. Am. Clin. Assoc. **120**, 403–412 (2009)
2. Bolick, D., Hackel, D., Reimer, K., Ideker, R.: Quantitative analysis of myocardial infarct structure in patients with ventricular tachycardia. Circulation **74**(6), 1266 (1986)
3. Ursell, P.C., Gardner, P.I., Albala, A., Fenoglio, J., Wit, A.L.: Structural and electrophysiological changes in the epicardial border zone of canine myocardial infarcts during infarct healing. Circ. Res. **56**, 436–451 (1985)
4. Janse, M.J., Wit, A.L.: Electrophysiological mechanisms of ventricular arrhythmias resulting from myocardial ischemia and infarction. Physiol. Rev. **69**(4), 1049–1169 (1989)
5. Bello, D., Fieno, D.S., Kim, R.J., et al.: Infarct morphology identifies patients with substrate for sustained ventricular tachycardia. J. Am. Coll. Cardiol. **45**(7), 1104–1110 (2005)
6. Pop, M., Ghugre, N.R., Ramanan, V., Morikawa, L., Stanisz, G., Dick, A.J., Wright, G.A.: Quantification of fibrosis in infarcted swine hearts by ex vivo late gadolinium-enhancement and diffusion-weighted MRI methods. Phys. Med. Biol. **58**(15), 5009 (2013)
7. Yan, A., Shayne, A., Brown, K., Gupta, S., Chan, C., Luu, T., Di Carli, M., Reynolds, H., Stevenson, W., Kwong, R.: Characterization of the peri-infarct zone by contrast-enhanced cardiac magnetic resonance imaging is a powerful predictor of post-myocardial infarction mortality. Circulation **114**, 32 (2006)
8. Wijnmaalen, A., van der Geest, R., van Siebelink, C.F.B.H., Wijnmaalen, H., Kroft, L., Bax, J., Reiber, J., Schalij, M., Zeppenfeld, K.: Head-to-head comparison of contrast-enhanced magnetic resonance imaging and electroanatomical voltage mapping to assess post-infarct scar characteristics in patients with ventricular tachycardias: real-time image integration and reversed registration. Eur. Heart J. **32**, 104 (2011)
9. Pop, M., Sermesant, M., Liu, G., Relan, J., Mansi, T., Soong, A., Peyrat, J.-M., Truong, M.V., Fefer, P., McVeigh, E.R., Delingette, H., Dick, A.J., Ayache, N., Wright, G.A.: Construction of 3D MR image-based computer models of pathologic hearts, augmented with histology and optical imaging to characterize the action potential propagation. Med. Image Anal. **16**(2), 505–523 (2012)
10. Verma, A., Kilicaslan, F., Schweikert, R., Tomassoni, G., Rossillo, A., Marrouche, N., Ozduran, V., Wazni, O., Elayi, S., Saenz, L., et al.: Short-and long-term success of substrate-based mapping and ablation of ventricular tachycardia in arrhythmogenic right ventricular dysplasia. Circulation **111**(24), 3209 (2005)
11. Desjardins, B., Crawford, T., Good, E., Oral, H., Chugh, A., Pelosi, F., Morady, F., Bogun, F.: Infarct architecture and characteristics on delayed enhanced MR imaging and electroanatomic mapping in patients with postinfarction ventricular arrhythmia. Heart Rhythm **6**(5), 644–651 (2009)
12. Haws, C.W., Lux, R.L.: Correlation between in vivo transmembrane APD and ARI from EGM. Circulation **81**(1), 281–288 (1989)
13. Pop, M., Ramanan, V., Yang, F., Zhang, L., Newbigging, S., Ghugre, N., Wright, G.A.: High resolution 3D T1* mapping and quantitative image analysis of the gray zone in chronic fibrosis. IEEE Trans. Biomed. Eng. **61**(12), 2930–2938 (2014)
14. Zhang, Y., Wang, H., Kovacs, A., Kanter, E.M., Yamada, K.A.: Reduced expression of Cx43 attenuates ventricular remodelling after myocardial infarction via impaired TBF-B signaling. Am. J. Physiol. Heart Circ. Physiol. **298**(2), H477–H487 (2010)
15. Jansen, J., van Veen, T.A.B., de Jong, S., van der Nagel, R., van Rijen, H.V.M., et al.: Reduced Cx43 expression triggers increased fibrosis due to enhanced fibroblast activity. Circ. Arrhythmia Electrophsiol. **5**, 380–390 (2012)

16. Codreanu, A., Odille, F., Aliot, E., et al.: Electro-anatomic characterization of post-infarct scars comparison with 3D myocardial scar reconstruction based on MR imaging. J. Am. Coll. Cardiol. **52**, 839–842 (2008)
17. Gepstein, L., Hayam, G., Ben-HAim, S.A.: Activation-repolarization coupling in the normal swine endocardium. Circulation **96**, 4036–4043 (1997)
18. www.vassarstats.net
19. Wrobleski, D., Houghtaling, C., Josephson, M.E., Ruskin, J., Reddy, V.: Use of electrogram characteristics during sinus rhythm to delineate the endocardial scar in a porcine model of healed myocardial infarction. J. Cardiovasc. Electrophysiol. **14**, 524–529 (2003)
20. Callans, J.D., Ren, J.-F., Michele, J., Marchlinski, F., Dillon, S.: Electroanatomic left ventricular mapping in the porcine model of healed anterior myocardial infarction. Correlation with intracardiac echocardiography and pathological analysis. Circulation **100**, 1744–1750 (1999)

Improving the Spatial Solution of Electrocardiographic Imaging: A New Regularization Parameter Choice Technique for the Tikhonov Method

Judit Chamorro-Servent[1,2,3](✉) [iD], Rémi Dubois[1,4,5], Mark Potse[1,2,3] [iD], and Yves Coudière[1,2,3] [iD]

[1] IHU Liryc, Electrophysiology and Heart Modeling Institute, Foundation Bordeaux Université, Pessac, Bordeaux, France
[2] CARMEN Research Team, Inria, Bordeaux, France
judit.chamorro-servent@inria.fr
[3] Univ. Bordeaux, IMB, UMR 5251, CNRS, INP-Bordeaux, Talence, France
[4] Univ. Bordeaux, CRCTB, U1045, Bordeaux, France
[5] INSERM, CRCTB, U1045, Bordeaux, France

Abstract. The electrocardiographic imaging (ECGI) inverse problem is highly ill-posed and regularization is needed to stabilize the problem and to provide a unique solution. When Tikhonov regularization is used, choosing the regularization parameter is a challenging problem. Mathematically, a suitable value for this parameter needs to fulfill the Discrete Picard Condition (DPC). In this study, we propose two new methods to choose the regularization parameter for ECGI with the Tikhonov method: (i) a new automatic technique based on the DPC, which we named ADPC, and (ii) the U-curve method, introduced in other fields for cases where the well-known L-curve method fails or provides an over-regularized solution, and not tested yet in ECGI. We calculated the Tikhonov solution with the ADPC and U-curve parameters for in-silico data, and we compared them with the solution obtained with other automatic regularization choice methods widely used for the ECGI problem (CRESO and L-curve). ADPC provided a better correlation coefficient of the potentials in time and of the activation time (AT) maps, while less error was present in most of the cases compared to the other methods. Furthermore, we found that for in-silico spiral wave data, the L-curve method over-regularized the solution and the AT maps could not be solved for some of these cases. U-curve and ADPC provided the best solutions in these last cases.

Keywords: Inverse problem · Regularization · Electrocardiographic imaging · Potentials · Tikhonov regularization · Ill-posed problems

1 Introduction

During the past decades, much progress has been made in solving the inverse problem of electrocardiography [1–7]. However, despite all the success of non-invasive electrocardiographic imaging (ECGI), the understanding and treatment of many cardiac diseases is not feasible yet without an improvement of the inverse problem solution [8].

© Springer International Publishing AG 2017
M. Pop and G.A. Wright (Eds.): FIMH 2017, LNCS 10263, pp. 289–300, 2017.
DOI: 10.1007/978-3-319-59448-4_28

Solution of the 'inverse problem' depends on the 'forward problem', i.e. on specification of the relationship between potential sources on the cardiac surface, Φ_E, and the body surface measured potentials, Φ_T.

The inverse problem (finding $\Phi = \Phi_E$ at epicardial surface Γ_E) is innately ill-posed. That means that is extremely sensible to: (i) noise on the measured potentials, (ii) uncertainty in the location of measurement sites with respect to the surface on which the sources are distributed, (iii) errors of segmentation of the geometries, and (iv) influence of cardiac motion.

Regularization incorporates additional knowledge in the inverse problem by applying constraints to the solution, to stabilize the problem yielding realistic and unique results. When regularization is applied, the weight of the constraints (regularization parameter) has to be determined to find a balance between solutions purely based on the body-surface potentials and solutions that are constrained too strictly. While the former may be severely distorted by ill-posedness, the latter may have too much bias. Given the ill-posedness, the regularization parameter choice has an important influence on the solution [8].

In this study we focused on the two-norm Tikhonov regularization technique for the method of fundamental solution (MFS), an homogeneous meshless method adapted to ECGI by Wang and Rudy [9, 10]. Specifically, we will focus on the choice of the regularization parameter. In many implementations, the Tikhonov regularization problem is solved by manually selecting the regularization parameter α. This is done using a sequence of regularization parameters and selecting the value that gives the best results, as judged by the user. Obviously, the procedure is subjective and time consuming. To overcome this problem, several automatic methods for selecting regularization parameters have been suggested over the years. These include: (i) Strategies based on the calculation of the variance of the solution, which requires prior knowledge of the noise, (ii) Strategies that do not need *apriori* information. For ECGI, we are interested in the latter.

Regarding the second group (ii) of automatic methods previously used in the MFS ECGI literature [9, 10], the regularization parameter is obtained by using the mean of the regularization parameters provided by the Composite Residual and Smoothing Operator (CRESO) technique [10]. The CRESO method has been found to perform comparably to the "optimal" regularization parameter that provides the minimum root-mean-square error (RMSE) between the computed epicardial potentials (Φ_E) and the measured ones [10]. When other numerical methods, such as the Boundary Element method (BEM), were used to solve the ECGI problem, the L-curve criterion has been highly used by the community to find the regularization parameter [8, 11].

While the efficacy of the CRESO and L-curve methods has been proven in the wide inverse problems literature [8–16], finding an automatic regularization parameter for Tikhonov regularization that is suitable for all ill-posed inverse problems is a challenging problem [12, 15] and CRESO and L-curve may over-regularize the solution or fail for some cases. This is especially important when we deal with medical problems such as cardiac pathologies, where different pathologies may need different regularization.

For this reason, the overarching goal of this paper is to show the feasibility of the U-curve method [17, 18], not used yet in cardiac inverse problems, to our knowledge; and to develop a new automatic method, ADPC, based on the Discrete Picard Condition (DPC), a mathematical condition that any suitable regularization parameter for Tikhonov regularization must fulfill [12].

2 Methods

2.1 Method of Fundamental Solutions and Tikhonov Regularization

MFS is a homogeneous meshless approach that was adapted to ECGI to overcome some of the issues of the classical meshes-based methods [9, 10]. MFS does not need the topological relations between nodes and so completely avoids disadvantages of accuracy degradation and complexity augmentation frequently encountered in classical numerical methods because of remeshing. Thus it avoids, negative effects introduced by errors in the segmentation process and/or singularities in the boundaries.

In the MFS, the potentials are expressed as a linear combination of Laplace fundamental solution over a discrete set of virtual source points (Dirac masses) placed outside of the domain of interest, Ω, where Ω is the volume conductor enclosed by the epicardial surface Γ_E and the body surface, Γ_T. Specifically the potential Φ for $x \in \Omega$ is sought as

$$\Phi(x) = a_0 + \sum_{j=1}^{NS} f(x - y_j) a_j, \tag{1}$$

where the $(y_j)_{j=1...N_S}$ are the N_S locations of the sources $(y_j \notin \Omega)$, and the $(a_j)_{j=1...N_S}$ are their coefficients. Here, f stands for the fundamental solution to the Laplace equation, $f(r) = \frac{1}{4\pi} \frac{1}{|r|}$ where r is the 3D Euclidean distance between point x and virtual source y. After discretization on collocation points on Γ_T, and fixed number N_S of source locations [9], this formulation yields a linear system of equations with matrix system M:

$$M = \begin{pmatrix} 1 & f(r_{11}) & \cdots & f(r_{1N_S}) \\ \vdots & & \ddots & \vdots \\ 1 & f(r_{N_T 1}) & \cdots & f(r_{N_T N_S}) \\ 0 & \partial_n f(r_{11}) & \cdots & \partial_n f(r_{1N_S}) \\ \vdots & & \ddots & \vdots \\ 0 & \partial_{n_{N_T}} f(r_{N_T 1}) & \cdots & \partial_{n_{N_T}} f(r_{N_T N_S}) \end{pmatrix},$$

where $r_{ij} = x_i - y_j, x_i \in \Gamma_T$ and $(y_j)_{j=1...N_S}, y_j \notin \Omega$. Hence, homogeneous Neumann conditions (no flux conditions, $\partial_n \Phi = 0$) and Dirichlet conditions ($\Phi = \Phi_T$, where Φ_T are the potentials acquired on the torso) on Γ_T are considered in an apparently equivalent manner in the standard MFS described in [9]. Therefore, only the sources coefficients are unknown, resulting in a quadratic minimizing problem (Tikhonov regularization problem), find the sources coefficients $a = (a_0, a_1, \ldots, a_{N_S})^T \in \mathbb{R}^{1+N_S}$ that minimize

$$J(a, \alpha) = ||Ma - b||^2 + \alpha ||a||^2 \tag{2}$$

Where $b = \begin{pmatrix} \Phi_T^* \\ 0 \end{pmatrix}$ is a $2N_T x1$ vector, given some potentials $\Phi_T^* = \left(\Phi_i^*\right)_{i=1,\cdots,N_T}$

recorded on N_T torso electrodes, and $\alpha > 0$ is the Tikhonov regularization parameter. Note that the first term $\|Ma - b\|^2$ measures the goodness-of-fit, i.e., how well the unknown sources coefficients a predicts the given boundary conditions at body surface. The second term $\|a\|^2$ measures the regularity of the sources coefficients a. It suppresses (most of) its large noise components by controlling its norm (since the naïve solution is dominated by high-frequency components with large amplitudes). The balance between both terms is controlled by the regularization parameter, α.

Once solved (2) and found the source coefficients $a = \left(a_0, a_1 \dots, a_{N_S}\right)^T \epsilon \mathbb{R}^{1+N_s}$, the Φ_E are found by using (1) for $x \in \Gamma_E$ and $\left(y_j\right)_{i=1..N_S}, y_j \notin \Omega$.

2.2 Discrete Picard Condition

Instead of the condition number of the matrix of a linear system, the rate of decrease of the singular values (SVs) is a better indication of the conditioning of the problem [12, 13]. Furthermore, the Discrete Picard Condition (DPC) [12] provides an objective assessment of the ill-posedness of the entire problem relating the matrix information (by means of the SVs) with the measurements [12].

The Tikhonov solution can be obtained by equalling the gradient of (2) to zero:

$$\nabla J_a(a, \alpha) = \nabla\left((Ma - b)^T(Ma - b) + \alpha Ia^T a\right) = (M^T M)a - M^T b + \alpha^2 Ia = 0, \text{ and } a_\alpha = \left(M^T M + \alpha^2 I\right)^{-1} M^T b.$$

Then, by decomposing the forward matrix of the MFS in terms of SV decomposition $(M = USV^T)$ and writing $I = VV^T$,

$$a_\alpha = \sum_{i=1}^{\min(2*N_T, N_s+1)} \frac{\sigma_i}{\sigma_i^2 + \alpha^2} u_i^T b v_i = \sum_{i=1}^{\min(2*N_T, N_s+1)} \frac{\sigma_i^2}{\sigma_i^2 + \alpha^2} \frac{u_i^T b}{\sigma_i} v_i \qquad (3)$$

where σ_i are the SVs in descending order, $\sigma_1 \geq \cdots \geq \sigma_{\min(2*N_T, N_s+1)}$ (the elements of the diagonal matrix S).

The DPC is satisfied if the data space coefficients $|u_i^T b|$, on average, decay to zero faster than the respective σ_i's. The representation of $|u_i^T b|$, σ_i, and the respective quotient in a same plot in logarithmic scale is known as a Picard plot.

In ill-posed problems, such as ECGI, there may be a point where the data become dominated by errors and the DPC fails. In other words, for larger values of the index i the solution coefficients $\dfrac{|u_i^T b|}{\sigma_i}$ start to increase and the computed solutions (3) are completely dominated by the SVD coefficients corresponding to the smallest SVs. In these cases, to compute a satisfactory solution by means of Tikhonov regularization, the DPC has to be fulfilled [12, 18]. That is, the σ_i above the regularization parameter α

(useful SVs) must decay to zero slower than the corresponding right hand side coefficients, $|u_i^T b|$. The DPC determines how well the regularized solution approximates the unknown, exact solution.

2.3 Existent Regularization Parameter Choice Techniques

Composite Residual and Smoothing Operator (CRESO)
CRESO was developed by Colli-Franzone et al. [14]. The method was presented as an empirical heuristic but has become widely accepted as the preferred method of parameter estimation in a large class of ill-posed bioelectric inverse problems [15]. The CRESO method chooses that parameter value which generates the first local maximum of the difference between the derivative of the constraint term and the derivative of the residual term

$$C(\alpha) = \left\{ \frac{d}{d(\alpha^2)} \left(\alpha^2 ||x(\alpha)||^2 \right) - \frac{d}{d(\alpha^2)} ||Mx(\alpha) - b||^2, \alpha > 0 \right\} \tag{4}$$

L-curve
The best-known method for estimating a value for the regularization parameter is defined in terms of the L-curve [15, 16]

$$L(\alpha) = \{(||Mx(\alpha) - b||, ||x(\alpha)||), \alpha > 0\} \tag{5}$$

To choose the L-curve regularization parameter we used the criterion proposed by Hansen and O'Leary [16], where the optimal α-value corresponds to the point on the log-log plot of the L-curve possessing maximum curvature.

U-curve
The U-curve method was proposed by Krawczyck-Stando and Rudnicki [17]. The U-curve is the plot of the sum of the inverse of the regularized solution norm $||x(\alpha)||$ and the corresponding residual error norm $||Mx(\alpha) - b||$, for $\alpha > 0$ on a log-log scale

$$U(\alpha) = \frac{1}{||Mx(\alpha) - b||^2} + \frac{1}{||x(\alpha)||^2} \tag{6}$$

The U-curve is a U-shaped function, where the sides of the curve correspond to regularization parameters for which either the solution norm or the residual norm dominates. The optimum regularization value is the value for which the U-curve has a minimum. U-curve is computationally efficient since it provides an *a priori* interval where to find this minimum [17, 18]. Furthermore, while its feasibility has not been tested yet for ECGI, it has been shown to perform well for some numerical and biomedical problems when the L-curve failed or gavean over-regularized solution [17, 18].

2.4 ADPC: A New Regularization Parameter Choice Method

As cited before, a suitable regularization parameter α for Tikhonov regularization must fulfill the DPC. This means that the σ_i above the suitable α must decay to zero slower than the corresponding $|u_i^T b|$, in order to prevent the computed Tikhonov solutions (3) from being completely dominated by the SVD coefficients corresponding to the smallest SVs. Therefore, we performed an automatic regularization parameter choice method based on this condition as follows:

1. We calculate the singular value decomposition of the MFS matrix, M, to obtain the left singular vectors (u_i) and the singular values (σ_i).

2. For each instant of time, $t_k(ms)$, we compute $\left|u_i^T b_{t_k}\right|$ and we calculate the coefficients for a polynomial $p(i)_{t_k}$ of degree from 5 to 7 that is a best fit (in a least-squares sense) for the log ($\left|u_i^T b_{t_k}\right|$). Therefore, we have: $p(i)_{t_1},\dots,p(i)_{t_{N_t}}$ polynomials, one for each instant $t_k, k = 1, \cdots, N_t$ instants of time.

3. For each $p(i)_{t_k}$, we find: $\alpha_{t_k} = \sigma_{max\{i\}}$, such that $\log(\sigma_i) \geq p(i)_{t_k}$, with σ_i in descending order.

4. The ADPC regularization parameter is: $\alpha = \text{median}(\alpha_{t_k})$.

Note that step 2 and 3 of this algorithm are empirically choosing the lower limit that any suitable Tikhonov regularization parameter can achieve to still fulfill the DPC. Both steps are looking for the last index i before the SVD coefficients (corresponding to the smallest SVs) start to dominate the solution. That means, before $\log(\sigma_i)$ starts to be smaller than log ($\left|u_i^T b_{t_k}\right|$). The fitting of the log ($\left|u_i^T b_{t_k}\right|$) by a polynomial is done to simplify the search for a suitable index i. Since the behavior of the SVs of ECGI MFS problem and the fact that ADPC parameter choice is based on the necessary DPC, our algorithm provides a suitable regularization parameter.

2.5 Simulated Data

To test the effect of the new regularization parameter choice method and to compare with previous ones, eight different activation patterns (1 single site pacing in the right ventricular (RV) free wall, 1 single site pacing in the left ventricular (LV) lateral endo-cardial wall, 1 single site pacing in the LV mid wall, 1 single pacing site in the LV lateral epi and 4 single spiral waves) were simulated [19]. Propagating activation was simulated with a monodomain reaction-diffusion model in a realistic 3D model of the human ventricles, with transmembrane ionic currents computed with the Ten Tusscher et al. model for the human ventricular myocyte [20]. The transmembrane currents from the monodomain problem were used to compute the extracellular potential distribution throughout the torso by solving a static bidomain problem in a heterogeneous, aniso-tropic torso model at 1 mm resolution [21]. The torso model used had heterogeneous conductivity, with anisotropic skeletal muscle, lungs, and intracavitary blood. The heart model consisted of left and right ventricles, with a 0.2 mm spatial resolution, and an anisotropic conduction derived from rule-based fiber orientation. Cardiac and thoracic anatomy was based on MRI data. The simulations provide the theoretical, in-silico Φ_T and Φ_E every 1 ms.

2.6 Analysis of the Reconstructed Potentials and Activations Time Maps: Statistical Approach

Once reconstructed the potentials on the epicardium, correlations coefficients (CC) and root mean square errors (RMSE) [9], and relative errors ($RE = ||\Phi_E - \widetilde{\Phi_E}||/||\Phi_E||$, where Φ_E are the simulated potentials and $\widetilde{\Phi_E}$ the reconstructed ones), were computed through time. Thus, we compared the reconstructed potentials on the epicardium against the simulated ones. Afterwards, we calculated the three quartiles of the resulting CCs, RMSEs and REs, (Median [1st quartile, 3rd quartile]).

Similarly, we calculated the activation time (AT) maps and the respective CCs and REs (Median, [min, max]).

3 Results

3.1 Ill-Posedness and Discrete Picard Condition

The DPC condition allows us to investigate the behavior of the SVD coefficients $|u_i^T b|$ (of the right hand side) and $\dfrac{|u_i^T b|}{\sigma_i}$ (of the solution (3)). A logarithmic plot of these coefficients, together with the respective SVs, σ_i (also in logarithmic scale), is often referred to as DPC plot [12]. Given $\sigma_1 \geq \cdots \geq \sigma_{\min(2*N_T,N_s+1)}$, lower indices i correspond to larger SVs and bigger indices correspond to smaller SVs.

The DPC plot of Fig. 2 is an example for an in-silico single spiral wave in a fixed instant of time chosen here randomly around $t_k = 100$ ms. This choice is based on the behavior of the body-surface potentials of Fig. 1 and in the literature [23].

The DPC plot above shows the ill-posedness of the inverse problem. We can see that after the index $i \sim 240$SVs (being $N_T = 252$), the data become dominated by errors/noise and the solution (3) starts to become instable and affected by them. This is due to the fact that for $i > 240$, the SVs (dot points), σ_i, start to decrease to zero faster than the right hand side coefficients (crosses) and the DPC fails for any regularization parameter below σ_{240}. Besides, by looking at the different color arrows for the different parameter choices hold on the DPC plot, we can observe the over-regularization provided by the L-curve method.

We want to remark that while here we only plot the DPC for a certain spiral wave data and for a fixed instant of time, t_k, (as example to explain the details of a DPC plot) the SVs for the ECGI MFS matrix as formulated in [9, 10] usually decrease slowly initially, and they start to highly decrease for larger values of the index i (commonly

$i \sim N_T$). The same SV decay behavior for ECGI MFS for other real torso-heart geometries can be observed in [24].

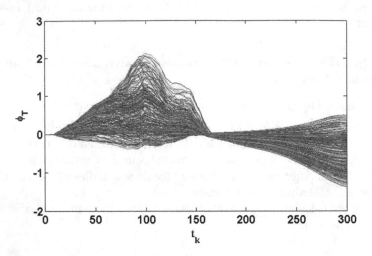

Fig. 1. In-silico body-surface potentials, Φ_T for $t_k = 1, \cdots, 300$ ms (corresponding to the first beat)
of the single spiral wave used for the DPC plot example of Fig. 2.

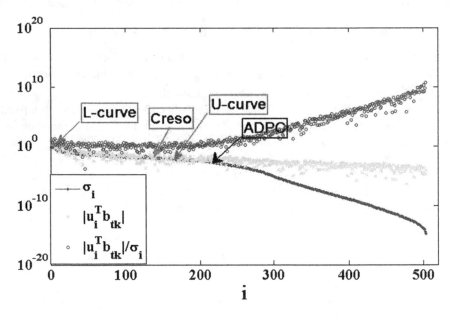

Fig. 2. DPC plot for a time using the regularization toolbox of Matlab [22] from PC. Hansen.
The dot points show the decay of σ_i, the crosses $|u_i^T b|$, and the circles the quotient between both.
The values of the different regularization parameter choices are located in the plot with different
color arrows and their respective methods' name.

Figure 3 shows the reconstructed potentials in a random location on the epicardium through the different time instants, t_k (ms) for the different regularization parameters together with the simulated ones. We observe that in this example the solution based on the L-curve is over-regularized.

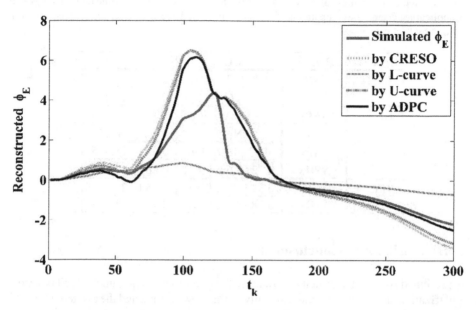

Fig. 3. Signal reconstructed in a point of the epicardium along the time for the different regularization parameters methods as legend indicates against the simulated one.

3.2   Correlation Coefficients and Relative Errors of Reconstructed Potentials in Time and Activation Time Maps

Table 1 presents the three quartiles of CCs, RMSEs and REs of the reconstructed potentials with the different regularization parameters found for the single site pacing

Table 1. Median [1st quartile, 3rd quartile] of the:CCs, RMSE and REs of the reconstructed potentials along the time.

Method	Dataset	CCs	RMSEs	REs (L₂-norm)
CRESO	SSPs	0.77 [0.49, 0.86]	0.05 [0.04, 0.06]	0.81 [0.62, 1.03]
	SSWs	0.74 [0.57, 0.86]	0.07 [0.05, 0.10]	0.77 [0.64, 0.96]
L-curve	SSPs	0.76 [0.51, 0.85]	0.05 [0.04, 0.06]	0.89 [0.67, 1.06]
	SSWs	0.70 [0.46, 0.82]	0.08 [0.05, 0.13]	0.91 [0.80, 1.08]
U-curve	SSPs	0.76 [0.53, 0.86]	0.06 [0.05, 0.08]	0.80 [0.60, 0.97]
	SSWs	0.76 [0.60, 0.87]	0.07 [0.05, 0.09]	0.75 [0.61, 0.93]
ADPC	SSPs	0.81 [0.70, 0.89]	0.06 [0.04, 0.08]	0.71 [0.54, 0.83]
	SSWs	0.81 [0.68, 0.90]	0.06 [0.04, 0.09]	0.69 [0.55, 0.87]

simulations (SSPs) and single spiral waves simulations (SSWs). Table 2 shows the minimum, median, and maximum of CCs and REs of the respective activation time (AT) maps. In both tables the best CCs, RMSEs and/or REs are outlined in red color.

Table 2. Median [min, max] of the CCs and of the REs for the activation time (AT) maps.NA*: Not applicable because the computation of the ATs is inhibited due to the over-regularized solution provided by L-curve.

Method	Dataset	CCs	REs (L_2-norm)
CRESO	SSPs	0.8 [0.75, 0.9]	0.57 [0.41, 0.68]
	SSWs	0.61 [0.60, 0.83]	0.24 [0.15, 0.38]
L-curve	SSPs	NA*	0.61 [0.4, 1]
	SSWs	NA*	1 [0.4, 1]
U-curve	SSPs	0.83 [0.77, 0.9]	0.44 [0.28, 0.54]
	SSWs	0.72 [0.68, 0.84]	0.26 [0.11, 0.37]
ADPC	SSPs	0.84 [0.81, 0.86]	0.35 [0.23, 0.46]
	SSWs	0.78 [0.42, 0.84]	0.25 [0.11, 0.37]

4   Discussion and Conclusions

We presented ADPC, a new method to choose the regularization parameter for Tikhonov regularization and we showed the feasibility of this new method and the existent U-curve method (never before used for ECGI problems). Our results showed the importance of the choice of the regularization parameter.

For the in-silico data used here, we found that the L-curve provided an over-regularized solution for the single site pacing (SSP) in the RV and for three of the single spiral waves simulations (SSWs), inhibiting the computation of the AT map for these SSWs. In Fig. 2, for one of the SSWs, we can clearly see how the L-curve regularization parameter provided is far above the other regularization parameters, as well as of the moment where the SVs start to decay to zero faster than the respective right hand side coefficients. For the cases that this happens, a few SVs are highly weighted in the Tikhonov solution (3), and this results in a highly over-regularized solution. In Fig. 3, we can visualize this fact in terms of the reconstructed potentials in a random epicardial point along the time (for a SSP in the RV dataset); when using the L-curve regularization parameter.

While CRESO seems to give lower RMSE in agreement with the work of Rudy [10] for the single site pacing simulations (Table 1), the U-curve and specially the ADPC provide higher CCs in terms of potentials (in time) and lower REs. Furthermore, the CCs of the ATs are also higher for the U-curve and the ADPC, especially for the SSWs(5–11% of improvement), while the REs are not affected for the SSWs and decreased for the SSPs in a 21% respect to the usual CRESO method (Table 2).

We would like to remark that this study provides results for the ECGI MFS problem, such as described in [9, 10]. However, it is well known that a suitable automatic regularization parameter choice method depends on the inverse problem treated [12].

Therefore, if the ECGI problem is numerically approximated by a different method such as the finite element method or the boundary element method, this study must be repeated for the specific problem in each case. DPC provides an interval that any Tikhonov regularization parameter should fulfill. The empirical choice of the lower threshold for the ADPC algorithm is working well for the ECGI MFS problem [9, 10] where the decay behavior of SVs does not have jumps and always decays slowly for the first SVs $i \sim NT$, $i \leq NT$, and the respective coefficients of the solution start to be unstable exactly at that moment. The empirical lower threshold of DPC (step 3), chosen in our ADPC algorithm, may be differently adapted for others approaches with different SV decayment behavior. More detail of DPC can be found in earlier work [12, 18] that may help to the reader to adapt it to a specific inverse problem. Equally in different inverse problems involving spatio-temporal solutions, the choice of the mode or mean instead of median could be more accurate for step 4 than the median proposed here. Nevertheless, we believe that, as shown for our specific problem, a good adaptation of ADPC for each different numerical approach of ECGI or other ill-posed problems should provide suitable results since ADPC is really focused on the mathematical DPC that any Tikhonov solution should fulfill.

Finally, it seems advisable to test the four regularization parameter choice methods with experimental data with different pathologies and study the different solutions in future work.

Acknowledgements. This study received financial support from the French Government under the "Investments of the Future" program managed by the National Research Agency (ANR), Grant reference ANR-10-IAHU-04 and from the Conseil Régional Aquitaine as part of the project "Assimilation de données en cancérologie et cardiologie". This work was granted access to the HPC resources of TGCC under the allocation x2016037379 made by GENCI.

References

1. Shah, A.: Frontiers in noninvasive cardiac mapping, an issue of cardiac electrophysiology clinics. Elsevier Health Sci. **7**(1), 1–164 (2015)
2. Rudy, Y.: Noninvasive electrocardiographic imaging of arrhytmogenic substrates in humans. Circ. Res. **112**, 849–862 (2013)
3. Wang, Y., et al.: Noninvasive electro anatomic mapping of human ventricular arrhythmias with electrocardiographic imaging. Sci. Transl. Med. **3**(98), 98ra84 (2011)
4. Ramanathan, C., et al.: Noninvasive electrocardiographic imaging for cardiac electrophysiology and arrhythmia. Nat. Med. **10**(4), 422–428 (2004)
5. Dubois, R., et al.: Non-invasive cardiac mapping in clinical practice: Application to the ablation of cardiac arrhythmias. J. Electrocardiol. **48**(6), 966–974 (2015)
6. Haissaguerre, M., et al.: Noninvasive panoramic mapping of human atrial fibrillation mechanisms: a feasibility report. J. Cardiovasc. Electrophysiol. **24**, 711–717 (2013)
7. Cochet, H., et al.: Cardiac arrythmias: multimodal assessment integrating body surface ECG mapping into cardiac imaging. Radiology **271**(1), 239–247 (2014)
8. Cluitmans, M.J.M., et al.: Noninvasive reconstruction of cardiac electrical activity: update on current methods, applications and challenges. Neth. Heart J. **23**(6), 301–311 (2015)

9. Wang, Y., Rudy, Y.: Application of the method of fundamental solutions to potential-based inverse electrocardiography. Ann. Biomed. Eng. **34**, 1272–1288 (2006)
10. Rudy, Y.: U.S. Patent No. 6,772,004. U.S. Patent and Trademark Office, Washington, DC (2004)
11. Milanič, M., et al.: Assessment of regularization techniques for electrocardiographic imaging. J. Electrocardiol. **47**(1), 20–28 (2014)
12. Hansen, P.C.: Discrete Inverse Problems: Insight and Algorithms, vol. 7. SIAM, Philadelphia (2010)
13. Tsai, C.C., et al.: Investigations on the accuracy and condition number for the method of fundamental solutions. Comput. Model. Eng. Sci. **16**(2), 103 (2006)
14. Colli-Franzone, P., et al.: A mathematical procedure for solving the inverse potential problem of electrocardiography. Analysis of the time-space accuracy from in vitro experimental data. Math. Biosci. **77**(1–2), 353–396 (1985)
15. Ruan, S., Wolkowicz, G.S.K., Wu, J. (eds.): Differential Equations with Applications to Biology, vol. 21. American Mathematical Society, Providence (1999)
16. Hansen, P.C., O'Leary, D.P.: The use of the L-curve in the regularization of discrete ill-posed problems. SIAM J. Sci. Comput. **14**(6), 1487–1503 (1993)
17. Krawczyk-Stańdo, D., Rudnicki, M.: Regularization parameter selection in discrete ill-posed problems—the use of the U-curve. Int. J. Appl. Math. Comput. Sci. **17**(2), 157–164 (2007)
18. Chamorro-Servent, J., et al.: Feasibility of U-curve method to select the regularization parameter for fluorescence diffuse optical tomography in phantom and small animal studies. Opt. Express **19**(12), 11490–11506 (2011)
19. Duchateau, J., Potse, M., Dubois, R.: Spatially coherent activation maps for electrocardiographic imaging. IEEE Trans. Biomed. Eng. (2016, in print)
20. Ten Tusscher, K.H.W.J., et al.: A model for human ventricular tissue. Am. J. Physiol. Heart Circ. Physiol. **286**(4), H1573–H1589 (2004)
21. Potse, M., et al.: Cardiac anisotropy in boundary-element models for the electrocardiogram. Med. Biol. Eng. Compu. **47**(7), 719–729 (2009)
22. Hansen, P.C.: Regularization tools version 4.0 for Matlab 7.3. Numer. Algorithms **46**, 189–194 (2007)
23. Ghodrati, A., et al.: Wavefront-based models for inverse electrocardiography. IEEE Trans. Biomed. Eng. **53**(9), 1821–1831 (2006)
24. Chamorro-Servent, J., et al.: Adaptive placement of the pseudo-boundaries improves the conditioning of the inverse problem. Comput. Cardiol. **43**, 705–708 (2016)

Statistical Atlases for Electroanatomical Mapping of Cardiac Arrhythmias

Mihaela Constantinescu[1]([✉]), Su-Lin Lee[1], Sabine Ernst[2],
and Guang-Zhong Yang[1]

[1] The Hamlyn Centre for Robotic Surgery, Imperial College London, London, UK
`mihaela.constantinescu12@imperial.ac.uk`
[2] The Royal Brompton and Harefield Hospital, London, UK

Abstract. Electroanatomical mapping is a mandatory time-consuming
planning step in cardiac catheter ablation. In practice, interventional
cardiologists target specific endocardial areas for mapping based on per-
sonal experience, general electrophysiology principles, and preoperative
anatomical scans. Effective fusion of all available information towards
a useful mapping strategy has not been standardised and achieving the
optimal map within time and space constraints is challenging. In this
paper, a novel framework for computing optimal endocardial mapping
locations in patients with congenital heart disease (CHD) is proposed.
The method is based on a statistical electroanatomical model (SEAM)
which is instantiated from preoperative anatomy in order to achieve an
initial prediction of the electrical map. Simultaneously, the anatomical
areas with the highest frequency of mapping among the similar cases in
the dataset are detected and a classifier is trained to filter these points
based on the electroanatomical data. The framework was tested in an
iterative process of adding mapping points to the SEAM and computing
the instantiation error, with retrospective clinical data of 66 CHD cases
available.

1 Introduction

Cardiac rhythm disorders are serious life-long comorbidities affecting patients
with surgical repair of congenital heart disease (CHD). These life-threatening
conditions are commonly treated by radiofrequency (RF) catheter ablation with
a high input from the clinician in terms of personalised electroanatomical mapping, RF energy delivery and follow-up. CHD electroanatomical mapping is additionally challenging due to the structural differences in anatomy and unusual
haemodynamics. All these physiological changes affect the electrical conduction
system and build the substrate for arrhythmias uncommon to normal hearts,
but specific to each CHD in particular [4].

Pre-procedural planning is a major factor in the success and duration of
cardiac catheter ablation. The state-of-the-art in intra-operative image guidance
systems such as CARTO (Biosense Webster, Diamond Bar, CA, USA) or EnSite
(St Jude Medical, St Paul, MN, USA) are able to reconstruct the anatomy

© Springer International Publishing AG 2017
M. Pop and G.A. Wright (Eds.): FIMH 2017, LNCS 10263, pp. 301–310, 2017.
DOI: 10.1007/978-3-319-59448-4_29

from the mapping catheter tip motion and the electrical activation from the catheter tip electrode. However, the catheter tip can only be in contact with the endocardium at sparse points and while a large number of points yields better mapping accuracy, this increases pre-procedural time. Emerging multi-electrode systems such as Rhythmia (Boston Scientific, Marlborough, MA, USA with their basket catheter configuration limit the reachability of narrow sites in CHD patients, despite being able to collect many mapping points in the same time [6]. Moreover, the construction of a clinically informative map is a skill of experienced clinicians, who are able to adapt general electrophysiology principles to the specific CHD and patient anatomy and to decide on the position of the mapping points.

As part of the procedural pre-planning and also for better understanding of the electrophysiology, several electromechanical models have been proposed [10,14]. The models were built and parameterised from a small number of measurements, thus limiting the instantiation ability at finer level of deformed anatomy and atypical activation caused for example by surgical scars. Other approaches focused on improving the electrophysiology model by coupling a generic equation of the anisotropic myocardial fibre orientation [7] and further enhancing it with ECG-derived measures [16]. However, these have proved unable to describe activation patterns measured intraoperatively [7].

Parameterisation of shape atlases has also been in extensive use in describing cardiac anatomy. Since their introduction [5], statistical shape models (SSM) have moved from simple shape descriptions on Riemannian manifolds to more complicated multi-dimensional spaces such as parameters of rigid transformations [3] and to combined statistical atlases of shape and texture [1] or shape and pose [11], thus showing their applicability outside the traditional point distribution models commonly implemented in cardiac shape analysis. Furthermore, combined inter- and intra-subject shape modelling has been used in the study of cardiac [9] and respiratory motion [15]. In CHD patients, shape analysis on the myocardium in Tetralogy of Fallot showed that disease-specific markers can be computed from medical images [8,17]. Moreover, the values differed significantly from the healthy subjects, thus encouraging cohort-specific statistical analysis.

In this paper, a novel approach for optimal electroanatomical mapping of CHD is proposed. Firstly, a statistical electroanatomical model (SEAM) is built for each disease and cardiac chamber separately. Secondly, the frequency of anatomical sites chosen as mapping points in the specific CHD anatomy is computed. Finally, the vertices of a new shape are classified into mapping and regular points based on the atlas electroanatomical knowledge and sorted in descending order of their mapping frequency across the anatomy-specific dataset. The framework was tested in 5 CHD groups, adding to 66 CHD electrophysiology studies, to propose subject-specific mapping points location and compute the error reduction in electrical feature instantiation of the SEAM, i.e. unipolar and bipolar voltages and local activation times (LAT). The instantiation errors from the proposed sequence of mapping points were compared against the instantiation errors from the retrospective sequence of mapping points acquired in CARTO.

The results showed a steeper reduction in the electroanatomical reconstruction error when the mapping points were selected with the proposed approach.

2   Methods

2.1   Data Acquisition

Electroanatomical data from CARTO 3 studies of 66 CHD anatomies was exported. Two CHD groups were represented: Tetralogy of Fallot (34 studies) and univentricular hearts repaired by Fontan procedure with total cavo-pulmonary connection (32 studies). In the Fallot group, there were 21 studies of right ventricle (RV) and 13 of right atrium (RA$_{Fallot}$), while in the Fontan group, there were 16 left atria (LA), 9 right atria (RA$_{Fontan}$), and 7 total cavo-pulmonary connections (TCPC).

Each CARTO study included the preoperative MRI, the fast electroanatomical map (FAM) created by the mapping catheter, the unipolar and bipolar voltages and the LAT at each FAM vertex, the list and position of the sparse mapping points, as well as the rigid transformation from the intraoperative manual registration of the MRI onto the FAM. The number of mapping points varied within the same anatomy and the same CHD, with 49 ± 35 points in RV, 35 ± 18 in RA$_{Fallot}$, 33 ± 21 in LA, 34 ± 23 in RA$_{Fontan}$, and 33 ± 22 in TCPC. The MRI meshes were smoothed in MeshLab [2].

For each of the five groups, an analysis inspired by mutual information was performed, in order to select as template the mesh that is closest to the group mean in terms of Cartesian distance and unipolar and bipolar voltages and LAT difference between pairwise vertices. This yielded 6206 vertices for RV, 3940 for RA$_{Fallot}$, 6508 for LA, 7973 for RA$_{Fontan}$, and 5086 for TCPC. The correspondences were chosen as the list of vertices on each template mesh and were propagated on the other meshes using landmark-free nonrigid registration [12]. In order to match the electrical values, the MRI meshes were registered nonrigidly on their corresponding FAM. All distances and electrical values were normalised within each case dataset.

2.2   Statistical Models

A statistical shape model was first built to fit a new shape **s** to the atlas described by the mean shape $\bar{\mathbf{s}}$ and the matrix of eigenvectors $\mathbf{P}_a$. The shape **s** was approximated by the SSM as $\hat{\mathbf{s}}_a$ from the set of parameters $\mathbf{b}_a$, the result of least square optimisation. This can be represented by:

$$\hat{\mathbf{s}}_a = \bar{\mathbf{s}} + \mathbf{P}_a \mathbf{b}_a \qquad (1)$$

Simultaneously, the correspondences on shape **s** built a subset of an instance in the statistical electroanatomical model (SEAM) defined by the mean electroanatomical vector $[\bar{\mathbf{s}}^T \ \bar{\mathbf{e}}^T]^T$ and the eigenvectors $\mathbf{P}_{ae}$. Again using least square optimisation, the current electroanatomical vector $[\mathbf{s}^T \ \mathbf{e}^T]^T$ was approximated as $[\hat{\mathbf{s}}_{ae}^T \ \hat{\mathbf{e}}^T]^T$, defined by the model through the parameters $\mathbf{b}_{ae}$ (Eq. (2)).

$$\begin{bmatrix} \hat{\mathbf{s}}_{ae} \\ \hat{\mathbf{e}} \end{bmatrix} = \begin{bmatrix} \bar{\mathbf{s}} \\ \bar{\mathbf{e}} \end{bmatrix} + \mathbf{P}_{ae}\mathbf{b}_{ae} \tag{2}$$

Due to least square optimisation forcing the approximated shape to converge to the original in both models, it can be assumed that $\mathbf{s} \approx \hat{\mathbf{s}}_a \approx \hat{\mathbf{s}}_{ae}$ and therefore $\mathbf{P}_a \ \mathbf{b}_a \approx \mathbf{P}_{ae,s}\mathbf{b}_{ae}$, whereby $\mathbf{P}_{ae,s}$ is the matrix formed by the rows of $\mathbf{P}_{ae}$ corresponding to the shape vector $\mathbf{s}$. Finally, the unknown parameter vector $\mathbf{b}_{ae}$ and subsequently the electrical values $\hat{\mathbf{e}}$ can be recovered as in Eqs. (3) and (4), where $\mathbf{P}_{ae,s}^+ = (\mathbf{P}_{ae}^T \ \mathbf{P}_{ae})^{-1} \cdot \mathbf{P}_{ae}^T$ is the Moore-Penrose pseudoinverse of $\mathbf{P}_{ae,s}$ and $\mathbf{P}_{ae,e}$ are the rows of matrix $\mathbf{P}_{ae}$ corresponding to the electrical value vector.

$$\mathbf{b}_{ae} \approx \mathbf{P}_{ae,s}^+\mathbf{P}_a\mathbf{b}_a \tag{3}$$

$$\hat{\mathbf{e}} \approx \bar{\mathbf{e}} + \mathbf{P}_{ae,e}\mathbf{b}_{ae} \tag{4}$$

Simultaneously to building the statistical models, the template meshes were further decimated until the number of vertices was below 200, thus clustering the vertices of each template mesh around sparse points, while still preserving the anatomy. The value of 200 was chosen empirically in order to cover the maximal number of mapping points per anatomy in the dataset (136 for one RV).

Each mapping vertex of the full mesh was then approximated to the nearest vertex of the decimated template mesh. This was performed for all subjects within the same CHD group. The mapping frequency of each vertex on the low-resolution mesh was defined as the sum of mapping vertices on the full-resolution mesh, across all subjects in the anatomy-specific dataset.

2.3   Classification

RUSBoost classification of the vertices on a new instantiated shape was performed in order to define target mapping areas (Algorithm 1). This particular boosting algorithm is suitable for classes with imbalanced numbers, i.e. regular vertices *vs.* mapping vertices [13]. The features on which the classifier is trained are the concatenated normalised coordinates of all shapes in the database $[\mathbf{x} \ \mathbf{y} \ \mathbf{z}]$ and their corresponding normalised electrical features $[\mathbf{uni} \ \mathbf{bi} \ \mathbf{LAT}]$. The anatomical features in the test set $[\mathbf{x}_{test} \ \mathbf{y}_{test} \ \mathbf{z}_{test}]$ are the normalised Cartesian coordinates of the current shape $\mathbf{s}$, while the unipolar (uni) and bipolar voltages (bi) and the local activation time (LAT) are the normalised electrical features estimated from the SEAM Eq. (4).

2.4   Iterative Addition of Mapping Points

In order to assess the performance of the proposed framework, the computed mapping points were added iteratively to the SEAM in decreasing order of their probability. In each iteration, the known shape vector $\mathbf{s}$ was enhanced with the

---

Data:
- $(\mathbf{y}_{\text{train}}, [\mathbf{x}_{\text{train}}\ \mathbf{y}_{\text{train}}\ \mathbf{z}_{\text{train}}\ \mathbf{uni}_{\text{train}}\ \mathbf{bi}_{\text{train}}\ \mathbf{LAT}_{\text{train}}]),\ \mathbf{y}_{\text{train}} \in \{0, 1\}$,
  where 0 denotes regular vertex and 1 mapping vertex.
- number of mapping vertices is significantly lower than the number of regular vertices, i.e. $n_1 \ll n_0$.
- $([\mathbf{x}_{\text{test}}\ \mathbf{y}_{\text{test}}\ \mathbf{z}_{\text{test}}\ \mathbf{uni}_{\text{test}}\ \mathbf{bi}_{\text{test}}\ \mathbf{LAT}_{\text{test}}])$
- weak learner,
  which does not necessarily yield a good initial classification.

Initialisation: $w_{1,i} = \frac{1}{n_{\text{train}}}, i = \overline{1, n_{\text{train}}}$, where $w_{k,i}$ is the weight of sample i in iteration k and n_{train} is the number of samples in the training set;
while *preset number of iterations not reached* **do**

1. subsample from the full set using the weights $w_{k,i}$, $i = \overline{1, n_{\text{train}}}$;
2. feed the subset and the weights to the learner;
3. learner estimates the labels of the training data;
4. update the weights with the classification error;

end
Result: $\mathbf{y}_{\text{test}}$

Algorithm 1. RUSBoost classification algorithm for computing mapping points on a given electroanatomical map. Adapted from [13].

electrical features of the computed mapping vertices. The electrical parameters of the remaining vertices were estimated as in Eqs. (3) and (4). The instantiation error was compared to the one obtained from the ground truth mapping points. For each electrophysiology study, the number of iterations was equal to the number of mapping points exported from CARTO. The iterative instantiation is presented comparatively in Algorithm 2.

3   Results

3.1   Statistical Models

Table 1 shows the first mode of variation of the SEAM for each anatomy. Among the noticeable features, the SEAM is able to describe the variation in the amount of septal activation in RV and the atrial dilatation, a common issue in CHD. Cross-validation on a leave-one-out basis was performed within the dataset of each CHD anatomy. The mean instantiation errors for shape and electrical properties were also computed. The shape instantiation error err_s was evaluated in terms of mean Cartesian distance between SEAM-computed vertices position and ground truth, while the error for the electrical values was computed as vertex-wise L1-norm between the true and estimated parameters.

The mapping frequency of the vertices on the low-resolution template mesh of each anatomy was computed according to Sect. 2.2 (Table 1). The atlas of mapping frequency was also built on a leave-one-out basis, as for SEAM validation.

Data:
- decimated template mesh and mapping frequency of each vertex, computed according to Sect. 2.2
- descending order of the vertices on the decimated mesh according to mapping frequency (cluster vertices)
- n_P number of CARTO mapping points

Initialisation: perform SEAM according to Eqs. (2–4);
for $i \leftarrow 1$ **to** n_P **do**

Proposed framework	Ground truth
1. classify vertices on instantiated electroanatomy; 2. select only vertices mapped to the cluster vertex with i-highest probability; 3. perform SEAM as in Eqs. (2–4);	1. add mapping vertices corresponding to the next chronological CARTO point; 2. perform SEAM as in Eqs. (2–4);

end

Algorithm 2. Iterative addition of mapping points for SEAM with points computed from the proposed framework and points added chronologically from the ground-truth CARTO point list.

The atlas was further used in ranking potential mapping points in decreasing order of their probability and added iteratively to refine the SEAM instantiation. The resulting colour-coded maps in Table 1 indicate the outflow tract in the RV and the interatrial septum as frequent mapping areas.

3.2 Classification

The adapted RUSBoost classifier was cross-validated CHD-specifically by training and testing it on normalised electroanatomical values within the same anatomical group. Overall, the accuracy averaged at 67.91 %, with a true positive rate (sensitivity) of 54.35 %. The high accuracy at low sensitivity in RA_{Fontan} indicates that the true negative rate is high, i.e. the model is reluctant to recommend a vertex as a mapping point if not enough previous cases are available.

3.3 Iterative Addition of Mapping Points

Mapping points selected by the classifier and ordered by their mapping frequency according to the anatomy-specific atlas were added iteratively to the SEAM to improve its performance and test the contribution of the proposed mapping points. Starting with no electrical information (original SEAM), the unipolar

Table 1. SEAM modes of variation and instantiation errors, probabilistic mapping atlas (second set of meshes), accuracy and sensitivity of the RUSBoost classifier and SEAM error reduction with the addition of mapping points computed by the proposed method (green) when compared with the addition of CARTO-exported mapping points (red). The mesh orientation is given by the superior-inferior axis (red), left-right axis (black), and anterior-posterior axis (green).

	RV	RA$_{Fallot}$	LA	RA$_{Fontan}$	TCPC
+2σ	[ms] 100, 87.5, 75, 62.5, 50	[ms] 100, 80, 60, 40, 20	[ms] 85, 75, 65, 55, 45	[ms] 50, 42.5, 35, 27.5, 20	[ms] 70, 55.5, 40, 25.5, 10
mean	[ms] 30, 17.5, 5, -7.5, -20	[ms] 5, -2.5, -10, -17.5, -25	[ms] 8.0, 3.0, -2.0, -7.0, -12.0	[ms] 14.0, 10.0, 6.0, 2.0, -2.0	[ms] 0, -7.5, -15, -22.5, -30
-2σ	[ms] -20, -40, -60, -80, -100	[ms] -50, -62.5, -75, -87.5, -100	[ms] -50, -57.5, -65, -72.5, -80	[ms] 0, -12.5, -25, -37.5, -50	[ms] -60, -67.5, -75, -82.5, -90
err$_s$ [mm]	3.69±2.75	3.29±3.17	2.42±2.44	1.96±2.08	5.58±4.63
err$_{uni}$ [mV]	1.51±1.02	0.44±0.23	0.89±1.72	1.11±1.39	1.04±1.73
err$_{bi}$ [mV]	1.22±1.22	0.37±0.21	0.84±1.50	0.74±0.87	1.25±2.60
err$_{LAT}$ [ms]	13.82±18.82	27.31±13.19	22.33±14.30	33.83±13.42	23.63±17.09
map prob	1, 0.75, 0.5, 0.25, 0	1, 0.75, 0.5, 0.25, 0	1, 0.75, 0.5, 0.25, 0	1, 0.75, 0.5, 0.25, 0	1, 0.75, 0.5, 0.25, 0
acc [%]	66.61	64.22	66.21	68.64	73.87
sens [%]	71.52	60.32	50.84	36.88	52.22
err$_{uni}$ [mV]	1.1 / 0.9	0.58 / 0.5	1.05 / 0.95	0.605 / 0.6	1.41 / 1.37
err$_{bi}$ [mV]	0.9 / 0.88	0.32 / 0.31	0.83 / 0.82	0.37 / 0.35	0.842 / 0.84
err$_{LAT}$ [ms]	120 / 80	280 / 240	145 / 132	89.0 / 84.0	145 / 137
	24 points	70 points	78 points	58 points	33 points

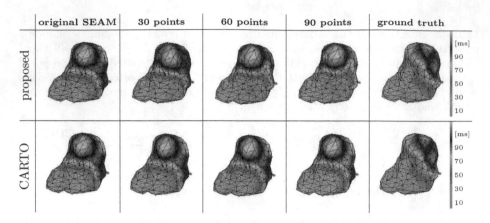

Fig. 1. Iterative addition of mapping points for a RV. Comparison of proposed combined SEAM-classification method with the chronological addition of mapping points as exported from CARTO. The electroanatomical maps show electrical propagation in terms of LAT. The ground truth is the CARTO-exported LAT map.

and bipolar voltages and LATs of vertices in the regional cluster with i-highest mapping probability were added in iteration i. Figure 1 shows snapshots of the improvement over 3 iterations for a RV. Table 1 also includes the curves of error reduction for one case of each CHD anatomy.

4   Discussion and Conclusion

Electroanatomical mapping as pre-procedural planning of cardiac catheter ablation is a patient-specific and time consuming step which requires high skills and knowledge from the electrophysiologist. In this paper, a novel combination of statistical electroanatomical mapping model instantiation and classification is employed in order to compute areas of potential interest based on previous cases of similar disease and on patient preoperative anatomical data.

The chain of methods relies on the instantiation of a pure shape model from known anatomy and the direct substitution into a combined SEAM to recover the electrical data. While initial results presented in Table 1 are promising, the method relies on the approximation that the shapes in the two models are equal. A quantitative analysis showed that they differ in reality by an average of 3 mm, which is in the range of the shape recovery error of the SEAM. The proposed framework was tested in an iterative addition of the computed mapping points to the SEAM. The error curves in Table 1 show good results for a large representative training dataset, e.g. RV (21 cases), but inconclusive results for a database smaller than 10 subjects (e.g. TCPC). Moreover, a statistical analysis on CHD electroanatomy was only possible due to the electrical activation pattern homogeneity, which needs further investigation in application to other patient groups, such as myocardial infarction survivors.

In conclusion, a novel method for objective identification of electroanatomical mapping areas was proposed. The framework can be regarded as a first step in computer-aided standardisation of pre-procedural mapping in cardiac catheter ablation and can be used to transfer expert knowledge to trainees. Moreover, targeted patient-specific electroanatomical mapping can help in reducing the overall intervention time and to effectively detect potential ablation sites.

References

1. Andreopoulos, A., Tsotsos, J.K.: Efficient and generalizable statistical models of shape and appearance for analysis of cardiac MRI. Med. Image Anal. **12**(3), 335–357 (2008)
2. Cignoni, P., Corsini, M., Ranzuglia, G.: Meshlab: an open-source 3D mesh processing system. In: ERCIM News, pp. 45–46
3. Constantinescu, M.A., Lee, S.L., Navkar, N.V., Yu, W., Al-Rawas, S., Abinahed, J., Zheng, G., Keegan, J., Al-Ansari, A., Jomaah, N., Landreau, P., Yang, G.Z.: Constrained statistical modelling of knee flexion from multi-pose magnetic resonance imaging. IEEE Trans. Med. Imaging **35**(7), 1686–1695 (2016)
4. Roy, K., Gomez-Pulido, F., Ernst, S.: Remote magnetic navigation for catheter ablation in patients with congenital heart disease: a review. J. Cardiovasc. Electrophysiol. **27**(Suppl 1), S45–S56 (2016)
5. Cootes, T.F., Taylor, C.J., Cooper, D.H., Graham, J.: Active shape models-their training and application. Comput. Vis. Image Underst. **61**(1), 38–59 (1995)
6. El Yaman, M.M., Asirvatham, S.J., Kapa, S., Barrett, R.A., Packer, D.L., Porter, C.B.: Methods to access the surgically excluded cavotricuspid isthmus for complete ablation of typical atrial flutter in patients with congenital heart defects. Heart Rhythm **6**(7), 949–956 (2009)
7. Krueger, M.W., Seemann, G., Rhode, K., Keller, D.U., Schilling, C., Arujuna, A., Gill, J., O'Neill, M.D., Razavi, R., Doessel, O.: Personalization of atrial anatomy and electrophysiology as a basis for clinical modeling of radio-frequency ablation of atrial fibrillation. IEEE Trans. Med. Imaging **32**(1), 73–84 (2013)
8. Mansi, T., Voigt, I., Leonardi, B., Pennec, X., Durrleman, S., Sermesant, M., Delingette, H., Taylor, A.M., Boudjemline, Y., Pongiglione, G., Ayache, N.: A statistical model for quantification and prediction of cardiac remodelling: application to tetralogy of fallot. IEEE Trans. Med. Imaging **30**(9), 1605–1616 (2011)
9. Perperidis, D., Mohiaddin, R., Rueckert, D.: Construction of a 4D statistical atlas of the cardiac anatomy and its use in classification. Med. Image Comput. Comput. Assist. Interv. **8**(Pt 2), 402–410 (2005)
10. Prakosa, A., Sermesant, M., Allain, P., Villain, N., Rinaldi, C.A., Rhode, K., Razavi, R., Delingette, H., Ayache, N.: Cardiac electrophysiological activation pattern estimation from images using a patient-specific database of synthetic image sequences. IEEE Trans. Biomed. Eng. **61**(2), 235–245 (2014)
11. Rasoulian, A., Rohling, R., Abolmaesumi, P.: Lumbar spine segmentation using a statistical multi-vertebrae anatomical shape+pose model. IEEE Trans. Med. Imaging **32**(10), 1890–1900 (2013)
12. Rueckert, D., Sonoda, L.I., Hayes, C., Hill, D.L., Leach, M.O., Hawkes, D.J.: Nonrigid registration using free-form deformations: application to breast MR images. IEEE Trans. Med. Imaging **18**(8), 712–721 (1999)

13. Seiffert, C., Khoshgoftaar, T.M., Van Hulse, J., Napolitano, A.: RUSBoost: a hybrid approach to alleviating class imbalance. IEEE Trans. Syst. Man Cybern. Part A Syst. Hum. **40**(1), 185–197 (2010)
14. Sermesant, M., Chabiniok, R., Chinchapatnam, P., Mansi, T., Billet, F., Moireau, P., Peyrat, J.M., Wong, K., Relan, J., Rhode, K., Ginks, M., Lambiase, P., Delingette, H., Sorine, M., Rinaldi, C.A., Chapelle, D., Razavi, R., Ayache, N.: Patient-specific electromechanical models of the heart for the prediction of pacing acute effects in CRT: a preliminary clinical validation. Med. Image Anal. **16**(1), 201–215 (2012)
15. Wilms, M., Ehrhardt, J., Handels, H.: A 4D statistical shape model for automated segmentation of lungs with large tumors. Med. Image Comput. Comput. Assist. Interv. **15**(Pt 2), 347–354 (2012)
16. Zettinig, O., Mansi, T., Georgescu, B., Kayvanpour, E., Sedaghat-Hamedani, F., Amr, A., Haas, J., Steen, H., Meder, B., Katus, H., Navab, N., Kamen, A., Comaniciu, D.: Fast data-driven calibration of a cardiac electrophysiology model from images and ECG. Med. Image Comput. Comput. Assist. Interv. **16**(Pt 1), 1–8 (2013)
17. Zhang, H., Wahle, A., Johnson, R.K., Scholz, T.D., Sonka, M.: 4-D cardiac MR image analysis: left and right ventricular morphology and function. IEEE Trans. Med. Imaging **29**(2), 350–364 (2010)

Prediction of Post-Ablation Outcome in Atrial Fibrillation Using Shape Parameterization and Partial Least Squares Regression

Shuman Jia[1(✉)], Claudia Camaioni[2], Marc-Michel Rohé[1], Pierre Jaïs[2], Xavier Pennec[1], Hubert Cochet[2], and Maxime Sermesant[1]

[1] Université Côte d'Azur, Asclepios Research Group, Inria, Sophia Antipolis, France
shuman.jia@inria.fr
[2] IHU Liryc, University of Bordeaux, Pessac, France

Abstract. To analyze left atrial remodeling may reveal shape features related to post-ablation outcome in atrial fibrillation, which helps in identifying suitable candidates before ablation. In this article, we propose an application of diffeomorphometry and partial least squares regression to address this problem. We computed a template of left atrial shape in control group and then encoded the shapes in atrial fibrillation with a large set of parameters representing their diffeomorphic deformation. We applied a two-step partial least squares regression. The first step eliminates the influence of atrial volume in shape parameters. The second step links deformations directly to post-ablation recurrence and derives a few principle modes of deformation, which are unrelated to volume change but are involved in post-ablation recurrence. These modes contain information on ablation success due to shape differences, resulting from remodeling or influencing ablation procedure. Some details are consistent with the most complex area of ablation in clinical practice. Finally, we compared our method against the left atrial volume index by quantifying the risk of post-ablation recurrence within six months. Our results show that we get better prediction capabilities (area under receiver operating characteristic curves $AUC = 0.73$) than left atrial dilation ($AUC = 0.47$), which outperforms the current state of the art.

Keywords: Atrial fibrillation · Catheter ablation · Post-ablation outcome · Left atrial remodeling · Statistical shape analysis · Partial least squares regression

1 Introduction

Atrial fibrillation (AF) is the most common type of cardiac arrhythmia [1], characterized by uncoordinated electrical activation and disorganized contraction of the atria. This condition is associated with life-threatening consequences, such as stroke and heart failure. Catheter ablation is an effective treatment for AF and may be recommended for drug refractory patients. Yet, for 30% of patients, AF redevelops, resulting in repeated interventions and higher risk.

© Springer International Publishing AG 2017
M. Pop and G.A. Wright (Eds.): FIMH 2017, LNCS 10263, pp. 311–321, 2017.
DOI: 10.1007/978-3-319-59448-4_30

In order to optimize the planning of AF treatment, a number of studies have been looking into predictors of recurrent arrhythmia. Previous studies reported that hypertension, Holter duration, left atrial (LA) volume and LA sphericity [2–5] were potentially related to post-ablation recurrence, but their underlying mechanism is still unclear, as is LA remodeling in AF.

In this exploratory work, we focus on shape variations as predictors. By studying principal deformation modes associated with a higher post-ablation recurrence rate, we expect to find cartographic/regional criteria that will help us stratify the risk of post-ablation recurrence. So far, LA volume index (LAVI) has been the only clinically accessible and reproductive index for identifying suitable candidates for ablation. However, its relation to post-ablation outcome has been constantly reported to be weak [6]. We therefore decided to explore shape features that are more comprehensive than LAVI. For example, using maps of the deformation modes, we can visualize both regional and global shape variations, both of which may indicate potential for recurrence.

Unlike in [7], we decided to quantify anatomical variability using a registration approach based on diffeomorphism. This allows for finer shape variations to be captured without the need for pre-defined markers. In statistical analysis, we used partial least squares (PLS) regression rather than principal component analysis (PCA) since PLS regression recognizes the components that are directly linked to recurrence. We included a two-step regression to compare shape features and volume index. Finally, we used computed tomography (CT) to ensure detailed and accurate anatomical descriptions.

2 Methods and Experiments

2.1 Data Preparation

Population. 40 paroxysmal AF (PAF) patients were studied. They had no previous ablation at the time of imaging. Average age was 59 ± 11 years old and $31/40$ of the patients are male. Post-ablation recurrence within six months was observed in $13/40$ patients. Meanwhile, we chose 24 control subjects without significant difference in age or gender from the PAF group.

Image Acquisition. All subjects were imaged before catheter ablation using a 64-slice Siemens SOMATOM CT scanner [8]. Multi-detector CT was performed, during the intravenous injection of iodinated contrast agent. The scans were ECG-gated for acquisition window to occur at end-systole, when LA motion is minimal. Temporal resolution was 66 ms. The trans-axial images were acquired with a slice thickness of 0.5 mm and reconstructed with a voxel size of $0.5 \times 0.4 \times 0.4 \, mm^3$. The protocol of this study was approved by the local research ethics committee.

Endocardium Segmentation. We segmented the LA endocardium from CT images using a region growing algorithm, as described in [9]. Semi-automatically drawn polygons isolated the left atrium and served as the boundary for succedent region growing process. A patient-specific intensity threshold, computed from tissue sampling, controlled the growing of blood pool during the iteration process until it touched the frontier of the endocardium. Pulmonary veins were cut several centimetres from the ostia. Based directly on image intensity, our segmentation can achieve high accuracy with such contrast CT, as well as identify PVs with several kinds of anatomies[1].

Mesh Generation. Next, we generated 3D volume tetrahedral meshes of the LA relaying on restricted Delaunay triangulation [11][2]. Range for size and angle of mesh elements were controlled via input parameters. To meet the requirements of implementation, we extracted surface triangular meshes of the LA from the volume meshes.

2.2 Quantification of Shape Variations

We encoded LA shape by deformation parameters in this step, adopting diffeomorphometry. The term was introduced recently and refers to the comparison of shapes and forms using a metric structure based on diffeomorphism. Methods under this framework have been proven to be efficient to qualify anatomical configurations and their differences for computational anatomy studies [12].

Mathematical Currents, Varifold Metric, Diffeomorphism. First, the shape S was represented by flux of 3D vector field/current w across the surface, that belongs to a Gaussian reproducing kernel Hilbert space (RKHS) W with kernel K_w. The current w was parameterized by a set of $\vec{\delta}_{c_k}$, attached to distinct points on surface S.

Then, the deformation ϕ was estimated using the large deformation diffeomorphic metric mapping (LDDMM) on surface, characterized by initial velocity $\vec{v}_0$. The velocity vector belongs to a Gaussian RKHS V with kernel K_V, and was weighted by a set of moment vectors $\vec{\alpha}_{c_k}$. Unless otherwise stated, $\{c_k\}_{k=1,...,N_{cp}}$ refers to the control points of deformation in the following text. At point x, velocity

$$v(x) = \sum_{k=1}^{N_{cp}} K_V(x, c_k)\vec{\alpha}_{c_k}. \tag{1}$$

Thus, for patient #i, the surface S_i could be represented as the sum of deformed template T and some residuals ϵ_i, as

$$S_i = \phi_i T + \epsilon_i. \tag{2}$$

For registration, we minimized the residuals ϵ, in other words, the distance between S_i and $\phi_i T$. For atlas construction, we minimized the sum of distances from the template to every mesh, resulting in a Fréchet mean of shape complex,

---

[1] We used the *MUSIC* software for endocardium segmentation [10].

[2] We used *CGAL* 3D mesh generation algorithm http://doc.cgal.org/latest/Mesh_3/index.html.

$$T = \arg\min_{T \in W} \sum_i d_W^2(T, S_i). \qquad (3)$$

The template and the deformations were estimated simultaneously.

In order to be independent of the surface orientation, the distance between surface meshes was defined based on varifold metric [13] as

$$d_W(S, S')^2 = ||S - S'||_{W*}^2 = \langle S, S \rangle_{W*} + \langle S', S' \rangle_{W*} + \langle S, S' \rangle_{W*} \qquad (4a)$$

$$\text{with } \langle S, S' \rangle_{W*} = \sum_p \sum_q K_W(c_p, c_q') \frac{(n_p^T n_q')^2}{|n_p^T||n_q'|} \qquad (4b)$$

where c (resp. c') refers to control points on surface S (resp. S') and n (resp. n') denotes normals related to controls points.

The line search strategy was used in optimization process. More details can be found in [14].

Remodeling. A rigid alignment of all meshes was performed in the first place, to reduce the impact of different origin in CT images. Iterative closest point algorithm was applied to calculate an optimized rigid registration in the least square sense.

Using methods described above, we computed a 3D template of the LA shape in the control group. The template offered a reference for anatomical invariant. Then, we warped this template towards each mesh of PAF patients and calculated its deformation moments. We illustrate the process in Fig. 1(a)[3].

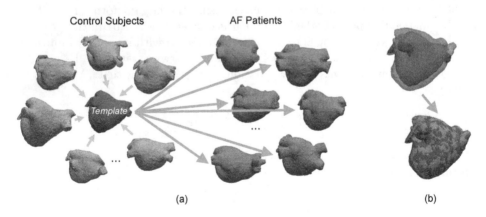

(a)   (b)

Fig. 1. Extraction of remodeling information vs. controls. (a) reference for atrial remodeling offered by control subjects; (b) the template in blue, before and after warping in a registration process toward the target in orange. (Color figure online)

---

[3] Atlas estimation and registration have been integrated in the *Deformetrica* software http://www.deformetrica.org/.

Deformations were parameterized as moment vectors $\vec{\alpha}_{c_k}$ attached to a same set of control points on the template for all patients. We use M_i to represent the deformation moments for patient #i

$$M_i = \left[\vec{\alpha}^i_{c_1} \; \vec{\alpha}^i_{c_2} \cdots \vec{\alpha}^i_{c_k} \cdots \vec{\alpha}^i_{c_{N_{cp}}} \right]^T, \tag{5}$$

where $\vec{\alpha}^i_{c_k} = \left[\alpha^i_{c_k 1} \; \alpha^i_{c_k 2} \; \alpha^i_{c_k 3} \right]^T$ is the deformation moment vector associated with control point #k for patient #i; N_{cp} is the total number of control points.

Kernel width parameters were set as: varifold kernel width $\sigma_W = 10\,\mathrm{mm}$ and deformation kernel width $\sigma_V = 10\,\mathrm{mm}$, ensuring respectively a suitable scale of shape variations and deformation to capture. $N_{cp} = 3952$ control points were placed near most variable parts on the template and helped in representing shape differences between the template and the meshes of PAF patients. With chosen parameters, atlas estimation and pairwise registration were performed efficiently with surfaces, as shown in Fig. 1(b).

Whitening Transform. Then, we used whitening transform to reduce the correlation between the deformation moments.

According to Eq. 1, the norm of the velocity vector for patient #i

$$\|v^i\|^2 = \left[\vec{\alpha}^i_{c_1} \; \vec{\alpha}^i_{c_2} \cdots \vec{\alpha}^i_{c_{N_{cp}}} \right] K \left[\vec{\alpha}^i_{c_1} \; \vec{\alpha}^i_{c_2} \cdots \vec{\alpha}^i_{c_{N_{cp}}} \right]^T = M_i^T K M_i, \tag{6}$$

where K is a positive definite matrix that defined a metric, with its blocks $K_{ij} = K_V(c_i, c_j) \otimes \mathbb{1}_3$. We used the Gaussian kernel $K_V(x, y) = exp(\frac{-|x-y|^2}{\sigma_V^2})$. Since the velocity vector v belongs to a RKHS space, the whitened moment $K^{1/2} M_i$ belongs to the Euclidean space with L_2 norm. This allows us to apply seamlessly standard statistical methods like PLS.

We arranged the whitened deformation moments $K^{1/2} M_i$ into a matrix as

$$\Lambda = \left[\vec{\lambda}_1 \; \vec{\lambda}_2 \cdots \vec{\lambda}_i \cdots \vec{\lambda}_{N_p} \right]^T, \tag{7}$$

where $\vec{\lambda}_i$ is a column vector that contains all the elements in $K^{1/2} M_i$; $N_p = 40$ is the total number of patient. Thus, the dimension of Λ turns out to be $N_p \times (3N_{cp})$.

To sum up, modeling complex geometries with mathematical currents avoided using pre-defined markers, and therefore has the potential to summarize any shape feature related to certain clinical factors. The finite dimensional approximation of shape as deformation moments was robust to detect subtle anatomical differences. Finally, whitening transform reduced the correlation due to the kernel metric and constructed a L_2 space for statistical analysis.

2.3 Statistical Modeling

PLS regression combines PCA with linear regression by projecting inputs X and Y to a new space. X is a $n \times m$ matrix of m predictors for n observations, and Y is a $n \times w$ matrix of w response variables for n observations. It relates predictors directly to response variables by finding multidimensional directions in X that explain the maximum variance in Y. Here, we use same notations as the tutorial [15].

We chose PLS regression so as to find the principle dimensions in deformation parameters that correlate with post-ablation recurrence.

Dependency Analysis with Left Atrial Volume. In the first PLS regression, the whitened deformation parameters were considered as predictors, while the LAVI was considered as response variable, as

$$X = \Lambda \quad \text{and} \quad Y = \left[LAVI_1 \, LAVI_2 \, \cdots \, LAVI_i \, \cdots \, LAVI_{N_p} \right]^T,$$

where Λ is defined in Eq. 7; $LAVI$ was calculated for each subject as $LAVI = V_{LA}/BSA$, based on the size of the atrium V_{LA} in units of mL and the body surface area BSA in units of m^2.

According to the percentage of variance explained shown in Table 1, the first mode spanned an optimal subspace that explained 84.61% of LAVI variance for the population under study. It was expected that this principle mode of variation would be linked to volume change. To analyze shape features which are complementary to the atrial size, we subtracted, for every patient, the components in deformation projected on this mode, as

$$\Lambda' = X - T(:, 1)P(:, 1)^T, \tag{8}$$

where Λ' is the matrix of volume-reduced deformation parameters; $T(:, 1)$ is the first column of X score matrix T; $P(:, 1)$ is the first column of X loading matrix P, referring to the first PLS component of deformation related to LAVI.

Table 1. Percentage of variance explained in the first partial least squares regression, for shape and left atrial volume index respectively.

PLS components	1st	2nd	3rd	4th	5th
Percentage of shape variance %	53.45	7.09	7.50	4.47	5.70
Percentage of LAVI variance %	83.61	5.45	2.03	2.07	0.96

We therefore created a new matrix of deformation parameters that were not linearly related to LA volume change.

Correlation with Post-ablation Recurrence. In the second PLS regression, we studied the correlation between atrial shape and post-ablation recurrence, applying discriminant analysis and leave-one-out prediction. The volume-reduced deformation parameters were considered as predictors, while the post-ablation outcome within six months was considered as response variable.

For patient #i, PLS regression was performed among all the other patients as

$$X = \Lambda'([1:i-1, \ i+1:end], \ :)$$
$$Y = \left[\, R_1 \ R_2 \cdots R_{i-1} \ R_{i+1} \cdots R_{N_p} \right]^T,$$

where Λ' was computed based on Eq. 8; R_i represents the post-ablation outcome within six months for patient #i, with $R_i = 1$ standing for with recurrence, $R_i = 0$ without recurrence.

Then, we projected the deformation parameters of patient #i on the subspace constructed by the first n PLS components, to calculate a predicted response for this new observation

$$SI_i = \Lambda'(i, \ :)P(:, 1:n)B(1:n, 1:n)Q(:, 1:n)^T, \tag{9}$$

where $\Lambda'(i, \ :)$ is the volume-reduced deformation parameters for patient #i; P refers to the X loading matrix; B is the diagonal matrix of coefficients b_h; Q is the Y loading matrix.

We repeated the leave-one-out regression and prediction, and thus obtained shape indices for every patient $\{SI_i\}_{i=1,2,\ldots N_p}$ which qualified their potential for recurrence.

3   Results

We present here the first three modes of deformation, with a higher signal-to-noise ratio, and their prediction capabilities. Beyond the third mode, components may be dominated by noise, since the percentage of variance explained by them was smaller than half of the gap between the percentage of variance explained by previous two successive modes. The percentage of variance explained is shown in Table 2.

Table 2. Percentage of variance explained in the second partial least squares regression, for shape and post-ablation recurrence respectively.

PLS components	1st	2nd	3rd	4th	5th
Percentage of shape variance %	21.44	9.02	14.30	9.71	5.69
Percentage of recurrence variance %	27.47	18.09	5.67	3.98	3.95

The principle modes of deformation related to post-ablation recurrence can be visualized and interpreted, shown in Fig. 2. Area strain [16] and the magnitude of displacement were mapped onto meshes to illustrate detailed variations in deformations. For area strain, red indicates enlargement of triangular elements, while blue indicates shrinking. For displacement, regions in red deform with a larger scale of displacement than regions in blue.

318   S. Jia et al.

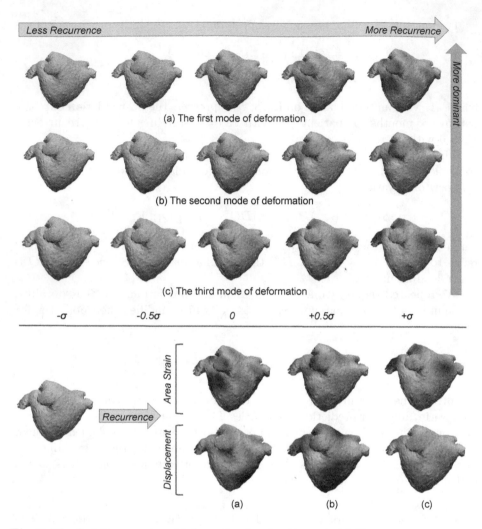

Fig. 2. The first three modes of deformation involved in post-ablation recurrence. σ represents the standard derivation on each mode in the population. (Color figure online)

Results show that the most dominant mode has an emphasis on regions underneath pulmonary veins, both with area strain and displacement of mesh elements. Clinical experts confirmed that this is one of the most complicated area to ablate in pulmonary vein isolation. Also, the remodelling around pulmonary veins seems to be an important aspect of AF. The second mode contains a twist of the upper-left part of the LA (from the posterior view, including roof, lateral and anterior segments) and also changes in orientation of pulmonary veins. The third mode, from less recurrence to more recurrence, reflects a slight change of roundness. These 3D deformation sequences can also be shown as videos in order to be more explicit. However, the subtle regional variations need further interpretation.

We compared prediction capabilities of shape indices with that of LAVI, using student's t-test and receiver operating characteristic (ROC) curves. The indices are shown in Table 3.

Table 3. Student's t-test to compare shape indices and left atrial volume index in two sub-groups and area under receiver operating characteristic curves.

PAF patients	Non-recurrence	Recurrence	P-value	AUC
SI - 1 PLS component	0.28 ± 0.20	0.41 ± 0.23	0.04	0.65
SI - 2 PLS components	0.29 ± 0.22	0.48 ± 0.26	0.01	0.73
SI - 3 PLS components	0.30 ± 0.21	0.51 ± 0.34	0.01	0.70
Volume index/mL/m_2	64.25 ± 19.34	64.97 ± 22.23	0.46	0.47

- Shape indices derived from PLS regression differed significantly between recurrence and non-recurrence groups, with p-value reaching $p = 0.01 < 0.05$ with the first two components, while LAVI did not, with $p = 0.46$.
- Shape indices show a better classification performance as the discrimination threshold varies. We draw ROC curves of the shape indices and the volume index in Fig. 3. The area under the ROC curve (AUC) using the first two components is 0.73, compared with 0.47 for LAVI.
- With discrimination threshold $t = 0.36$ for shape index with the first two components, we predicted recurrence with 0.77 sensitivity and 0.67 specificity.

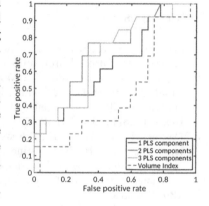

Fig. 3. Receiver operating characteristic curves of 6-month post-ablation recurrence prediction.

From the differences, we can conclude that the shape analysis discovered extra anatomical features compared with the volume index for the understanding of post-ablation recurrence. Meanwhile, the principle modes of deformation revealed in this study turned out to be clinically meaningful.

4 Conclusion

We adapted a shape-based statistical model, extended from mathematical currents and diffeomorphism, to address the problem of post-ablation recurrence. We eliminated the impact of atrial size in shape analysis. Then, from regressions, we obtained shape indices better predicting post-ablation recurrence, when compared with dilatation. Our shape analysis approach summarizes all shape variations at a scale that is greater than the kernel width. It is also more robust, with respect to training samples, than using only one shape parameter, as is done with the volume index.

From a clinical viewpoint, we revealed the principal LA deformation modes that were related to a higher post-ablation recurrence rate. We visualized them and some anatomical details were consistent with the complexity of ablation in clinical practice. These features also bring new insights on how a shape evolves during LA remodeling, apart from volume change.

Future work includes using a larger database to reduce random effects even further, as well as combining shape variations with other factors, such as age, sex and electrocardiography, to stratify the risk of recurrence.

Acknowledgments. Part of the research was funded by the *Agence Nationale de la Recherche (ANR)*/ERA CoSysMed SysAFib and ANR MIGAT projects. The authors would like to thank Alan Garny, Côme Le Breton and Marco Lorenzi for their great support.

References

1. Zoni-Berisso, M., Lercari, F., Carazza, T., Domenicucci, S., et al.: Epidemiology of atrial fibrillation: European perspective. Clin. Epidemiol. **6**, 213–220 (2014)
2. Berruezo, A., Tamborero, D., Mont, L., Benito, B., Tolosana, J.M., Sitges, M., Vidal, B., Arriagada, G., Méndez, F., Matiello, M., et al.: Pre-procedural predictors of atrial fibrillation recurrence after circumferential pulmonary vein ablation. Eur. Heart J. **28**(7), 836–841 (2007)
3. Dagres, N., Kottkamp, H., Piorkowski, C., Weis, S., Arya, A., Sommer, P., Bode, K., Gerds-Li, J.-H., Kremastinos, D.T., Hindricks, G.: Influence of the duration of holter monitoring on the detection of arrhythmia recurrences after catheter ablation of atrial fibrillation: implications for patient follow-up. Int. J. Cardiol. **139**(3), 305–306 (2010)
4. Shin, S.-H., Park, M.-Y., Oh, W.-J., Hong, S.-J., Pak, H.-N., Song, W.-H., Lim, D.-S., Kim, Y.-H., Shim, W.-J.: Left atrial volume is a predictor of atrial fibrillation recurrence after catheter ablation. J. Am. Soc. Echocardiogr. **21**(6), 697–702 (2008)
5. Bisbal, F., Guiu, E., Calvo, N., Marin, D., Berruezo, A., Arbelo, E., Ortiz-Pérez, J., Caralt, T.M., Tolosana, J.M., Borràs, R., et al.: Left atrial sphericity: a new method to assess atrial remodeling. Impact on the outcome of atrial fibrillation ablation. J. Cardiovasc. Electrophysiol. **24**(7), 752–759 (2013)
6. Marrouche, N.F., Wilber, D., Hindricks, G., et al.: Association of atrial tissue fibrosis identified by delayed enhancement mri and atrial fibrillation catheter ablation: the decaaf study. JAMA **311**(5), 498–506 (2014)
7. Varela, M., Bisbal, F., Zacur, E., Berruezo, A., Aslanidi, O., Mont, L., Lamata, P.: Novel computational analysis of left atrial anatomy improves prediction of atrial fibrillation recurrence after ablation. Frontiers Physiol. **8**, 68 (2017)
8. Labarthe, S., Coudière, Y., Henry, J., Cochet, H.: A semi-automatic method to construct atrial fibre structures : a tool for atrial simulations. In: CinC 2012 - Computing in Cardiology, vol. 39, pp. 881–884 (2012)
9. Jia, S., Cadour, L., Cochet, H., Sermesant, M.: STACOM-SLAWT challenge: left atrial wall segmentation and thickness measurement using region growing and marker-controlled geodesic active contour. In: Mansi, T., McLeod, K., Pop, M., Rhode, K., Sermesant, M., Young, A. (eds.) STACOM 2016. LNCS, vol. 10124, pp. 211–219. Springer, Cham (2017). doi:10.1007/978-3-319-52718-5_23

10. Cochet, H., Dubois, R., Sacher, F., Derval, N., Sermesant, M., Hocini, M., Montaudon, M., Haïssaguerre, M., Laurent, F., Jaïs, P.: Cardiac arrythmias: multimodal assessment integrating body surface ecg mapping into cardiac imaging. Radiology **271**(1), 239–247 (2013)
11. Jamin, C., Alliez, P., Yvinec, M., Boissonnat, J.-D.: CGALmesh: a generic framework for delaunay mesh generation. ACM Trans. Math. Softw. (TOMS) **41**(4), 23 (2015)
12. Miller, M.I., Younes, L., Trouvé, A.: Diffeomorphometry and geodesic positioning systems for human anatomy. Technology **2**(01), 36–43 (2014)
13. Charon, N., Trouvé, A.: The varifold representation of nonoriented shapes for diffeomorphic registration. SIAM J. Imaging Sci. **6**(4), 2547–2580 (2013)
14. Durrleman, S., Prastawa, M., Charon, N., Korenberg, J.R., Joshi, S., Gerig, G., Trouvé, A.: Morphometry of anatomical shape complexes with dense deformations and sparse parameters. NeuroImage **101**, 35–49 (2014)
15. Geladi, P., Kowalski, B.R.: Partial least-squares regression: a tutorial. Anal. Chim. Acta **185**, 1–17 (1986)
16. Kleijn, S.A., Aly, M.F.A., Terwee, C.B., van Rossum, A.C., Kamp, O.: Three-dimensional speckle tracking echocardiography for automatic assessment of global and regional left ventricular function based on area strain. J. Am. Soc. Echocardiogr. **24**(3), 314–321 (2011)

Adjustment of Parameters in Ionic Models Using Optimal Control Problems

Diogène Vianney Pongui Ngoma[1], Yves Bourgault[2(✉)],
Mihaela Pop[3], and Hilaire Nkounkou[1]

[1] Department of Mathematics, Marien Ngouabi University,
Brazzaville, Republic of Congo
[2] Department of Mathematics and Statistics,
University of Ottawa, Ottawa, Canada
ybourg@uottawa.ca
[3] Sunnybrook Research Institute, University of Toronto, Toronto, Canada

Abstract. We developed numerical methods to optimally adjust the parameters in cardiac electrophysiology models, using optimal control and non-differentiable optimization methods. We define three optimal control problems to capture the main features of the cardiac action potential (AP). The first two control problems adjust parameters in single-cell models to recover the duration of the various phases of the AP or the trans-membrane potential at a given cell recorded over time. A third control problem is defined to adjust the conductance in the monodomain model to recover the conduction speed of an AP wave. The methodology is used to adjust parameters in the monodomain model with Mitchell-Schaeffer ion kinetics to recover the phase durations and conduction velocity in three cardiac tissues. Error on the phase durations lies within 1–3%, except for the depolarization time. The Aliev-Panfilov and Mitchell-Schaeffer model are adjusted to experimental recording of the trans-membrane potential obtained through optical fluorescence imaging. The Mitchell-Schaeffer model achieves a better fit to the data.

Keywords: Ionic models · Optimal control · Non-differentiable optimization · Parameter identification

1 Introduction

Several ionic models are available to describe the evolution of the electrical potential across cardiac cell membranes. These models usually read as a systems of coupled highly nonlinear differential equations with many adjustable parameters. The adjustment of parameters becomes increasingly important to be able to personalise these models using medical data (see for instance [9,10]) or to compare models with each other in the best possible way. It is not easy to study the combined effect of varying the parameters and the literature is usually not too explicit on the way the parameters are adjusted in ionic models. Moreover, methods are not

© Springer International Publishing AG 2017
M. Pop and G.A. Wright (Eds.): FIMH 2017, LNCS 10263, pp. 322–332, 2017.
DOI: 10.1007/978-3-319-59448-4_31

available to check in a systematic way if two models could provide a similar solution (e.g. same recording of the potential at a cell over time), for instance when parameters are well adjusted in one model to fit the other model.

Parameter adjustment is possible with simpler ionic models using asymptotic formula connecting the parameters with the phase durations [7,10,11]. The conductance can be adjusted to the conduction velocity (CV) using Eikonal equations as in [10]. Very few attempts have been made to address the adjustment of the ionic model parameters or the conductance using fully nonlinear models. We are aware of the very recent paper [5] where a genetic algorithm was used to build a cell-specific cardiac electrophysiology model and [6] where simulated annealing is used to compare two ionic models.

This paper proposes numerical methods to optimally adjust the parameters in ionic models, in particular when these models involve terms (ionic currents or gating source terms) that are not continuous or stiff in the state variables. Our method is based on the numerical solution of an optimal control problem with a least-square objective function. Three types of least-square functions will be used. The first one attempts to fit the main features of the cardiac action potential (AP), namely the action potential duration (APD), the depolarization time (DT), recovery time (RT), etc. The second function attempts to fit the trans-membrane potential predicted by the model to experimental recording on a single cell. The third least-square function fits the CV predicted by the monodomain model to an experimental value.

We will illustrate the efficiency of the method for the Mitchell-Schaeffer model [7], which is a simple two variables ionic model with a limited set of parameters and one discontinuity in the r.h.s. of the ODE for the gating variable. Our methodology is not limited to this model. Numerical results are presented, in particular model fitting to experimental AP measurements obtained through an optical fluorescence imaging technique.

2   Mathematical Models

2.1   Mitchell-Schaeffer Model

As one particular example where the proposed parameter identification technique can be applied, we consider the Mitchell-Schaeffer (MS) two-variable model [7]. This model describes the dynamics of the trans-membrane potential u in the myocardium and a gating variable v representing in a lumped way the opening and closing of ionic channels controlling the passage of ions across the cell membranes. Here we will consider two situations, either the 0D model for a single cell (no space dependence of the variables u and v) or the monodomain model where spatial propagation is assumed in the myocardium.

Single-Cell Model: The dependent variables $u = u(t)$ and $v = v(t)$, $t > 0$, are solutions of:

$$\frac{du}{dt} = f(u,v) + I_{stim}(t), \quad \text{with} \quad f(u,v) = \frac{1}{\tau_{in}}vu^2(1-u) - \frac{1}{\tau_{out}}u, \quad (1)$$

$$\frac{dv}{dt} = g(u,v), \quad \text{with} \quad g(u,v) = \begin{cases} \dfrac{1-v}{\tau_{open}}, & \text{if} \quad u < u_{gate}, \\[3mm] \dfrac{-v}{\tau_{close}}, & \text{if} \quad u \geq u_{gate}. \end{cases} \tag{2}$$

The trans-membrane current $f(u,v)$ is the sum of the gated inward current $vu^2(1-u)/\tau_{in}$ with time scale τ_{in} that tends to depolarize the cardiac cell and the ungated current $-u/\tau_{out}$ that tends to repolarize the cardiac cell with time scale τ_{out}. Finally, I_{stim} represent an external current produced by a stimulation electrode. The dynamics of the gating variable v depends on the threshold potential u_{gate} for the initiation of an action potential, and on two time constants, τ_{open} and τ_{close}, respectively controlling the opening and closing of the gate. We set $\tau = [\tau_{in}, \tau_{out}, \tau_{open}, \tau_{close}]$ to simplify notations. The functions f and g depend on the parameter τ. Eqs. (1)–(2) requires initial conditions $u(0) = u_0$ and $v(0) = v_0$, where $u_0, v_0 \in [0,1]$ are given.

Monodomain Model: The dependent variables $u = u(x,t)$ and $v = v(x,t)$, $x \in \Omega$, $t > 0$, are solutions of:

$$\frac{\partial u}{\partial t} - \nabla \cdot (\sigma \nabla u) = f(u,v) + I_{stim}(t), \tag{3}$$

$$\frac{\partial v}{\partial t} = g(u,v). \tag{4}$$

The functions f and g are defined as for the single-cell MS model, except that the unknowns u and v depends also on the space variable x. The cardiac tissue constitutes the domain Ω where the equations are solved. The parameter σ is the conductance of the cardiac tissue, usually taken as a 2-tensor to represent the anisotropic conduction properties of the myocardium. In our test cases, we consider the 1D monodomain model where propagation is assumed to be in one spatial direction only (e.g. planar waves). In this case, σ is a scalar constant (e.g. conductance along the fibers). Eqs (3)–(4) come with initial and boundary conditions, here taken as homogeneous Neumann boundary conditions.

2.2 Optimal Control Problems

We introduce three different control problems. The first problem adjusts the parameter τ for the single-cell MS model so that the durations of the four phases are close to known values, while the second one fits the trans-membrane potential u from the model to a recorded experimental potential $\tilde{u} = \tilde{u}(t)$. The third problem adjusts the conductance σ of the monodomain model to match the speed of propagation of the AP.

Least Square Fit on the Phase Durations: Consider the phase durations ΔT_i^*, $i = 1, 2, 3, 4$, (i.e. DT, APD, RT and recovery phase duration) obtained experimentally. To identify the parameters τ in the model (1)–(2), we introduce an optimization problem whose goal is to reduce the gap between the phase durations ΔT_i predicted by the model and the target durations ΔT_i^*.

This optimization problem reads as: Find τ^* minimizing the following least square function

$$J(\tau) = \frac{1}{2} \sum_{i=1}^{4} \omega_i (\Delta T_i - \Delta T_i^*)^2, \tag{5}$$

where $\omega_i \geq 0$, $i = 1, 2, 3, 4$, are weight constants for varying the relative importance of each variable to adjust, u and v are solution of (1)–(2), the times $T_i = T_i(\tau)$, $T_1 < T_2 < \ldots < T_5$, are such that

$$\begin{cases} u(T_i) = \gamma_i, & i = 1, 3, 4, \quad \gamma_i \quad \text{thresholds given,} \\ u(T_2) = \max_t (u(t)), \\ v(T_5) = \gamma_5, \quad \gamma_5 \quad \text{given.} \end{cases} \tag{6}$$

We set $\Delta T_i = T_{i+1} - T_i$.

The thresholds γ_i are characteristic values of the potential u (or v for γ_5) indicating the beginning or the end of the phases. The values of the thresholds $\gamma_1 = 0.13$, $\gamma_3 = 0.5$, $\gamma_4 = 0.05$ and $\gamma_5 = 0.9$ are used for all our test cases. For instance, $\gamma_1 = 0.13$ since $u_{gate} = 0.13$ is the threshold potential to initiate a depolarization. The maximum of the potential is reached at time t_2, hence ΔT_1 is the duration of the depolarization phase (DT). The threshold γ_3 flags the end of the plateau, hence ΔT_2 is the duration of the plateau (more or less the APD). The threshold γ_4 is close to the equilibrium or rest value of the potential u, hence ΔT_3 is the repolarization time. The threshold γ_5 is used to flag the return of the recovery variable to equilibrium when the cell is excitable again. The time t_5 thus corresponds to the end of the refractory period. We call ΔT_5 the recovery period, which is nothing but the refractory period minus the APD.

A natural choice for making each square dimensionless in the function J is given by

$$\omega_i = \left(\frac{1}{\Delta T_i^*} \right)^2, i = 1, 2, 3, 4. \tag{7}$$

Least Square Fit of the Potential: We change the least square function to adjust the potential $u = u(t)$ predicted by the single-cell MS model to a given potential $\tilde{u} = \tilde{u}(t)$, $t \in [0, T]$ (measured or obtained with an other ionic model). This optimization problem reads as: Find τ^* minimizing

$$J(\tau) = \frac{1}{2} \int_0^T | u(s, \tau) - \tilde{u}(s) |^2 \, ds, \tag{8}$$

where u and v are solution of (1)–(2) with parameters τ. It is no longer needed to set thresholds γ_i, recover phase durations and wave speed. The connection between the parameter τ and the objective function J is more direct, but still subject to the lack of regularity induced by the discontinuity in the function g. We successfully solved this optimization problem to fit various ionic models among each other (up to 3 variables and 8 parameters – not shown here). It is possible to add other least square terms to fit other variables.

Least Square Fit on the Wave Speed: Consider the wave speed c^* obtained experimentally. To identify the conductance σ in the model (3)–(4), we introduce an optimization problem whose goal is to reduce the gap between the speed of the wave c predicted by the model and the target wave speed c^*. It is known from an asymptotic argument presented in [11] that the parameters in the 1D MS model are in one-to-one relation with the durations ΔT_i^* and wave speed c^*. It thus makes sense to try to identify the conductance σ by matching the wave speed.

This optimization problem reads as: Find σ^* minimizing the following least square function

$$J(\sigma) = \frac{1}{2}(c - c^*)^2, \tag{9}$$

where u and v are solutions of (3)–(4), and the wave speed $c = c(\sigma)$ is considered a function of σ only since the parameters τ are assumed known from minimizing either (5) or (8). We calculate the wave speed from the solution of the model using

$$c = \frac{x_2 - x_1}{t_2 - t_1},$$

where $t_1 < t_2$ are the passage times of the wave at the points $x_1 < x_2$ given in the domain where the AP wave is propagated. At each point x_i, the passage of the wave can be flagged by finding the smallest time $t_i > 0$ such that $u(x_i, t_i) = \gamma_1$, with γ_1 the threshold for the initiation of the AP given above.

2.3   Numerical Methods

The Eqs (1)–(2) are solved using the function `ode` in `Scilab`, which implements the Adams predictor-corrector method. The Eqs (3)–(4) are solved by discretizing with finite difference formulae in space and the second order semi-implicit backward differentiation formulae (SBDF2) in time [4]. Eq. (6) are solved using linear interpolation within time steps to ensure the accuracy of the times T_i.

The function g is discontinuous in u, which eventually leads to a lack of regularity of the solution (u, v) of (1)–(2) (similarly for (3)–(4)) and consequently of the function $J = J(\tau)$. The derivatives of J with respect to τ may not be well defined. Attempts with numerical differentiation of J and plots of the square terms in J showed that numerical derivatives do not converge when the steps in τ are refined. Direct computations of the sensitivities δu and δv with respect to τ by solving numerically the sensitivity ODEs for a regularized version of the MS model were not more successful. To avoid the computations of the sensitivities and the gradient of J with respect to τ (or σ), we use non-differentiable optimization methods [2]. A two-step strategy was required to identify the parameters τ^* and σ^*. First, we solve the optimization problem (1)–(2)+(5)–(6) for τ^* using the Nelder-Mead method. Setting the parameters τ^* from this first step, we then solved the optimization problem (1)–(2)+(9) for σ^* using the Golden Section method. We used the function `fminsearch` in `Scilab` that implements the Nelder-Mead method. We implemented our own script for the Golden Section method in `Scilab`.

Changing from one to an other least square function presented above is possible at no cost since the Nelder-Mead method requires only values of J at given iterates τ_k, and no derivatives of this function.

The optimization method used introduces a sensitivity to the initial guess of the parameters (or the initial interval for the Golden section method). If convergence is reached (e.g. measured in distance between consecutive iterates) but the value of the least-square function J is not small for the final iterate, the minimum is likely to be local only and new initial guesses must be attempted in the hope of getting a better fit. Asymptotic formula from [11] could be used to obtain initial guesses for the parameters τ in the MS model. We used our experience with the fitted models as well as trial and error to find good initial guesses.

3   Numerical Results

3.1   Validation

To validate our approach, we tried to recover with the help of the Nelder-Mead method the value of the parameters $\tau^* = [0.3, 6, 130, 150]$ that correspond to $J(\tau^*) = 0$ for $\Delta T^* = [6.71, 251.48504, 34.311947, 270.42669]$, the phase durations obtained by solving the 0D MS with this τ^*. We studied the behavior of the method varying the tolerance Tol and the initial parameter values τ_0, where the convergence criteria is given by

$$\|\tau_{k+1} - \tau_k\| < Tol,$$

for two successive iterates τ_k and τ_{k+1}. Table 1 shows the impact of varying Tol. NIter et NEval are the number of iterations of the Nelder-Mead method and the number of function evaluations, respectively. The global minimizer is reached with an accuracy of 10^{-2}, while the global minimum of J is accurate at 10^{-10}. Starting from $\tau_0 = [0.37, 7, 140, 160]$, the best value of τ^* achieved is $\tau_{final} = [0.3413, 5.707, 132.41, 169.48]$ with $J(\tau_{final}) = 0.7975322$ (table not shown here). Hence initial values τ_0 must be chosen carefully to ensure convergence to the global minimum. The value of $J(\tau_{final})$ is a good measure of the fit (the closer to zero, the better!).

A similar validation test case was carried for recovering σ^* with the Golden Section method, with similar conclusions on the sensitivity to the interval chosen to initialize the method.

Table 1. Sensitivity of Nelder-Mead method to Tol. $\tau_0 = [0.27, 5.8, 127, 140]$

Tol	NIter	NEval	τ_{final}	$J(\tau_{final})$
10^{-4}	203	352	[0.3002057, 5.9988204, 130.01025, 150.08348]	$1.239D - 09$
10^{-6}	225	409	[0.3002056, 5.9988201, 130.01024, 150.08346]	$5.382D - 10$
10^{-8}	242	458	[0.3002056, 5.9988201, 130.01024, 150.08346]	$4.100D - 10$

3.2   Application to Three Cardiac Tissues

We proceed now with an application of the methodology introduced above. Table 2 presents commonly accepted values of the phase durations and wave speed (see for instance [3]). Note that the recovery period corresponds in our approach to phase between the time when the potential u goes back to rest (end of the repolarization) and the time where the gating variable v goes back to rest (end of the refractory period). Table 3 shows the parameters τ_{final} obtained by solving the control problem for the 0D model with the Nelder-Mead method, the value of the minimum $J(\tau_{final})$ and the phase durations ΔT_i predicted by the MS model for these parameter values. Table 4 shows the conductance σ_{final} obtained by solving the control problem for the 1D model with the Golden Section method. Since we have a range of experimental conduction speeds, we obtained a range of possible conductances. This table also presents the conduction velocities and phase durations predicted by the 1D MS model using the parameters fitted with the two-step identification method. For the Purkinje fibers, we provide only values for $c = 2\,\text{m/s}$. Larger speeds require a very large domain with a very fine numerical resolution for the fit to be reliable because of the fast moving and highly spread AP wave.

The fit is exceptionally good in both parameter identification steps, between the experimental phase durations and the ones predicted by both the 0D and 1D MS models. The same remark is valid for the conduction speed. A joint

Table 2. Experimental durations (ms) and wave speed (m/s) for three tissues

Tissues	ΔT_1^*	ΔT_2^*	ΔT_3^*	ΔT_4^*	$\hat{c}^*$ (m/s)
Left ventricle (LV)	8	250	30	260	$0.3 - 0.5$
Purkinje fibers (PF)	8	380	65	320	$2 - 4$
Right atria (RA)	$4 - 5$	100	20	250	$0.3 - 0.5$

Table 3. Results from the parameter identification in the 0D MS model

Tissue	ΔT_1	ΔT_2	ΔT_3	ΔT_4	τ_{final}	$J(\tau_{final})$
LV	6.41	249.96	30.28	260.00	[0.276, 4.92, 126.4, 161.5]	0.1812972
PF	8.14	379.98	64.97	320.00	[0.397, 13.3, 152.2, 168.7]	0.0101284
RA	3.9	100.00	19.99	250.00	[0.180, 4.23, 116.7, 53.4]	0.0756264

Table 4. Results from the parameter identification in the 1D MS model

Tissues	σ_{final}	c	ΔT_1	ΔT_2	ΔT_3	ΔT_4
LV	$0.303 - 5.08$	$0.30 - 0.50$	$8-7.6$	$247.96-247.54$	$30.70-30.71$	$259.13-259.37$
PF	26.729	2.00	9.9	377.95	66.22	318.86
RA	$0.0980 - 0.418$	$0.30 - 0.50$	$5.4-5.0$	$98.45-98.70$	$20.59-20.52$	$249.43-249.43$

identification of τ and σ could probably improve the fit, but the difficulty of getting initial values that lead to convergence of the Nelder-Mead method is avoided with the two-step method with a limited impact on the quality of the fit.

3.3   Adjusting Models to Experimental Data

Data Acquisition. For experimental validation and mathematical model parameterization, action potentials waves were recorded using voltage-based optical fluorescence imaging, as described in [8]. Briefly, fluorescence dye (di4-ANEPPS) and mechanical uncoupler (to block contraction) were injected into coronary circulation of a healthy explanted swine heart connected to a Langendorff perfusion system. The optical dye was excited with green light ($530 \pm 20 \, nm$) via 150 W halogen source lamps. The emitted signals from the heart were filtered ($> 610 \, nm$) and captured by a high-speed dual CCD system (MICAM02, Brain-Vision Inc. Japan) with 3.91 ms temporal resolution (256 frames/second). The field of view was 184×124 pixels ($12 \times 10 \, cm$), yielding an approximately 0.7mm spatial resolution. The temporal change in fluorescence signal intensity recorded at each pixel, gives directly the action potential waves. For fitting the models, we use the AP recorded at one pixel selected from an area in the left ventricle (LV) where tissue was homogeneously illuminated, and also both fluorescence signal and tissue perfusion were homogeneous. Notably, we did not average the optical fluorescence signal over a selected ROI in purpose, since this would result in a smoother AP wave form, particularly a smoother up-stroke (which can result in incorrect model parameters).

Model Fitting. The AP recorded by optical fluorescence were fitted with two different models, the MS model presented above and the Aliev-Panfilov (A-P) model [1]. The A-P model has 2 variables (a potential u and a gating variable v) and 5 adjustable parameters. The goal is to compare the quality of fit of the two models.

A single AP was isolated from a sequence of recorded AP. This AP was renormalized between 0 and 1 to obtain the potential $\tilde{u} = \tilde{u}(t)$, $t \in [0, 1400]$ (in ms) required in the least square function (8). The Nelder-Mead method was used to solve the 0D parameter fitting problem (step 1 only). A single AP is triggered with different stimulation currents for the two models: a current lasting $50ms$ and starting at $t = 100$ ms is used for for the MS model, and a current lasting $20ms$ and starting at $t = 105$ ms for the A-P model. These stimulation currents

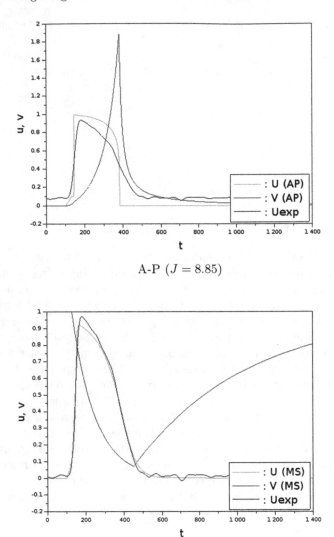

Fig. 1. A-P and MS models fitted to AP recorded by optical fluorescence

were adjusted manually to ensure the best fit of the models. Figure 1 shows the solution (u and v) and the experimental potential. The MS model achieves a much better fit than the A-P model. This is seen by comparing the graphs and the value of the least square function that is 32 times smaller for the MS model.

No efforts were made to control the evolution of the variable v for both models, since this variable is not included in the least square function J. The result is that the duration of the recovery (and refractory) period is different for the two models. The A-P model gives a cell that is excitable again after

1000 ms (roughly), while the MS model predicts that more 1400 ms are required for this to occur. This situation could be improved by adding a term to control the variable v in the least square function, assuming that experimental data on the refractory period are available.

4   Conclusions and Perspectives

We provided a new framework for fitting electrophysiology model parameters based on control theory. Three least-square functions are proposed, allowing the adjustment of parameters in simple single-cell ionic models to fit either the durations of the various phases of the AP or the shape of the trans-membrane potential to experimental recording. The methodology can be extended to adjust the conductance in the monodomain model in order to fit an experimentally recorded CV. Differentiability of the ionic model is not required as no derivatives are computed. The method is thus potentially applicable to the many ionic models that lack differentiability (e.g. through jump functions included in those models). In its current form, the methodology is capable of fitting models with a modest number of parameters. So far, it has been tested and worked for models with up to 8 parameters (not shown here), but it may eventually work for models with a larger number of parameters as long as good initial guesses are available.

The accuracy of the fit is easily below 1% on the fitted AP phase durations in the single-cell MS model, except for the depolarization phase which is very short and thus subject to larger relative error ($2-20\%$ – still below 1.5 ms in absolute error). The two-step approach allows a perfect fit of the CV while introducing a minimal extra error on the phase durations. The least-square function (8) gave a good fit of the trans-membrane potential predicted by the model to the potential recorded over time on a single cell. However, the fit for this latter is as good as a given ionic model can represent the data. For instance, the A-P model did not turn out to be as flexible as the MS model in representing the potential recorded by optical fluorescence imaging. The approach can definitely be used to sort out models in terms of their capability to reproduce a given AP, be it experimental or from an other ionic model, while using the optimal value in the parameter space.

Future applications of the methodology include fitting the CV in more complex situations and for adjusting the restitution properties of ionic models. The fact that no derivatives are required in the optimization method is particularly appealing for the latter. We will have to be careful in adjusting the conductance with spatially distributed data (e.g. with our optical imaging), as there is a limitation in the number of parameters that can be adjusted with the current approach. For instance, fitting the conductance at the 256×256 pixels of our optical fluorescence image is out of reach of the proposed method. We acknowledge that the model parameters might be different in the right ventricle, RV (e.g. shorter APD). Thus, in the future we will investigate these parameters using optical signals recorded from RV. Furthermore, we will perform data fitting on 26 AHA segments (17 in the LV and 9 in the RV) as in [9].

Acknowledgments. This work was funded by a NSERC Discovery Grant from the Canadian Government, the Agence Univesitaire de la Francophonie (AUF) through the program Inter-Regional Project, Horizons Francophones and a scholarship from the African Institute of Mathematical Sciences (AIMS), Research of Africa Project.

References

1. Aliev, R.R., Panfilov, A.V.: A simple two-variable model of cardiac excitation. Chaos, Solitons and Fractals **7**(3), 293–301 (1996)
2. Chong, E.K.P., Zak, S.H.: An Introduction to Optimization, 3rd edn. Wiley, Hoboken (2008)
3. Djabella, K., Landau, M., Sorine, M.: A two-variable model of cardiac action potential with controlled pacemaker activity and ionic current interpretation. In: 46th IEEE Conference on Decision and Control, pp. 5186–5191 (2007)
4. Ethier, M., Bourgault, Y.: Semi-implicit time-discretization schemes for the bidomain model. SIAM J. Numer. Anal. **46**, 2443–2468 (2008)
5. Groenendaal, W., et al.: Cell-specific cardiac electrophysiology models. PLOS Comput. Biol., 22 p. (2015)
6. Lombardo, D.M., Fenton, F.H., Narayan, S.M., Rappel, W.-J.: Comparison of detailed and simplified models of human atrial myocytes to recapitulate patient specific properties. In: PLOS Comput. Biol., 15 p. (2016)
7. Mitchell, C.C., Schaeffer, D.G.: A two-current model for the dynamics of cardiac membrane. Bull. Math. Biol. **65**(5), 767–793 (2003)
8. Pop, M., Sermesant, M., Lepiller, D., Truong, M.V., McVeigh, E.R., Crystal, E., Dick, A., Delingette, H., Ayache, N., Wright, C.A.: Fusion of optical imaging and MRI for the evaluation and adjustment of macroscopic models of cardiac electrophysiology: a feasibility study. Med. Image Anal. **13**(2), 370–380 (2009)
9. Relan, J., Pop, M., Delingette, H., Wright, G., Ayache, N., Sermesant, M.: Personalisation of a cardiac electrophysiology model using optical mapping and MRI for prediction of changes with pacing. IEEE Trans. Biomed. Eng. 10(10), 11 p. (2011)
10. Relan, J., et al.: Coupled personalization of cardiac electrophysiology models for prediction of ischaemic ventricular tachycardia. Interface Focus **1**, 396–407 (2011)
11. Rioux, M., Bourgault, Y.: A predictive method allowing the use of a single ionic model in numerical cardiac electrophysiology. ESAIM Math. Modell. Numer. Anal. **47**, 987–1016 (2013)

Smoothed Particle Hydrodynamics for Electrophysiological Modeling: An Alternative to Finite Element Methods

Èric Lluch[1,2(✉)], Rubén Doste[1], Sophie Giffard-Roisin[3], Alexandre This[2,4],
Maxime Sermesant[3], Oscar Camara[1], Mathieu De Craene[2],
and Hernán G. Morales[2]

[1] PhySense, ETIC, Universitat Pompeu Fabra, Barcelona, Catalonia
[2] Medisys, Philips Research, Paris, France
eric.lluch@philips.com
[3] Université Côte d'Azur, Inria, Nice, France
[4] Inria, Paris, France

Abstract. Finite element methods (FEM) are generally used in cardiac 3D-electromechanical modeling. For FEM modeling, a step of a suitable mesh construction is required, which is non-trivial and time-consuming for complex geometries. A meshless method is proposed to avoid meshing. The smoothed particle hydrodynamics (SPH) method was used to solve an electrophysiological model on a left ventricle extracted from medical imaging straightforwardly, without any need of a complex mesh. The proposed method was compared against FEM in the same left-ventricular model. Both FEM and SPH methods provide similar solutions of the models in terms of depolarization times. Main differences were up to 10.9% at the apex. Finally, a pathological application of SPH is shown on the same ventricular geometry with an added scar on the heart wall.

Keywords: SPH · Meshless · FEM · Cardiac electrophysiology

1 Introduction

Patient-specific modeling has become an interesting research topic in the cardiac electrophysiology community because it can help to understand the electrical propagation and its pathologies [2]. FEM is a well-established numerical approach often used to investigate the electro-mechanics of the human heart [3,13]. The generation of complex meshes is necessary. Meshing is one of the main bottlenecks for the clinical translation of cardiac modeling tools since it is difficult to have a streamlined and automated pipeline to generate accurate FE simulations from imaging data [10]. Another non-trivial step of FEM in electro-mechanics is the coupling between electrophysiology and mechanics when meshes with different resolution for both problems are used. It is expected that a way to overcome these difficulties could be through a meshless approach.

© Springer International Publishing AG 2017
M. Pop and G.A. Wright (Eds.): FIMH 2017, LNCS 10263, pp. 333–343, 2017.
DOI: 10.1007/978-3-319-59448-4_32

Various meshless methods have demonstrated the ability to provide a computational feasible model for cardiac electrophysiology simulations, without burden mesh generation [2,5,15]. In this paper, SPH [8] is proposed to numerically solve the Mitchell-Schaeffer (MS) electrophysiological model [7] on the electrical depolarization of the left ventricle. To evaluate the accuracy of this approach, a comparison with a FEM implementation [6] was conducted. Finally, a scar was added to the ventricular myocardium to show the potential use of the proposed meshless approach in a pathological case. The goal is to explore the accuracy, speed and limitations of SPH with respect to FEM, as a first step towards a potential full electro-mechanical heart model using a meshless approach.

2   Method

In this section, the electrophysiological model and the SPH discretization scheme are explained. For further details of the FEM approach, refer to [6].

2.1   Electrophysiological Model

In this paper, the macroscopic biophysical mono-domain model Mitchel-Schaeffer together with a diffusion term [13] was used to model the cardiac electrophysiology. This model was chosen because it captures the action potential duration (APD) (Fig. 1), considers fiber orientation in the diffusion term and is only governed by 6 parameters, which might facilitate a more precise model personalization since less parameters need to be fitted to given data.

$$
\begin{cases}
\partial_t v = \dfrac{w v^2 (1-v)}{\tau_{\text{in}}} - \dfrac{v}{\tau_{\text{out}}} + I_{\text{app}} \\
\partial_t w = 2 \begin{cases} \dfrac{1-w}{\tau_{\text{open}}} & \text{if } v < v_{\text{gate}} \\ \dfrac{-w}{\tau_{\text{close}}} & \text{if } v > v_{\text{gate}}. \end{cases}
\end{cases}
\tag{1}
$$

When considering the geometry, a diffusion term $\text{div}(C\nabla v)$ is required to the first Eq. (1) [13]. $C \in \mathbb{R}^{3,3}$ is the connectivity tensor defined as

$$
C = (\tau \otimes \tau (1 - ar) + Id \cdot ar) \cdot c,
\tag{2}
$$

with $\tau \in \mathbb{R}^3$ being the vector corresponding to the fiber orientation, $\otimes$ the tensor product, $Id \in \mathbb{R}^{3,3}$ the identity matrix, $ar \in \mathbb{R}$ the anisotropic ratio and $c \in \mathbb{R}$ the conductivity coefficient that controls the propagation velocity. ar controls the conduction velocity in the fiber orientation with respect to the transverse plane, e.g. in the case $ar = 1$, the fiber orientation is not anymore taken into account, hence reducing the model to the isotropic case.

The parameters $\tau_{\text{in}}, \tau_{\text{out}}, \tau_{\text{open}}, \tau_{\text{close}} \in \mathbb{R}$ control the duration of the four stages of the APD. The depolarization phase is controlled by $w \in \mathbb{R}$ and $v_{\text{gate}} \in \mathbb{R}$ defines at which point the APD starts. $I_{\text{app}} \in \mathbb{R}$ corresponds to the first stimulus of the transmembrane potential $v \in \mathbb{R}$.

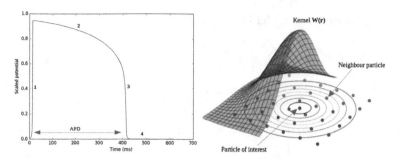

Fig. 1. Left: Example of the 4 stages of the cardiac action potential: initiation (**1**), plateau (**2**), decay (**3**), and recovery (**4**). Right: Example of an SPH Kernel for 2D.

2.2 SPH Discretization

SPH is a meshless Lagrangian method, where each solid particle carries its own properties such as density, conductivity, etc. Given a continuous function $f :$ $\mathbb{R}^3 \to \mathbb{R}$ representing a particle property at the spatial position r, it can be approximated with a delta Dirac function (3):

$$f(r) = (f * \delta)(r) = \int_{\mathbb{R}^3} f(r')\delta(r - r')\mathrm{d}r' \approx \int_{\mathbb{R}^3} f(r')W(r - r', h)\mathrm{d}r'. \quad (3)$$

Notice that the delta Dirac function was approximated with a kernel function $W(r - r', h)$, where $h \in \mathbb{R}$ is the so-called smoothing length (Fig. 1). For Eq. (3) to hold, W must fulfill the following

$$\int_{\mathbb{R}^3} W(r - r', h)\mathrm{d}r' = 1 \text{ and } \lim_{h \to 0} W(r - r', h) = \delta(r - r'). \quad (4)$$

The integral in (4) is approximated as a finite sum, where the density ρ_j and the mass m_j are obtained by replacing the infinitesimal volume $\mathrm{d}r'$ by the finite volume (5).

$$f(r) = \lim_{h \to 0} \int_{\mathbb{R}^3} \frac{f(r')}{\rho(r')} W(r - r', h)\rho(r')\mathrm{d}r'$$

$$\propto \sum_{j=1}^{N} m_j \frac{f_j}{\rho_j} W(r_i - r_j, h) = f_i, \quad (5)$$

where f_i is the approximated value of the function f at the position r, i.e. at the particle of interest i. Due to the previous formulation (5) the derivative of the function f in the same position r can be approximated as a derivative of the kernel function (6)

$$\nabla f_i = \sum_j m_j \frac{f_j}{\rho_j} \nabla W(r_i - r_j, h). \quad (6)$$

The electrophysiological model (1) in the SPH formulation reads:

$$\partial_t v_i = C_i \circ \sum_{j=1}^{n_i} m_j \frac{v_j - v_i}{\rho_j} \nabla^2 W(\boldsymbol{r_i} - \boldsymbol{r_j}, h)$$

$$+ \nabla v_i \text{div}(C_i) + \frac{w_i v_i^2 (1 - v_i)}{\tau_{\text{in}}} - \frac{v_i}{\tau_{\text{out}}} + I_{\text{app},i} \tag{7}$$

$$\partial_t w_i = 2 \begin{cases} \frac{1-w_i}{\tau_{\text{open}}} & \text{if } v_i < v_{\text{gate}} \\ \frac{-w_i}{\tau_{\text{close}}} & \text{if } v_i > v_{\text{gate}}, \end{cases}$$

where $\circ$ is an element wise multiplication (Hadamard product), $\nabla^2 W \in \mathbb{R}^{3,3}$ is the Hessian matrix of the kernel including all the Hessian derivatives and

$$\nabla v_i \text{div}(C_i) = \sum_{j=1}^{n_i} m_j \nabla W(\boldsymbol{r_i} - \boldsymbol{r_j}, h) \frac{C_j - C_i}{\rho_j} \cdot \sum_{j=1}^{n_i} m_j \frac{v_j - v_i}{\rho_j} \nabla W(\boldsymbol{r_i} - \boldsymbol{r_j}, h),$$

$$\tag{8}$$

where n_i is the number of neighbors of the particle p_i.

Boundary conditions are difficult to handle in SPH, even when simple boundary conditions such as symmetric surface boundary are required. This is due to the truncation of the particle neighborhood near a boundary, which results in a truncation of the integral of Eq. (5) [4]. Moreover, when it is assumed that the system connectivity only changes because the particles lose or gain connectivity through a boundary (no-flux), there is no need to place special conditions on the gradient of the potential function v near the boundary [4,8]. In other words, if all the boundaries fulfill the no-flux condition, then the symmetry of the SPH ensures that the system conserves its flux because the particles interact amongst themselves. On top of this, a corrective smoothed particles method (CSPM) was implemented to overcome the lack of particles in the boundaries while enhancing the solution accuracy inside the domain [4]. After applying CSPM, the discretization scheme has an accuracy of $O(h^2)$ for interior points and $O(h)$ for points near or on the boundary, where h is the distance between particles. The distance depends on the choice of the spatial resolution.

The cubic B-spline kernel (9) was used here since its first derivatives are positive for neighbor particles close to the particle of interest [4]

$$W(\boldsymbol{x} - \boldsymbol{x}', h) = \frac{\alpha_d}{h^3} \begin{cases} 1 - \frac{3}{2}q^2 + \frac{3}{4}q^3 & \text{for } q < 1 \\ \frac{1}{4}(2 - q)^3 & \text{for } 1 \leq q < 2 \\ 0 & \text{elsewhere}, \end{cases} \tag{9}$$

where in order to fulfill (4), the coefficient $\alpha_d = \frac{1}{\pi}$ and q is defined as

$$q = \frac{|\boldsymbol{x} - \boldsymbol{x}'|}{h} = \frac{\boldsymbol{r}}{h}. \tag{10}$$

Regarding the time integration scheme, a forward explicit Euler method was used for SPH whose accuracy is $O(h)$, h being the time step. For FEM, the modified Crank-Nicholson/Adams-Bashforth (MCNAB) was used [13].

3 Experiments

To evaluate the proposed SPH-based electrophysiological model, the same model was solved with a FEM scheme and results were compared for the electrical depolarization. An image-based left-ventricular geometry was evaluated in this work to have a preliminary comparison between the methods. The two methods are labeled as:

- **FEM$_{MS}$**: Mitchell-Schaeffer model discretized with FEM [7].
- **SPH$_{MS}$**: Mitchell-Schaeffer model discretized with SPH.

For FEM and SPH approaches, two different resolutions were considered. The low resolution had 18667 nodes and the high resolution had 51037 nodes for both approaches. For **FEM$_{MS}$**, a tetrahedral mesh was computed from the segmented left ventricle. For the proposed approach **SPH$_{MS}$**, a set of equidistributed points from the same anatomy was used. Each SPH particle has a density of 1053kg/m^3, corresponding to reported myocardium density [14]. The mass of each particle was computed as the product of the density times the volume of the cubic cell defined between the particle of interest and the neighboring particles. To understand the impact of key intrinsic parameters of the **SPH$_{MS}$**, additional experiments were conducted for several kernel sizes in these two resolutions. In order to evaluate the accuracy of these experiments, an analysis of the L^2 differences between FEM and SPH activation times, as well as the computational time, were investigated (Table 1).

For all simulations, myocardial fiber orientation was included in each of the nodes to achieve a physiological behavior. Fibers were assigned following the rule-based model angles described by D. Streeter et al. [12]. Regarding the parameters, an initial electrical impulse $I_{app} = -580000 \frac{mV}{s}$ was imposed in a set of points on the apex surface corresponding to 80 mm^2 during 4 ms so that in the first time step with an integration time of $dt = 10^{-4}$ s an initial potential of $v = -58$ mV was obtained. Time variables were $\tau_{open} = 120$ ms, $\tau_{close} = 150$ ms, $\tau_{in} = 0.3$ ms, $\tau_{out} = 6$ ms, following [7]. An anisotropic ratio $ar = 0.16$ was used.

Moreover, a scar was added in the myocardium of the same left ventricle to show how SPH handles a pathological example. In particular, the electrical activation was simulated during one second for both healthy and pathological scenarios. It was assumed that the heart rate was 75 bpm, i.e. the heart period was 0.8 s. Under this assumption, three activation phases were observed within one simulated second: the depolarization phase, where particles get activated; the repolarization phase, where particles get deactivated; the second depolarization phase, where particles get activated again. The scar was placed in the septal-anterior region close to the base. Shape and location of the induced scar are shown in Fig. 2. The high resolution (51037 particles) model with a kernel size of 3 mm was used for both healthy and pathological simulations. The scar tissue was applied to 1621 particles while 5891 particles were treated as gray tissue (tissue near the scar). The rest of the particles were considered as healthy tissue. Two different pathological experiments with different model parameters were simulated. In the first pathological simulation, denoted as 'pathological with low conductivity', particles in the gray zone and in the scar were modified according to [1], in such a way that their conductivity coefficient is reduced but not null. In the second pathological case, denoted as 'pathological with zero conductivity', particles in the scar are assumed to have zero conductivity.

Fig. 2. Scar regions in black, Grey zones in grey and healthy zones in red. (Color figure online)

To visualize the discretized domains using both discretization schemes, the structure of the meshes is shown in Fig. 3. In the case of SPH, a 3D Delaunay filter from Paraview (Kitware Inc., Clifton Park, USA) was used to enhance the volume visualization, since unconnected points in 3D do not provide a good visual 3D representation.

4   Results and Discussion

Results section is structured as follows: first, a sensitivity analysis of the impact of key intrinsic SPH parameters is presented. Then, a qualitative comparison against the FEM solution for depolarization is presented. Finally, experiments to show the SPH_{MS} applicability in a pathological case with a scar are shown. Depolarization and repolarization phases were compared between healthy and pathological cases using SPH.

4.1   Sensitivity Analysis

A sensitivity analysis of particle resolution, as well as the kernel size was conducted for SPH and compared against FEM results. Five kernel sizes ranging from 0.003 to 0.007 m were evaluated. A kernel size < 0.0025 m fails due to insufficient number of neighbors, while a kernel size > 0.007 m has so many neighbors that the computational time was excessive for the potential gain in accuracy.

Table 1 shows that the L^2 error is not linear neither over the kernel size nor over the resolution. A reduction of the difference between SPH and FEM was observed when the number of degrees of freedom was incremented for the evaluated kernel sizes. In terms of computational time, it increases linearly over the kernel size and faster over the degrees of freedom. The FEM implementation was faster than the SPH one. A GPU implementation (relatively easy with SPH

Table 1. $L^2(\mathbb{R})$ norm of the difference of depolarization times between SPH and FEM simulations in the endocardium and computational time (in brackets) of 150 ms with a 4 processor Intel computer for both SPH and FEM.

Number of particles	SPH kernel size					FEM
	0.003	0.004	0.005	0.006	0.007	
18667	10.294	9.792	9.767	9.500	10.911	–
	(37 s)	(1 m 13 s)	(3 m 13 s)	(6 m 43 s)	(10 m 43 s)	(9.92 s)
51037	4.258	4.209	4.723	5.103	6.082	–
	(4 m 52 s)	(11 m 03 s)	(21 m 02 s)	(38 m 41 s)	(57 m 26 s)	(27.41 s)

formulation) could overcome this disadvantage of the SPH approach [9]. The choice of a kernel size of 3 mm and a resolution of 51037 particles is a good balance between kernel size and number of degrees of freedom since the L^2 error with respect to FEM is very small and it was relatively fast to compute. For the rest of the results, this choice of kernel size and resolution was used.

4.2 Qualitative Comparison

Depolarization times were first qualitatively compared using a discrete colormap divided by ten isochronous on the ventricle. In all simulations the electrical activation started from the apex until the septal base. **SPH$_{\text{MS}}$** and **FEM$_{\text{MS}}$** show the same range of depolarization times and a similar activation pattern (Fig. 3). Moreover, it is observed that all particles in the endocardium get activated after 123 ms.

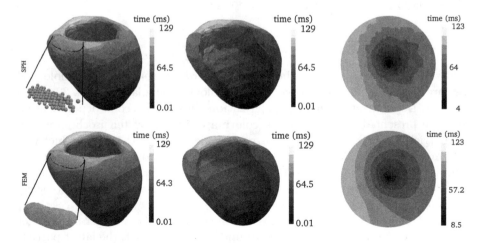

Fig. 3. Left: Contour color map of depolarization times for the left ventricle model. Middle: a longitudinal cross section of the ventricle. Right: projected endocardium onto a disk for both SPH and FEM simulations.

To evaluate the propagation with both approaches, a picture of the activation times for the same cross section in both approaches is shown in Fig. 3 (middle column). In general, a visual comparison of this figure shows that the behavior over the whole ventricular volume is similar for both FEM and SPH methods. The left ventricle endocardium was then mapped into a disk by the use of Quasi-conformal mapping (QCM) [11] to better visualize differences in all regions. The mapped results are shown in the bullseye plots of Fig. 3.

To insight into the differences between $\mathbf{SPH_{MS}}$ and $\mathbf{FEM_{MS}}$, the absolute differences between the depolarization times of these two numerical approaches were computed on the endocardium. These differences were projected on a home-omorphic disk as shown in Fig. 4. In this figure, it is observed that the highest differences of depolarization times in the endocardium occur near the apex with a peak difference of 13.5 ms. For most of the domain, the differences are less than 4 ms as can be seen in the histogram of Fig. 4.

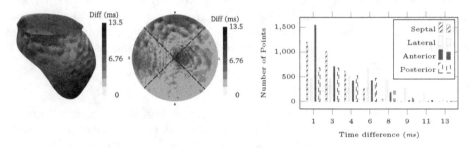

Fig. 4. Left: Difference map of depolarization times between $\mathbf{SPH_{MS}}$ and $\mathbf{FEM_{MS}}$. Middle: projection of the differences onto a disk divided into septal (**S**), lateral (**L**), anterior (**A**) and posterior (**P**) regions. Right: histogram of number of points per region with respect to the time difference.

4.3 Pathological Case

Results of the experiments described in Sect. 3 for both healthy and pathological cases are shown in this subsection. The pathological simulations took 22 m 45 s for 99 ms. Depolarization times, repolarization times and second depolarization times are presented using a discrete colormap divided by ten isochronous on the ventricle (Fig. 5). Results for the pathological case with zero conductivity were thresholded at 0.66 s to avoid having particles at two repolarization phases simultaneously, which facilitates the comparison among the first phases for the three experiments (healthy, pathological with low conductivity and pathological with zero conductivity).

In pathological cases, some regions of the ventricle take a longer time to get activated due to the low conductivity around the scar. In fact, the latest particle gets activated at 169.97 ms (low conductivity) and 656.12 ms (zero conductivity) for the two pathological scenarios, whereas in the healthy case, the latest particle does it after 129 ms. However it was observed for all three cases, that

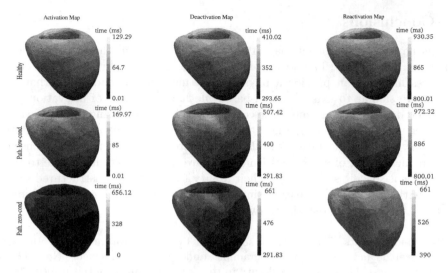

Fig. 5. Different phases for each of the three simulations (from above to below): healthy, pathological with low conductivity and pathological with zero conductivity on the scar).

the ventricle gets repolarized from the apex to the base. In the two pathological cases, the activation pattern goes around the geometry of the scar. In the middle column of Fig. 5, the depolarization phase is depicted. For all cases, the heart starts its depolarization from the apex as it should be. Nevertheless, the depolarization times for the pathological cases are higher due to the lower conductivity in the gray and scar zones. Similarly to the repolarization time, the pattern for the deactivation in the pathological cases takes the shape of the scar into account.

Finally, the second repolarization is shown on the right of Fig. 5. In a heart without arrhythmia, the second repolarization phase should have the same pattern as the first repolarization phase when the electrical impulse is given in the apex again after 0.8 s. Nevertheless, for the pathological case with zero conductivity, particles in the gray zone get reactivated before 0.8 s due to their low (but not zero) conductivity. In particular, the first particle in the gray zone to be activated (at 390 ms) forces the apex to reactivate much earlier than 0.8 s. This implies that the heart gets reactivated before it should, which is known as 'reentry arrhythmia' and has been observed in patients with scars.

Finally, as part of the limitation of this study, full heart geometries will be considered in the future to evaluate the robustness of the proposed SPH_{MS}. Moreover, the validation of SPH for electrophysiology should be performed by comparing it with patient data or with a higher number of validated synthetic geometries. The impact of the particle distribution on the results needs to be revised as well, especially when particle motion is taken into account.

5   Conclusions

In this paper, it was shown that SPH is an alternative method to model cardiac electrophysiology. This work has not only demonstrated that the presented meshless method can provide a physiological meaningful model, but that the results are similar to existing mesh-based methods in terms of activation patterns and depolarization times. The comparison shows promising results towards a proper validation of the method and accuracy assessment against real data. Moreover, a pathological case was also investigated to show the potential use of SPH in the present of a scar. SPH methods are a promising alternative to produce patient-specific simulations. Their ability to import an unstructured set of points without any mesh makes the integration of sparse imaging data (including anatomy and velocities) straightforward.

Acknowledgements. The work is supported by the European Union Horizon 2020 research and innovation programme under grant agreement No 642676 (CardioFunXion). The authors would like to thank the organizers of this project: Bart Bijnens and Mathieu De Craene. Finally, the authors would also like to thank David-Soto Iglesias for all the help provided with the conformal mapping of the endocardium.

References

1. Cabrera Lozoya, R.: Radiofrequency ablation planning for cardiac arrhythmia treatment using modeling and machine learning approaches. Theses, Université Nice Sophia Antipolis, September 2015
2. Campos, J.O., Oliveira, R.S., dos Santos, R.W., Rocha, B.M.: Lattice Boltzmann method for parallel simulations of cardiac electrophysiology using GPUs. J. Comput. Appl. Math. **295**, 70–82 (2016). VIII Pan-American Workshop in Applied and Computational Mathematics
3. Chabiniok, R., Wang, V.Y., et al.: Multiphysics and multiscale modelling, data-model fusion and integration of organ physiology in the clinic: ventricular cardiac mechanics. Interface Focus **6**(2), 20150083 (2016)
4. Chen, J.K., Beraun, J.E., Carney, T.C.: A corrective smoothed particle method for boundary value problems in heat conduction. Int. J. Numer. Method Eng. **46**(2), 231–252 (1999)
5. Chinchapatnam, P., Rhode, K., Ginks, M., Nair, P., Razavi, R., Arridge, S., Sermesant, M.: Voxel based adaptive meshless method for cardiac electrophysiology simulation. In: Ayache, N., Delingette, H., Sermesant, M. (eds.) FIMH 2009. LNCS, vol. 5528, pp. 182–190. Springer, Heidelberg (2009). doi:10.1007/978-3-642-01932-6_20
6. Marchesseau, S., Delingette, H., et al.: Fast parameter calibration of a cardiac electromechanical model from medical images based on the unscented transform. Biomech. Model. Mechanobiol. **12**(4), 815–831 (2013)
7. Mitchell, C.C., Schaeffer, D.G.: A two-current model for the dynamics of cardiac membrane. Bull. Math. Biol. **65**(5), 767–793 (2003)
8. Monaghan, J.J.: Smoothed particle hydrodynamics. Rep. Prog. Phys. **68**(8), 1703 (2005)
9. Nishiura, D., Furuichi, M., Sakaguchi, H.: Computational performance of a smoothed particle hydrodynamics simulation for shared-memory parallel computing. Comput. Phys. Commun. **194**, 18–32 (2013)

10. Smith, N., de Vecchi, A., et al.: euHeart: personalized and integrated cardiac care using patient-specific cardiovascular modelling. Interface Focus **1**(3), 349–364 (2011)
11. Soto-Iglesias, D., Butakoff, C., et al.: Integration of electro-anatomical and imaging data of the left ventricle: an evaluation framework. Med. Image Anal. **32**, 131–144 (2016)
12. Streeter, D.D., Spotnitz, H.M., et al.: Fiber orientation in the canine left ventricle during diastole and systole. Circ. Res. **24**(3), 339–347 (1969)
13. Talbot, H., Marchesseau, S., et al.: Towards an interactive electromechanical model of the heart. Interface Focus 3(2) 2013
14. Yipintsoi, T., Scanlon, P.D., et al.: Density and water content of dog ventricular myocardium. Proc. Soc. Exp. Biol. Med. **141**(3), 1032–1035 (1972)
15. Zhang, H., Wang, L., Hunter, P.J., Pengcheng, S.: Meshfree framework for image-derived modelling. In: 2008 5th IEEE International Symposium on Biomedical Imaging: From Nano to Macro, pp. 1449–1452. IEEE, May 2008

A Rule-Based Method to Model Myocardial Fiber Orientation for Simulating Ventricular Outflow Tract Arrhythmias

Rubén Doste[1]([✉]), David Soto-Iglesias[1], Gabriel Bernardino[1],
Rafael Sebastian[2], Sophie Giffard-Roisin[3], Rocio Cabrera-Lozoya[3],
Maxime Sermesant[3], Antonio Berruezo[4], Damián Sánchez-Quintana[5],
and Oscar Camara[1]

[1] Physense, Universitat Pompeu Fabra, Barcelona, Spain
ruben.doste@upf.edu
[2] Computational Multiscale Simulation Lab (CoMMLab),
Department of Computer Science, Universitat de Valencia, Valencia, Spain
[3] Asclepios Research Group, Inria, Sophia-Antipolis, France
[4] Arrhythmia Section, Cardiology Department, Thorax Institute, Hospital Clinic,
University of Barcelona, Barcelona, Spain
[5] Department of Anatomy and Cell Biology, Faculty of Medicine,
University of Extremadura, Badajoz, Spain

Abstract. Myocardial fiber orientation determines the propagation of electrical waves in the heart and the contraction of cardiac tissue. One common approach for assigning fiber orientation to cardiac anatomical models are Rule-Based Methods (RBM). However, RBM have been developed to assimilate data mostly from the Left Ventricle. In consequence, fiber information from RBM does not match with histological data in other areas of the heart, having a negative impact in cardiac simulations beyond the LV. In this work, we present a RBM where fiber orientation is separately modeled in each ventricle following observations from histology. This allows to create detailed fiber orientation in specific regions such as the right ventricle endocardium, the interventricular septum and the outflow tracts. Electrophysiological simulations including these anatomical structures were then performed, with patient-specific data of outflow tract ventricular arrhythmias (OTVA) cases. A comparison between the obtained simulations and electro-anatomical data of these patients confirm the potential for in silico identification of the site of origin in OTVAs before the intervention.

Keywords: Fiber orientation · Rule-based method · Electrophysiological simulations · Arrhythmias · Outflow tracts

1 Introduction

Outflow tract ventricular arrhythmias (OTVAs) are a type of arrhythmia in which the site of origin (SOO) of the ectopic beat is located in one of the two

© Springer International Publishing AG 2017
M. Pop and G.A. Wright (Eds.): FIMH 2017, LNCS 10263, pp. 344–353, 2017.
DOI: 10.1007/978-3-319-59448-4_33

outflow tracts. In order to treat this disease, clinicians need to locate and ablate its SOO by Radio Frequency Ablation (RFA). Usually, the ectopic focus is identified after visual inspection of the electrocardiogram (ECG) by experienced observers, which is confirmed during the intervention with Electro-Anatomical Mapping (EAM) data. Unfortunately, sometimes the SOO cannot be properly determined from the ECG since it can be located in regions where both OTs are very close. In those cases, RFA interventions can last several hours if several EAMs need to be acquired to identify the SOO.

Personalized electrophysiological simulations of the heart have recently shown promising results to support clinical decisions [1]. Nevertheless, there are some limitations in the existing models for helping to determine the SOO in OTVAs. Most simulation studies in the literature have focused on the left ventricle (LV) due to the complexity of obtaining accurate data of the right ventricle (RV), especially on fiber orientation (e.g. with Diffusion Tensor Magnetic Resonance Imaging [2]). Even though biventricular geometries are often considered in the models, they do not include the outflow tracts, but an artificial basal plane well below the valves. One of the main reasons for that practice is that Rule-Based Methods (RBM) usually employed to generate myocardial fiber orientation do not include specific information about the RV or the outflow tracts, preventing the use of cardiac simulations with OTVA data.

Since electrical wave propagation is determined by fiber orientation [3,4], a proper fiber configuration is needed in order to obtain more accurate and reliable simulation results. We present here an adaptation of an existing RBM [5] that includes specific fiber orientation in cardiac regions relevant for OTVAs, following observations from histological data. According to this data, fiber orientation in the RV sub-endocardium has a longitudinal direction from the apex towards the pulmonary and tricuspid valves, as illustrated in Fig. 1. Moreover, fiber orientation in the sub-endocardium and sub-epicardium of the OT have a longitudinal and circumferential direction, respectively. In addition, the developed RBM processes both ventricles independently, which gives more flexibility to generate different fiber configurations. Septal fiber orientation can also be independently modified, allowing the study of its discontinuity, which is still under debate: whereas some studies indicate the existence of a continuous septum [6], others show fiber discontinuity and evidences of a bilayered septum [7,8].

In this work, fiber configurations generated with the proposed RBM are compared with state-of-the-art RBM in a simple heart geometry without OTs for verification purposes. Several electrophysiological simulations were performed to study the effect of using different fiber configurations on the heart electrical propagation. Finally, the proposed RBM was applied to several patient-specific geometries showing OTVA in which electrophysiological simulations with different SOOs were run. The resulting in silico isochrones around the RV earliest activated point were then compared with patient-specific EAM data for validation purposes.

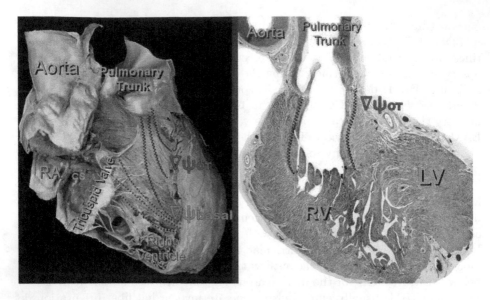

Fig. 1. Histological data of the heart. Left: Fiber configuration in the RV sub-endocardium, with longitudinal directions to the pulmonary and tricuspid valves (dashed red and green lines, respectively). Right: RVOT slice showing longitudinal direction in the sub-endocardium wall. R: Right; L: Left; A: atrium; V: ventricle; CS: coronary sinus; $\nabla\Psi_{basal}$: apico-basal direction; $\nabla\Psi_{OT}$: apico-OT direction (Color figure online)

2 Methodology

The developed RBM was inspired by the work of Bayer et al. [5], which is mainly based on Streeter's observations [9] from the left ventricle. We have improved that RBM by:

(i) Extending fiber orientation up to the outflow tracts.

(ii) Modifying the fibers in the RV endocardium according to histological studies. These fibers have two main longitudinal directions, one from the apex to the pulmonary valve and other from the apex to the tricuspid valve (see Fig. 1).

(iii) Allowing septal discontinuities by modifying the angle between the septal fibers of LV and RV.

2.1 Mesh Generation

The geometries used in this work were patient-specific tetrahedral meshes built from the processing of computed tomography (CT) images. A bi-ventricular model including the outflow tracts was built from semi-automatic segmentation performed by region growing techniques available in 3DSlicer[1]. Subsequently,

---

[1] https://www.slicer.org.

surface meshes were generated applying the classical Marching Cubes method, which was followed by some post-processing steps (e.g. smoothing, labelling), performed in Blender[2]. Finally, tetrahedral meshes (~80000 nodes and ~400000 elements) were created using iso2mesh[3].

2.2 Fiber Generation Algorithm

A scheme of this process can be visualized in Fig. 2. The first step for assigning fiber orientation was to generate a local orthonormal reference system at each node of a given personalized heart geometry. The axes of the local orthonormal reference system were the longitudinal ($\hat{e}_l$) transmural ($\hat{e}_t$) and circumferential ($\hat{e}_c$) directions. Transmural and longitudinal directions were defined by solving the Laplace equation using the corresponding geometrical surfaces as Dirichlet boundary conditions and computing the gradient of the resulting distribution map; the geometrical surfaces were: RV and LV endocardium; epicardium; apex; and the tricuspid, mitral, pulmonary and aortic valves. The circumferential direction was computed from the cross product of transmural and longitudinal directions. Fiber orientation was then obtained by rotating the obtained vector $\hat{e}_c$ around $\hat{e}_t$ by a given angle α.

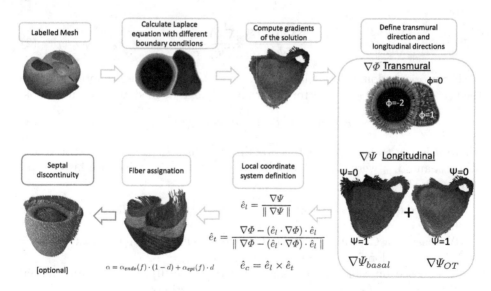

Fig. 2. Schematic showing the different steps of the presented RBM. $\nabla\Psi$: longitudinal gradient; $\nabla\Psi_{basal}$: apico-basal gradient; $\nabla\Psi_{OT}$: apico-OT gradient; $\nabla\Phi$: transmural gradient; $\hat{e}_l$: longitudinal axis; $\hat{e}_t$: transmural axis; $\hat{e}_c$: circumferential axis.

Specifically, transmural direction ($\nabla\Phi$) was defined solving the Laplace equation between the endocardium and epicardium of each ventricle independently,

---

[2] https://www.blender.org.
[3] http://iso2mesh.sourceforge.net.

and subsequently computing the gradient of the Laplace solution. The Laplace equation was simultaneously computed for both ventricles, assigning a negative value to the LV endocardium and a positive to the RV one, allowing independent fiber configurations for both ventricles. Longitudinal direction ($\nabla\Psi$) was also defined separately in each ventricle. This direction was the result of a weighted sum of the apico-basal gradient ($\nabla\Psi_{basal}$) and the apico-OT gradient ($\nabla\Psi_{OT}$), defined individually in both ventricles: $\nabla\Psi_{basal}$ considered the mitral or tricuspid valves and $\nabla\Psi_{OT}$ the aortic or pulmonary valves, for LV and RV, respectively. These two main directions were already described by Greenbaum et al. [10] and can be visualized in Fig. 1 (dashed lines). The previous sum was weighted by an intra-ventricular interpolation function f, which was computed by solving the Laplace equation between the pulmonary-tricuspid valve and the aortic-mitral valve in the RV and LV, respectively. In this way, we obtained a distribution of values allowing to control the smoothness in fiber changes near the OT in different geometries. The resulting longitudinal direction for each ventricle was:

$$\nabla\Psi = \nabla\Psi_{basal} \cdot f + \nabla\Psi_{OT} \cdot (1 - f) \tag{1}$$

Using the previously calculated gradients ($\nabla\Psi$, $\nabla\Phi$), the local coordinate system was set up for each vertex, which is fully described with the following vectors:

$$\hat{e}_l = \frac{\nabla\Psi}{\| \nabla\Psi \|} \qquad \hat{e}_t = \frac{\nabla\Phi - (\hat{e}_l \cdot \nabla\Phi) \cdot \hat{e}_l}{\| \nabla\Phi - (\hat{e}_l \cdot \nabla\Phi) \cdot \hat{e}_l \|} \qquad \hat{e}_c = \hat{e}_l \times \hat{e}_t \tag{2}$$

Fiber orientation was estimated in every vertex by rotating counterclockwise the vector $\hat{e}_c$ around $\hat{e}_t$ the vector $\hat{e}_c$ with an angle α, defined in each ventricle as:

$$\alpha = \alpha_{endo}(f) \cdot (1 - d) + \alpha_{epi}(f) \cdot d \tag{3}$$

where d is the transmural depth normalized from 0 to 1. The different values of α_{endo} and α_{epi} were chosen to replicate observations from several histological studies [9–11]:

- LV (based on Streeter's observations [9]): $\alpha_{endo}(f = 1) = -60°$; $\alpha_{epi}(f = 1) = 60°$
- RV (based on Greenbaum [10] and Sanchez-Quintana [11]): $\alpha_{endo}(f = 1) = 90°$ (same as longitudinal direction); $\alpha_{epi}(f = 1) = -25°$
- OTs (based on Sanchez-Quintana [11]): $\alpha_{epi}(f = 0) = 0°$ (circumferential direction); $\alpha_{endo}(f = 0) = 90°$ (longitudinal direction)

2.3   Septal Discontinuity

The presented RBM is able to include fiber angle discontinuity in the septum. This can be done since we define transmural direction ($\nabla\Phi$) independently in each ventricle. The first step was to determine the contribution of each ventricle to the septal wall; some studies [7] state that two thirds of the septal wall belong to the LV. Therefore, we applied modified Dirichlet conditions when computing the Laplace equation: the negative values assigned to the LV were the double of the positive ones assigned to the RV surfaces. In this way, we could identify a border surface between the positive/negative values and use it for guiding the interpolation of the fiber angles in the septum. Therefore, the expression for assigning the fiber angle in the septum remained as:

$$\alpha = \alpha_{endo}(f) \cdot (1 - d) + \alpha_{septal}(f) \cdot d \tag{4}$$

where α_{septal} is the fiber angle at the border of the ventricles. Continuity or a certain angle discontinuity in fiber orientation could easily be forced by modifying this angle.

3   Results

3.1   Angle Difference

The performance of the developed method was analyzed setting up an experiment to estimate fiber angle differences with respect to Bayer's RBM [5] and several configurations of the presented RBM. However, Bayer's RBM is not able to reproduce fiber orientation above the basal plane since it is used as a boundary condition, making the comparison of the OT fibers unfeasible. Thus a bi-ventricular geometry of OTVA patients was cut below the outflow tracts so that all the existing RBM could be run. As expected, the main angle differences between both RBM were found in the RV endocardium and near the RVOT (see upper row of Fig. 3), where we forced longitudinal directions from the apex towards the pulmonary and tricuspid valves in our method.

3.2   Electrical Activation

The impact of the proposed RBM in the electrical activation of the heart was evaluated running several electrophysiological simulations with fiber configurations provided by the existing and proposed RBMs. Simulations were performed with the SOFA software[4], using the Mitchell-Schaeffer model [12]. Using one of the available bi-ventricular geometries (shown in Fig. 3), simulations were run placing the ectopic focus in the RV apex. In each simulation only fiber orientation was changed, keeping unaltered the remaining parameters of the model such as the conductivity anisotropy ratio (in our simulations conduction velocity in the fiber direction is 2.5 larger than in the transverse plane). Lower row of the Fig. 3 shows the electrical activation isochrones obtained with the continuous and discontinuous septum versions of the proposed RBM as well as with Bayer's RBM (left, center and right in the lower row of the figure, respectively). It can be observed in Fig. 3 that the electrical activation patterns are quite similar except in the septum and the RV endocardium (see white arrows), using

---

[4] An Open Source medical simulation software http://www.sofa-framework.org.

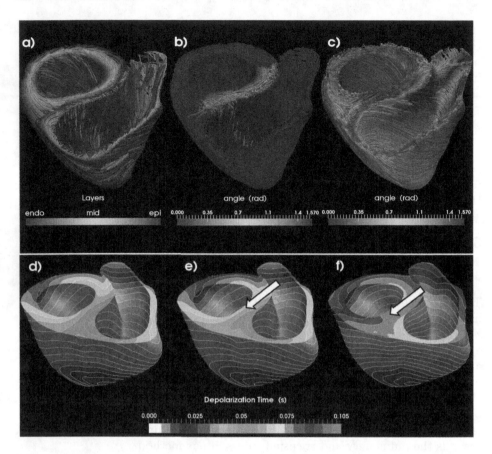

Fig. 3. Upper row:(a) Fiber orientation provided by the proposed RBM with a continuous septum; (b) angle differences (in radians) with fibers in a discontinuous septum; (c) and with Bayer's RBM method. Lower row: Electrical activation isochrones (in seconds) provided by different models: the proposed RBM with a continuous (d) and discontinuous (e) septum; (f) Bayer's RBM. White arrows point towards the main differences found in the septal activation pattern.

the continuous septum version of the proposed RBM as reference. The larger delay in the basal septum showed in panel C comparing to the ones from our method (panels A and B) is due to the RV septal fiber difference. In our code, an apex-base direction was forced ($\nabla \Psi_{basal}$), as a consequence, the impulse travels faster in our geometry than in the obtained by Bayer's RBM, which has more circumferential fibers.

These results can be evaluated in a more quantitative way by calculating the mean local activation time (LAT) differences for each geometry. Taking the continuous septum version of the proposed RBM as reference, LAT differences were: 2 ± 2 ms (mean $\pm$ std) with the discontinuous septum version; and 6 ± 5 ms compared with Bayer's RBM. A more localized analysis in the septal wall showed larger LAT differences: 4 ± 2 ms with the discontinuous septum version; and 8 ± 6 ms with Bayer's RBM.

3.3 OTVA Simulations

Lastly, we have also run electrophysiological simulations with fiber orientation computed with the proposed RBM on heart geometries of different OTVA patients. The obtained simulated electrical propagation waves were then compared with patient-specific EAM data. In particular, we analyzed the characteristics of the isochrones around the earliest activated point in the RV endocardium since they provide useful information about the SOO [13,14]: if the long axis of

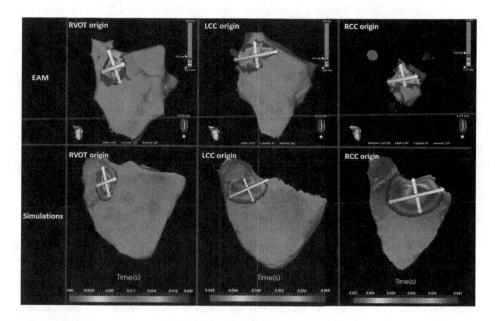

Fig. 4. Isochronal maps obtained from EAMs after 20 ms from earliest activated point in the RV (top) and corresponding electrophysiological simulations (bottom) in cases with different SOO. Orange and blue arrows indicate the longitudinal and perpendicular axis of the isochrones respectively. RVOT: right ventricular outflow tract. LCC and RCC: left and right coronary cusp, respectively. (Color figure online)

Table 1. Area (mm^2) and longitudinal/perpendicular axis ratio of the 10 ms and 20 ms isochrones obtained from simulations and EAM data from three cases.

Isochrones	RVOT origin		LCC origin		RCC origin	
	10 ms	20 ms	10 ms	20 ms	10 ms	20 ms
Area Exp	490 ± 80	990 ± 140	260 ± 90	1340 ± 150	500 ± 100	2600 ± 250
Area Sim	160 ± 30	540 ± 40	270 ± 30	880 ± 50	650 ± 40	2110 ± 70
Long/Perp ratio Exp	2.04 ± 0.7	1.5 ± 0.3	0.5 ± 0.14	0.87 ± 0.13	0.56 ± 0.14	0.92 ± 0.1
Long/Perp ratio Sim	1.8 ± 0.2	1.6 ± 0.1	0.45 ± 0.04	0.6 ± 0.03	0.71 ± 0.04	0.72 ± 0.03

the isochrones is more longitudinal (i.e. follows the apico-basal axis) the SOO should be in the RVOT (following the fibers in the OT); while LVOT origins create more isotropic isochrones or with a larger perpendicular axis [13,14].

Simulations were performed in heart geometries from three OTVA patients, each one with a different SOO according to clinical diagnosis: RVOT; Right Coronary Cusp (RCC) and Left Coronary Cusp (LCC) in the LVOT. The area and the axis ratio (longitudinal axis/perpendicular axis) of the simulated isochrones were compared in each case with patient data, as shown in Fig. 4 and Table 1. In agreement with clinical observations, parameters obtained from simulations confirmed that when the ectopic focus was originated in the RVOT the longitudinal diameter of the isochrones was larger than the cross-fiber one. By contrast, LVOT origins were associated with a larger perpendicular diameter, making the diameter ratio smaller. Table 1 shows how the obtained ratios of the RVOT isochrones are bigger than 1 (longer longitudinal axis) whereas the LVOT isochrones had a ratio smaller than 1. Nevertheless, we found substantial differences in isochronal areas, that could be caused by the difficulties for measuring them and the effect of far field in EAM data.

4   Discussion and Conclusions

We have presented a RBM that includes fiber information specific to the RV, interventricular septum and both OTs of the ventricles, to replicate histological observations. The proposed model allows running electrophysiological simulations in applications when these regions are important such as in OTVA patients, which was not possible before. The method has been indirectly validated comparing the results of OTVA simulations (using fiber orientation provided by the proposed RBM) with patient EAM data, showing good agreement. Nevertheless, further studies are required to better determine the accuracy of the proposed RBM, such as a more quantitative comparison between RBM-derived fibers and the ones observed in histology. Additionally, OTVA simulations with RBM-derived fibers and their validation with EAM data need to be performed on a larger database of cases. It would also be desirable to compare RBM results with in-vivo cardiac diffusion imaging, which starts to provide very promising results [15,16]. This would eventually allow to overcome the existing RBM limitations to reproduce patient-specific alterations of the myofiber distribution due to infarcts or different pathologies. Future work will be devoted to analyze the impact of the new fiber configurations in mechanical simulations of the heart.

References

1. Arevalo, H.J., et al.: Arrhythmia risk stratification of patients after myocardial infarction using personalized heart models. Nat. Commun. **7**, 11437 (2016)
2. Lombaert, H., et al.: Human atlas of the cardiac fiber architecture: study on a healthy population. IEEE Trans. Med. Imaging **31**(7), 1436–1447 (2012)

3. Young, R.J., Panfilov, A.V.: Anisotropy of wave propagation in the heart can be modeled by a Riemannian electrophysiological metric. Proc. Natl. Acad. Sci. U.S.A. **107**(34), 15063–15068 (2010)
4. Hooks, D.A., et al.: Laminar arrangement of ventricular myocytes influences electrical behavior of the heart. Circ. Res. **101**(10), 103–113 (2007)
5. Bayer, J.D., et al.: A novel rule-based algorithm for assigning myocardial fiber orientation to computational heart models. Ann. Biomed. Eng. **40**(10), 2243–2254 (2012)
6. Agger, P., et al.: Insights from echocardiography, magnetic resonance imaging, and microcomputed tomography relative to the mid-myocardial left ventricular echogenic zone. Echocardiography **33**(10), 1546–1556 (2016)
7. Boettler, P., et al.: New aspects of the ventricular septum and its function: an echocardiographic study. Heart **91**(10), 1343–1348 (2005)
8. Kocica, M.J., et al.: The helical ventricular myocardial band: global, three-dimensional, functional architecture of the ventricular myocardium. Eur. J. Cardiothoracic Surg. **29**(SUPPL.), 1 (2006)
9. Streeter, D.D., et al.: Fiber orientation in the canine left ventricle during diastole and systole. Circ. Res. **24**(3), 339–347 (1969)
10. Greenbaum, R.A., et al.: Left ventricular fibre architecture in man. Br. Heart J. **45**(1980), 248–263 (1981)
11. Sanchez-Quintana, D., et al.: Anatomical basis for the cardiac interventional electrophysiologist. BioMed. Res. Int. Ao. **2015** (2015)
12. Talbot, H., et al.: Towards an interactive electromechanical model of the heart. Interface Focus **3**(2) (2013)
13. Acosta, J., et al.: Impact of earliest activation site location in the septal right ventricular outflow tract for identification of left vs right outflow tract origin of idiopathic ventricular arrhythmias. Heart Rhythm **12**(4), 726–734 (2015)
14. Herczku, C., et al.: Mapping data predictors of a left ventricular outflow tract origin of idiopathic ventricular tachycardia with V3 transition and septal earliest activation. Circ. Arrhythm. Electrophysiol. **5**(3), 484–491 (2012)
15. Toussaint, N., et al.: In vivo human cardiac fibre architecture estimation using shape-based diffusion tensor processing. Med. Image Anal. **17**(8), 1243–1255 (2013)
16. Mekkaoui, C., et al.: Diffusion tractography of the entire left ventricle by using free-breathing accelerated simultaneous multisection imaging. Radiology, 152613 (2016)

Biomechanics: Modelling and Tissue Property Measurements

Feasibility of the Estimation of Myocardial Stiffness with Reduced 2D Deformation Data

Anastasia Nasopoulou[✉], David A. Nordsletten, Steven A. Niederer, and Pablo Lamata

Division of Imaging Sciences and Biomedical Engineering,
Department of Biomedical Engineering, King's College London, London, UK
anastasia.nasopoulou@kcl.ac.uk

Abstract. Myocardial stiffness is a useful diagnostic and prognostic bio-marker, but only accessible through indirect surrogates. Computational 3D cardiac models, through the process of personalization, can estimate the material parameters of the ventricles, allowing the estimation of stiffness and potentially improving clinical decisions. The availability of detailed 3D cardiac imaging data, which are not routinely available for the conventional cardiologist, is nevertheless required to constrain these models and extract a unique set of parameters. In this work we propose a strategy to provide the same ability to identify the material parameters, but from 2D observations that are obtainable in the clinic (echocardiography). The solution combines the adaptation of an energy-based cost function, and the estimation of the out of plane deformation based on an incompressibility assumption. In-silico results, with an analysis of the sensitivity to errors in the deformation, fibre orientation, and pressure data, demonstrate the feasibility of the approach.

Keywords: Cardiac mechanics · Myocardial stiffness · Energy-based cost function · Parameter estimation · 2D images

1 Introduction

Left ventricular (LV) stiffness has been identified as a useful biomarker in the diagnosis and monitoring of heart failure (HF) with preserved ejection fraction, a syndrome which is associated with high morbidity and mortality rates and poor prognosis [1,2]. The assessment of myocardial stiffness in vivo is a complex task which can be tackled with the use of cardiac biomechanical models. In these models material stiffness is determined from the choice of parameter values in the employed material constitutive equations, and therefore the problem of stiffness estimation is posed as a parameter estimation problem. A major limitation in this approach is the parameter coupling in common material models [3], resulting in multiple parameter combinations corresponding to equivalent solutions in the optimization. This inability to uniquely constrain parameters limits the correlation of changes in material parameters to possible changes in patient pathophysiology. The problem of parameter coupling was recently addressed by

© Springer International Publishing AG 2017
M. Pop and G.A. Wright (Eds.): FIMH 2017, LNCS 10263, pp. 357–368, 2017.
DOI: 10.1007/978-3-319-59448-4_34

a novel energy-based cost function (CF), which was applied to detailed 3D geometry and deformation data available from magnetic resonance imaging (MRI) [4].

However, MRI scanners are expensive and scarce and the reliance of parameter estimation pipelines on the availability of 3D data limits their possible impact in the large scale. Conversely, planar ultrasound images are ubiquitous in clinical practice. Aiming to translate the recent advances in parameter estimation from 3D clinical models into the clinic and the constraints of everyday practice, this paper investigates the possibility to assess myocardial parameters from 2D geometry data, such as would be available from long axis echocardiographic views of the LV.

2   Materials and Methods

Parameter estimation in this study relies on recent work [4] where a' CF was proposed based on the energy conservation of the myocardium during diastolic filling. This CF solves the problem of coupling between the scaling and bulk exponential parameters in material models and was applied on a popular transversely isotropic model proposed by Guccione et al. [5].

To investigate the feasibility of myocardial stiffness estimation, a synthetic data set was generated in order to provide ground truth (GT) solutions to material parameters and knowledge of 'real' 3D deformations (Sect. 2.2). The information from this model was subsequently modified in order to serve two scenarios. In the first, geometrical information is available on a 2D plane only, but the full 3D displacements and displacement gradients of the material points on this plane are assumed to be known. In the second, only 2D information on LV geometry and deformation is available. To assess the effect of data noise and certain material modelling assumptions, a sensitivity study was performed highlighting the robustness of each of the two cases of data availability (Sect. 2.3).

A new version of the energy-based CF [4] is introduced in this study for the estimation of the exponential parameter, and the different strategies employed for the CF evaluation depending on the available data set are explained in Sect. 2.5. In the last paragraph of this section, we complete the parameter estimation by presenting a method to estimate the remaining coupled scaling parameter (Sect. 2.6).

2.1   Material Model

The myocardium was modelled as a transversely isotropic material [5] with a strain energy density function (Eqs. (1) and (2)) expressed as a function of a scaling parameter (C_1) and a bulk exponential parameter (α) [6]. The parameters r_f, r_t, r_{ft} scale the Green-Lagrange strain tensor (E), where E is expressed in fiber coordinates. The subscripts f, s, n denote the tensor components in the fiber, sheet and sheet normal directions. The fiber orientation in the LV myocardial wall was assumed to follow an idealised $-60°/+60°$ distribution from the epi- to the endo-cardium [7].

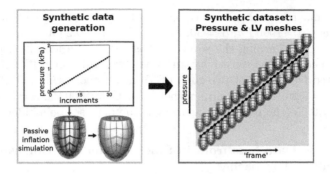

Fig. 1. Passive inflation simulation of a prolate spheroidal finite element (FE) model of the LV to generate the original pressure - 3D geometry data.

$$\Psi = \frac{1}{2}C_1(e^Q - 1) \tag{1}$$

$$Q = \alpha[r_f E_{ff}{}^2 + r_{ft}(E_{fs}{}^2 + E_{sf}{}^2 + E_{fn}{}^2 + E_{nf}{}^2)$$
$$+ r_t(E_{ss}{}^2 + E_{nn}{}^2 + E_{sn}{}^2 + E_{ns}{}^2)] \tag{2}$$

As the main direction of coupling in the Guccione law is along C_1 and α, which scale stiffness along all directions [6], in this study we focus on C_1-α estimation and assume the anisotropy parameters are fixed (see Sect. 2.2 for the chosen GT values), following a common approach [4,8].

2.2   Ground Truth: 3D in Silico Model

For the generation of the 3D GT data set, from which the synthetic 2D 'images' will be derived, the left ventricle (LV) was modelled as a truncated ellipsoid of revolution of human dimensions. It was passively inflated to 1500 Pa end diastolic pressure under fully constrained basal displacements to generate the original pressure - 3D geometry data. The constructed finite element (FE) mesh consisted of 320 elements (4 circumferential, 4 transmural, 4 longitudinal and 16 in the apical cap) and 9685 nodes. The pressure was applied in 30 increments of 50 Pa each and basal plane was fixed by prescribing the Dirichlet boundary conditions directly at the basal nodes of the mesh.

The deformation was found by solving the linearised total potential energy equations using the *CHeart* nonlinear mechanics solver [9], using a split $\boldsymbol{u}$-p formulation outlined in [10]. The initial geometry ($\boldsymbol{X}$), deformed geometry ($\boldsymbol{x}$) and fiber orientation ($\boldsymbol{f}, \boldsymbol{s}, \boldsymbol{n}$) vector fields were interpolated through cubic-Lagrange shape functions. To avoid locking phenomena, the 'hydrostatic pressure' field (p) approximation needs to be of a lower order [11], and in this case a Linear-Lagrange interpolation scheme was used for computational economy (reduction of stiffness matrix size and use of the already available linear mesh). The meshes and simulation outputs were visualised with *cmGui*[1].

---

[1] http://www.cmiss.org/cmgui.

The Guccione material parameters chosen for the generation of the synthetic data set were: $C_1 = 1000\,Pa$, $\alpha = 30$, $r_f = 0.55$, $r_{ft} = 0.25$ and $r_t = 0.2$ [4]. For clarifying aspects regarding the severity of errors in the identified parameters as well as aspects of the sensitivity analysis, two additional 3D GT data sets were generated (see also last two rows of Table 2), one with $C_1 = 300$, $\alpha = 100$, $r_f = 0.55$, $r_{ft} = 0.25$, $r_t = 0.2$ and another with $C_1 = 1000\,Pa$, $\alpha = 30$, $r_f = 0.85$, $r_{ft} = 0.1$, $r_t = 0.05$ (the remaining aspects of the data set being identical).

2.3 Synthetic Data Sets from in Silico Model and Corresponding Modelling Approaches

Case 1: Synthetic Data Set of 2D LV Geometry - 3D Deformation. The first data set that will be examined consists of cavity pressure measurements and geometry and 3D deformation of a long axis plane of the myocardium across 31 'frames' corresponding to the simulation increments (reference mesh and 30 simulation outputs) of the initial 3D synthetic data set (Sect. 2.2). In this case the long axis plane was chosen to 'cut' the existing 3D mesh into 2 symmetric halves and coincides with the XZ plane in the undeformed configuration (see Fig. 2).

Case 2: Synthetic Data Set of 2D LV Geometry - 2D Deformation. The second case of the synthetic data set consists of the pressure measurements, 2D geometry and 2D deformation across the 31 'frames' from the original simulation (Sect. 2.2). The synthetic 'image' of the LV is again in the XZ plane as in the first case. The difference here is that only 2D information of the deformation field is provided (Eq. 3), which is achieved by projecting the 'real' deformation field into the 'imaging' plane (XZ) (see Fig. 2).

The estimation of the energy-based CF in Sect. 2.5 requires the estimation of the Green-Lagrange strain tensor in the 3D fiber coordinate system (Eq. (2)),

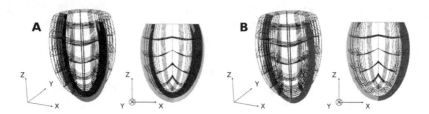

Fig. 2. Synthetic 'images' for the 3D simulations (Fig. 1). **A.** Data set 1 (3D deformation information): undeformed geometry is the slice in black and deformed geometry is shown in grey, where a small twist can be observed. **B.** Data set 2 (2D deformation information): the 'real' deformed geometry (used in data set 1) is shown in grey as in panel A, and in blue is the geometry resulting from the projection of the 'real' deformation on the XZ plane (this is used in data set 2) - note that the twist is now removed. The mesh in both panels corresponds to the original 3D undeformed geometry. (Color figure online)

which in turn requires the description of deformation in three dimensions. In the case of the limited, in-plane deformation data this requires an assumption to be made about the off-plane deformation patterns. For this purpose, two different approaches were followed. In the first, plane strain is assumed and the 3D displacement and deformation gradient are taken as $\boldsymbol{u}_{3D}$ and $\boldsymbol{F}_{p.s.}$ in Eq. (4). In the second, the off-plane shear is again assumed to be zero but the term $\frac{\partial y}{\partial Y}$ is estimated from the isovolumic deformation assumption for the myocardium (the 3D description of the displacement is again $\boldsymbol{u}_{3D}$ but the deformation gradient is now $\boldsymbol{F}_{iso}$ in Eq. (4)).

$$\boldsymbol{u}_{2D} = \begin{bmatrix} u_x \\ u_z \end{bmatrix}, \quad \boldsymbol{F}_{2D} = \begin{bmatrix} \frac{\partial x}{\partial X} & \frac{\partial x}{\partial Z} \\ \frac{\partial z}{\partial X} & \frac{\partial z}{\partial Z} \end{bmatrix} \tag{3}$$

$$\boldsymbol{u}_{3D} = \begin{bmatrix} u_x \\ 0 \\ u_z \end{bmatrix}, \quad \boldsymbol{F}_{p.s.} = \begin{bmatrix} \frac{\partial x}{\partial X} & 0 & \frac{\partial x}{\partial Z} \\ 0 & 1 & 0 \\ \frac{\partial z}{\partial X} & 0 & \frac{\partial z}{\partial Z} \end{bmatrix}, \quad \boldsymbol{F}_{iso} = \begin{bmatrix} \frac{\partial x}{\partial X} & 0 & \frac{\partial x}{\partial Z} \\ 0 & \frac{1}{det(\boldsymbol{F}_{2D})} & 0 \\ \frac{\partial z}{\partial X} & 0 & \frac{\partial z}{\partial Z} \end{bmatrix} \tag{4}$$

2.4 Evaluation of Performance

Sensitivity Study. To provide an estimation of the severity of data quality and model-data discrepancies on the estimated parameter values in 2D, a sensitivity study was performed. For estimating effects of miscalibration in pressure measurements, synthetic data sets with modified pressure traces (pressure offset by $\pm10\%$ of mean pressure value) were used. To examine the effect of the imaging data on the accuracy of the analysis two additional data sets were used, where white Gaussian noise was inserted to the deformation field. Specifically, for each component of the deformation gradient F_{ij} ($i, j = 1, \dots, N$, where $N = 9$ for the 3D deformation data set and $N = 4$ for the 2D) its mean value ($\bar{F}_{ij}$) over the Gauss points (GPs) of the in-plane surface mesh elements was estimated (432 GPs used in total, equivalent to 3 GPs used per element direction). A normal distribution of random numbers was generated (independently for each frame and F_{ij}), was weighted by either 1% or 2% of $\bar{F}_{ij}$ and subsequently added to each GP of the mesh at each frame generating two sets of noisy deformation data used instead of the 1st and 2nd data sets of Sect. 2.3.

Displacement Errors and Stress-Strain Curves. The insilico tests will report the estimated parameters (α and C_1), together with an analysis of the impact of the error in these parameters with regard to two aspects: (1) the difference in the displacements (L^2 displacement norm), and (2) the difference in the stress-strain relationship in a simple fibre-stretch model. Specifically, forward simulations on the original 3D mesh (Sect. 2.2) were performed using the identified parameters and the resulting L^2 displacement norm ($|\Delta\boldsymbol{u}|$), comparing the deformed geometry of the simulation with the identified parameters and the GT, was estimated [4]. The stress-strain plots refer to an idealised scenario of stretch along the fiber direction of an incompressible cube with the specified α-C_1 parameters. They show the Cauchy stress along the fiber direction (σ_f) minus the indeterminate 'hydrostatic pressure' p term (see Sect. 2.5), in the absence of specified boundary conditions, with respect to fiber stretch (λ_f) (Fig. 3).

To contextualise this analysis, two additional sets of $C_1 - \alpha$ parameters are selected which lie on the $C_1 - \alpha$ principal parameter coupling line at ± 20 of the GT α value (α_{GT}). This line represents parameter combinations that effectively reproduce the same deformation in a model. In our analysis the coupling line was obtained from an exponential fit to the parameter combinations that minimize the $|\Delta u|$ residual (following parameter sweeps over C_1 and α on the original 3D simulation, see Sect. 2.2) and agreed well with the proposed coupling direction by [6] using the mean value of Q (Fig. 3).

2.5   Estimation of the Exponential Parameter α from the Reformulated Energy-Based CF

This study is based on previous work [4], where unique parameter estimation was achieved with the use of an energy-based CF that allowed determination of the α parameter in Eq. (2). This CF was based on energy conservation, dictating that the work of internal stresses inside the tissue stored as elastic energy (W_{int}) and the external work of cavity pressure (W_{ext}) are equal.

Here we re-formulate the energy-based CF based on the principle of virtual work, which expresses the linearisation of the energy balance with respect to an arbitrary displacement field δu, called the virtual displacement field. Assuming quasi-static loading and absence of residual active tension in the diastolic window of relevance allows the internal (δW_{int}) and external (δW_{ext}) components of the virtual work to be expressed as in Eqs. (5) and (6) respectively. In Eq. (5), Ψ is given by (1), and p, J and C denote the 'hydrostatic pressure', the determinant of the deformation gradient and the right Cauchy Green tensor respectively. Equation (6) is based on a pull-back expression of δW_{ext} based on Nanson's formula [12], where p_C, dA, δu, F denote the cavity pressure, infinitesimal area vector in the material configuration, virtual displacement, and deformation gradient. The terms $\frac{\partial \Psi}{\partial E}$ and $DE[\delta u]$ are given by Eqs. (7), (8) and (9) [6,12]. The virtual displacement field in these equations can be arbitrary provided that the boundary conditions of the problem are respected. Here we have employed a virtual field equal to the measured displacements from the data.

Note that although we refer to p as the tissue 'hydrostatic pressure', it is not equal to $1/3 tr(\boldsymbol{\sigma})$ due to the non use of the distortional formulation of Ψ ($\hat{\Psi}$), see [12]. The estimation of p from the data is not necessary, as the contribution of the term pJC^{-1} on the virtual work is zero (see Eq. (10) and [12] for the derivation), as long as the virtual displacement field employed respects the incompressibility constraint (so that $DJ[\delta u] = 0$). The latter holds in our approach, since the virtual displacements used are equal to the real (incompressible) ones (see Sect. 2.2).

$$\delta W_{int} = \int_V \left(\frac{\partial \Psi}{\partial E} + pJC^{-1} \right) : DE[\delta u] dV. \tag{5}$$

$$\delta W_{ext} = \int_A p_C J \delta \mathbf{u} \cdot \left(\mathbf{F}^{-T} d\mathbf{A} \right) \tag{6}$$

$$\frac{\partial \Psi}{\partial \mathbf{E}} = C_1 \alpha e^Q \begin{bmatrix} r_f & r_{ft} & r_{ft} \\ r_{ft} & r_t & r_t \\ r_{ft} & r_t & r_t \end{bmatrix} \circ \begin{bmatrix} E_{ff} & E_{fs} & E_{fn} \\ E_{sf} & E_{ss} & E_{sn} \\ E_{nf} & E_{ns} & E_{nn} \end{bmatrix} \tag{7}$$

$$D\mathbf{E}[\delta \mathbf{u}] = \frac{1}{2} \left(D\mathbf{F}[\delta \mathbf{u}]^T \mathbf{F} + \mathbf{F}^T D\mathbf{F}[\delta \mathbf{u}] \right) \tag{8}$$

$$D\mathbf{F}[\delta \mathbf{u}] = \boldsymbol{\nabla} \delta \mathbf{u} \mathbf{F} \tag{9}$$

$$p J \mathbf{C}^{-1} : D\mathbf{E}[\delta \mathbf{u}] = p J tr \left(\boldsymbol{\nabla} \delta \mathbf{u} \right) = p D J[\delta \mathbf{u}] \tag{10}$$

Expressing the principle of virtual work over two diastolic frames 1 and 2, the modified energy-based CF can then be expressed as in Eq. (11). Since both the numerator and denominator of the δW_{int} ratio in Eq. (11) contain the parameter C_1, it is evident that it cancels out and thus f allows for the unique estimation of α. As the α parameter in Eq. (11) lies within e^Q only, (the product $C_1 \alpha$ in $\frac{\partial \Psi}{\partial \mathbf{E}}$ cancels out), the use of higher strains increases the nonlinearity of the CF and therefore its identifiability. For this purpose the two last 'frames' of the synthetic data set (Frames 29 and 30, corresponding to the last two simulation increments of Sect. 2.2) were chosen as diastolic frames 1 and 2 respectively.

$$f = \left| \frac{\delta W_{ext}^1}{\delta W_{ext}^2} - \frac{\delta W_{int}^1}{\delta W_{int}^2} \right| \tag{11}$$

2.6   Estimation of the Scaling Parameter C_1

To complete the parameter estimation, C_1 parameter must also be identified. In [4] following α estimation from the energy-based CF, an additional geometric CF was used for assessing C_1. In this work, an alternative method is applied, where the principle of virtual work ($\delta W_{int} = \delta W_{ext}$) is used for the estimation of C_1 according to Eq. (12). C_1^f denotes the assumed C_1 value in Eq. (11) for the estimation of f ($C_1^f = 1000$ Pa). This method was applied to the original 3D data set, and was able to approximate the GT value ($C_1 = 989$ Pa).

$$C_{1\,sol} = \frac{\delta W_{ext}}{\left(\frac{\delta W_{int}}{C_1^f} \right)} \tag{12}$$

3   Results

The estimated parameters are reported in Table 1, together with the L^2 norm of error in deformation. The impact of parameter errors in the stress-strain curve is illustrated in Fig. 3. Note that the choice of other coupled parameters (taking $\alpha = 10$ or $\alpha = 50$) lead to a small error in deformation but to a large error in the strain-stress relationship.

Table 1. Estimated parameters by the three alternative combinations of data an assumptions, together with their $|\Delta u|$ residual. For reference, the residual with two alternative ('coupled') sets of parameters is provided (see also Sect. 2.4 and Fig. 3).

	Estimated parameters			Coupled parameters	
	3D F	2D F-pl.str.	2D F-isovol.	$-20\ \alpha_{GT}$	$+20\ \alpha_{GT}$
α	34	64	31	10	50
C_1	690	672	764	4100	50
$|\Delta u|$ (mm)	0.54	1.57	0.65	0.26	0.33

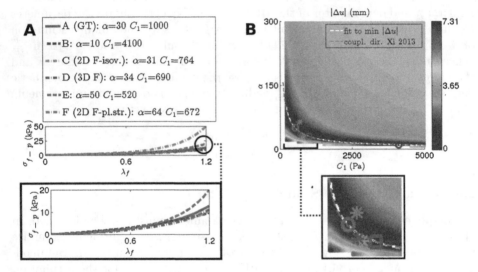

Fig. 3. A. Impact of parameter errors in the stress-strain relationship by comparison of the six parameter sets (ground truth, estimated parameters from the 2D analysis and two coupled sets of parameters - see also Sect. 2.4 and Table 1). Note that the mean and maximum stretch ratio along the fiber direction (λ_f) in the original 3D data set of Sect. 2.2 is 1.09 and 1.17 respectively. **B.** The position in parameter space of the ground truth (marked as square), the identified parameters from 3D and 2D deformation data (marked as asterisks) and the 'coupled' parameters at $\pm20\ \alpha_{GT}$ (marked as circles) with respect to the coupled parameter line (See Sect. 2.4).

The cumulative results from the processing of both data sets (availability of 3D deformation vs. 2D deformation) and both analysis strategies (plane strain vs. isovolumic) are shown in Table 2.

Table 2. Default and sensitivity analysis results for all data sets (see Sects. 2.2 and 2.4).

Data set / analysis:		3D F		2D F: Plane str.		2D F: Isovol.	
GT	Change to GT	α_{sol}	C_{1sol} (Pa)	α_{sol}	C_{1sol} (Pa)	α_{sol}	C_{1sol} (Pa)
$\alpha=30$ $C_1=1000^a$	None	34	690	64	672	31	764
	Pressure: $+10\%$ $\bar{p}_C$ offset	31	832	59	801	29	885
	-10% $\bar{p}_C$ offset	36	600	68	580	34	629
	F noise: STD 1% $\bar{F}_{ij}$	19	1522	108	258	86	103
	STD 2% $\bar{F}_{ij}$	9	3600	73	520	162	7
	Fibers: $\pm$ 90 ^{o}	33	752	49	832	32	773
	$(r_f,r_{ft},r_t)=$ $(0.85,0.1,0.05)$	32	1052	69	1810	29	1018
	$(0.34,0.33,0.33)$	30	647	50	543	30	649
$\alpha=100$ $C_1=300$ ^{b}	None	104	215	176	235	95	239
$r_f=0.85$ ^{c}	Fibers: $\pm$ 90 ^{o}	24	1180	28	1518	24	1111

Material parameters used in each GT data set: $^{a.}$ $C_1 = 1000$, $\alpha = 30$, $r_f = 0.55$, $r_{ft} = 0.25$, $r_t = 0.2$. $^{b.}$ $C_1 = 300$, $\alpha = 100$, $r_f = 0.55$, $r_{ft} = 0.25$, $r_t = 0.2$. $^{c.}$ $C_1 = 1000$ Pa, $\alpha = 30$, $r_f = 0.85$, $r_{ft} = 0.1$, $r_t = 0.05$. (see also Sect. 2.2)

4   Discussion

This study provides the initial evidence of the feasibility of the unique material parameter estimation of a transversely isotropic material model from 2D long axis images. The proposed strategy uses a CF based on energy conservation [4] that addresses the problem of parameter coupling in myocardial material laws [3,6]. In this contribution a modified version of this method based on the expression of virtual work was employed and extended, not only to cope with the lack of 3D data, but to also allow the estimation of the complete set of parameters without the need of mechanical simulations.

Two different types of synthetic long axis data sets were tested against this framework: one which includes the 3D deformation of the myocardium along the imaging plane (Case 1) and one where only 2D deformation is available (Case 2). For the second data set two different assumptions regarding the out-of-plane deformation of the myocardium (plane strain and isovolumetric deformation) were tested. The results showed that the exponential α parameter was sufficiently identified in the case of 3D deformation data availability, and, in the absence of out-of-plane deformation, when the incompressibility assumption was used to approximate part of the unknown deformation gradient terms. In both these cases the errors were small and not proportional to the α value (compare the two GT models: $\alpha = 30$, $C_1 = 1000$ Pa vs. $\alpha = 100$, $C_1 = 300$ Pa in Table 2). Conversely, the plane strain assumption did not perform well. The superior performance of the isovolumic assumption is due to its better approximation of the 'real' 3D data deformation field. Specifically, the deformation gradient from the isovolumic assumption matched better to the original 3D deformation gradient provided by the first data set, as the shear out-of-plane components ($\partial y/\partial X$, $\partial y/\partial Z$, $\partial x/\partial Y$, $\partial z/\partial Y$) were small (in the order of 10^{-3}–10^{-2}), while the mean $\partial y/\partial Y$ component is about 1.10. Therefore assuming ($\partial y/\partial Y) = 1$ according to the plane strain assumption creates an error of at least an order of magnitude bigger comparing to the error introduced by neglecting the off-plane shear terms which occurs in both plane strain and isovolumetric assumptions.

In the recovery of the full set of parameters, errors were larger for estimating C_1, as it is affected by any error in the estimated α paremeter (Eq. (12)). However, these errors did not lead to large discrepancies in the resulting stresses, where there was good agreement with the behaviour of the GT α-C_1 pair (see Fig. 3). Moreover, a comparison of the global displacement error $|\Delta u|$ from forward simulations of the original 3D model (Sect. 2.2) with the identified parameters to simulations with parameter sets that lie on the 'principal' coupling line [6] demonstrates the good performance of the identified parameters in terms of deformation prediction (Table 1). On the contrary, not addressing the parameter coupling problem can lead to an error in the predicted developing stresses within the range of cardiac deformation (Fig. 3) [4].

The sensitivity analysis showed that the biggest source of error in parameters was the noise in the deformation field, which is an anticipated result given the exponential nature of the material law and the formulation of the proposed CF (Eq. (11)) with the additional term $E: DE\,[\delta u]$ compared to the previously proposed energy based CF [4]. However, this result does not discredit the proposed methodology, as the type of white Gaussian noise is unrealistic (it is common practice that image derived deformation data are regularised [13]) and was used only in lack of a more suitable approach to model real image registration noise. The effect of pressure offsets in the data, altered fiber field and prescribed anisotropy parameter ratios in the model on the estimated α values was small and comparable to the 3D data analysis results [4], although a higher sensitivity to the fiber field in the case of a highly anisotropic myocardium should be noted (last row of Table 2).

In this work we have examined the feasibility of parameter estimation from 2D imaging data in a synthetic data set, where the use of a perfect symmetric long axis plane promote a good performance of our proposed parameter estimation strategies. However, the benefit of symmetry is lost in real data and additional information may be required for estimating parameters from clinical images. Moreover, the good performance of the analysis with the isovolumic assumption may not hold when processing real cardiac images, as a higher amount of out-of-plane shear components (such as torsion) is expected and also the assumption of incompressibility can be challenged due to the changes in myocardial volume caused by perfusion.

Further work is required to investigate the possibility of reducing parameter errors (especially in the C_1 material parameter) and the feasibility of this approach with clinical data, addressing the challenges in modelling real cardiac data [4,6,8], and to define the optimal strategy combining data acquisition and model analysis.

5   Conclusions

This study provides preliminary in-silico evidence of the feasibility to provide a unique parameter estimation with tolerable errors from a partial 2D observation of the 3D anatomy.

References

1. Burkhoff, D.: Mortality in heart failure with preserved ejection fraction: an unacceptably high rate. Eur. Heart J. **33**(14), 1718–1720 (2012)
2. Westermann, D., Kasner, M., Steendijk, P., Spillmann, F., Riad, A., Weitmann, K., Hoffmann, W., Poller, W., Pauschinger, M., Schultheiss, H.-P., Tschöpe, C.: Role of left ventricular stiffness in heart failure with normal ejection fraction. Circulation **117**(16), 2051–2060 (2008)
3. Gao, H., Li, W.G., Cai, L., Berry, C., Luo, X.Y.: Parameter estimation in a Holzapfel-Ogden law for healthy myocardium. J. Eng. Math. **95**(1), 231–248 (2015)
4. Nasopoulou, A., Shetty, A., Lee, J., Nordsletten, D., Rinaldi, C.A., Lamata, P., Niederer, S.: Improved identifiability of myocardial material parameters by an energy-based cost function. Biomech. Model. Mechanobiol. (2017)
5. Guccione, J.M., McCulloch, A.D., Waldman, L.K.: Passive material properties of intact ventricular myocardium determined from a cylindrical model. J. Biomech. Eng. **113**(1), 42–55 (1991)
6. Xi, J., Lamata, P., Niederer, S., Land, S., Shi, W., Zhuang, X., Ourselin, S., Duckett, S.G., Shetty, A.K., Rinaldi, C.A., Rueckert, D., Razavi, R., Smith, N.P.: The estimation of patient-specific cardiac diastolic functions from clinical measurements. Med. Image Anal. **17**(2), 133–146 (2013)
7. Streeter, D.D., Spotnitz, H.M., Patel, D.P., Ross, J., Sonnenblick, E.H.: Fiber orientation in the canine left ventricle during diastole and systole. Circ. Res. **24**(3), 339–347 (1969)
8. Hadjicharalambous, M., Chabiniok, R., Asner, L., Sammut, E., Wong, J., Carr-White, G., Lee, J., Razavi, R., Smith, N., Nordsletten, D.: Analysis of passive cardiac constitutive laws for parameter estimation using 3D tagged MRI. Biomech. Model. Mechanobiol. **14**(4), 807–828 (2014)
9. Lee, J., Cookson, A., Roy, I., Kerfoot, E., Asner, L., Vigueras, G., Sochi, T., Deparis, S., Michler, C., Smith, N.P., Nordsletten, D.A.: Multiphysics computational modeling in CHeart. SIAM J. Sci. Comput. **38**(3), C150–C178 (2016)
10. Hadjicharalambous, M., Lee, J., Smith, N.P., Nordsletten, D.A.: A displacement-based finite element formulation for incompressible and nearly-incompressible cardiac mechanics. Comput. Methods Appl. Mech. Eng. **274**(100), 213–236 (2014)
11. Zienkiewicz, O., Taylor, R., Zhu, J.: The Finite Element Method: Its Basis and Fundamentals. Elsevier, Amsterdam (2010)

12. Bonet, J., Wood, R.D.: Nonlinear Continuum Mechanics for Finite Element Analysis, 2nd edn. Cambridge University Press, New York (2008)
13. Rueckert, D., Sonoda, L.I., Hayes, C., Hill, D.L., Leach, M.O., Hawkes, D.J.: Non-rigid registration using free-form deformations: application to breast MR images. IEEE Trans. Med. Imaging **18**(8), 712–21 (1999)

Analysis of Coronary Contrast Agent Transport in Bolus-Based Quantitative Myocardial Perfusion MRI Measurements with Computational Fluid Dynamics Simulations

Johannes Martens[1(✉)], Sabine Panzer[1], Jeroen P.H.M. van den Wijngaard[2], Maria Siebes[2], and Laura M. Schreiber[1]

[1] Chair for Cellular and Molecular Imaging, Comprehensive Heart Failure Center, University Hospital Wuerzburg, Wuerzburg, Germany
martens_j@ukw.de
[2] Department of Biomedical Engineering and Physics, Academic Medical Center, Amsterdam, Netherlands

Abstract. Aim of the project is the analysis of contrast agent dispersion in bolus-based quantitative myocardial perfusion MRI measurements. 3D-models are extracted from high-resolution cardiovascular cryomicrotome imaging data and subsequently meshed with computational grids. Computational fluid dynamics simulations are performed to solve Navier-Stokes equations for blood flow and the advection-diffusion equation for contrast agent transport to obtain bolus dispersion in epicardial vessels, i.e. bolus duration is increased which results in a systematic underestimation of myocardial blood flow. The dispersion of the injected bolus is observed at different positions along the passage through the cardiovascular vessel geometry and is quantified by means of the variance and transit times of the vascular transport function. We find multi-faceted influences on bolus shape from length of traversed vessels to bifurcation angles and vessel curvature. Therefore, depending on the exact anatomical region systematic errors of blood flow measurements are prone to spatial variance.

Keywords: Computational fluid dynamics · Myocardial blood flow · Contrast agent dispersion

1 Introduction

In dynamic contrast-enhanced magnetic resonance perfusion imaging the amount of intravenously injected contrast agent (CA) in tissue at different times is monitored and analyzed. Using appropriate physiologic models, myocardial blood flow (MBF) can be quantified on the basis of these CA concentration time curves. To enable this, knowledge of the shape of CA wash-in through upstream epicardial vessels is required, the arterial input function (AIF). For technical reasons this cannot be quantified directly in the supplying vessels and is thus measured in

© Springer International Publishing AG 2017
M. Pop and G.A. Wright (Eds.): FIMH 2017, LNCS 10263, pp. 369–380, 2017.
DOI: 10.1007/978-3-319-59448-4_35

the left ventricle (LV), which introduces the risk of systematic errors in MBF quantification due to bolus dispersion in coronary vessels.

The influence of epicardial vessels on CA bolus dispersion and resulting systematic errors in MBF have been investigated in several studies [1–5] using computational fluid dynamics (CFD) simulations. These analyses show systematic underestimation of MBF values due to various parameters (e.g. flow velocity, length, curvature). Furthermore, they suggest decreasing influence of smaller vessels on dispersion and prompt the existence of a limiting vessel generation, after which influence of smaller vessels vanishes. We expect this to happen around vessel generation nine or ten (vessel radii ~ 0.1mm).

Aim of this project is the conduction of CFD simulations on coronary 3D-geometries including arteries down to vessel generations six and seven. With the help of these simulations we target a more accurate estimation of errors in MBF quantification, the overall aim of the project. Moreover, a better understanding of myocardial perfusion in general is expected from the findings of this work.

2 Materials and Methods

2.1 Imaging Data and Computing Facilities

The analysis is based on an imaging data set of a healthy *ex vivo* porcine heart at a resolution of 160μm, generated by an imaging cryomicrotome at Amsterdam Medical Center [6, 7].

The simulations presented in this work are performed at a local server on one graphical processing unit (NVIDIA Tesla K80) as well as the high performance computing cluster *elwetritsch* at University Kaiserslautern, parallelized on 128 processors. The simulation of blood flow and CA transport (cf. Sect. 2.3) took up to two weeks per computed model.

2.2 Model Creation and Meshing

With the help of dedicated software packages (VMTK[1], SimVascular[2]), 3D-models of epicardial vessels are extracted from a high resolution imaging data set (cf. Sect. 2.1). For a systematic study of CA bolus dispersion dependence on vessel generation, two models with varying level of detail have been created (cf. Fig. 1).

Subsequently, in a semi-automatic procedure these geometries are meshed with computational grids of preferentially hexahedral type, which has been shown to reduce numerical diffusion errors in CFD, [8]. The two models in Fig. 1 are meshed with different hexahedral meshing formalisms. For Fig. 1(a) software cfMesh[3] by creativeFields is used, while the larger model Fig. 1(b) is meshed

---

[1] www.vmtk.org

[2] www.simvascular.org

[3] cfmesh.com, open source meshing library for OpenFOAM.

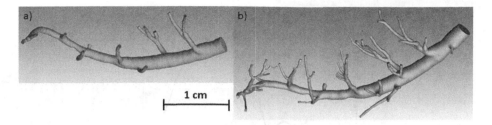

Fig. 1. Models to study vessel dependence of CA bolus dispersion, starting at first diagonal artery (vessel generation 3) branching off from LAD (inlet diameter $\sim$ 2mm). Model (a) includes generations 3–5 and (b) generations 3–7.

with ICEM-CFD[4] by ANSYS Inc. These packages differ in the way that cfMesh allows more automatization and thus faster meshing with mainly hexahedral cells ($< 1\%$ of non-hexahedral type), which are not aligned with predominant flow direction. This meshing process took $\sim$ 10min and only requires manual naming of individual parts beforehand. Meshing with ICEM-CFD as applied in model Fig. 1(b) allows orientation of the hexahedral grid with the expected flow direction, but requires much more user intervention before a computational grid of sufficient quality can be created. Depending on the model's complexity this can take days to weeks. Prior to these simulations an analysis comparing the two utilized mesh-types has been undertaken on an idealized geometry (analogous to [9]) and shows good agreement.

2.3 Governing Equations and Implementation in CFD Simulations

Following this, CFD simulations are performed in two steps with software Open-FOAM[5] and RapidCFD[6]. First Navier-Stokes equations for blood flow

$$\rho\left(\frac{\partial \mathbf{v}}{\partial t} + \mathbf{v} \cdot \nabla \mathbf{v}\right) = -\nabla p + \mu \Delta \mathbf{v} + \mathbf{b}, \tag{1}$$

are solved for one cardiac cycle and resulting velocity and pressure fields are stored as interim results. Here $\rho, \mathbf{v}, p$ and μ denote physical density, velocity, pressure and viscosity of the fluid, respectively. $\mathbf{b}$ stands for body forces. At the model inlet the pressure-time curve depicted in Fig. 2 has been applied. It is taken from [3] and scaled by 50% according to [10] to account for pressure decrease between the LAD in [3] and the smaller diagonal artery in this work.

The boundary condition at the outlets is a resistance model computed according to the so-called structured tree [11,12]. The considered number of vessels in the structured tree is calculated following [13] down to a minimal radius of 50μm at arteriolar level. Combined with the outflowing blood volumes these resistances

---

[4] Proprietary meshing software package usable in OpenFOAM.

[5] www.openfoam.org

[6] https://github.com/Atizar/RapidCFD-dev, OpenFOAM fork for execution on GPU.

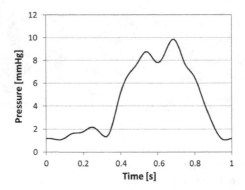

Fig. 2. Applied inlet pressure time curve from [3] scaled by 50%.

provide outlet pressures needed for the solution of the Navier-Stokes-equations. Application of realistic *in vivo* blood pressure profiles at the model inlet would require inclusion of the complete epicardial vasculature with capillary arterioles, venoles and veins and is not practicable in this study. However, since the solutions of the Navier-Stokes equations only depend on the difference between the applied inlet pressure and the correspondingly computed outlet pressures, the physiologically unrealistic pressure range in Fig. 2 can be ignored. At the vessel walls, which are modelled rigid and inelastic, a "no-slip" condition is applied.

For the subsequent computation of CA transport the stored physical fields of one cardiac cycle are repeatedly read in to solve the advection-diffusion equation

$$\rho\frac{\partial c}{\partial t} + \mathbf{v} \cdot \nabla - D\Delta c = 0, \tag{2}$$

over several cardiac cycles, where c and D represent CA concentration and diffusion coefficient, respectively. As inlet boundary condition a gamma-variate function is used [14,15] $Y_{CA}(t) = a(t - t_0)^b e^{-c(t-t_0)}$, with $a = 1.013 \cdot 10^{-4}, b = 2.142, c = 0.454\mathrm{s}^{-1}$. The diffusion coefficient was set as $D = 2.98 \cdot 10^{-10}\mathrm{m}^2\mathrm{s}^{-1}$ for Gd-DOTA [16]. The concentration time curves obtained at the model outlets then represent the real dispersed $AIF_{\mathrm{disp}}(t)$ in the region-of-interest to be used for MBF quantification by use of an adequate perfusion model.

2.4   Analysis and Quantification of CA Bolus Dispersion

According to [5] in a linear system that fulfils conservation of mass the obtained dispersed AIFs at the model outlets are related to the AIF in the LV via the convolution

$$AIF_{\mathrm{disp}}(t - t_0) = AIF_{\mathrm{LV}}(t) \otimes VTF(t - t_0), \tag{3}$$

where VTF denotes the vascular transport function, which includes dispersion effects, AIF_{LV} the measured signal in the LV and AIF_{disp} as above the dispersed AIF at the considered outlet. As a measure for bolus dispersion mean vascular

transit times $(MVTT)$ and VTF's variance σ^2 can be calculated without deconvolution of (3) by using $0^{\text{th}}, 1^{\text{st}}$ and 2nd moments of the AIFs as outlined in
(4) and (5).

$$MVTT = \frac{AIF_{\text{disp}}^{(1)}}{AIF_{\text{disp}}^{(0)}} - \frac{AIF_{\text{LV}}^{(1)}}{AIF_{\text{LV}}^{(0)}}, \tag{4}$$

$$\sigma^2 = \frac{AIF_{\text{disp}}^{(2)}}{AIF_{\text{disp}}^{(0)}} - \frac{AIF_{\text{LV}}^{(2)}}{AIF_{\text{LV}}^{(0)}} + \left(\frac{AIF_{\text{LV}}^{(1)}}{AIF_{\text{LV}}^{(0)}}\right)^2 - \left(\frac{AIF_{\text{disp}}^{(1)}}{AIF_{\text{disp}}^{(0)}}\right)^2, \tag{5}$$

where the integral momentum $f^{(i)} = \int_0^\infty t^i f(t) \mathrm{d}t$ is approximated by a Riemann-
sum over all N computed time-steps t_k with distance Δt

$$f^{(i)} = \sum_{k=0}^{N} (t_k)^i f(t_k) \Delta t. \tag{6}$$

According to the central volume theorem of indicator-dilution theory for infinitely short bolus lengths under conditions of a well-mixed single compartment
distribution of the CA [17], tissue blood flow TBF is given by the ratio of tissue
blood volume TBV and mean transit time MTT of CA within the tissue volume
under consideration

$$TBF = \frac{TBV}{MTT} = \frac{TBV}{\int_0^\infty R(t)\mathrm{d}t}, \tag{7}$$

where $R(t)$ denotes the residue function, which describes the proportion of CA
remaining in the tissue volume at time t, following injection of a true, i.e. a
δ-shaped CA bolus. $R(t)$ only depends on the structure of the vascular bed and
is related to the time-dependent concentration of CA in tissue by $R_{\text{measured}} =
R \otimes AIF$. MTT can be considered a function of σ^2, as defined above and thus
proves highly relevant in the evaluation of vascular dispersion effects.

Comparison of MBF values calculated conventionally using the undispersed
$AIF_{\text{LV}}(t)$ show substantial MBF underestimation up to $\sim 40\%$ when dispersion
is neglected [2,3].

3   Results

Figure 3(a) shows the computed results for CA bolus dispersion for all 12 outlets
of the small model shown in Fig. 1 in comparison to the AIF applied at the inlet.
Figure 3(b) shows the obtained values for variance σ^2 as a function of MVTT,
calculated as described in [5] for both considered models (Fig. 1). The values
of MVTT and σ^2 ranged between $0.34 - 0.97\,\mathrm{s}$ (mean value $(0.63 \pm 0.20)\mathrm{s}$) and
$0.07 - 1.12\,\mathrm{s}^2$ (mean value $(0.52 \pm 0.34)\mathrm{s}^2$) respectively in the small model and
between $0.39 - 1.63\,\mathrm{s}$ (mean value $(0.83 \pm 0.27)\mathrm{s}$) and $0.03 - 1.81\,\mathrm{s}^2$ (mean value
$(0.56 \pm 0.49)\mathrm{s}^2$) respectively in the large model. The obtained results for AIF_{disp}
at the model outlets as well as the computed variances and MVTT show reduced
dispersion effects on CA bolus disperson than what is observed in [3,5] alike.

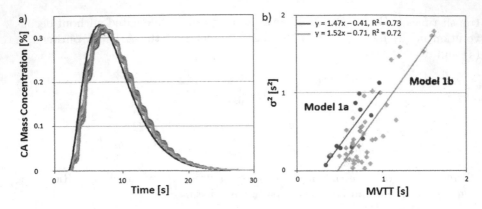

Fig. 3. Results for dispersion, MVTT and σ^2. (a) shows AIF_{disp} at all outlets of model Fig. 1(a), with the applied inlet concentration time curve AIF_{LV} (black). (b) depicts σ^2 as a function of MVTT for both models. The results at the outlets of model Fig. 1(a) are shown in red, with a linear fit (red line, standard errors for slope and intercept: ±0.28, ±0.19, resp.). Analogously the results for model Fig. 1(b) are depicted in yellow (standard errors ±0.15, ±0.13). Error margins of the graphs overlap, however, they do not yield information on vessel generation dependence of dispersion. (Color figure online)

The data presented in Fig. 3 give information about the shape of the AIF at the different model outlets and their respective σ^2 and MVTT. However, a clear understanding of the influence of different vessel segments and generations on CA bolus dispersion remains unclear. Figure 4(a) shows locations in model Fig. 1(b) between inlet and exemplary outlets, where additional information about CA concentration time curves are extracted. These allow for the analysis of variance σ^2 and MVTT of VTF depending on the distance from the model inlet as a measure for CA bolus dispersion in the epicardial vessels.

Figure 4(b) and (c) show the dependence of MVTT and variance at discrete positions in the pathway of the CA bolus to the outlets marked by arrows in Fig. 4(a). The different branches individually show very different behavior. Branch 8, the most unidirectional branch of the analysed bifurcations in the model, roughly shows the expected saturation effects in bolus broadening with increasing vessel generation until the CA bolus reaches the region marked in blue in Fig. 4. Depending on bifurcation angles and vessel curvature the other branches show both dispersion as well as narrowing effects alike. In particular, results obtained in branches 2, 6 and 7 show that CA bolus broadening can be reduced until no additional dispersion actually remains at the distal ends. Branch 1, 2, 3, 4 and 5 show bolus dispersion effects just behind the bifurcation with 2, 3 and 4 with strongly reduced dispersion later in the pathway. The remaining branches 6 and 7 differ in their behaviour, overall reducing obtained bolus dispersion.

To further investigate the behaviour of CA bolus dispersion in the vessels branching off from the central vessel, Fig. 5 shows CA flow through a model

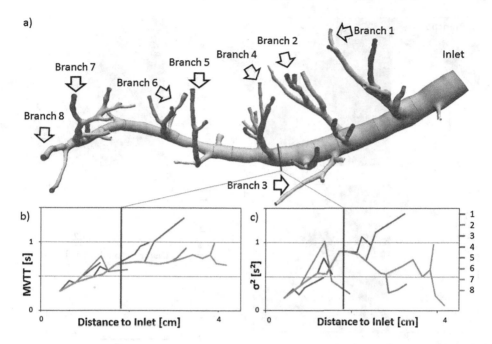

Fig. 4. Position of cross sectional planes and analysis of MVTT and σ^2) dependence on distance from model inlet. The vessel branches numbered in (a) are analyzed in more detail. (b) and (c) show the evolution of MVTT and σ^2 in the course of the numbered branches at discrete positions (red cross sections marked in (a)). The straight lines connecting the values solely serve to enhance visual presentation and do not hold any physical meaning. Close to the inlet both quantities increase. Travelling further along the considered vessels the behaviour varies strongly comprising dispersion as well as narrowing effects. Until the region marked in blue in the middle of the vascular model both quantities describe the expected monotonously increasing behavior. (Color figure online)

segment including branches 2 and 3 at different points in time. These two vessels branch off in different angles from the central branch 8, both draw bends of different curvature downstream and themselves branch into further smaller vessels. Branch 2 with a steeper bifurcation angle shows more dispersion than branch 3 in the first segment, however, also more dispersion reduction afterwards in its more pronounced bend.

Figure 6 shows velocity streamlines in the two branches analyzed in detail in Fig. 5. The two images allow for a more vivid picture of how CA is transported by convection from the main branch through bifurcations into downstream vessels. Streamlines indicate that depending on their orientation daughter vessels are "fed" by different cross sectional areals of the mother vessel. Accordingly the varying distribution of CA in the main branch will have an impact on observed dispersion in branching vessels.

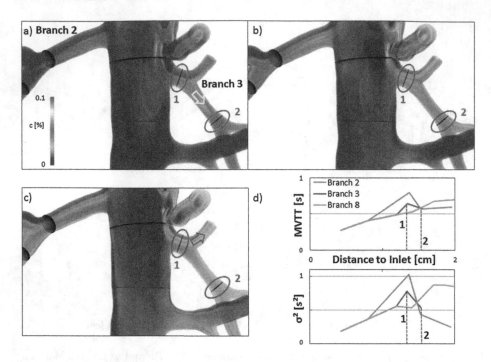

Fig. 5. CA flow through model segment and corresponding MVTT and σ^2. (a,b,c) show CA flow of a shortened bolus (for better visualization) at different points in time. In (d) MVTT and σ^2 in branches 2 and 3 are plotted depending on the travelled distance. As can be seen in a,b,c the CA bolus takes long to completely pass region 1, resulting in increased MVTT and σ^2 (cf. (d)). While in (a) and (b) most CA passing region 1 enters region 2 (yellow arrow), in (c) CA mostly flows into the steep bifurcation to the side (blue arrow), thus reducing MVTT and σ^2 in region 2 (cf. (d)). Similar but more pronounced behaviour is observed in branch 3, with a bifurcation angle of nearly $90°$. (Color figure online)

4   Discussion

The computed AIF_{disp} depicted in Fig. 3(a) and the corresponding values for MVTT and σ^2 suggest reduced influence of increasingly smaller vessels on CA dispersion in coronary arteries as compared to previous studies [1–5,9,14]. Both vascular models show a similar correlation between MVTT and VTF's variance. However, only considering CA concentration time curves at the model outlets does not yield specific information about the influence of vessels of different generations and their length on CA bolus dispersion. This requires a more detailed analysis with regard to the covered distance within the considered vessel sections.

The calculated results for variance σ^2 and MVTT along the different branches marked in Fig. 4(a) vary strongly from one another. Effects both leading to increasing dispersion (broadening of the VTF, i.e. increase of σ^2) as well as reducing dispersion (i.e. decrease of σ^2) are observed together with distinct fluc-

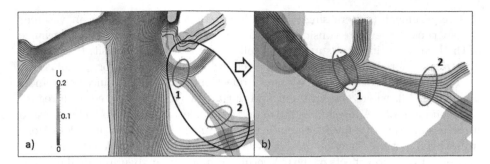

Fig. 6. Streamlines in branches 2 and 3 at 0.65s of one cardiac cycle (cf. Fig. 2), same timestep as Fig. 5(a). (a) Seed of the plotted streamlines is a horizontal 2D line source in the image plane ahead of the two branches at the upper end of the depicted model section. The sector encircled in black is represented enlarged in (b). The velocity scale (units ms^{-1}) in (a) applies for both graphics. (b) The 2D line source used as seed is indicated in blue. The two branches and subsequent daughter vessels are "fed" from different cross sectional areals of the main branch. CA flow and velocities in the main branch thus strongly influence CA concentration time curves in branching vessels. (Color figure online)

tuations in MVTT (cf. Fig. 4). The only branch showing the expected reduced influence of increasingly smaller vessels on CA dispersion in coronary arteries is the central (most unidirectional) branch 8 of the model. However, this branch also underlies fluctuations in MVTT and σ^2.

The observation of an apparently reduced dispersion in the curved bifurcating vessels is in fact a consequence of the inhomogeneous distribution of CA across the branching vessel. In earlier studies similar observations are made in curved vessels, where CA distribution is also severely inhomogeneous across the vessel lumen, and where the variance increased in and behind the stenosis [1,9,18]. After a few millimeters behind the stenosis the variance fell significantly but towards a value higher than before the stenosis. In analogy to those observations we see a transitional increase of the variance before it returns to a lower value.

The exact amount and temporal variation of CA which enters the branching vessel depends on a number of different factors. First, CA is transported by convective flow as suggested by the streamlines in Fig. 6. The amount of CA entering the branching vessel depends strongly on the (inhomogeneous) CA distribution within the main vessel, and the exact position of the ostium of the branching vessel. Moreover, diffusion of the contrast agent modifies this transport.

Strictly speaking, the concept of defining a mean vascular transit time in Eq. (3) may not be completely adequate: a general assumption of indicator dilution theory is that the intraluminal vessel space is a well-mixed single compartment. This assumption is not fulfilled because of the inhomogeneity of CA across the vessel lumen. Therefore, quantification of MBF on the basis of Eq. 7 is not feasible. Conservation of mass is only given if there is no other sink between the inlet and the vessel positions investigated. As the CA travels along the vessels, this requirement is increasingly not fulfilled. In our study the mean vascular transit time is therefore considered a parameter to easily calculate the variance of the bolus, but it is not supposed to reflect the true mean transit. It is mainly a parameter for simple analysis with a yet to be determined error.

Furthermore, the vascular system in question needs to be a linear system as defined in [19], where system input and output are linearly related with regard to the amount of CA injected at the main vessel inlet. It still needs to be assessed if the assumption of a linear system is fulfilled in the presented work. Deviations from this hypothesis may be present because of the non-negligible concentration dependent diffusion effects of the CA resulting in appearingly unphysical findings of reduced MVTT and σ^2 with increasing vessel generation.

However, the evolution of both quantities in the main branch of the considered model until CA reaches the region marked by the blue line in Fig. 4(a) shows the expected monotonous increase. This suggests that in the upstream region before branch 4 the above conditions for a linear system are satisfied at least in a first approximation. Behind this point distinct stenoses in the main vessel can be identified just after branch 4 and 5, and furthermore branch 8 itself becomes smaller and also underlies more curvature. In addition, more CA leaves the main branch at bifurcations and as a result the requirements for a linear system may not hold perfectly.

Verification of these results with the help of an alternative and more generally valid definition of MVTT and σ^2 as a measure for dispersion is still necessary. Nevertheless, the observations made in this work underline the complexity of CA transport and diffusion due to vessel curvature, tapering and branching.

5   Conclusion

Surprisingly, the obtained CA bolus dispersion results do not reinforce the hypothesis of asymptotically decreasing bolus dispersion effects in higher vessel generation models. On the contrary, bifurcations and orientation of branching vessels even lead to reduced dispersion of the CA bolus in some vessel branches, which will eventually result in a heterogeneous reduction of errors in subsequent MBF quantification. Due to the complex interplay of these different influences, in certain anatomical regions systematic underestimation in MBF quantification caused by bolus dispersion effects might after all not be as pronounced as expected.

The presented results put the assumption of the considered vascular model to be a linear system in doubt. They show that a generally valid AIF must be replaced with a localized input function for every branching vessel. According to the explanations above, the used formulae should hold in vessel segments between bifurcations, which will be analyzed in the next step. However, the influence of diffusion effects on their validity also remains questionable.

To better understand and investigate the complex impact of ever smaller vessels originating from their tapering, curvature and branching as well as downstream alignment of daughter vessels on CA bolus dispersion, an exhaustive investigation is currently being prepared on a model including vessel generations nine to ten. The ratio of vessel segment volume to volume flow through vessel segments without bifurcations represents an interesting indicator to further enhance the analysis of bolus dispersion effects performed in this study. Most importantly, an alternative indicator other than MVTT to analyze and assess dispersion and transit times in cardiovascular models with one inlet and several outlets is currently sought after and will be used for verification of the findings described above.

The presented work represents a methodical approach to better understand and analyze transport processes in the coronary vasculature and their effects on CA dispersion and later MBF quantification. An exhaustive study on patient and animal data to verify and statistically substantiate the described findings is still necessary and envisaged.

Acknowledgments. We acknowledge financial support of German Ministry of Education and Research (BMBF, grants: 01EO1004, 01E1O1504). We acknowledge University Kaiserslautern for access to HPC-cluster *elwetritsch*. Special thanks to Dr. Karsten Sommer and Regine Schmidt (University Mainz).

References

1. Graafen, D., Münnemann, K., Weber, S., Kreitner, K.-F., Schreiber, L.M.: Quantitative contrast-enhanced myocardial perfusion magnetic resonance imaging: Simulation of bolus dispersion in constricted vessels. Med. Phys. **36**(7), 3099–3106 (2009)
2. Sommer, K., Bernat, D., Schmidt, R., Breit, H.-C., Schreiber, L.M.: Contrast agent bolus dispersion in a realistic coronary artery geometry: Influence of outlet boundary conditions. Ann. Biomed. Eng. **42**(4), 787–796 (2013)
3. Sommer, K., Schmidt, R., Graafen, D., Breit, H.-C., Schreiber, L.M.: Resting myocardial blood flow quantification using contrast-enhanced magnetic resonance imaging in the presence of stenosis: A computational fluid dynamics study. Med. Phys. **42**(7), 4375–4384 (2015)
4. Calamante, F., Yim, P.J., Cebral, J.R.: Estimation of bolus dispersion effects in perfusion MRI using image-based computational fluid dynamics. NeuroImage **19**, 341–353 (2003)
5. Calamante, F., Willats, L., Gadian, D.G., Connelly, A.: Bolus delay and dispersion in perfusion MRI: Implications for tissue predictor models in stroke. Magn. Reson. Med. **55**, 1180–1185 (2006)
6. van den Wijngaard, J.P.H.M., Schwarz, J.C.V., van Horssen, P., van Lier, M.G.J.T.B., Dobbe, J.G.G., Spaan, J.A.E., Siebes, M.: 3D Imaging of vascular networks for biophysical modeling of perfusion distribution within the heart. Biomech **46**, 229–239 (2012)
7. Spaan, J.A.E., ter Wee, R., van Teeffelen, J.W.G.E., Streekstra, G., Siebes, M., Kolyva, C., Vink, H., Fokkema, D.S., VanBavel, E.: Visualisation of intramural coronary vasculature by an imaging cryomicrotome suggests compartmentalization of myocardial perfusion areas. Med. Biol. Eng. Comput. **43**, 431–435 (2005)
8. De Santis, G., Mortier, P., De Beule, M., Segers, P., Verdonck, P., Verhegghe, B.: Patient-specific computational fluid dynamics: Structured mesh generation from coronary angiography. Med. Biol. Eng. Comput. **48**, 371–380 (2010)

9. Schmidt, R., Graafen, D., Weber, S., Schreiber, L.M.: Computational fluid dynamics simulations of contrast agent bolus dispersion in a coronary bifurcation: Impact on MRI-based quantification of myocardial perfusion. Comput. Math. Methods Med. 2013 (2013)
10. Endspurt - die Skripten fürs Physikum - Physiologie 1, Georg Thieme Verlag KG (2011)
11. Olufsen, M.S., Peskin, C.S., Kim, W.Y., Pedersen, E.M., Nadim, A., Larsen, J.: Numerical simulation and experimental validation of blood flow in arteries with structured-tree outflow conditions. Ann. Biomed. Eng. 28, 1281–1299 (2000)
12. Cousins, W., Gremaud, P.A.: Boundary conditions for hemodynamics: The structured tree revisited. Comput. Phys. 231, 6086–6096 (2012)
13. Adam, J.A.: Blood vessel branching: Beyond the standard calculus problem. Math. Mag. 84, 196–207 (2011)
14. Graafen, D., Hamer, J., Weber, S., Schreiber, L.M.: Quantitative myocardial perfusion magnetic resonance imaging: The impact of pulsatile flow on contrast agent bolus dispersion. Phys. Med. Biol. 56, 5167–5185 (2011)
15. Mischi, M., den Boer, J.A., Korsten, H.H.M.: On the physical and stochastic representation of an indicator dilution curve as a gamma fit. Physiol. Meas. 29, 281–294 (2008)
16. Wieseotte, C., Wagner, M., Schreiber, L.M.: An estimate of Gd-DOTA diffusivity in blood by direct NMR diffusion measurement of its hydrodynamic analogue Ga-DOTA. In: Conference Paper, ISMRM Annual Meeting (2014)
17. Ewing, R.J., Bonekamp, D., Barker, P.B.: Clinical Perfusion MRI - Techniques and Applications. Cambridge University Press, Cambridge (2013)
18. Graafen, D.: Diploma Thesis. Johannes Gutenberg University Mainz, Untersuchung der Blutströmung in Herzkranzarterien mittels Computational Fluid Dynamics (2008)
19. King, R.B., Deussen, A., Raymond, G.M., Bassingthwaighte, J.B.: A vascular transport operator. Am. J. Physiol. 265, H2196–H2208 (1993)

Microstructurally Anchored Cardiac Kinematics by Combining *In Vivo* DENSE MRI and cDTI

Luigi E. Perotti[1,2(✉)], Patrick Magrath[1,2], Ilya A. Verzhbinsky[1],
Eric Aliotta[1,3], Kévin Moulin[1], and Daniel B. Ennis[1,2,3]

[1] Department of Radiological Sciences, University of California, Los Angeles, USA
[2] Department of Bioengineering, University of California, Los Angeles, USA
luigiemp@ucla.edu
[3] Department of Biomedical Physics IDP, University of California, Los Angeles, USA

Abstract. Metrics of regional myocardial function can detect the onset of cardiovascular disease, evaluate the response to therapy, and provide mechanistic insight into cardiac dysfunction. Knowledge of local myocardial microstructure is necessary to distinguish between isotropic and anisotropic contributions of local deformation and to quantify myofiber kinematics, a microstructurally anchored measure of cardiac function. Using a computational model we combine *in vivo* cardiac displacement and diffusion tensor data to evaluate pointwise the deformation gradient tensor and isotropic and anisotropic deformation invariants. In discussing the imaging methods and the model construction, we identify potential improvements to increase measurement accuracy. We conclude by demonstrating the applicability of our method to compute *myofiber strain* in five healthy volunteers.

Keywords: Cardiac kinematics · Diffusion tensor imaging · Cardiac deformation invariants · Myofiber strain

1 Introduction

Heart disease, including myocardial infarction (MI) and hypertrophic cardiomyopathy (HCM), heterogeneously impacts left ventricular (LV) structure and function. Global metrics of myocardial function, such as ejection fraction (EF), however, may be preserved despite the presence of disease, and this masks regional functional deficits that portend a poor clinical outcome.

Displacement ENcoding with Stimulated Echoes (DENSE) magnetic resonance imaging (MRI) encodes time-resolved Eulerian cardiac displacements directly into the MRI signal phase [1,20]. The displacement field derived from DENSE MRI can be used to calculate several regional measures of cardiac function, including radial (E_{rr}) and circumferential (E_{cc}) strain. Deformation-based biomarkers show clinical promise and regional sensitivity for the diagnosis of, for example, ischemic cardiac pathologies (e.g., MI) and inherited cardiac pathologies (e.g., HCM and Duchenne muscular dystrophy).

© Springer International Publishing AG 2017
M. Pop and G.A. Wright (Eds.): FIMH 2017, LNCS 10263, pp. 381–391, 2017.
DOI: 10.1007/978-3-319-59448-4_36

Aletras *et al.* [2] employed DENSE to identify depressed systolic E_{cc} in regions of Late Gadolinium Enhanced (LGE) defined focal and diffuse fibrosis in HCM, as well as in hypertrophic regions for which LGE MRI findings were negative. The authors speculate that functional deficits absent positive findings of fibrosis may be explained by microstructural remodeling of myofiber geometry. In fact, considerable attention has been given to the link between MRI based strain measures and local microstructure, e.g., the work of Tseng et al. [15] or the recent work of Wang et al. [17] where *in vivo* displacement data and *ex vivo* cardiac Diffusion Tensor Imaging (cDTI) data were combined to study myofiber strains and reorientation.

Recent advances in cDTI [3,14] enable *in vivo* measurements of local myofiber orientation and microstructural rearrangement during the cardiac cycle [9]. Combining cDTI with DENSE displacement data permits computing microstructurally anchored myofiber deformation to characterize myocyte performance *in vivo*. Here, in agreement with the current literature, we use the term *myofiber* orientation and deformation. However, we emphasize that we compute the averaged myofiber deformation since the acquired *in vivo* cDTI data provide averaged (or preferential) myofiber orientation within a voxel, not single cardiomyocyte orientation.

Building on these recent advances in cDTI and DENSE imaging techniques, our objectives were to: (1) Construct a computational framework to evaluate *in vivo* myofiber (anisotropic) and extracellular matrix (isotropic) deformation metrics based on single-slice DENSE and cDTI data; (2) Evaluate sources of error in our modeling framework using an analytic computational deforming phantom (CDP) of cardiac motion that provides ground truth kinematics; (3) Demonstrate the applicability of our framework in healthy volunteers ($N = 5$) and compute *in vivo* myofiber and matrix deformation.

2   Methods

A long term goal is to apply our framework in a clinical setting under the constraint of limited acquisition time. Therefore, we base our computational framework on single-slice DENSE and cDTI data. Validation of our computational framework and evaluation of the error involved with our assumptions requires the knowledge of a ground truth, which is provided here by a cardiac CDP (Sect. 2.3).

2.1   Single-Slice Model Construction with cDTI and DENSE

The epicardial and endocardial borders of the LV myocardium at each cardiac phase were defined using cubic spline interpolation and a motion guided segmentation algorithm [13] (Fig. 1A). The region between the borders defining the first cardiac phase recorded in the DENSE sequence after the QRS trigger was meshed with linear triangular finite elements (FE) (Fig. 1E).

Using an open-source post-processing tool [12], myocardial (3D) Lagrangian displacements u_i (Fig. 1C) at each cardiac phase were computed by unwrapping the displacement encoded phase data. Subsequently, the myocardial displacements u_i computed at the center of each image voxel were linearly interpolated to the nodes of the FE mesh (Fig. 1E). Here and in the following $i = 1$, 2, and 3 represents, respectively, the component along the X_1, X_2, and X_3 coordinate and we assume, without loss of generality, that the single-slice is initially in the $X_1 - X_2$ plane and the LV long-axis is coaxial with X_3.

All cDTI images were registered to DENSE using a rigid translation and rotation in the $X_1 - X_2$ plane, followed by a non-rigid transformation step. Rotation and rigid translation were determined by matching the two RV insertion points in the cDTI and DENSE images. Subsequently the cDTI magnitude images were segmented using cubic splines as was previously done for the DENSE magnitude images. The cubic spline DENSE and cDTI segmentations were then transformed into binary masks and registered using the b-spline based registration algorithm outlined in [11]. The computed non-rigid registration did not include a rotation component, as the proper rotation was already taken into account by matching the RV insertion points. The same rigid translation and rotation, and non-rigid registration based on the magnitude images were applied to all cDTI images encoding different diffusion directions (Fig. 1B).

cDTI acquired diffusion tensors were interpolated using linear invariant interpolation [6] to the location of the FE quadrature points where the deformation invariants were computed (Fig. 1D). At these locations, the myofiber preferential orientation was computed as the principal eigenvector $\mathbf{f}$ of the interpolated diffusion tensors. Finally, $\mathbf{f}$ was rotated around X_3 using the previously computed rigid rotation to account for the change in LV orientation between cDTI and DENSE images.

2.2   Calculation of the Deformation Gradient and Invariants

The deformation gradient tensor $\mathbf{F}$ was computed using FE interpolation functions and the displacements u_i calculated from DENSE at each FE node with initial position X_i. In the reference configuration corresponding to the first DENSE image, we assume that all nodes have the same X_3. For displacements acquired only within a single short-axis slice, we cannot directly compute the third column of the deformation gradient $\partial \varphi_i / \partial X_3$, where $\varphi_i = X_i + u_i$. In order to evaluate $\mathbf{F}$ using only single-slice data, we make the following two assumptions: (1) the myocardium is incompressible, i.e., $\det \mathbf{F} = 1$; and (2) intra-slice torsion is very small, therefore u_1 and u_2 are not a function of X_3. Consequently, we evaluate $\mathbf{F}$ as:

$$\mathbf{F} = \begin{bmatrix} \partial \varphi_1 / \partial X_1 & \partial \varphi_1 / \partial X_2 & 0 \\ \partial \varphi_2 / \partial X_1 & \partial \varphi_2 / \partial X_2 & 0 \\ \partial \varphi_3 / \partial X_1 & \partial \varphi_3 / \partial X_2 & 1/(F_{11}F_{22} - F_{21}F_{12}) \end{bmatrix}$$

Subsequently, we compute the right Cauchy-Green deformation tensor $\mathbf{C} = \mathbf{F}^{\top}\mathbf{F}$ and a selection of its isotropic (I_1, R_1, R_2, R_3) and anisotropic invariants

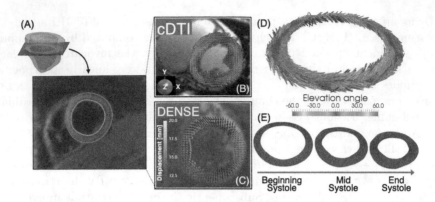

Fig. 1. (A) LV short-axis single-slice (B) including myofiber preferential orientations (principal eigenvectors) from cDTI and (C) Lagrangian displacement vectors from beginning to end systole. (D) Interpolated myofiber preferential orientations at FE quadrature points after cDTI-DENSE registration and (E) FE-based mesh deformation through systole driven by the DENSE measured displacement field.

(I_4, I_5) to characterize, respectively, extracellular matrix and myofiber kinematics. In particular: $I_1 = \text{tr}(\mathbf{C})$, $R_1 = \sqrt{\mathbf{C} : \mathbf{C}}$, $R_2 = \sqrt{3/2}\sqrt{\bar{\mathbf{C}} : \bar{\mathbf{C}}}/R_1$ where $\bar{\mathbf{C}} = \mathbf{C} - 1/3\,\text{tr}(\mathbf{C})\mathbf{I}$, $R_3 = 3\sqrt{6}\det(\bar{\mathbf{C}}/\sqrt{\bar{\mathbf{C}} : \bar{\mathbf{C}}})$, $I_4 = \text{tr}\left(\mathbf{C}(\mathbf{f} \otimes \mathbf{f})\right)$, and $I_5 = \text{tr}\left(\mathbf{C}^2(\mathbf{f} \otimes \mathbf{f})\right)$. Further details on the construction of these sets of invariants are discussed in [4,7]. One of the simplest choices to characterize myocardial deformation consists of adopting I_1 (isotropic) and I_4 (anisotropic) invariants, which measure, respectively, the sum of the squares of the principal stretches and the the square of the myofiber stretch. Additionally, I_4 is related to the Green-Lagrange myofiber strain E_{ff} through $E_{\text{ff}} = 0.5 \cdot (I_4 - 1)$ while the more common circumferential strain E_{cc} is computed by projecting $\mathbf{E}$ in the circumferential direction $\mathbf{c}$, i.e., $E_{\text{cc}} = 0.5 \cdot (\text{tr}\left(\mathbf{C}(\mathbf{c} \otimes \mathbf{c})\right) - 1)$

2.3   Computational Deforming Phantom

In order to evaluate the model's accuracy for computing myocardial deformation, we constructed a 3D Computational Deforming Phantom (CDP) in which ground truth strain values were known by construction. The motion of the CDP and the corresponding exact deformation gradient tensor are determined analytically by the functions described in Appendix B of [8]. Importantly, the analytic mapping presented in [8] allows decoupling the slice displacements into in-plane deformation and intra-slice torsion. Using the 3D CDP, we evaluated the effect of computing the invariants of $\mathbf{C}$ from a single short-axis slice of data and our model (see Sec. 2.2) using the following tests:

Test 1 *Effect of using a single slice model in absence of intra-slice torsion.* We evaluate the effect of analyzing 3D CDP motion using only single slice data.

Test 2 *Effect of FE interpolation and including intra-slice torsion.* Based on the analytical functions in [8] and including intra-slice torsion, we assigned the exact displacements to the nodes of a regular grid as it would be sampled by the DENSE experiment. Subsequently, we interpolated the displacements to the nodes of our FE mesh.

Test 3 *Effect of pipeline processing.* We computed the Eulerian displacements at a fixed short-axis slice location in the CDP. Based on the Eulerian displacement components and a chosen encoding strength k_e, we assigned a phase value to the center of each image voxel (uniform 2×2mm grid). These constructed DENSE phase images corresponding to the motion of the CDP were subsequently processed identically to *in vivo* images. The computed displacements were interpolated to the nodes of the FE mesh. This test reproduced the entire pipeline from DENSE image processing to the calculation of deformation invariants, including image segmentation and displacement interpolation from Eulerian to Lagrangian in the DENSE processing toolbox.

We report the mean (μ) and standard deviation (σ) for the analytically (μ_{an}, σ_{an}) and numerically (μ_{num}, σ_{num}) computed invariants, together with the percent error of the means ($\frac{(\mu_{an} - \mu_{num}) \cdot 100}{\mu_{an}}$) and the pointwise percent error ($\frac{(I_i^{an} - I_i^{num}) \cdot 100}{I_i^{an}}$), where I_i (or R_i) represents one of the computed invariants. We included myofiber preferential orientation data in the CDP in order to compare computed and analytic myofiber stretch. The modeled myofiber helix angle varies linearly from $39.5°$ (endocardial) to $-53.5°$ (epicardial) as measured in [5].

2.4 *In Vivo* Image Acquisition

Using an IRB approved protocol, a single mid-ventricular short-axis slice was acquired in healthy volunteers ($N = 5$) using DENSE and cDTI at 3T (Prisma, Siemens) after obtaining informed consent. All imaging used navigator triggered free breathing. DENSE: balanced 4-point encoding, $2.5 \times 2.5 \times 8$ mm, TE/TR = $1.04/15$ ms, k_e=0.06 cycles/mm, $N_{avg} = 3$, spiral interleaves=10, T_{scan}=5 min. CODE [3] cDTI: $2 \times 2 \times 5$ mm, TE/TR= $74/4000$ ms, b-value=0,350s/mm^2, $N_{avg} = 10$, $N_{dir} = 12$, $T_{scan} = 4$ min.

3 Results

We first evaluated our modeling framework using the tests listed in Sect. 2.3 and subsequently demonstrated its application in healthy volunteers.

3.1 Model Evaluation: From 3D Motion to a 2D Short-Axis Slice

Table 1 compares the deformation invariants computed analytically and numerically when intra-slice torsion is set to zero in the CDP and exact displacements

Table 1. Exact (analytical) and computed (numerical) deformation invariants at peak systole neglecting intra-slice torsion (Test 1).

	I_1	R_1	R_2	R_3	I_4	I_5
$\mu_{\mathrm{an}} \pm \sigma_{\mathrm{an}}$	3.46 ± 0.05	2.27 ± 0.06	0.58 ± 0.02	0.84 ± 0.11	0.79 ± 0.05	1.31 ± 0.12
$\mu_{\mathrm{num}} \pm \sigma_{\mathrm{num}}$	3.46 ± 0.05	2.27 ± 0.05	0.58 ± 0.02	0.84 ± 0.12	0.79 ± 0.05	1.31 ± 0.12
% error	$2 \cdot 10^{-3}$	0.01	0.03	0.13	-0.01	-0.02

are applied to the nodes of the FE mesh without interpolating from a regular grid (Test 1). The mean percent error between analytically and numerically computed invariants is small ($< 1\%$).

Subsequently, we compared the analytical and numerical deformation invariants including: (1) intra-slice torsion; and (2) interpolation of the displacement field from a uniform grid to the FE nodes (Test 2). The mean percent errors between analytically and numerically computed invariants remain small (Table 2), but the pointwise comparison (Fig. 2) highlights regions of larger discrepancies.

Table 2. Exact (analytical) and computed (numerical) deformation invariants at peak systole including intra-slice torsion and displacement interpolation to FE nodes (Test 2).

	I_1	R_1	R_2	R_3	I_4	I_5
$\mu_{\mathrm{an}} \pm \sigma_{\mathrm{an}}$	3.46 ± 0.04	2.28 ± 0.05	0.58 ± 0.02	0.83 ± 0.13	0.80 ± 0.08	1.33 ± 0.19
$\mu_{\mathrm{num}} \pm \sigma_{\mathrm{num}}$	3.46 ± 0.05	2.28 ± 0.06	0.58 ± 0.02	0.84 ± 0.12	0.79 ± 0.05	1.31 ± 0.14
% error	0.04	0.02	-0.03	-1.50	1.10	1.36

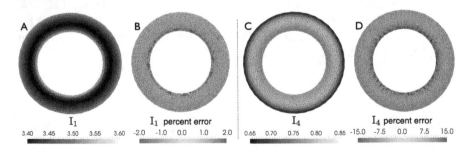

Fig. 2. Isotropic (I_1) and anisotropic (I_4) deformation invariants: analytic invariants (A and C) and pointwise % error between analytical and numerical values (B, D) computed in Test 2.

3.2   Model Validation: Effect of Pipeline Processing

Table 3 summarizes the error introduced by the entire post-processing pipeline in the calculation of the deformation invariants (Test 3). Figure 3 shows the point-wise comparison for I_1 (isotropic) and I_4 (anisotropic) deformation invariants. Percent error of mean values are smaller for I_1 and I_4 among the isotropic and anisotropic invariants, respectively.

Table 3. From simulated imaging data to deformation invariants in the single slice model (Test 3).

	I_1	R_1	R_2	R_3	I_4	I_5
$\mu_{an} \pm \sigma_{an}$	3.46 ± 0.05	2.28 ± 0.06	0.58 ± 0.02	0.82 ± 0.12	0.80 ± 0.05	1.33 ± 0.14
$\mu_{num} \pm \sigma_{num}$	3.34 ± 0.11	2.13 ± 0.13	0.51 ± 0.09	0.73 ± 0.35	0.82 ± 0.06	1.26 ± 0.10
% error	3.61	6.5	13.16	11.01	-2.51	5.20

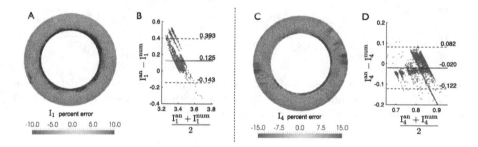

Fig. 3. Pointwise comparison and Bland-Altman plot for I_1 (A, B) and I_4 (C, D) computed analytically and using the single slice model with DENSE simulated data (Test 3). The biases in the Bland-Altman plots are 0.125 for I_1 (B) and -0.020 for I_4 (D).

3.3   *In Vivo* Results

We demonstrated the applicability of our framework *in vivo* by combining DENSE and cDTI data from a single-slice. The modeling framework was used to compute I_1 from DENSE displacements alone and myofiber deformation I_4 from co-registered *in vivo* DENSE and cDTI (Fig. 4). We also computed the time evolution of averaged I_1 and I_4 deformation invariants in each healthy volunteer (Fig. 5). Intersubject average I_4 at peak systole is equal to 0.74 ± 0.03, which corresponds to average myofiber strain equal to $-12.8\% \pm 1.6\%$.

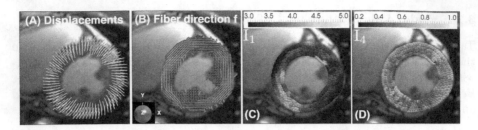

Fig. 4. (A) Displacement vectors from beginning to end systole in a healthy volunteer. (B) cDTI myofiber vectors **f** acquired at mid-systole. (C,D) Peak systolic deformation invariants I_1 and I_4.

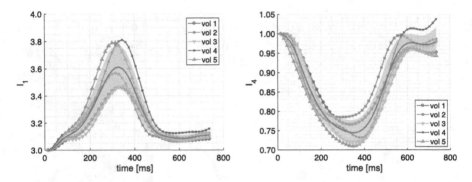

Fig. 5. Time evolution of averaged I_1 (isotropic) and I_4 (anisotropic) deformation invariants in healthy volunteers ($N = 5$) together with intersubject mean (red line) and standard deviation (blue shaded region). (Color figure online)

4 Discussion

We have implemented a computational modeling framework to calculate microstructurally anchored deformation invariants from DENSE and cDTI imaging data. The comparison of exact (analytic) and numerically computed deformation invariants permits evaluating sources of error in the modeling framework and possible remedies.

We observed that the 2D model approximation (Test 1) led to deformation invariants close to the exact analytical values (error $< 1\%$) if intra-slice torsion is neglected (Table 1).

The agreement, however, between exact and numerical deformation invariants worsens when intra-slice torsion is included in the CDP to provide more realistic cardiac motion (Table 2, Test 2). Lower agreement is expected since correctly representing torsion requires data from parallel short-axis slice locations and our acquired data was purposefully restricted to a single short-axis slice. Indeed, one of our assumptions in Sect. 2.2 was the presence of a negligible intra-slice torsion. This assumption, of course, can be relaxed by acquiring data at two adjacent short-axis slices.

Lastly, we compared the analytic deformation invariants to the ones computed after the entire processing pipeline, starting from simulated DENSE images (Table 3, Test 3). Other studies have evaluated the accuracy of strain measures computed from DENSE displacements. Wehner et al. [18] report that measured DENSE strains show good agreement with those computed from Tagged MRI in human volunteers. Young et al. [19] echo these results, and also demonstrate that measured DENSE strains in a rotating deforming gel phantom show good agreement with those computed analytically for the phantom. Our results complement these studies by comparing computed deformation invariants with *analytic values* available through the CDP. This comparison with exact analytic values enables determining which deformation invariants are less affected by the assumptions and discretization steps included in our modeling pipeline.

To understand the spatial distribution of the differences between analytical and numerical deformation invariants, we compared I_1 and I_4 pointwise (Figs. 2 and 3). Although their average difference is small, local differences can be much larger, especially in the epicardial and endocardial regions where the effect of intra-slice torsion appears to be larger and the correction outlined above may be most useful. We notice that the results in Fig. 3(B,D) are not uniformly scattered but partially clustered along distinct "bands". This may be due to the inhomogeneous error in I_1 and I_4 across the myocardial wall as illustrated in Fig. 3(A,C).

After evaluating several sources of error in our model, we showed its applicability to *in vivo* acquired data (Figs. 4 and 5). Leveraging the measured patient-specific cDTI myofiber orientations permits estimating a measure of myofiber function (I_4), which is inherently microstructurally anchored. The averaged $I_4(\approx 0.75)$ values across all volunteers (Fig. 5) is consistent with values of myofiber strain reported in the literature [16]. *In vivo* averaged I_1 (≈ 3.6) is consistent with the analytic values computed from the CDP ($I_1^{an} \approx 3.5$ - Table 3). In this study, we focused on comparing I_1 and I_4 because these deformation invariants are commonly used to formulate material energy laws describing passive cardiac mechanics, e.g., [7,10].

In this work, we have evaluated the model assumptions for interpolating displacements from a single short-axis slice of cine DENSE data to compute the deformation invariants. These approximations can have a significant impact on the computed values since the displacement field is differentiated to calculate $\mathbf{F}$, and this operation can amplify experimental noise. Accurate calculation of $\mathbf{F}$ is central to computing myofiber strain and deformation invariant biomarkers. Future validation of our computational model will include evaluating the effect of non-rigid registration of DENSE and cDTI data and the effect of experimental noise on both DENSE and cDTI data. We foresee that a higher order FE scheme or a meshless method, together with a more sophisticated displacement interpolation method, may be necessary to decrease the effect of experimental noise.

An additional step in validating our single slice framework consists in repeating the tests presented herein using a different CDP to reproduce different cardiac motion, especially if specific cardiac phases are of particular interest. For example, to determine an optimal strategy to analyze passive cardiac kinematics, we want to test our framework using a CDP that closely reproduces the motion during diastole, since different numerical assumptions may lead to a different amount of error in different cardiac phases.

Finally, we emphasize that the proposed framework may be easily extended to the full 3D case, in which several short-axis DENSE and cDTI slices are combined to construct a 3D finite element model. In the 3D case, no assumption about intraslice torsion is necessary, thereby avoiding one major source of error in the current framework.

In conclusion, a computational framework was developed to evaluate extracellular matrix and myofiber kinematics based on single-slice DENSE and cDTI data. Our preliminary results suggest that I_1 (isotropic) and I_4 (anisotropic) invariants have the lowest average error among the invariants considered and consequently may form a robust basis for describing systolic cardiac kinematics and formulating strain energy laws for the heart.

Acknowledgements. We gratefully acknowledge the funding support from the Department of Radiological Sciences at UCLA, the Center for Duchenne Muscular Dystrophy at UCLA pilot grant under NIH P30AR05723, NIH R01HL131975 grant, and NIH K25HL135408 grant.

References

1. Aletras, A.H., Ding, S., Balaban, R.S., Wen, H.: DENSE: Displacement encoding with stimulated echoes in cardiac functional MRI. J. Magn. Reson. **137**(1), 247–252 (1999)
2. Aletras, A.H., Tilak, G.S., Hsu, L.Y., Arai, A.E.: Heterogeneity of intramural function in hypertrophic cardiomyopathy. Circ. Cardiovasc. Imaging **4**(4), 425–434 (2011). Clinical Perspective
3. Aliotta, E., Wu, H.H., Ennis, D.B.: Convex optimized diffusion encoding (CODE) gradient waveforms for minimum echo time and bulk motion-compensated diffusion-weighted MRI. Magn. Reson. Med. **77**(2), 717–729 (2016)
4. Ennis, D.B., Kindlmann, G.: Orthogonal tensor invariants and the analysis of diffusion tensor magnetic resonance images. Magn. Reson. Med. **55**(1), 136–146 (2006)
5. Ennis, D.B., Nguyen, T.C., Riboh, J.C., Wigström, L., Harrington, K.B., Daughters, G.T., Ingels, N.B., Miller, D.C.: Myofiber angle distributions in the ovine left ventricle do not conform to computationally optimized predictions. J. Biomech. **41**(15), 3219–3224 (2008)
6. Gahm, J.K., Wisniewski, N., Kindlmann, G., Kung, G.L., Klug, W.S., Garfinkel, A., Ennis, D.B.: Linear invariant tensor interpolation applied to cardiac diffusion tensor MRI. In: Ayache, N., Delingette, H., Golland, P., Mori, K. (eds.) MICCAI 2012. LNCS, vol. 7511, pp. 494–501. Springer, Heidelberg (2012). doi:10.1007/978-3-642-33418-4_61
7. Holzapfel, G.A., Ogden, R.W.: Constitutive modelling of passive myocardium: A structurally based framework for material characterization. Philos. Trans. R. Soc. Lond. A: Math. Phys. Eng. Sci. **367**(1902), 3445–3475 (2009)

8. Moghaddam, A.N., Saber, N.R., Wen, H., Finn, J.P., Ennis, D.B., Gharib, M.: Analytical method to measure three-dimensional strain patterns in the left ventricle from single slice displacement data. J. Cardiovasc. Magn. Reson. **12**(1), 33 (2010)

9. Nielles-Vallespin, S., Khalique, Z., Ferreira, P.F., de Silva, R., Scott, A.D., Kilner, P., McGill, L.A., Giannakidis, A., Gatehouse, P.D., Ennis, D., et al.: Assessment of myocardial microstructural dynamics by in vivo diffusion tensor cardiac magnetic resonance. J. Am. Coll. Cardiol. **69**(6), 661–676 (2017)

10. Perotti, L.E., Ponnaluri, A.V., Krishnamoorthi, S., Balzani, D., Ennis, D.B., Klug, W.S.: Method for the unique identification of hyperelastic material properties using full field measures. Application to the passive myocardium material response. Int. J. Numer. Methods Biomed. Eng. (2017)

11. Rueckert, D., Sonoda, L.I., Hayes, C., Hill, D.L., Leach, M.O., Hawkes, D.J.: Non-rigid registration using free-form deformations: Application to breast MR images. IEEE Trans. Med. Imaging **18**(8), 712–721 (1999)

12. Spottiswoode, B.S., Zhong, X., Hess, A., Kramer, C., Meintjes, E.M., Mayosi, B.M., Epstein, F.H.: Tracking myocardial motion from cine DENSE images using spatiotemporal phase unwrapping and temporal fitting. IEEE Trans. Med. Imaging **26**(1), 15–30 (2007)

13. Spottiswoode, B.S., Zhong, X., Lorenz, C.H., Mayosi, B.M., Meintjes, E.M., Epstein, F.H.: Motion-guided segmentation for cine DENSE MRI. Med. Image Anal. **13**(1), 105–115 (2009)

14. Stoeck, C.T., Von Deuster, C., Genet, M., Atkinson, D., Kozerke, S.: Second-order motion-compensated spin echo diffusion tensor imaging of the human heart. Magn. Reson. Med. (2015)

15. Tseng, W.Y.I., Dou, J., Reese, T.G., Wedeen, V.J.: Imaging myocardial fiber disarray and intramural strain hypokinesis in hypertrophic cardiomyopathy with MRI. J. Magn. Reson. Imaging **23**(1), 1–8 (2006)

16. Tseng, W.Y.I., Reese, T.G., Weisskoff, R.M., Brady, T.J., Wedeen, V.J.: Myocardial fiber shortening in humans: Initial results of MR imaging. Radiology **216**(1), 128–139 (2000)

17. Wang, V.Y., Casta, C., Zhu, Y.M., Cowan, B.R., Croisille, P., Young, A.A., Clarysse, P., Nash, M.P.: Image-based investigation of human in vivo myofibre strain. IEEE Trans. Med. Imaging **35**(11), 2486–2496 (2016)

18. Wehner, G.J., Suever, J.D., Haggerty, C.M., Jing, L., Powell, D.K., Hamlet, S.M., Grabau, J.D., Mojsejenko, W.D., Zhong, X., Epstein, F.H., et al.: Validation of in vivo 2D displacements from spiral cine DENSE at 3T. J. Cardiovasc. Magn. Reson. **17**(1), 5 (2015)

19. Young, A.A., Li, B., Kirton, R.S., Cowan, B.R.: Generalized spatiotemporal myocardial strain analysis for DENSE and SPAMM imaging. Magn. Reson. Med. **67**(6), 1590–1599 (2012)

20. Zhong, X., Spottiswoode, B.S., Meyer, C.H., Kramer, C.M., Epstein, F.H.: Imaging three-dimensional myocardial mechanics using navigator-gated volumetric spiral cine DENSE MRI. Magn. Reson. Med. **64**(4), 1089–1097 (2010)

A Patient-Specific Computational Fluid Dynamics Model of the Left Atrium in Atrial Fibrillation: Development and Initial Evaluation

Alessandro Masci[1], Martino Alessandrini[1(✉)], Davide Forti[2],
Filippo Menghini[2], Luca Dedé[2], Corrado Tommasi[3], Alfio Quarteroni[2],
and Cristiana Corsi[1]

[1] Department of Electrical, Electronic and Information Engineering,
University of Bologna, Bologna, Italy
martino.alessandrini@gmail.com
[2] Chair of Modelling and Scientific Computing,
Swiss Federal Institute of Technology (EPFL), Lausanne, Switzerland
[3] Hospital "S. Maria Delle Croci", Ravenna, Italy

Abstract. Atrial fibrillation (AF) is associated to a five-fold increase
in the risk of stroke and AF strokes are especially severe. Stroke risk is
connected to several AF related morphological and functional remodeling mechanisms which favor blood stasis and clot formation inside the
left atrium. The goal of this study was therefore to develop a patient-specific computational fluid dynamics model of the left atrium which
could quantify the hemodynamic implications of atrial fibrillation on a
patient-specific basis. Hereto, dynamic patient-specific CT imaging was
used to derive the 3D anatomical model of the left atrium by applying a
specifically designed image segmentation algorithm. The computational
model consisted in a fluid governed by the incompressible Navier-Stokes
equations written in the Arbitrary Lagrangian Eulerian (ALE) frame of
reference. In this paper, we present the developed model as well as its
application to two AF patients. These initial results confirmed that morphological and functional remodeling processes associated to AF effectively reduce blood washout in the left atrium, thereby increasing the
risk of clot formation. Our analysis is a step forward towards improved
patient-specific stroke risk stratification and therapy planning.

Keywords: Atrial fibrillation · Left atrial appendage · Stroke · Risk
stratification · Patient-specific analysis · Computational fluid dynamics

1 Introduction

Atrial fibrillation (AF) is the most common form of arrhythmia worldwide [2]. Its
prevalence is expected at least to double in the next 50 years as the population
ages [2]. AF is associated to a five-fold increase in the risk of stroke [2] and AF
related strokes are more severe and likely to be fatal [2]. Clinical scores used

© Springer International Publishing AG 2017
M. Pop and G.A. Wright (Eds.): FIMH 2017, LNCS 10263, pp. 392–400, 2017.
DOI: 10.1007/978-3-319-59448-4_37

to stratify the risk of stroke, such as the recommended CHA_2DS_2-VASc score [2], are based on very generic empirical factors such as age, hypertension and diabetes mellitus. As a result, their predictive power remains low [3].

In this context, several clinical studies suggested that stroke risk stratification could be improved by using hemodynamic information on the left atrium (LA) and left atrial appendage (LAA) [3,6]. These studies are based on the consideration that AF effectively alters the intra-atrial blood flow dynamics in such a way to yield blood stasis and, therefore, clot formation and embolism. One factor is anatomical remodeling, which consists in the progressive LA enlargement and increase in the LAA elongation [6]. A second factor is given by the altered mechanical function. Atrial contraction becomes completely chaotic and therefore ineffective during the AF episodes. Moreover, there is a progressive loss of atrial function even in sinus rhythm [6].

Yet, the exact way these complex mechanisms interplay cannot be assessed experimentally and remains largely unknown. In this respect, computational fluid dynamics (CFD) represents a unique tool to test different boundary conditions on a complex fluid dynamics system, such as intra-cardiac hemodynamics, in a non-invasive fully controllable and reproducible way. The aim of this study was therefore to develop a patient-specific CFD model of the LA in AF which could help elucidate the role of the key anatomical and functional features of AF on the LA blood flow. By that, the provided tool might be used for improving personalized stroke risk stratification and therapy planning.

Related work: Despite the relevance of the problem, the literature on CFD LA modeling in AF is very scarce [8,16]. Due to the higher priority, the majority of modeling efforts focused ventricular fluid-dynamics. Moreover, existing atrial models consider almost exclusively electrophysiology. Few coupled atrioventricular CFD models exist but still with main focus on the ventricle [14].

The first CFD study of the LA in AF was the one by Zhang and coll. [16]. They performed a purely theoretical (i.e. not patient-specific) study making use of a simple toy model of LA anatomy and motion. Inflow boundary conditions were also accounted for by simplistic wave-forms. More recently, Koizumi [8] used a geometry segmented from a healthy volunteer. Hence, anatomical remodeling effects were not taken into account. A trivial motion model was used consisting of simple contraction/expansion in the direction from the mitral valve (MV) to the LA roof. Fixed pressure values taken from the literature were used as boundary conditions and the MV was simulated as an on-off switch.

Statement of the contribution: In this paper, we present what to our knowledge represents the first patient-specific model of atrial blood flow in AF. The main original contributions consist of: (1) LA anatomy was obtained from an AF patient, hence, anatomical remodeling was accounted for; (2) Atrial motion was extracted from patient data by image tracking; (3) We employed realistic inflow boundary conditions obtained from real Doppler measurements on AF patients; (4) we used a state-of-the art library for CFD simulations (LifeV[1]) which has

---

[1] https://cmcsforge.epfl.ch/projects/lifev/.

been extensively validated and optimized for fluid dynamics studies and cardiac applications [11].

In this paper, we present the development of our computational model as well as an initial evaluation on two AF patients, where the model is used to study the hemodynamic implications of an AF episode.

2   Materials and Methods

The project's flowchart is summarized in Fig. 1.

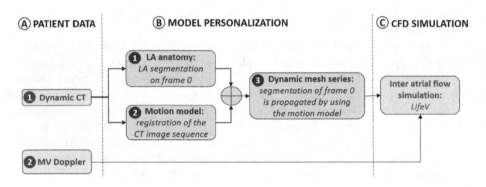

Fig. 1. Project flowchart.

2.1   Patient Data for Model Personalization

Clinical data was selected retrospectively from two AF patients. CT data (block A.1 in Fig. 1) was acquired from a Philips Brilliance 64 CT scanner. Ten volumes (170 axial slices, pixel size 0.4 mm, slice thickness 1 mm) spanning one cardiac cycle from the end of (ventricular) diastole were reconstructed by retrospective ECG gating.

Moreover, we used PW Doppler to set realistic inflow boundary conditions. Specifically, PW Doppler at the MV was used to constrain, indirectly, the inflow at the four pulmonary veins, as detailed in Sect. 2.3. Unfortunately, these meauserements were not available for the two considered patients. Representative MV velocity profiles were therefore taken from the literature [8] (block A.2 in Fig. 1).

2.2   Patient Specific Anatomy and Motion Model

Segmentation of the First Frame - The 3D LA anatomy, inclusive of pulmonary veins (PV's) and left atrial appendage (LAA), was segmented from the first CT volume (block B.1 in Fig. 1). Each slice was processed independently, as follows.

For each slice, an initial segmentation was computed by intensity threshold-ing. A set of morphological opening and closing were then applied to remove spurious regions. Finally, a curvature based level set was used to regularize the contours [9].

Further processing was then required to make the 3D model compatible with the CFD simulator. This included smoothing via the HC Laplacian algo-rithm [15] and manual cutting of the inflow (i.e. the 4 PV's) and outflow (i.e. the MV) planes. Finally, a volumetric tetrahedral mesh was generated and the inflow/outflow surfaces were labeled with the VMTK library[2]. A final mesh quality check was performed. An example of segmentation in provided in Fig. 2.

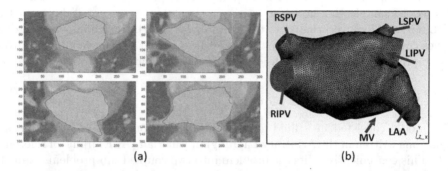

Fig. 2. Reconstructed anatomy for patient 1: (a) segmentation result on four slices; (b) final tetrahedral mesh. The principal anatomical structures are annotated in (b): right inferior PV (RIPV), right superior PV (RSPV), left inferior PV (LIPV), left superior PV (LSPV), LAA and MV. The final mesh contained 170,428 points and 1,042,766 tetrahedra.

Motion Model - We performed image registration on the full time sequence in order to extract the patient specific LA motion over the cardiac cycle (block B.2 in Fig. 1). Hereto, the displacement $\mathbf{d}_{i \to i+1}(\mathbf{x})$ between two successive CT volumes $I_i(\mathbf{x})$ and $I_{i+1}(\mathbf{x})$ was computed by elastic image registration [13], using the mean squared difference as a similarity metric.

The global displacement with respect to the reference volume at time 0 was then computed by accumulating the successive inter-frame estimates by means of the recursive formula $\mathbf{d}_{i \to 0}(\mathbf{x}) = \mathbf{d}_{i \to i-1}(\mathbf{x}) \circ \mathbf{d}_{i-1 \to 0}(\mathbf{x})$, with $\mathbf{d}_{0 \to 0}(\mathbf{x}) = \mathbf{0}$. To control error accumulation, the tracking was performed in the forward direction (i.e., with respect to frame 0) for half of the frames and in the backward direction (i.e., with respect to frame $N-1$, being N the total number of frames) for the remaining half.

The displacement field was then used to propagate the initial mesh segmented at time 0 over the full cardiac cycle (block B.3 in Fig. 1). By calling $\mathbf{x}_0$ the

---

[2] http://www.vmtk.org/.

position of a mesh node at time 0, we computed its position $\mathbf{x}_i$ at time i by sampling the computed displacement field, i.e. $\mathbf{x}_i = \mathbf{x}_0 + \mathbf{d}_{i\to0}(\mathbf{x}_0)$. As such, a set of N tetrahedral meshes representing the instantaneous position of the LA on each available CT volume were obtained. Image registration and the sampling of the displacement field were all performed in Elastix [7]. All parameters were tuned to optimize tracking quality, as assessed visually.

Temporal Interpolation - A continuous interpolation of the discrete mesh node positions was used to increase temporal resolution. Hereto, Fourier series interpolation was used, in agreement with the physiological periodicity of the heart motion. From a computational perspective, the continuous representation allowed to sample the model at a proper time step to ensure stability of the CFD solver. Moreover, thanks to the periodicity, an arbitrary number of cardiac cycles could be simulated, which was necessary in order to remove influence from the unphysiological initial condition on the fluid velocity.

2.3   The Computational Model

The model consisted in a fluid governed by the incompressible Navier-Stokes equations written in the Arbitrary Lagrangian Eulerian (ALE) frame of reference [12]. This conveniently splits the problem into two coupled sub-problems, namely the fluid problem, describing the fluid-dynamics, and the geometry problem, which attains to the motion of the computational domain. The latter determines the displacement of the fluid domain which, in turn, defines the ALE map.

For spatial discretization of the model, we consider a Galerkin finite element approximation using $\mathbb{P}1 - \mathbb{P}1$ Lagrange polynomials with a suitable VMS-SUPG stabilization scheme [5,10] while for temporal discretization, we applied a second order semi-implicit backward differentiation formula scheme [4,5].

Inflow boundary conditions at each PV were assigned as follows. First, we considered a representative MV flow rate Q_0. The latter was computed as as $Q_O = v_{MV}A_{MV}$, being v_{MV} mean velocity profile v_{MV} (cf. Sect. 2.1) and A_{MV} the MV cross sectional area.

The flow rate at each PV was then obtained by mass balance: $Q_1^{pv} + Q_2^{pv} + Q_3^{pv} + Q_4^{pv} + Q^O + Q_{wall} = 0$, where $Q_i^{pv}, i = \{1, 2, 3, 4\}$ are the flow rates at each PV and Q^O is the desired MV flow rate. Q_{wall} is the flux associated to the LA volume variation, i.e. $Q_{wall} = dV/dt$, being V the instantaneous LA volume.

Having defined $Q_{tot}^{pv} = Q_1^{pv} + Q_2^{pv} + Q_3^{pv} + Q_4^{pv}$, the individual PV flow rates were then computed as: $Q_l^{pv} = (A_l/A_t) Q_{tot}^{pv} - Q_l^w$, where A_l is the sectional area and A_t is the sum of all PVs sectional areas. Q_l^w represents the flow due to the mesh velocity for each PV considering that the PVs sections move and their area changes during the cardiac cycle. Moreover, we chose to impose a parabolic velocity profile at each PV. Finally, we applied natural boundary conditions to penalize backflow at the MV [1]. The CFD simulations were performed through the LifeV library[1] (block C of Fig. 1) in a parallel setting.

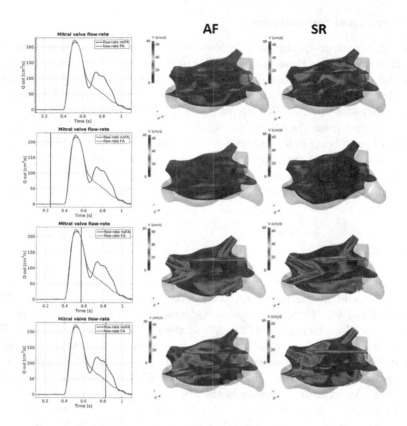

Fig. 3. Computed LA blood flow in one patient. First column reports the computed flow rate at the MV cut plane in SR (blue) and AF (red). Each row correspond to a different time in the cardiac cycle, denoted by the vertical black marker: end of diastole; LV systole; LV filling (E-wave); atrial systole (A-wave). The computed flow velocity fields are reported in the second (AF) and third (SR) column. (Color figure online)

Table 1. Number of particles which remained in the LAA at the end of each cardiac cycle for the two considered patients in SR and AF. The percentage indicates the current number of particle with respect to the initial umber. Cardiac cycle = 0 indicates the beginning of the simulation.

Cardiac cycle	Patient 1		Patient 2	
	SR	AF	SR	AF
0	500	500	500	500
1	500	500	500	499
2	478	480	455	465
3	130 (26%)	228 (45%)	195 (39 %)	251 (50.2 %)

2.4 Numerical Simulations

We ran two patient-specific simulations from two sets of patient data. Each simulation was run for three cardiac cycles to remove influence from the unphysical initial conditions on the fluid velocities. The results on the 3rd simulated cycle will be therefore reported. For each patient, we run one simulation in sinus rhythm (SR) and one in AF. Hereto, SR employed the patient specific motion model extracted from the CT volumes, as described previously. AF was instead simulated by applying independently to each node a random displacement, that was kept small ($\approx$0.01 mm) to avoid numerical issues arising from an excessive worsening of the mesh quality. Moreover, in AF, we redefined the inflow boundary conditions by considering a representative MV flow rate by removing the A-wave.

3   Results

The simulated LA blood velocity is displayed in Fig. 3 for the first patient. We note that at end of diastole (first row) zero velocity is measured at MV, as expected, and PV's. During LV systole (second row) MV velocity is zero. However, we noticed an increment to 20 cm/s at the four PVs (20 cm/s) in SR due to atrial diastole. Due to the random motion model, this was not the case in AF, where PV velocities remained mostly null. During LV filling (third row), the MV flow increased (E-wave) to a peak value of about 65/70 cm/s in both SR and AF simulations. An expected increase of the PV velocity was also confirmed by the simulation results (mean value 40 cm/s). During atrial systole (fourth row) we saw the expected A-wave at the MV (peak velocity between 35 and 40 cm/s) while the A-wave was missing in AF. Besides peak and mean values, the model allows to access the full complexity of the LA blood flow, cf. Fig. 3. In particular, detailed flow features, such as small vortexes, were represented by the model even if they are not discussed here due to the lack of space.

Blood stasis in the LAA might result in the formation of blood clots [6]. As a measure of blood stasis, we populated the LAA with 500 point particles, and counted how many remained inside the LAA after each cardiac cycle. The results for the two patients are reported in Table 1. After 3 cycles, 26% of the particles remained in the LAA in SR, while 45.6% remained in AF for patient 1. For patient 2, 39% remained in SR and 50% in AF. In both cases, although more incisively for patient 1, AF implied an expected reduced washout of the LAA which in the long term might be indicative of the generation of blood clots.

4   Conclusions

We presented what to our knowledge represents the first patient-specific model of atrial hemodynamics in AF. An initial evaluation of the model was also presented to study the effect of an AF episode on the intra-atrial blood flow.

In this initial testing, the model returned realistic blood flow patterns both at SR and during AF. The model confirmed that AF episodes result in a reduced washout of the LAA which can lead to the formation of thrombi.

One limitation is that the model was not benchmarked against experimental flow measurements. Moreover, a comparison against a healthy atrium is necessary to understand the hemodynamic implications of AF better. Finally, an evaluation on more patients including a diverse range of anatomies and blood flow profiles has to be performed.

This is our initial effort towards the development of a better tool for stroke risk assessment and therapy delivery in the atrial fibrillation patient.

Acknowledgment. This project has received funding from the European Union's Horizon 2020 research and innovation programme under the Marie Sklodowska-Curie grant agreement No 659082.

References

1. Bazilevs, Y., Gohean, J., Hughes, T., Moser, R., Zhang, Y.: Patient-specific isogeometric fluid-structure interaction analysis of thoracic aortic blood flow due to implantation of the jarvik 2000 left ventricular assist device. Comput. Methods Appl. Mech. Eng. **198**(45), 3534–3550 (2009)
2. Camm, A.J., Kirchhof, P., Lip, G.Y., Schotten, U., Savelieva, I., Ernst, S., et al.: Guidelines for the management of atrial fibrillation. Eur. Heart J. **31**(19), 2369–2429 (2010)
3. Fluckiger, J.U., Goldberger, J.J., Lee, D.C., Ng, J., Lee, R., Goyal, A., Markl, M.: Left atrial flow velocity distribution and flow coherence using four-dimensional FLOW MRI: A pilot study investigating the impact of age and Pre- and Postintervention atrial fibrillation on atrial hemodynamics. J. Magn. Reson. Imaging **38**(3), 580–587 (2013)
4. Forti, D.: Parallel algorithms for the solution of large-scale fluid-structure interaction problems in hemodynamics. Ph.D. thesis, École Politechnique Fédérale de Lausanne (2016)
5. Forti, D., Dedè, L.: Semi-implicit BDF time discretization of the Navier-Stokes equations with VMS-LES modeling in a high performance computing framework. Comput. Fluids **117**, 168–182 (2015)
6. Gupta, D.K., Shah, A.M., Giugliano, R.P., Ruff, C.T., et al.: Left atrial structure and function in atrial fibrillation: Engage af-timi 48. Eur. Heart J. **35**(22), 1457–1465 (2014)
7. Klein, S., Staring, M., Murphy, K., Viergever, M.A., Pluim, J.P.W.: elastix: A toolbox for intensity-based medical image registration. IEEE Trans. Med. Imaging **29**(1), 196–205 (2010)
8. Koizumi, R., Funamoto, K., Hayase, T., Kanke, Y., Shibata, M., Shiraishi, Y., Yambe, T.: Numerical analysis of hemodynamic changes in the left atrium due to atrial fibrillation. J. Biomech. **48**(3), 1–7 (2014)
9. Osher, S., Sethian, J.A.: Fronts propagating with curvature-dependent speed: Algorithms based on hamilton-jacobi formulations. J. Comput. Phys. **79**(1), 12–49 (1988)
10. Quarteroni, A.: Numerical Models for Differential Problems, vol. 2. Springer, Heidelberg (2010)

11. Quarteroni, A., Lassila, T., Rossi, S., Ruiz-Baier, R.: Integrated heart: Coupling multiscale and multiphysics models for the simulation of the cardiac function. Comput. Methods Appl. Mech. Eng. **314**, 345–407 (2017). special Issue on Biological Systems Dedicated to William S. Klug
12. Reymond, P., Crosetto, P., Deparis, S., Quarteroni, A., Stergiopulos, N.: Physiological simulation of blood flow in the aorta: Comparison of hemodynamic indices as predicted by 3-D FSI, 3-D rigid wall and 1-D models. Med. Eng. Phys. **35**(6), 784–791 (2013)
13. Rueckert, D., Sonoda, L.I., Hayes, C., Hill, D.L.G., Leach, M.O., Hawkes, D.J.: Nonrigid registration using free-form deformations: Application to breast MR images. IEEE Trans. Med. Imaging **18**(8), 712–721 (1999)
14. Vedula, V., George, R., Younes, L., Mittal, R.: Hemodynamics in the left atrium and its effect on ventricular flow patterns. J. Biomech. Eng. **137**, 8 (2015)
15. Vollmer, J., Mencl, R., Mller, H.: Improved laplacian smoothing of noisy surface meshes. Comput. Graph. Forum **18**(3), 131–138 (1999)
16. Zhang, L.T., Gay, M.: Characterizing left atrial appendage functions in sinus rhythm and atrial fibrillation using computational models. J. Biomech. **41**(11), 2515–2523 (2008)

Assessment of Atrioventricular Valve Regurgitation Using Biomechanical Cardiac Modeling

R. Chabiniok[1,2,3]($\boxtimes$), P. Moireau[1,2], C. Kiesewetter[3], T. Hussain[4],
Reza Razavi[3], and D. Chapelle[1,2]

[1] Inria, Université Paris-Saclay, Palaiseau, France
radomir.chabiniok@inria.fr
[2] LMS, Ecole Polytechnique, CNRS, Université Paris-Saclay, Paris, France
[3] Division of Imaging Sciences and Biomedical Engineering, St Thomas' Hospital,
King's College London, London, UK
[4] Department of Pediatrics, UT Southwestern Medical Center, Dallas, TX, USA

Abstract. In this work we introduce the modeling of atrioventricular valve regurgitation in a spatially reduced order biomechanical heart model. The model can be fast calibrated using non-invasive data of cardiac magnetic resonance imaging and provides an objective measure of contractile properties of the myocardium in the volume overloaded ventricle, for which the real systolic function may be masked by the significant level of the atrioventricular valve regurgitation. After demonstrating such diagnostic capabilities, we show the potential of modeling to address some clinical questions concerning possible therapeutic interventions for specific patients. The fast running of the model allows targeting specific questions of referring clinicians in a clinically acceptable time.

Keywords: Cardiac modeling · Reduced-order model · Atrioventricular regurgitation · Model-based diagnosis assistance · Therapy planning

1 Introduction

Mitral or tricuspid regurgitation (MR, TR) also named mitral (tricuspid) valve insufficiency or incompetent mitral (tricuspid) valve, is a condition in which the atrioventricular (AV) valve allows backward flow (regurgitation) during systole due to anatomical or functional defects. The AV valve regurgitation (AV-R) is typically quantified by imaging techniques – echocardiography or magnetic resonance imaging (MRI) – with classification into mild (regurgitation fraction RF being below 30%, and regurgitation volume RVol < 30 ml), moderate (RF 30–40% and RVol 30–45 ml), moderate-to-severe (RF 40–50%, RVol 45–60 ml) and severe (RF above 50% or RVol above 60 ml). The ventricular volume overload due to a moderate or severe AV valve insufficiency makes the assessment of the true systolic function of ventricle rather difficult. The indicators practically considered are the ventricular ejection fraction (EF), ventricular end-diastolic

© Springer International Publishing AG 2017
M. Pop and G.A. Wright (Eds.): FIMH 2017, LNCS 10263, pp. 401–411, 2017.
DOI: 10.1007/978-3-319-59448-4_38

volume (EDV) indexed to body surface area (iEDV) representing the level of ventricular dilatation, the percentage of RF, the regurgitation volume RVol and the morphology of the leaking valve. Putting all these available data into a clear clinical picture is not obvious and is one of the potentials of biomechanical modeling.

Capturing the anatomy of valve leaflets coupled with a realistic 3D heart model is mathematically extremely challenging and computationally intensive [1], with a limited potential to be used at bedside. The role of such detailed models would rather be in testing new valve implants, and the intensive computations would be run in the prosthetic valve development. Our goal in this paper is to present a relatively easy approach of incorporating the AV-R into a biomechanical heart model with the capability to provide a fast quantification of the actual contractile property of the ventricle, which would allow to objectively assess the level of overload the patient's heart is facing. The abnormal contractility might play a very significant role in the mismatch between the clinical assessment of a valve insufficiency and the actual outcome of a therapeutic intervention on the valve [7] – a phenomenon not fully understood to date. We aim at providing the clinical information in a relatively short time period, typically within hours – the time scale compatible with data analysis and reporting for an MRI exam of heart disease patients – so that the added indicators could be directly taken into consideration in clinical conclusions. Secondly, once the model is set up using patient-specific data, it can be used to predict the effect of possible valve repair or other therapeutic interventions – an information of invaluable importance prior to deciding about the type of therapy.

The valve insufficiency has already been introduced in the cardiac modeling community as for instance for the pulmonary valve. In [6] a 3D mechanical heart model was constrained by external forces from processed image data to allow regurgitation flows. The model was used to estimate the constitutive properties of myocardium and a possible outcome of a complete valve repair was predicted. In our case, the global mechanical model – including the component of AV-valve insufficiency – that we are proposing allows to obtain more comprehensive predictions. Examples thereof include the investigation of the effect of a partial improvement by invasive or non-invasive techniques – as is often the case in correcting the AV valve insufficiency – or modifications of the physiological properties of the cardiovascular system by a number of interventions, pharmacological or others. We will exemplify this variety of predictions with two clinical cases and several associated treatment scenarios.

2   Methods

2.1   Clinical Data

Two AV-R patients and one healthy control were analyzed in this pilot study. The data used were the non-invasively acquired magnetic resonance imaging data (MRI), together with peripheral pressure cuff measurement. MRI data contained cine images in short axis (the stack covering whole ventricles and part

Table 1. Patients' information obtained during cardiac MRI exam.

Patient #	Age (years)	EDV (ml)	iEDV (ml/m²)	EF (%)	RF (%)	RVol (ml)	type of defect
Healthy control	30	130	70	55	0	0	N/A
1	17	246	119	59	36	45	MR
2	17	308	169	55	26	43	TR

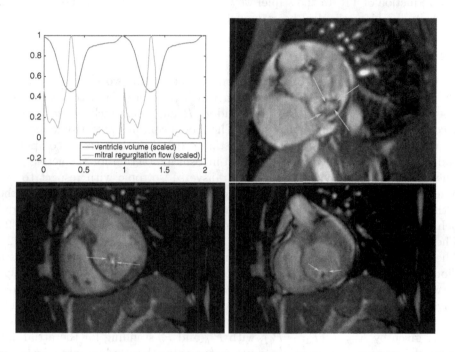

Fig. 1. Top: Mitral valve defect in Patient 1 with holosystolic regurgitant flow through the defect (the opening at end-systole is marked by arrows). Bottom: Tricuspid valve defect in Patient 2 present during early systole (left) and closing up during late systole (right).

of the atria) and stacks of long axis orientations (typically 5–9 slices for 2-, 3- and 4-chamber views) in order to well visualize the AV-valve defect in several orientations throughout the cycle, and phase contrast images of flow through aortic and pulmonary valves (the outflows of the ventricles).

The first patient has a moderate-to-severe MR with RF of 36% and RVol 45 ml, see Table 1. MR was caused by an anatomical defect in the posterior valve leaflet leading to MR throughout the systole, which can be appreciated in the top line of Fig. 1. The second patient is a TR patient with dilated RV – which is connected to the systemic circulation and has outflow to the aorta (the so-called systemic RV) – and in addition has a complete atrioventricular block (i.e. the atria and ventricles beat independently with a very low ventricular rate of

~40 bpm). The defect in the tricuspid valve is central and disappears when the RV volume falls below 200 ml (due to a decrease in size of the tricuspid valve ring), as assessed from the detailed short and long axis cine MRI stacks (see bottom line in Fig. 1).

While for the first patient mitral valve surgery might be indicated, the clinicians' main question about the second patient is whether or not to implant a pacemaker/CRT to correct the AV-block, as increasing the heart rate might lead to a reduction of TR. In the sequel we will address these clinical questions with the help of a biomechanical heart model.

2.2   Biomechanical Heart Model

The biomechanical model of heart function used in this work was detailed in [2]. It is a reduced-order (0D) model, in which the ventricle of concern is represented by a sphere with the inner radius R and wall thickness d adjusted according to the patient's image data. The model contains anisotropic mechanics with Holzapfel-Ogden-type hyperelastic potential in the passive component and chemically controlled model of actin-myosin interaction for the active part. The constitutive behavior is taken as transverse isotropic with respect to the fiber direction, and the fibers are assumed to be isotropically distributed in the orthoradial plane through the wall thickness. Consequently, the resulting overall behavior is different in the radial vs. orthoradial directions, as detailed in [2]. The circulation system is represented by a 2-stage Windkessel model (see Fig. 2). The phases of the cardiac cycle and opening/closure of valves are handled via a diode system as is described in [9]. In particular the AV-R is incorporated via a two-direction diode which is controlled by the phase of cardiac cycle. Under physiological conditions, the value of the AV valve conductance is set to a high value K_{open} when the valve is open with the resulting atrium-to-ventricle flow being given by $K_{open} \cdot (P_{at} - P_v)$, with P_{at} and P_v standing for the atrial and ventricle pressures. During the closure of the valve the conductance is set to a very low value K_{close} as a penalty for the negligible flow $K_{close} \cdot (P_v - P_{at})$. By increasing the coefficient K_{close} the AV-R can be introduced and tuned to the observed level of RF.

The complete heart-circulation model – i.e. the 0D cavity with the connected Windkessel model – is calibrated sequentially. The order for calibrating each component is given by the available data so that the calibrated part of a given module is carried further into a more complete model. First, the geometry is adjusted according to the data (reference volume and wall thickness of the ventricle [5]). Taking the advantage of having the measured flows leaving the ventricles at hand, the parameters of Windkessel model (representing the circulation, we recall Fig. 2) are tuned by directly imposing the measured flow at the inlet and using the peak and minimal arterial pressure taken by cuff measurement. The passive parameters are adjusted according to the clinically predicted atrial pressure. Finally, the active contractility is calibrated in order to obtain the peak aortic flow velocity as in the data. The main mechanical parameters of the model are summarized in Table 2.

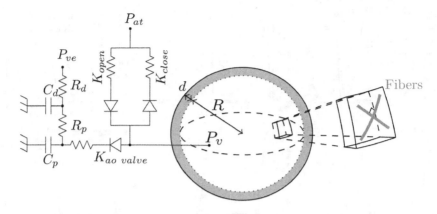

Fig. 2. Schematics of the model including the dual-direction diode for the AV-valve.

Concerning the calibration of the AV valve insufficiency, we kept the coefficient K_{close} constant for Patient 1, while for Patient 2 we made K_{close} linearly dependent on the volume of ventricle V_v with no regurgitation for the ventricular volume below the limit observed in the image data ($V_v^{competent} = 200$ ml). In detail,

$$K_{close} = \begin{cases} \alpha K_{close}^{incompetent\ valve} + (1-\alpha)K_{close}^{competent} & \text{when } V_v > V_v^{competent} \\ K_{close}^{competent} & \text{when } V_v \leq V_v^{competent}, \end{cases}$$

with linear interpolation factor $\alpha = \frac{V_v - V_v^{competent}}{V_v^{EDV} - V_v^{competent}}$.

3   Results

Figure 3 shows the level of accuracy obtained by our model representing the data in the measured key indicators. Table 1 summarizes data for the subjects in the study as obtained by a routine clinical analysis of the MR exam. Note that the ventricular size of Patient 1 is moderately increased (normal iEDV being below $100 \, \text{ml/m}^2$) and the ventricle of Patient 2 is severely dilated. According to the standard clinical assessment of systolic function, both patients fall into normal range (EF $\geq$ 55%). When comparing the level of ventricular contractility as estimated by the model (see Table 2), both patients have the contractility above the normal population (60–70 kPa, as exemplified by our healthy control case). The level of contractility in Patient 1 is mildly increased, and the contractility value of Patient 2 is already at moderately elevated inotropic level, even though the patient is at rest.

To better appreciate the patients' heart function, we plot in Fig. 4 the cardiac output as a function of the preload. We can see that in both cases the current working point is on a rather flat part of these "Starling curves", relatively independent of the current preload. Both patients have no signs of damaged

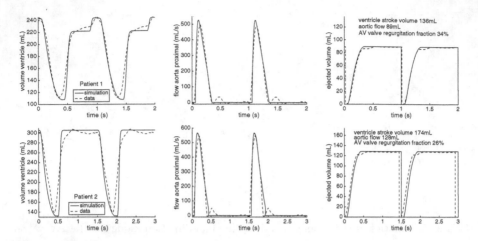

Fig. 3. Indicators of the calibrated models compared with the available data for Patient 1 (top row) and Patient 2 (bottom row).

Table 2. Main parameters of the model.

Patient #	Relat. passive stiffness	Contractility (kPa)	Periph. resistance ($Pa\ s/m^3$)	Periph. capacitance (m^3/Pa)
Healthy control	1.0	60	$1.55 \cdot 10^8$	$1.1 \cdot 10^{-8}$
1	1.13	92	$1.1 \cdot 10^8$	$1.5 \cdot 10^{-8}$
2	1.0	125	10^8	$2.2 \cdot 10^{-8}$

myocardium (in particular, no post-infarction scarring was present in gadolinium late-enhancement MRI), and we would therefore expect the level of contractility to be able to increase to a level of 180 kPa – in our experience the level obtained by using a moderate dose of inotropic drug in stress studies performed in clinically indicated cases [11]. The red cardiac output curve in Fig. 4 shows a prediction of such an inotropic effect. While the cardiac output of Patient 1 could significantly increase, the predicted cardiac output increase for Patient 2 is rather limited. This plot suggests that Patient 2 is more urgent to treat, as the contractile reserve of the patient's heart (which would be available to compensate for instance an acute destabilization of the patient's state) is limited, even though the level of RF is significantly less than for Patient 1.

Finally, Figs. 5 and 6 demonstrate the effect of therapies in question for the two patients. By correcting the mitral valve in Patient 1, while the inotropic level is kept at the same level, the model predicts that the heart will be working on smaller volumes (see blue curves in Fig. 5). This could be beneficial for the level of tissue stress and may induce a reverse remodeling of the dilated heart and bring the heart volumes from moderate dilatation to normal size. Alternatively, to reach the cardiac output as at baseline (5.3 l/min), the myocardial contractility could decrease by 15% – which would have positive effects on the heart energy requests.

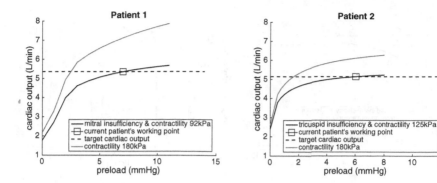

Fig. 4. Heart function presented as patient-specific cardiac output-vs-preload curves at the current physiological state (black solid line), and as predicted if the contractility was increased to a high level of 180 kPa (red line). (Color figure online)

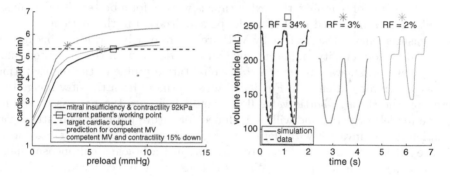

Fig. 5. Predicted effect of mitral valve repair in Patient 1: Patient's heart could work on the same level of contractility but on a lower preload (blue plot), or on a lower contractility level (red), assuming the repair would practically completely resolve the MR. (Color figure online)

For Patient 2, the model predicts a significant reduction of the regurgitation fraction when increasing the heart rate from 40 bpm to the normal value of 60 bpm after implanting a pacemaker, thanks to a reduction of the ventricular volume (see blue curves in Fig. 6). Even by reducing the the the contractility by 10% represented by red curves – which would definitely be of benefit for the patient, recall Fig. 4 – RF would still be around 14% (considered as only a trivial-to-mild level of TR).

4   Discussion and Future Work

A detailed visualization of the AV valve by a high number of cine slices in superior temporal resolution was key to adjust the AV-valve conductance in the central-TR defect, which was closing up for the smaller ventricular volumes. Although acquiring the additional ~20 slices in long axis orientations represents

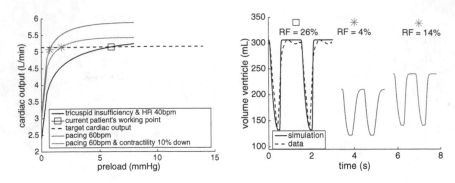

Fig. 6. Predicted effect of pacing on the tricuspid valve regurgitation fraction in Patient 2.

extra 10–20 min of scanning, this extra time is worthy for a better visual analysis and also for an increased level of model personalization in this patient.

Measurements of the AV flow by MRI are not routinely performed because of a limited reproducibility and accuracy. Our experience from the presented pilot study, however, shows that even a solely qualitative profile of the regurgitation jet (as shown for Patient 1 in Fig. 1, obtained actually from the flow image targeted on the through-aortic valve flow) would facilitate the process of setting up the model correctly, and will be included for future patients. Our study was completely non-invasive with data acquired during a routine congenital heart disease MRI exam. The obtained pressure was just a brachial cuff measurement and we tuned our model (i.e. the Winkessel resistances and capacitances R_p, R_d, C_p and C_d, see Fig. 2) to the corresponding two measured pressure points (minimum diastolic and maximum systolic) as if these points were present directly in the proximal aorta. We are aware of the difference between the central and peripheral pressure measurements, but the error in the range of 5–10 mmHg is acceptable for our purpose (as a comparison, in the cohort of patients followed, 10–20 % error in RF does not necessarily mean any difference in a standardized therapy). We do however plan to perform a sensitivity analysis with respect to the measured peripheral pressure for a group of patients with available central aortic pressure. The uncertainty of all parameters could be studied independently for each component of the model (Winkessel, passive and active properties), and remains to be done in future.

Our model suggests that correcting the MR in Patient 1 would be beneficial, and also directly gives some complementary therapeutic guides, e.g. the effect of decreasing contractility, or volume of circulating fluid which is adjusted by intrinsic physiological mechanisms of "pressure driven diuresis/natriuresis" [4] possibly supported by antidiuretic drugs. In this work, we have not dealt with the important effect of coupling the cardiac function with venous return as in [3,8], even though this would allow predicting using the antidiuretics, and we plan to extend our study also in this respect.

Our predictions for Patient 2 suggest a beneficial effect of pacing. We have to point out, however, that our model only captures the immediate effects and does not include possible negative effects caused by artificially fast rate. The validity of prediction of immediate effect of pacing, namely the changes of end-diastolic and stroke volume indicators, is a subject in our current clinical study in which the patients with MR conditional CRT devices are included. Instead of MRI, the echocardiography data processed into 0D signals of ventricular volumes and aortic flows could be used with the advantage of a possible validation on a much larger cohort by avoiding the contraindication of MRI in the majority of CRT patients.

Patient 2 has a congenital disease called congenitally-corrected transposition of great arteries (ccTGA), which is a very rare pathology. However, we believe that it is a group of patients who might extremely benefit from modeling. First, reduced order models – as the one used in the present work – can allow a quick assessment of the contractile properties of the remodeled systemic right ventricle – typically associated with some degree of TR – and predict the effect of possible pacing, as the AV block is commonly present in ccTGA. Secondly, the inter- and intra-ventricular delays (also common in ccTGA) may be directly treated by using a biventricular pacing (CRT) with a possibility of employing more detailed 3D models, which are able to capture the propagation of electrical activation throughout the myocardial tissue, with the aim to optimize the CRT [10]. Modeling can change the management of these patients since their early ages.

All the simulations performed in the present work rely on the assumptions of reduced order (0D) mechanical modeling [2] and therefore possess some limitations compared to 3D modeling. The 0D approach cannot rigorously capture the inhomogenities of material parameters, such as alterations of passive myocardial stiffness and active contractility in the infarcted heart. The effective spatial averaging of these parameters may lead to errors, e.g. false negativity if trying to discover an increased level of contractility at rest. Similarly, the propagation of electrical activation – often pathological in the AV-R patients with dilated ventricles – can be taken into account only by imposing the duration of propagation of depolarization wave according to the measured QRS duration in patients' ECGs. Local branch/partial blocks therefore cannot be directly included. We expect that out of these two limitations, the former one has a larger impact, and a fully 3D approach should therefore be considered in the infarcted hearts.

The fast run of 0D simulations, however, allows investigations of a number of virtual scenarios in a very short time. Indeed, one of the objectives in carrying out this work was to allow medical doctors to directly perform such simulations and hypothetical scenarios by themselves, in parallel with reporting on patients' MRI scans. Our experience from the presented work and other projects [8] suggests that the complexity of the model is such that this aim is realistic, indeed.

5 Conclusion

This paper shows a promising pathway in employing simplified models addressing the clinical questions for a given patient to extend the capabilities of imaging techniques. Although the pathology in question is related to the valves, the consequences on the myocardium itself are what really matters. Including modeling in clinical decision making has the potential to push further the understanding of the state of patients' hearts and predicting the effect of therapeutical interventions.

Acknowledgments. This work was supported by the National Institute for Health Research (NIHR) Healthcare Technology Co-operative for Cardiovascular Disease at Guy's and St Thomas' NHS Foundation Trust within the project entitled *"Quantification of regional myocardial contractility under stress conditions by using biophysical modelling and parameter estimation framework"*, and by NIHR Biomedical Research Centre at Guy's and St Thomas' NHS Foundation Trust in partnership with King's College London. The views expressed are those of the author(s) and not necessarily those of the NHS, the NIHR or the Department of Health.

References

1. Astorino, M., Gerbeau, J.-F., Pantz, O., Traoré, O.: Fluid-structure interaction and multi-body contact: Application to aortic valves. Comput. Methods Appl. Mech. Eng. **198**, 3603–3612 (2009)
2. Caruel, M., Chabiniok, R., Moireau, P., Lecarpentier, Y., Chapelle, D.: Dimensional reductions of a cardiac model for effective validation and calibration. Biomech. Model Mechanobiol. **13**(4), 897–914 (2014)
3. Chapelle, D., Felder, A., Chabiniok, R., Guellich, A., Deux, J.-F., Damy, T.: Patient-specific biomechanical modeling of cardiac amyloidosis – A case study. In: van Assen, H., Bovendeerd, P., Delhaas, T. (eds.) FIMH 2015. LNCS, vol. 9126, pp. 295–303. Springer, Cham (2015). doi:10.1007/978-3-319-20309-6_34
4. Guyton, A.C., Hall, J.E.: Textbook of Medical Physiology, 10th edn. Saunders, Philadelphia (2010)
5. Klotz, S., Hay, I., Dickstein, M.L., Yi, G.-H., Wang, J., Maurer, M.S., Kass, D.A., Burkhoff, D.: Single-beat estimation of end-diastolic pressure-volume relationship: A novel method with potential for noninvasive application. Am. J. Physiol. Heart Circ. Physiol. **291**, H403–H412 (2006)
6. Mansi, T., André, B., Lynch, M., Sermesant, M., Delingette, H., Boudjemline, Y., Ayache, N.: Virtual pulmonary valve replacement interventions with a personalised cardiac electromechanical model. In: Magnenat-Thalmann, N., Zhang, J.J., Feng, D.D. (eds.) Recent Advances in the 3D Physiological Human, pp. 201–210. Springer, Heidelberg (2009)
7. Ross Jr., J.: Afterload mismatch in aortic and mitral valve disease: Implications for surgical therapy. JACC **5**(4), 811–826 (1985)
8. Ruijsink, B., Moireau, P., Pushparajah, K., Wong, J., Hussain, T., Razavi, R., Chapelle, D., Chabiniok, R.: Biomechanical modeling of the dobutamine stress response in exercise induced failure of the Fontan circulation. Interface focus (2017). Submitted

9. Sainte-Marie, J., Chapelle, D., Cimrman, R., Sorine, M.: Modeling and estimation of the cardiac electromechanical activity. Comput. Struct. **84**, 1743–1759 (2006)
10. Sermesant, M., Chabiniok, R., Chinchapatnam, P., Mansi, T., Billet, F., Moireau, P., Peyrat, J.M., Wong, K., Relan, J., Rhode, K., Ginks, M., Lambiase, P., Delingette, H., Sorine, M., Rinaldi, C.A., Chapelle, D., Razavi, R., Ayache, N.: Patient-specific electromechanical models of the heart for the prediction of pacing acute effects in CRT: A preliminary clinical validation. Med. Image Anal. **16**(1), 201–215 (2012)
11. Wong, J., Pushparajah, K., de Vecchi, A., Ruijsink, B., Greil, G.F., Hussain, T., Razavi, R.: Pressure-volume loop-derived cardiac indices during dobutamine stress: A step towards understanding limitations in cardiac output in children with hypoplastic left heart syndrome. Int. J. Cardiol. **230**, 439–446 (2017)

In Silico Analysis of Haemodynamics in Patient-Specific Left Atria with Different Appendage Morphologies

Andy L. Olivares[1], Etelvino Silva[2], Marta Nuñez-Garcia[1], Constantine Butakoff[1],
Damián Sánchez-Quintana[3], Xavier Freixa[4], Jérôme Noailly[1], Tom de Potter[2],
and Oscar Camara[1(✉)]

[1] Department of Information and Communication Technologies, Universitat Pompeu Fabra,
Barcelona, Spain
oscar.camara@upf.edu
[2] Arrhythmia Unit, Department of Cardiology, Cardiovascular Center, Aalst, Belgium
[3] Department of Anatomy, University of Extremadura, Badajoz, Spain
[4] Department of Cardiology, Hospital Clinic de Barcelona, Universitat de Barcelona,
Barcelona, Spain

Abstract. The influence of the left atrial appendage (LAA) and its different possible morphologies in atrial haemodynamics and thrombus formation is not fully known yet. The main goal of this work is to analyse blood flow characteristics in relation with LA/LAA morphologies to better understand conditions that may lead to thrombus formation. We constructed several patient-specific computational meshes of left atrial geometries from medical imaging data. Subsequently, Computational Fluid Dynamics (CFD) methods were run with boundary conditions based on pressure and velocity measurements from literature. Relevant indices characterizing the simulated flows such as local maps of vorticity were related to simple LAA shape parameters. Our in silico study provided different 3D haemodynamics patterns dependent on the patient-specific atrial geometry. It also suggests that areas near the LAA ostium and with presence of lobes are more prone to coagulation due to the presence of low velocities and vortices.

Keywords: Left atrial appendage · Shape analysis · Haemodynamics · CFD · Vorticity · Thrombus formation

1 Introduction

Atrial fibrillation (AF) is the most common sustained heart rhythm disorder and is one of the main causes of cerebral strokes [1]. In patients with non-valvular atrial fibrillation (NVAF), around of 90% of thrombi leading to stroke are originated in the left atrial appendage [2], which is an interesting anatomical structure with large inter-subject morphological variability. Percutaneous left atrial appendage (LAA) closure is a promising approach for stroke prevention, which is recommended for patients with contra-indications to anticoagulant therapy. Due to the large inter-subject variability in LAA, an accurate morphological analysis of its geometry is crucial in selecting the appropriate device size during preoperative planning.

© Springer International Publishing AG 2017
M. Pop and G.A. Wright (Eds.): FIMH 2017, LNCS 10263, pp. 412–420, 2017.
DOI: 10.1007/978-3-319-59448-4_39

The left atrial appendage acts as a decompression chamber during left ventricular systole and in periods of high left atrial pressure [3], but its functioning and the influence its morphology are still unclear. Studies relating blood flow dynamics with LA configurations (e.g. number, location, size and orientation of pulmonary veins, volume) and LAA morphologies (e.g. chicken wing, cactus, cauliflower, windsock [4]) may shed some light on the relevance of this structure, his role in the cardiovascular system and the long-term effect of its occlusion.

Advanced imaging techniques such as Computational Tomography (CT) or 3D Rotational Angiography (3DRA) provide accurate 3D reconstructions of LAA geometries. Blood flow information is usually measured from Transesophageal Echocardiographic (TEE) images (including Doppler acquisitions) after LAA occlusion interventions to evaluate if the device is correctly positioned, but they only provide partial flow data on a plane or a given point of view. A more complete haemodynamics characterization is possible using novel 4D-flow Magnetic Resonance Imaging (MRI) [5, 6], but it is still in its infancy. On the other hand, Computational Fluid Dynamics (CFD) techniques can create 3D blood flow simulations to study haemodynamics patterns in different LA/LAA configurations. A few number of papers [7–10] can be found in the literature with CFD studies on LA. They are basically performed on synthetic geometries or in 1-2 patient-specific geometries from CT/MR imaging, but without jointly analysing haemodynamics and morphological indices to investigate the risk of thrombus formation.

The main objective of this study was evaluating haemodynamics parameters in different patient-specific LA/LAA morphologies to identify characteristics potentially related to high risk of thrombus formation. Four geometrical models of the LA, each one with a different type of LAA morphology, were built from patient-specific 3DRA data. Simple morphological parameters parameterizing the LAA geometry (e.g. size, ostium diameters, number of lobes, volumes, lengths) were jointly analysed with indices characterizing blood flow patterns obtained from CFD simulations (e.g. velocities, vorticity, streamlines, Reynolds number, oscillating shear index, relative residence time), providing a complete overview of each LA/LAA configuration.

2 Materials and Methods

2.1 Image Processing and Meshing

The computational pipeline for generating the set of CFD simulations started with the segmentation of patient-specific 3DRA images to obtain LA/LAA geometries. Left atrial meshes were obtained from imaging data of four AF patients (OLV Hospital, Aalst, Belgium). The ethical committee approved this study and informed consent was obtained from every patient. 3D Rotational Angiography images were acquired with an Innova 3D system (GE Healthcare, Chalfont St Giles, UK) and reconstructed with the scanner workstation, providing isotropic 3D images with 0.23 mm or 0.45 mm volumetric pixel size, for 512 or 256 pixels per dimension, respectively. Segmentation of the left atria was achieved with semi-automatic thresholding and region-growing algorithms available at the scanner console. From the resulting binary masks, surface meshes (with triangular elements) of the LAA were built with the classical Marching Cubes method. A Taubin filter [11] was applied to smooth the surface mesh while preserving the original volume. The catheter, always present in the reconstructed mesh (see Fig. 1), was manually removed using MeshLab[1]. Subsequently, inlet and outlet surfaces were added using MeshMixer[2]: four synthetic pulmonary veins (i.e. tubes) were added as inlets whereas the mitral valve area was defined as outlet. Location and physical dimensions of these surfaces were determined following measurements from the 3DRA images. From the triangulated surface, the tetrahedral mesh was created with Gmsh[3]. Figure 1 illustrates some intermediate results during the left atrial meshing pipeline.

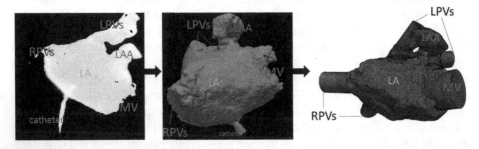

Fig. 1. Generation of left atrial meshes. Left: Segmentation of the left atria from 3DRA images. Middle: surface mesh after applying Marching cubes. Right: surface mesh after re-meshing, smoothing and adding appropriate inlet (PVs) and outlet (MV) surfaces. L: left; R: right; MV: mitral valve; PV: pulmonary vein. LAA: left atrial appendage.

2.2 Left Atrial Blood Flow Simulations

A second order implicit unsteady formulation was used for the solution of the momentum equations in the CFD simulations, in conjunction with a standard partial discretization for the pressure (under-relaxation factors of 0.3 and 0.7 for the pressure and momentum, respectively). Blood flow in the LA was modelled with the incompressible Navier–Stokes and continuity equations. Residuals of mass and momentum conservation equations lower than 0.001 were considered as absolute convergence criteria. Blood was modelled as an incompressible Newtonian fluid with density $\rho = 1060$ kg/m^3 [12]. The dynamic viscosity of blood in large vessels and the heart at normal physiological conditions was set to $\mu = 0.0035$ Pa-s [12]. Simulations were run using a laminar flow hypothesis under isothermal and non-gravitational effects. All walls were simulated rigid and with no-slip conditions, replicating the worst AF scenario (e.g. chronic AF), when atrial contraction is not possible anymore. The differential equations were solved using a time-step $\Delta t = 0.01$ s. At the inlets of the LA model (i.e. the four PVs) a time-varying blood flow function was applied (see Fig. 2a), following clinical observations from Fernandez-Perez et al. [13]. At the outlet (i.e. the mitral valve) a pressure of 8 mmHg [14] was imposed during the ventricular diastolic phase. In ventricular systole, the MV

---

[1] http://www.meshlab.net.

[2] http://www.meshmixer.com.

[3] http://gmsh.info/.

is closed; this was simulated as a wall boundary. The systolic and diastolic phase lasted 0.4 s and 0.65 s, respectively [13].

2.3   Geometrical Characterization of LA and LAA

Several indices were studied to characterise LAA shape (see Fig. 2b, c). They were inspired on measurements used by clinicians to determine the optimal LAA occluder dimensions and intervention planning: the maximum diameters (d1, d2), area and height of the ostium (equivalent to the landing zone concept); LAA depth was estimated as the length of a straight line (H0) from the middle of the LAA height (H) to the LAA apex (furthest point); and an index estimating the length of the LAA, $\tau = H/2 + H0$ (higher values of τ indicate longer blood flow pathways, from the ostium to the apex of the LAA).

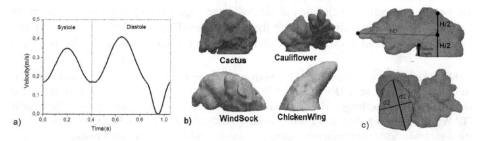

Fig. 2. (a) Generic blood velocity waveform applied to the pulmonary veins during ventricular systolic and diastolic phases. (b) the four typical LAA morphologies studied. (c) geometrical parameters of LAA, including its length and depth (H0 + H/2 and H, respectively) as well as ostium maximum diameters (d1 and d2).

2.4   Haemodynamics Indices Related to Thromboembolic Risk

The local Reynolds number was estimated from CFD simulations to determine the flow regime of blood towards the left ventricle. Mean flows and pressure waves were retrieved for a whole cardiac cycle at the inlets and outlets of the LA (e.g. PVs and MV) as well as in the LAA. LAA velocities < 0.55 m/s have been associated with a higher risk of stroke, with decreasing velocities of < 0.20 m/s being linked to the identification of thrombus within the LAA and a higher incidence of thromboembolic events [15]. Velocity streamlines of simulated flows were used to visually analyse the fluid dynamics profiles coming from left or right PVs. Vorticity maps, estimated from the second variant of the velocity gradient tensor (Q-criterion), were also computed to identify regions with high risk of thrombus formation. Thresholds on the Q-criterion (200-300 s^{-2}) were applied for vortex visualization purposes, similar to Otani et al. [10], to identify vortices associated to blood rouleaux or coagulation [3]. Paraview[4] was used for visualization and post-processing of CFD simulations.

---

[4] http://www.paraview.org.

3 Results and Discussion

Table 1 summarises the main geometrical and haemodynamics parameters characterising the four studied LA/LAA morphologies. LAA morphologies were classified into the usual four categories following the criteria of De Biase et al. [4]: case 1, cactus (CACT); case 2, cauliflower (CF); case 3, windsock (WS); and case 4, chicken wing (CW). Table 1 shows the variability of the studied morphologies, including large differences in LA volumes (from 131 to 220 cm^3), PV areas (from 14.9 to 29.1 cm^2), LAA ostium areas (estimated from diameters d1 and d2, from 3.13 to 6.15 cm^2), LAA volumes (from 7.7 to 14.8 cm^3), LAA landing zones (H from 1.27 to 2.35 cm), LAA lengths (from 3.35 to 4.45 cm) and number of lobes (from none to seven), among others. These measurements corresponded to the end diastolic phase, when the contrast in 3DRA acquisitions was filling the atria.

A simple verification of the obtained CFD simulations was achieved by comparing the peak of transmitral velocities in diastole (see Fig. 3a) with the ones provided by Fernandez-Perez [13], since we used the same inlet (PV) boundary conditions. Even though LA morphologies were quite different in both studies, transmitral velocities were

Table 1. LA/LAA geometrical and haemodynamics parameters. Volumes (V), areas (A), left and right pulmonary veins (LPV and RPV), mitral valve (MV), ostiumarea and depth (OsA and OsD), LAA depth and length (H, H0, τ), number of lobes (Nlobes), time-average wall shear stress (TAWSS), oscillating shear index (OSI), relative residence time (RRT), Reynolds number (Re). Bold values indicate the maximum/minimum (red/blue colours) among the 4 cases.

Geometrical Parameters	1	2	3	4	Haemodynamic Parameters	1	2	3	4
V_{LA} (cm^3)	131	147	220	157	E/A ratio (-)	2.05	2.43	2.34	2.33
V_{LAA} (cm^3)	13.9	8.7	7.7	14.8	Mean Blood flow Volume at the MV (cm^3)	72.3	101.0	108.0	85.8
LPV (cm^2)	1.53	2.08	2.52	1.49	TAWSS$_{LA}$ (Pa)	0.55	0.45	0.72	0.26
RPV (cm^2)	2.36	2.13	2.91	2.06	OSI$_{LA}$ (-)	0.35	0.31	0.27	0.36
MV (cm^2)	6.04	7.05	7.97	8.16	RRT$_{LA}$ (Pa^{-1})	5.9	5.7	3.1	14.0
OsA (cm^2)	3.77	3.13	4.96	6.15	TAWSS$_{LAA}$ (Pa)	0.12	0.06	0.05	0.03
H (cm)	2.35	1.80	1.27	1.95	OSI$_{LAA}$ (-)	0.35	0.33	0.31	0.23
H0 (cm)	2.71	3.16	2.72	3.58	RRT$_{LAA}$ (Pa^{-1})	26	51	53	56
τ (cm)	3.88	4.06	3.35	4.55	Re peak (-)	225	234	342	155
OsD (cm)	0.55	0.48	0.25	0.80	Case 1: Cactus (CACT); Case 2: Cauliflower (CF);			
Nlobes	3	7	3	0	Case 3: Windsock (WS); Case 4: Chicken Wing (CW).			

in the same range: 0.27–0.48 m/s and 0.33 m/s in our cases and Fernandez-Perez [13], respectively.

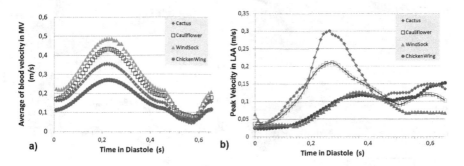

Fig. 3. Distribution of average blood velocity in the mitral valve (a) and the LAA ostium (b) during diastole.

The E/A ratio is a marker of diastolic performance of the heart, which is computed from the relation between early (E) to late (A) ventricular filling velocities. As expected, all E/A ratios estimated from CFD simulations are larger than two (E>>A), implying a restrictive and irreversible pattern in the LA [13]; in our case this is due to serious changes in LA elasticity and relaxation caused by AF. We also analysed peak diastolic (emptying) blood velocities in the LAA ostium (see Fig. 3b) since low values (<0.20 m/s) have been associated with the identification of thrombus [15]. In our experiments, case 1 (CACT, 0.30 m/s) and case 2 (CF, 0.21 m/s) presented a lower risk of thromboembolic events than case 3 (WS, 0.13 m/s) and case 4 (CW, 0.12 m/s), following the before mentioned criteria.

Figure 4 shows CFD-derived flow streamlines in two cases (case 1 and case 4, with cactus and chicken wing morphologies, respectively). It can be observed that blood flow enters the LAA mainly during systole, when the MV is closed. This shows the decompression role of the LAA in this cardiac phase, where higher pressures are found in the left atria. On the other hand, blood flow scarcely enters the LAA at the diastolic phase. Moreover, the LAA is preferentially filled in from flow coming from the LPV, while the RPV flow is mainly directed to the MV, in agreement with simulations from Vedula et al. [8].

Figure 5 displays the CFD-based vorticity maps of the four LA/LAA cases. Maps were thresholded within the range of 200–300 s^{-2} in the Q-criterion for visualization purposes, similarly to Otani et al. [10] (200 s^{-2} and 800 s^{-2} in their two cases); these maps are coloured with the magnitude of the blood flow velocities. Vortices were only created in the LAA during systole (closed MV) and highly depended on the underlying LA/LAA morphologies. Case 2, which had a Cauliflower LAA, was the one associated with more blood flow vortices and with a large number of secondary lobes. This is in agreement with the hypothesis relating the number of lobes with blood stasis and risk of thrombus formation. On the other hand, the Chicken Wing LAA (case 4) did not generated vortices, confirming clinical studies observing a lower risk of thrombosis in these morphologies. Furthermore, vortices with low blood flow velocities (dark blue in Fig. 5) were

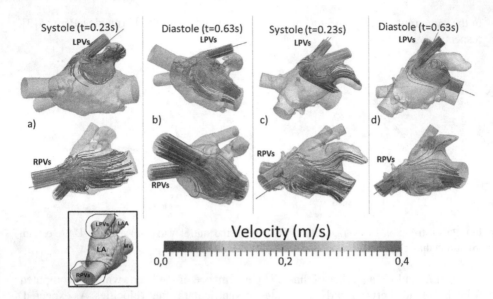

Fig. 4. Blood flow velocity streamlines with LPV (top) or RPV (bottom) origin at two time-points in systole and diastole. (a–b) Case 1 (cactus); (c–d) Case 4 (chicken wing).

consistently located around the LAA ostium or in small secondary lobes, potentially forming thrombus if blood flow is continuously slowed down in these areas.

The peak Reynolds (Re) numbers for the whole LA (see Table 1) remained in the blood laminar flow regime for the four cases, having a maximum and a minimum for the windsock (case 3) and chicken wing (case 4) morphologies, respectively. Case 3 had three secondary lobes and a very large LA (and PVs) compared with the other cases. This is probably caused by atrial remodelling due to AF, leading to an increased blood flow volume that may explain the largest Reynolds number. On the other hand, the chicken wing morphology (case 4) only had a single smooth main lobe, inducing low Re values.

Table 1 also summarises several indices related to CFD-derived wall shear stress such as the time-average wall shear stress (TAWSS), the oscillatory shear index (OSI) and the relative residence time (RRT), already used to assess the risk of thrombosis [9]. Case 4 (chicken wing) showed the lowest value for TAWSS in the LAA, which coincided with having only a main smooth lobe and the largest LAA volume. At the same time, this LAA morphology was associated with the largest RRT, implying the presence of areas where stagnation of blood flow is promoted. The largest value of the OSI index, associated to WSS vector deflection from blood flow predominant direction, was found in case 1; this was the cactus morphology, which had the most tortuous LAA, with its apex in the same plane than the ostium. Case 3 (windsock) also presented a relatively small TAWSS in the LAA even though it had the smallest LAA volume. One of the main characteristics of this case was its substantially larger LA and PV dimensions, leading to an increased blood flow volume (and higher TAWSS in the LA). This shows the relevance of analysing LAA haemodynamics considering the geometrical characteristics of the whole LA (and PV).

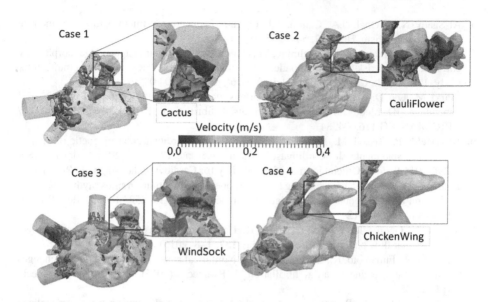

Fig. 5. Vorticity maps coloured by blood flow velocities in systole. Vortex values have been thresholded for visualization purposes (Q-criterion (200–300 s^{-2})). (Color figure online)

Case 3 will arguably be the hardest to implant a LAA occluder device due to the small landing zone area and an elevated presence of flow vortices.

4   Conclusions

Haemodynamics from CFD simulations and morphometric parameters from medical images were analysed in four patient-specific LA/LAA geometries corresponding to AF patients. A high number of secondary lobes and regions around the ostium were related with flow vortices and low velocities, which has been associated to high risk of thrombosis. Nevertheless, a deeper and more quantitative joint analysis of blood flow and geometrical indices is still required to better understand this process. Future work will focus on the validation of simulations with echocardiographic images and pressure measurements. Additionally, motion information will be included into the CFD simulations accounting for global and local mechanically derived changes (e.g. overall LA volume reduction and mitral valve ring displacement) in different phases of the cardiac cycle.

References

1. Wolf, P.A., Abbott, R.D., Kannel, W.B.: Original contributions atrial fibrillation as an independent risk factor for stroke: the framingham study. Stroke **22**, 983–988 (1991). doi: 10.1161/01.STR.22.8.983
2. Blackshear, J.L., Odell, J.A.: Appendage obliteration to reduce stroke in cardiac surgical patients with atrial fibrillation. Ann. Thorac. Surg. **61**, 755–759 (1996)

3. Al-Saady, N.M., Obel, O.A., Camm, A.J.: Left atrial appendage: structure, function, and role in thromboembolism. Heart **82**, 547–555 (1999)
4. Di Biase, L., Santangeli, P., Anselmino, M., et al.: Does the left atrial appendage morphology correlate with the risk of stroke in patients with atrial fibrillation? Results from a multicenter, study. J. Am. Coll. Cardiol. (2012). doi:10.1016/j.jacc.2012.04.032
5. Markl, M., Lee, D.C., Furiasse, N., et al.: Atrial structure and function left atrial and left atrial appendage 4D blood flow dynamics in atrial fibrillation, 1–10 (2016). doi:10.1161/ CIRCIMAGING.116.004984
6. Dyverfeldt, P., Bissell, M., Barker, A.J., et al.: 4D flow cardiovascular magnetic resonance consensus statement. J. Cardiovasc. Magn. Reson. **17**, 72 (2015). doi:10.1186/ s12968-015-0174-5. Official journal of the Society for Cardiovascular Magnetic Resonance
7. Zhang, L.T., Gay, M.: Characterizing left atrial appendage functions in sinus rhythm and atrial fibrillation using computational models. J. Biomech. **41**, 2515–2523 (2008). doi:10.1016/ j.jbiomech.2008.05.012
8. Vedula, V., George, R., Younes, L., Mittal, R.: Hemodynamics in the left atrium and its effect on ventricular flow patterns. J. Biomech. Eng. **137**, 1–8 (2015). doi:10.1115/1.4031487
9. Koizumi, R., Funamoto, K., Hayase, T., et al.: Numerical analysis of hemodynamic changes in the left atrium due to atrial fibrillation. J. Biomech. (2015). doi:10.1016/j.jbiomech. 2014.12.025
10. Otani, T., Al-Issa, A., Pourmorteza, A., et al.: A computational framework for personalized blood flow analysis in the human left atrium. Ann. Biomed. Eng. **44**, 3284–3294 (2016). doi: 10.1007/s10439-016-1590-x
11. Taubin, G.: Curve and surface smoothing without shrinkage. In: Proceedings of IEEE International Conference on Computer Vision, pp. 852–857 (1995). doi:10.1109/ICCV. 1995.466848
12. Ku, D.N.: Blood flow in arteries. Annu. Rev. Fluid Mech. **29**, 399–434 (1997). doi:10.1146/ annurev.fluid.29.1.399
13. Fernandez-Perez, G.C., Duarte, R., Corral de la Calle, M., et al.: Analysis of left ventricular diastolic function using magnetic resonance imaging. Radiologia **54**, 295–305 (2012). doi: 10.1016/j.rx.2011.09.018
14. Nagueh, S.F., Appleton, C.P., Gillebert, T.C., et al.: Recommendations for the evaluation of left ventricular diastolic function by echocardiography. Eur. J. Echocardiogr. **22**, 165–193 (2009). doi:10.1093/ejechocard/jep007
15. Beigel, R., Wunderlich, N.C., Ho, S.Y., et al.: The left atrial appendage: anatomy, function, and noninvasive evaluation. JACC: Cardiovascular Imaging (2014). doi:10.1016/j.jcmg. 2014.08.009

Identification of Transversely Isotropic Properties from Magnetic Resonance Elastography Using the Optimised Virtual Fields Method

Renee Miller[1,3]($\boxtimes$), Arunark Kolipaka[2], Martyn P. Nash[3,4],
and Alistair A. Young[1,3]

[1] Department of Anatomy and Medical Imaging, University of Auckland,
Auckland, New Zealand
renee.miller@auckland.ac.nz
[2] Department of Radiology, The Ohio State University Wexner Medical Center,
Columbus, USA
[3] Auckland Bioengineering Institute, University of Auckland,
Auckland, New Zealand
[4] Department of Engineering Science, University of Auckland,
Auckland, New Zealand

Abstract. Magnetic resonance elastography (MRE) has been used to estimate myocardial stiffness. However, inversion methods typically introduce unrealistic assumptions. The virtual fields method (VFM) has been proposed for estimating material stiffness from image data. This study applied the optimised VFM to identify transversely isotropic material properties from both simulated harmonic displacements in a left ventricular (LV) model with added Gaussian noise and isotropic phantom MRE data. Two material model formulations were implemented, estimating three and five material properties. In the LV model, mean estimated moduli were more accurate from the five-parameter estimation than the three-parameter estimation. In the isotropic phantom experiment, where the material was assigned an arbitrary fibre orientation, results accurately revealed an isotropic material ($G_{12} = G_{13}$) and estimated shear moduli were close to reference values. This preliminary investigation showed the feasibility and limitations of the VFM to identify transversely isotropic material properties from MRE.

Keywords: Magnetic resonance elastography · Transverse isotropy · Optimised virtual fields

1 Introduction

Myocardial stiffness is an important determinant of cardiac function, and significant increases in global stiffness are thought to be associated with diastolic heart failure [21]. However, myocardial stiffness is not widely measured clinically or in research studies due to the invasiveness of measurements. Therefore,

© Springer International Publishing AG 2017
M. Pop and G.A. Wright (Eds.): FIMH 2017, LNCS 10263, pp. 421–431, 2017.
DOI: 10.1007/978-3-319-59448-4_40

a non-invasive estimate of myocardial stiffness may be useful to characterise cardiovascular disease. Magnetic resonance elastography (MRE) has developed over the past two decades as a non-invasive method of estimating stiffness of biological tissue. Numerous inversion algorithms exist (e.g. manual, direct inversion of the Helmholtz wave equation, local frequency estimation, etc.) which, for the most part, assume that the tissue is isotropic and infinite. Myocardial stiffness, however, is anisotropic due to its fibrous architecture and can be modelled by a transversely isotropic (TI) material law with the greatest stiffness in the fibre direction.

In the case of TI, a linearly elastic material can be fully described by five independent parameters. However, all five parameters can be difficult to estimate from elastography displacements since three parameters depend on accurate estimation of the longitudinal (dilatational) wave speed (due to compressibility), which is on the order of 300 times greater than the shear (distortional) wave speed. Therefore, other studies have reduced the number of estimated parameters to either two (e.g. [3]) or three parameters (e.g. [16]), avoiding estimation of compressibility. Only one paper (to our knowledge) has estimated all five independent parameters from MRE displacements [15].

The virtual fields method (VFM), based on the variational formulation of the equilibrium equations, is an inverse method that has previously been applied to MRE displacement data to estimate isotropic shear moduli (e.g. [12,14]). An optimised VFM has also been implemented, that minimises the impact of Gaussian noise on the estimated shear modulus [4]. In this study, the optimised method was adapted for the estimation of TI material properties from simulated LV harmonic displacements with added Gaussian noise and from isotropic phantom MRE data. Two material model formulations were tested, estimating both five and three material parameters.

2   Methods

2.1   Left Ventricular Simulations

Simulations of steady-state harmonic motion were performed in an anatomically realistic canine LV geometry. Physiologically realistic helical fibers [10] measured from histology were embedded in the geometric finite element model by interpolating nodal parameters. A loading test was carried out where 64 different loading combinations of x, y and z prescribed displacement on two surfaces, one apical and one anterior, were investigated by simulating steady-state harmonic motion. The loading conditions which resulted in the best identification of TI properties were then used subsequently for further analyses with noise. For conciseness, the results from all 64 tests are not shown in this paper. The best resulting load combinations for the three-parameter estimation method was a harmonic displacement at 80 Hz prescribed to the anterior surface nodes in the [1, 1, 1] direction as well as an apical displacement applied in the [0, 1, 1] direction. For the five-parameter estimation method, the best results occurred with

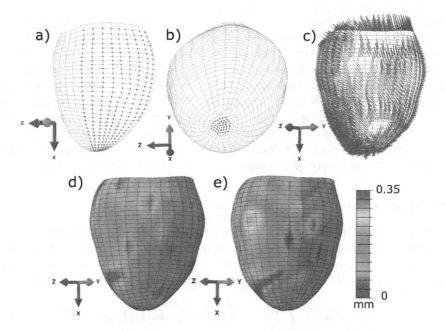

Fig. 1. LV finite element model illustrating (a) anterior surface nodes, (b) apical surface nodes, (c) fibre field measured from histology, (d) reference displacement field #1, anterior nodal displacement = [1, 1, 1], apical nodal displacement = [0, 1, 1] and (e) reference displacement field #2, anterior nodal displacement = [0, 0, 1], apical nodal displacement = [0, 1, 1].

a prescribed load on the anterior surface in the [0, 0, 1] direction and on the apical surface in the [0, 1, 1] direction (Fig. 1).

Reference stiffness values were defined based on cardiac anisotropic shear moduli measured from ultrasound elastography [6]. The fiber direction was assigned a Young's modulus (E_3) of 10.5 kPa, moduli in the transverse directions (E_1, E_2) were set to 6.5 kPa and the fiber shear moduli (G_{13}, G_{23}) were set to 2.5 kPa. The structural damping coefficient was set to 0.1; the Poisson's ratio was set to 0.4999 as cardiac tissue is largely incompressible; and a density of 1.06 g/cm^3 was assumed. Gaussian noise ($\sigma_{noise} = 15\% \cdot \sigma_{|disp|}$) was added to the displacements prior to material parameter estimation. A Monte-Carlo simulation was run (30 repeated simulations) with random Gaussian noise re-generated at each run.

2.2   Isotropic Phantom

Magnetic resonance elastography images of a PVC cylindrical gel phantom were obtained using a 3T MR scanner (Tim Trio, Siemens Healthcare, Erlangen, Germany) with gradients of 27 mT/m (2.7 G/cm) and a slew rate of 164 μs (TE/TR = 21.27/25 ms). A pneumatic driver system (Resoundant Inc.,

Rochester, MN) was used to apply a harmonic load to the bottom surface of the phantom at 60 Hz. Phase-contrast images (native resolution = 128 × 64 voxels, reconstructed resolution = 256 × 256 voxels, slice thickness = 5 mm, FOV = 250 mm × 250 mm) were collected at 16 longitudinal locations in the mid-region of the phantom. The cylindrical phantom had a radius of 76.2 mm and a height of 127 mm. At each location, 12 images were collected that encoded phase in three orthogonal directions at four phase offsets relative to the induced harmonic motion. A discrete Fourier transform was used to fit a sinusoid to the four phase offsets at each pixel in each direction. A finite element mesh was developed to represent the geometry of the imaged portion of the cylindrical phantom.

2.3 Transversely Isotropic Optimised VFM

The virtual fields method (VFM) utilises the principle of virtual work, which generally states that "a continuous body is at equilibrium if the virtual work of all forces acting on the body is null for any kinematically admissible virtual displacement." [13] The principle of virtual work is mathematically written as:

$$- \int_V \sigma : \epsilon^* dV + \int_S T \cdot u^* dS + \int_V b \cdot u^* dV = \int_V \rho a \cdot u^* dV. \qquad (1)$$

The test function (u^*), in this case a complex valued harmonic displacement field, was set to zero on the boundaries, eliminating the boundary traction term ($\int_S T \cdot u^* dS$). Body forces (b) were assumed to be negligible and the forcing frequency was assumed to be the same as the resulting displacement frequency. Acceleration (a) can be written in terms of the angular frequency (ω) and displacement (u). Thus, (1) was simplified to:

$$- \int_V \sigma : \epsilon^* dV = \int_V \rho \omega^2 u \cdot u^* dV, \qquad (2)$$

where σ is the internal stress, ϵ^* is the virtual strain field, ρ is the material density, ω is the loading frequency and u^* is the virtual displacement field.

In the first TI model formulation, all five independent parameters of the elasticity matrix, C_{11}, C_{33}, C_{44}, C_{66} and C_{13}, were estimated. The internal stress was written as a function of these five unknown parameters multiplied by the measured strains. Since five parameters were estimated, applying five independent virtual displacements fields (u^*) led to a set of five linear equations and five unknowns. Then, the optimised VFM [2] was implemented. By assuming an analytic model of noise, which is based solely on the virtual displacement field, the method obtained the optimal virtual field which minimised the variance in the estimated parameters due to Gaussian noise in the measured displacements. Then, the material constants: E_1, E_3, G_{12}, G_{13} and ν_{13}, were directly calculated from the inverse of the elasticity matrix.

In the second formulation, the TI model was rewritten in terms of four parameters: κ, G_{12}, G_{13} and τ [8].

$$C_{11} = \kappa + \frac{8}{9}G_{12} + \frac{4}{9}\tau \qquad C_{13} = \kappa + \frac{2}{9}G_{12} - \frac{8}{9}\tau$$
$$C_{33} = \kappa - \frac{4}{9}G_{12} + \frac{16}{9}\tau \qquad C_{12} = \kappa - \frac{10}{9}G_{12} + \frac{4}{9}\tau \qquad (3)$$
$$C_{44} = G_{12} \qquad\qquad C_{66} = G_{13},$$

where $\tau = G_{12} \cdot E_3/E_1$. This formulation lent itself to separating the longitudinal wave motion (dominated by κ) from the shear wave motion. Specialised conditions were applied to the numeric virtual fields to ensure that the term multiplied by κ was zero, thus, only three parameters were estimated: G_{12}, G_{13} and τ. Assuming incompressibility ($\nu = 0.5$), the Young's moduli were then calculated from the three estimated parameters.

The anisotropic optimised VFM required estimated material parameters to construct the minimisation matrix (which estimates variance of parameters due to Gaussian noise). Therefore, the method was iterative, during which, the parameters were updated on each calculation of the virtual displacement fields. The steps were repeated until the relative change in estimated parameters was less than 0.1% between two consecutive iterations. The maximum number of iterations was set to 100.

3   Results

All results from the five- and three-parameter estimation methods applied to the LV simulated harmonic displacements and phantom MRE data are shown in Table 1.

3.1   Left Ventricular Simulations

Five parameters were estimated for the entire region of the LV from simulated harmonic displacement fields without and with added Gaussian noise for both loading cases. In the first loading case with Gaussian noise, 20 out of 30 Monte-Carlo simulations did not converge within 100 iterations. In the second loading case, 22 out of 30 simulations converged. Results of the remaining simulations are shown below in Fig. 2. In the three-parameter estimation of both loading cases, all Monte-Carlo simulations converged to a solution. The three-parameter estimation method consistently overestimated all parameters. In comparison, the mean parameter estimates of the simulations with noise resulting from the five-parameter method were more accurate on average. However, five-parameter estimates showed large variation compared with the three-parameter method.

Table 1. Results from five- and three-parameter optimised VFM estimation for the LV model and phantom MRE data.

	G_{12} (kPa)	G_{13} (kPa)	E_1 (kPa)	E_3 (kPa)
LV reference parameters	1.923	2.5	6.5	10.5
Phantom reference parameters	5.45	5.45	16.35	16.35
5 parameter estimation				
LV (Load case #1)	1.768	2.333	6.201	12.321
LV (Load case #1 + Noise)	1.752 ± 0.136	2.351 ± 0.115	6.337 ± 0.735	12.173 ± 1.167
LV (Load case #2)	1.937	2.439	6.629	10.502
LV (Load case #2 + Noise)	2.028 ± 0.242	2.528 ± 0.216	8.484 ± 3.553	11.597 ± 3.857
Phantom	5.388 ± 1.418	5.390 ± 0.183	14.684 ± 30.663	2.411 ± 27.260
3 parameter estimation				
LV (Load case #1)	2.135	2.741	7.249	11.991
LV (Load case #1 + Noise)	2.076 ± 0.058	2.694 ± 0.055	7.094 ± 0.189	12.176 ± 0.620
LV (Load case #2)	2.656	3.301	8.962	14.315
LV (Load case #2 + Noise)	2.573 ± 0.107	3.165 ± 0.076	8.724 ± 0.294	14.363 ± 0.641
Phantom	5.959 ± 1.685	5.341 ± 0.183	17.442 ± 3.308	18.079 ± 3.378

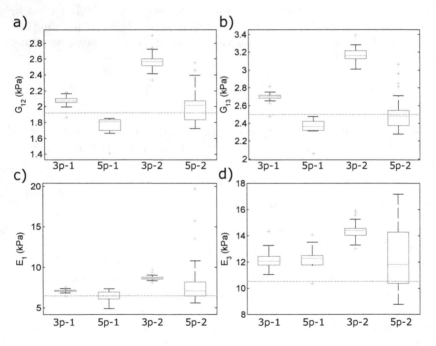

Fig. 2. Resulting estimated (a) transverse shear modulus (G_{12}), (b) fibre shear modulus (G_{13}), (c) transverse Young's modulus (E_1) and (d) fibre Young's modulus (E_3) for converged Monte-Carlo simulations from the five-parameter (5p) and three-parameter (3p) estimation methods for both loading cases (−1 and −2). Reference parameters are shown by red dotted lines. (Color figure online)

3.2   Isotropic Phantom

The five- and three-parameter inversion methods were applied to the isotropic phantom MRE data and the stiffness was estimated for 18 subzones. The mean shear moduli were close to those estimated by a finite element model update (FEMU: 5.55 kPa) and multi-modal direct inversion (MMDI: 5.45 kPa) method. Since the phantom is isotropic, an arbitrary material orientation was assigned in the [0, 0, 1] direction and the optimised VFM accurately estimated equivalent mean shear moduli (G_{12} and G_{13}). In the five-parameter estimation, the estimated Young's moduli (E_1 and E_3) showed wide variation, from -70 kPa to 80 kPa. In contrast, in the three-parameter estimation method, the calculated Young's moduli showed much smaller variation within physically realistic ranges. Estimated parameters for all phantom subzones estimated by each method are shown in Fig. 3.

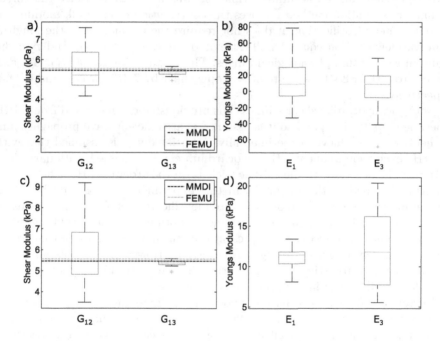

Fig. 3. Resulting estimated (a)/(c) shear moduli (G_{12} and G_{13}) and (b)/(d) Young's moduli (E_1 and E_3) for all subzones from the five-parameter (a and b) and three-parameter (c and d) estimation methods. The shear moduli measured by the MMDI and FEMU methods are shown by the black and red dotted lines, respectively in (a) and (c).

4   Discussion and Conclusions

The five-parameter optimised VFM did not converge to a solution in 20 noise cases and 8 noise cases for the first and second loading cases, respectively. In the three-parameter estimation method, all values of G_{12}, G_{13}, E_1 and E_3 were overestimated, but were more accurate in the first loading case. All Monte-Carlo simulations converged to a solution in the three-parameter method, showing that this method was more robust to noise than the five-parameter method. The results clearly show that the accurate identification of material parameters is dependent on the loading configuration and material law formulation.

Shear moduli of the phantom were accurately estimated from both the five- and three-parameter methods. However, with the current loading configuration and arbitrary material orientation, the Young's moduli estimated for each sub- zone varied greatly and were not identified accurately. In the five-parameter method, accurate estimation of Young's moduli depends on the measurement of the longitudinal (compressional) wavelength. The dimensions of the imaged region of the phantom were much smaller than the approximate longitudinal wavelength and thus an accurate estimation was not feasible. Since the three- parameter estimation method removes the contribution of the bulk modulus in the parameter identification, it does not require the estimation of the longitu- dinal wavelength. Consequently, all estimated values for G_{12}, G_{13} and τ in the phantom were within physiological ranges. The mean estimates of G_{12}, G_{13} and τ erred from the FEMU estimate of isotropic shear by 7.37%, 3.77% and 8.29%, respectively.

Isotropic myocardial shear stiffness has previously been measured from MRE experiments [7,9,11,19] and so it is assumed that sufficient wave propagation in the heart can be achieved experimentally. Additionally, it is assumed that with an 80 Hz excitation frequency, the myocardium can be assumed stationary since motion of the heart due to the cardiac cycle is on the order of 1 Hz. The loading conditions applied to the LV model in this experiment were chosen based on the loading which gives the most accurate identification of TI parameters using both the three-parameter and five-parameter optimised virtual fields method. The loading does not necessarily represent the true harmonic motion that would occur in the left ventricle during cardiac MRE. However, this was an initial test of feasibility to estimate TI properties in a realistic LV geometry with a physiological fibre field in the presence of noise.

Current methods for estimating myocardial stiffness in-vivo require invasive pressure measurements to estimate hyperelastic non-linear quasi-static material parameters by matching modelled inflation to the cardiac geometry measured from MR images given the pressure loading conditions [1,20]. MR elastography provides a non-invasive means of measuring linear dynamic cardiac tissue stiff- ness, eliminating the need for invasive LV pressure measurement. This study shows the feasibility of non-invasively estimating linearly elastic TI material properties from the harmonic displacement field in an LV model.

Two recent studies [17,18] estimated TI properties of tissue from MRE using simulated data. In one study [17], error in a parameter representing the anisotropic tensile ratio ($\zeta = E_3/E_1 - 1$) was typically at least 50% without added noise. Therefore, our study achieved improved identification of tensile parameters compared with previous MRE anisotropic inversion methods.

Overall, these initial results show the feasibility of estimating TI material properties from simulated harmonic displacements in the LV model as well as from MRE displacements measured from an isotropic phantom using the anisotropic optimised VFM. Unlike other inversion methods, the optimised VFM does not assume that the medium is isotropic and infinite. Additionally, the three-parameter method bypasses the need to estimate the longitudinal wavelength and appears to be more robust in reaching a converged solution in the presence of noise.

However, values estimated with the three-parameter method were consistently overestimated. At this point, the reason for the systematic error is not clear. Previous results from similar testing in a beam geometry (not published) did not exhibit this systematic overestimation. Additionally, it was not present in the phantom results. One possible cause may be the limited mesh resolution in the LV model. A previous study [5] reported that the accuracy of isotropic parameter estimation using the optimised VFM is dependent on the number of elements per wavelength. Eight elements per wavelength resulted in approximately 5% error in the estimation of an isotropic shear modulus. As can be visually estimated from the displacement field in Fig. 1e, in some places in the LV model, there are as few as four elements per wavelength. Additionally, as stated previously, any nodes on the boundary of the FE model were prescribed a virtual displacement of zero since traction forces were assumed to be unknown. Thus, in the LV model, there were only three "free" nodes in the transmural direction with which to estimate the wavelength, which may not be sufficient.

More work needs to be done investigating the effect of mesh resolution and loading configurations to gain a clearer understanding of their impact on estimated parameters in both the three-parameter and five-parameter material formulations. In the future, this method will also be applied to in-vivo cardiac MRE displacement data.

Acknowledgements. This research was supported by an award from the National Heart Foundation of New Zealand, American Heart Association 13SDG14690027, NHLBI R01HL124096 and The Royal Society of New Zealand Marsden Fund. The authors wish to acknowledge NeSI high performance computing facilities (https:// www.nesi.org.nz) for their support of this research.

References

1. Augenstein, K.F., Cowan, B.R., LeGrice, I.J., Young, A.A.: Estimation of cardiac hyperelastic material properties from MRI tissue tagging and diffusion tensor imaging. In: Larsen, R., Nielsen, M., Sporring, J. (eds.) MICCAI 2006. LNCS, vol. 4190, pp. 628–635. Springer, Heidelberg (2006). doi:10.1007/11866565_77
2. Avril, S., Grédiac, M., Pierron, F.: Sensitivity of the virtual fields method to noisy data. Comput. Mech. **34**(6), 439–452 (2004)
3. Chatelin, S., Charpentier, I., Corbin, N., Meylheuc, L., Vappou, J.: An automatic differentiation-based gradient method for inversion of the shear wave equation in magnetic resonance elastography: specific application in fibrous soft tissues. Phys. Med. Biol. **61**(13), 5000–5019 (2016)

4. Connesson, N., Clayton, E.H., Bayly, P.V., Pierron, F.: The effects of noise and spatial sampling on identification of material parameters by magnetic resonance elastography. Mech. Biol. Syst. Mater. **5**, 161–168 (2013)
5. Connesson, N., Clayton, E.H., Bayly, P.V., Pierron, F.: Extension of the optimised virtual fields method to estimate viscoelastic material parameters from 3D dynamic displacement fields. Strain **51**(2), 110–134 (2015)
6. Couade, M., Pernot, M., Messas, E., Bel, A., Ba, M., Hagege, A., Fink, M., Tanter, M.: In vivo quantitative mapping of myocardial stiffening and transmural anisotropy during the cardiac cycle. IEEE Trans. Med. Imag. **30**(2), 295–305 (2011)
7. Elgeti, T., Knebel, F., Hättasch, R., Hamm, B., Braun, J., Sack, I.: Shear-wave amplitudes measured with cardiac MR elastography for diagnosis of diastolic dysfunction. Radiology **271**(3), 681–687 (2014)
8. Feng, Y., Okamoto, R.J., Namani, R., Genin, G.M., Bayly, P.V.: Measurements of mechanical anisotropy in brain tissue and implications for transversely isotropic material models of white matter. J. Mech. Behav. Biomed. Mater. **23**, 117–132 (2013)
9. Kolipaka, A., McGee, K.P., Araoz, P.A., Glaser, K.J., Manduca, A., Ehman, R.L.: Evaluation of a rapid, multiphase MRE sequence in a heart-simulating phantom. Magn. Reson. Med. **62**(3), 691–698 (2009)
10. LeGrice, I.J., Smaill, B.H., Chai, L.Z., Edgar, S.G., Gavin, J.B., Hunter, P.J.: Laminar structure of the heart: ventricular myocyte arrangement and connective tissue architecture in the dog. Am. J. Physiol. **269**(2), H571–H582 (1995)
11. Mazumder, R., et al.: In vivo quantification of myocardial stiffness in hypertensive porcine hearts using MR elastography. J. Magn. Reson. Imaging **45**(3), 813–820 (2017)
12. Pierron, F., Bayly, P.V., Namani, R.: Application of the virtual fields method to magnetic resonance elastography data. In: Proul, T. (ed.) Application of Imaging Techniques to Mechanics of Materials and Structures. CPSEMS, vol. 4, pp. 135–142. Springer, New York (2013)
13. Pierron, F., Grediac, M.: The Virtual Fields Method: Extracting Constitutive Mechanical Parameters from Full-field Deformation Measurements. Springer, New York (2012)
14. Romano, A.J., Shirron, J.J., Bucaro, J.A.: On the noninvasive determination of material parameters from a knowledge of elastic displacements: theory and numerical simulation. IEEE Trans. Ultrason. Ferroelectr. Freq. Control **45**(3), 751–759 (1998)
15. Romano, A., Guo, J., Prokscha, T., Meyer, T., Hirsch, S., Braun, J., Sack, I., Scheel, M.: In vivo waveguide elastography: effects of neurodegeneration in patients with amyotrophic lateral sclerosis. Magn. Reson. Med. **72**(6), 1755–1761 (2014)
16. Schmidt, J., Tweten, D., Benegal, A., Walker, C., Portnoi, T., Okamoto, R., Garbow, J., Bayly, P.: Magnetic resonance elastography of slow and fast shear waves illuminates differences in shear and tensile moduli in anisotropic tissue. J. Biomech. **49**(7), 1042–1049 (2016)
17. Tweten, D.J., Okamoto, R.J., Bayly, P.V.: Requirements for accurate estimation of anisotropic material parameters by magnetic resonance elastography: a computational study. Magn. Reson. Med. (2017)
18. Tweten, D.J., Okamoto, R.J., Schmidt, J.L., Garbow, J.R., Bayly, P.V.: Estimation of material parameters from slow and fast shear waves in an incompressible, transversely isotropic material. J. Biomech. **48**, 4002–4009 (2015)

19. Wassenaar, P.A., Eleswarpu, C.N., Schroeder, S.A., Mo, X., Raterman, B.D., White, R.D., Kolipaka, A.: Measuring age-dependent myocardial stiffness across the cardiac cycle using MR elastography: a reproducibility study. Magn. Reson. Med. **75**, 1586–1593 (2015)
20. Xi, J., Lamata, P., Niederer, S., Land, S., Shi, W., Zhuang, X., Ourselin, S., Duckett, S.G., Shetty, A.K., Rinaldi, C.A., Rueckert, D., Razavi, R., Smith, N.P.: The estimation of patient-specific cardiac diastolic functions from clinical measurements. Med. Image Anal. **17**(2), 133–146 (2013)
21. Zile, M.R., Baicu, C.F., Gaasch, W.H.: Diastolic heart failure-abnormalities in active relaxation and passive stiffness of the left ventricle. N. Engl. J. Med. **350**(19), 1953–1959 (2004)

Longitudinal Parameter Estimation in 3D Electromechanical Models: Application to Cardiovascular Changes in Digestion

Roch Mollero[1]([✉]), Jakob A. Hauser[2], Xavier Pennec[1], Manasi Datar[3],
Hervé Delingette[1], Alexander Jones[2], Nicholas Ayache[1], Tobias Heimann[3],
and Maxime Sermesant[1]

[1] Université Côte d'Azur, Inria Sophia Antipolis, Asclepios Research Project,
Sophia Antipolis, France
rochmollero@hotmail.com
[2] Centre for Cardiovascular Imaging, Institute of Cardiovascular Science,
University College London, London, UK
[3] Imaging and Computer Vision, Siemens Corporate Technology,
Erlangen, Germany

Abstract. Computer models of the heart are of increasing interest for clinical applications due to their discriminative and predictive abilities. However the number of simulation parameters in these models can be high and expert knowledge is required to properly design studies involving these models, and analyse the results. In particular it is important to know how the parameters vary in various clinical or physiological settings. In this paper we build a data-driven model of cardiovascular parameter evolution during digestion, from a clinical study involving more than 80 patients. We first present a method for longitudinal parameter estimation in 3D cardiac models, which we apply to 21 patient-specific hearts geometries at two instants of the study, for 6 parameters (two fixed and four time-varying parameters). From these personalised hearts, we then extract and validate a law which links the changes of cardiac output and heart rate under constant arterial pressure to the evolution of these parameters, thus enabling the fast simulation of hearts during digestion for future patients.

1 Introduction

The main function of the heart is to create the necessary blood flow through the cardiovascular system, so that the oxygen supply of all the organs meets their needs. When an organ or a part of the body needs more energy (such as the muscles during exercise, or the digestive system during digestion), the heart rate and the blood flow increase because the overall demand in oxygen is higher.

The main changes in the cardiac function leading to an increase of the cardiac output are an increased heart rate, a decreased action potential duration and an increased contractility (positive inotropy). When the cardiac output increase is small (such as digestion or a mild exercice), the systolic pressure usually

© Springer International Publishing AG 2017
M. Pop and G.A. Wright (Eds.): FIMH 2017, LNCS 10263, pp. 432–440, 2017.
DOI: 10.1007/978-3-319-59448-4_41

increases but the diastolic pressure is constant, the latter being a consequence of the dilation of the arteries which lowers the arterial resistance [1]. Those qualitative changes are well-known, but are rarely quantified in the context of 3D cardiac electromechanical models, in part because most studies only involve personalisations on a single beat only (see [2] for a complete review).

A clinical study was performed in [3] to assess the cardiovascular response to a food stress protocol, involving the ingestion of a high-energy meal after fasting for 12 h. From the data of this study, we propose a consistent estimation of patient-specific 3D cardiac electromechanical models at two different instants of the protocol (pre-ingestion and t+1h). We first calibrate both the biomechanical parameters which are constant in time (such as the myocardial fibre directions) and time-varying (such as the arterial resistance) from the pre-ingestion measurements and heart motion extracted from the MRI. Then, we re-estimate values of the time-varying parameters (contractility and haemodynamics parameters) to reproduce changes in cardiac output and blood pressure at the second instant.

From these personalised simulations, we analyse the trends of the estimated parameters in relation to known physiological changes during mild exercise [4,5]. Finally, we build a law of evolution of the biomechanical parameters which leads to arbitrary changes of both the simulated cardiac output and stroke volume, while maintaining the same mean and diastolic pressure. The good accuracy of this law, which we validate with cross-validation over the 21 patients, then opens the door to the fast simulation of hearts during digestion in future patients.

2   Clinical Study and Data

More than 80 patients participated to a clinical study to assess the cardiovascular response after the ingestion of a high-energy (1635 kcal), high-fat (142 g) meal after fasting for 12 h, following the stress protocol in [3]. Informed consent was obtained from the subjects and the protocol was approved by the local Research Ethics Committee. An objective of the study was to analyze the evolution of blood flow toward the various organs of the body. In particular, a short axis cardiac cine MRI sequence was acquired before the ingestion, as well as measurements of the stroke volume, systolic, diastolic and mean cuff pressures at several time points within 1 h of the ingestion of the meal. Two instants are considered in particular: T_1 which is before the meal ingestion, and the latest measurement time T_2 around 1 h after ingestion, which also corresponds to the peak of the increased cardiac activity.

Overall (see Table 1), an increase of both the Heart Rate (HR) and the Cardiac Output (CO) of around 17% was observed. There were no significative changes in the values of the Systolic, Diastolic and Mean cuff pressure (SP, DP, MP) during the 1 h process of digestion (beyond the intra-patient variability of the measurements). Finally the Stroke Volume (SV) was constant on average but the measurement showed a high inter-patient variability of the evolution (11%).

Additionally, we tracked the boundaries of the endocardium over the entire cine MRI sequence acquired at T_1, then extracted from this sequence a point at

Table 1. Statistics of the measurements and their evolution Δ between T_1 and T_2 (in percentage of the value at T_1). Systolic, Diastolic and Mean cuff Pressure (SP, DP, MP), Stroke Volume (SV), Cardiac Output (CO) and Heart Rate (HR).

	SP (mmHg)	DP (mmHg)	MP (mmHg)	SV (mL)	CO (L/min)	HR (bpm)
Mean	117.13	60.95	84.16	92.11	6.17	67.65
Std	9.99	6.45	5.98	19.69	1.34	10.25
Mean Δ (%)	–	–	–	−0.10	**17.58**	**17.76**
Std. Δ (%)	–	–	–	**11.43**	17.74	13.19

the apex of the left ventricle and one at the top of the left ventricle septum. This was used to calculate the *Septal Shortening* (SS) as the maximal shortening of the distance between these two points during the cycle. It has an average value of −17% and a standard deviation of 3.7% across the population.

3   Patient-Specific Cardiac Modelling

3.1   3D Electromechanical Cardiac Model

We performed 3D cardiac modelling for 21 of these patients. A high-resolution biventricular tetrahedral mesh of the patient's heart morphology was extracted as in [6] from the pre-ingestion MRI at T_1, made of around 15 000 nodes. On this mesh, a *myocardial fibre direction* can be defined at each node of the mesh (see Fig. 1a), by varying the elevation angles of the fibre across the myocardial wall from α_1 on the epicardium to α_2 degrees on the endocardium. In this paper, α_2 is set at the default value of 90° and α_1 is a variable parameters in our experiments.

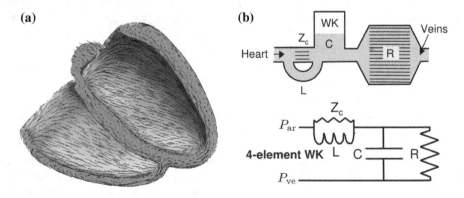

Fig. 1. (a): 3D heart geometry with myocardial fibres directions, (b): Schema and rheological model and of the Windkessel model (figure from [7])

The depolarization times across the myocardium were computed with the Multi-front Eikonal method [8]. The APD is set from the Heart Rate with classical values of the restitution curve and default values of conductivities are used as in [9]. Myocardial forces are computed based on the Bestel-Clement-Sorine model as detailed in [10]. It models the forces as the combination of an active contraction force in the direction of the fibre, in parallel with a passive anisotropic hyperelasticity driven by the Mooney-Rivlin strain energy. In this paper, we only consider two main parameters of the model: the *Maximal Contractility* σ and the *Passive Stiffness* c_1. Finally for the haemodynamics, the pressure in the cardiac chambers are described by global values, and the mechanical equations are coupled with a circulation model implementing the 4 phases of the cardiac cycle [11].

In particular the pressure of the aortic artery P_{ar} (cardiac after-load) is modeled with a 4-parameter Windkessel model [7], which describes the evolution of arterial blood pressure with the second-order equation of an electric circuit (see Fig. 1b). The blood inertia is modeled by the inductance L, the arterial compliance by a capacity C and the proximal and distal (peripheral) resistances respectively by a resistance Z_C and R (see Fig. 1b). Finally, the venous pressure P_{ve} models the mean pressure in the venous system. In the following, Z_C and L are fixed at a default value (see [11]) while C, R and P_{ve} are variable parameters.

3.2 Longitudinal Parameter Estimation

After building the heart mesh geometry, *parameter estimation* is the next step in order to have model simulations which reproduce the available data. Considering a set of simulated quantities called the "outputs" O (such as the Stroke Volume or the Mean Pressure for example), and a set of model parameters P, it consists in finding adequate values $\mathbf{x}$ of the parameters such that the output values $O(\mathbf{x})$ in the 3D model simulation fit the "target values" $\widehat{O}$ available in the data. This is done by performing an optimization of the parameter values $\mathbf{x}$ in order to minimize a distance $S(x, \widehat{O}) = ||O(x) - \widehat{O}||_S$ between the simulated values $O(\mathbf{x})$ and the target values $\widehat{O}$ (normalised to compare quantities with different units).

For each patient, we have here measurements of different *varying* quantities at the two instants T_1 and T_2 (such as the stroke volume and the heart rate), so we need to estimate different values for some cardiac model parameters (in particular the haemodynamic parameters) at these two instants. On the other hand, during the time-scale of the study (1 h on average), some parameters of the cardiac model can be considered *constant*. This is the case of the Epicardial Fibre Elevation Angle α_1 for example, or the cardiac stiffness c_1. In order to have consistent sets of estimated parameters at these two different instants, we need to use the same values for these parameters at these two instants.

To that end, we perform a two-step parameter estimation. First, we estimate values of both the fixed and varying parameters from the data at T_1. Then we reuse the estimated values of the fixed parameters for T_2 and estimate new values for the varying parameters only, from the data at T_2. As summarized in Table,

Table 2. Estimated parameters and target outputs in the parameter estimations at T_1 and T_2. Constant parameters whose values are reused for the estimation at T_2 are outlined in bold. The heart rate in the simulations for the estimation at T_1 (resp T_2) correspond to the measured value at T_1 (resp T_2).

Estimated parameters at T_1	Target outputs at T_1
Passive Stiffness c_1	**Septal Shortening**
Epicardial Fibre Elevation Angle α_1	Stroke volume at T_1
Maximal Contractility σ	Aortic Diastolic Pressure
Aortic Peripheral Resistance R	Aortic Mean Pressure
Aortic Compliance C	
Venous Pressure P_{ve}	
Estimated parameters at T_2	Target Outputs at T_2
Maximal Contractility σ	Stroke volume at T_2
Aortic peripheral Resistance R	Aortic Diastolic Pressure
Aortic Compliance C	Aortic Mean Pressure
Venous Pressure P_{ve}	

we then have two distinct *Parameter Estimation problems*: the estimation of 6 parameters values in order to fit 4 target output values at T_1 (with the heart rate of the simulations set to its value at T_1). Then the estimation of 4 parameters values in order to fit 3 target output values at T_2 (with the heart rate at T_2) (Table 2).

The optimisation was performed with an extended version of the framework described in [12]: the main algorithm is the CMA-ES genetic algorithm, which asks at each iteration for the score of a high number of 3D simulations. Instead of actually computing all these 3D simulations, we only compute a few within the parameter space ($2N + 1$ where N is the number of estimated parameters). Then we build a "low-fidelity" surrogate model [13] from these simulations which allows to approximate the outputs of the 3D simulations for many successive iterations of the algorithm, without performing all the 3D simulations. This robust and efficient "multifidelity optimization" allows a very fast exploration of large parameter sets with a low computational cost. In particular for the two problems at T_1 and T_2, we performed the optimization for the 21 patients simultaneously and the convergence was reached in around two days.

4   Exploitation of Estimated Parameters

4.1   Analysis of Parameter Trends in the Population

Across the 21 patients and the two estimations, the average *fit error* on the target ouput values are 1.9 mL for the Stroke Volume, 1% for the Septal Shortening, and 0.1 mmHg for both the mean and diastolic pressures, with few outliers. As a consequence of this step, we now have a population of 21 personalised patient hearts at two instants. For each parameter, we report in Table 3 the mean and

Table 3. Statistics of the estimated parameters and of the difference Δ between estimated parameters at T_1 and T_2

	c_1 (kPa)	α_1 (°)	σ (MPa)	P_{ve} (mmHg)	R (MPa.m^3.s)	C (MPa^{-1}.m^{-3})
Mean	$54.2e^1$	-58.7	82.6	48.3	47.4	$6.23e^{-3}$
Std	$27.7e^1$	2.94	34.0	12.9	17.2	$1.98e^{-3}$
Mean Δ	-	-	-1.52%	6.93%	$\mathbf{-14.2\%}$	-7.30%

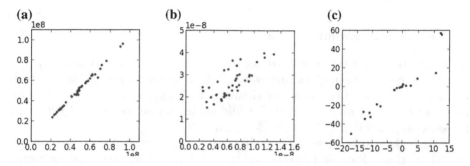

Fig. 2. (a): (MP-P_{ve})/CO as a function of R, (b): SV/(MP-DP) as a function of C, (c): $\Delta\sigma$ (%) as a function of ΔSV (%)

standard deviation of its estimated values at T_1 across the 21 patients, as well as the mean of its *evolution* Δ between the instants T_1 and T_2 (difference between the values estimated at T_2 and T_1).

The first remark is that on average, the parameters R which models the arterial peripheral resistance decreases by 14%. This was expected and corresponds to findings in [3]. In a clinical setting the peripheral resistance is indeed computed as the ratio between the blood flow and the blood pressure, and a similar relationship can be derived in the model: as shown in Fig. 2a the ratio (MP-P_{ve})/CO is almost exactly equal to the peripheral resistance R in our simulations. Across the population, since the cardiac output CO increases by around 17% but the pressures are constant, the peripheral resistance has to decrease by a close number (14.2% here) on average.

We then notice both an average increase of the venous pressure P_{ve} and decrease of the arterial compliance C. These two trends can be explained as to compensate the decrease of the resistance and avoid a drop in the mean blood pressure. Indeed, in the model, a decrease of R leads to a decrease of the "characteristic time" $\tau = RC$ at which the blood pressure decreases from the systolic pressure to the "asymptotic pressure" P_{ve}. A decrease of R only leads then to a decrease of the mean pressure. On the other hand, a decrease of C leads to an increase of the "pulse pressure" (difference between systolic and diastolic pressure) since C links an increase of arterial volume to an increase of arterial pressure with the formula $C\Delta P = \Delta V$ (a less compliant aorta has a higher pulse pressure for the same stroke volume). This contributes to the increase of the mean pressure (see Fig. 2b), and it is also the case of the increase of P_{ve}. Interestingly, we can note that these two trends (decrease of the arterial compliance and increase of venous pressure) in parameters correspond to actual cardiovascular phenomena which are commonly observed during exercise [4,5].

Finally, we can also observe a high correlation between changes in the Maximal Contractility σ and changes in the ejected volume, as shown in Fig. 2c. This is also a known phenomenon in cardiac dynamics, in particular at the core of the Starling Effect.

4.2   Parameter Evolution Law

From this data and the estimated parameters, we then build a law f which, from a given simulation, gives *variations of the electromechanical parameters* σ, P_{ve}, *R and C* which leads to a new simulation with *prescribed changes in heart period (HP) and stroke volume (SV)* while having **same mean and diastolic pressures**: $f(\Delta HP, \Delta SV) = (\Delta\sigma, \Delta P_{ve}, \Delta R, \Delta C)$

This is done by computing a multivariate regression between the changes (in %) in Heart Rate and Stroke Volume and the changes in the estimated parameters values at the two instants T_1 and T_2, for the 21 patients. We report in Table 4 the coefficients of this multivariate regression:

The predicted variations of parameters with the variations of the heart period (ΔHP) are consistent with the mean variations across the population described earlier. Interestingly with the coefficients of the second row (ΔSV), we can also note how the parameters have to change for an increase in Stroke Volume only with constant pressures.

We finally tested the accuracy of this law with a *leave-one-out* approach: for each patient, we computed the regression f from the data and estimated parameters of all the others patients. Then we changed the baseline parameters (at T_1) of this patient with the parameters predicted from f, and simulated the Pressure and Stroke Volume values at T_2. The obtained results were accurate: on average, the target Stroke Volume at T_2 was predicted within 1.9 mL and the mean absolute variations in Diastolic and Mean Pressure were within 2.1 mmHG, which is beyond the variability of both the intra-patient and population variabilities.

Table 4. Coefficient of the multivariate regression f

	$\Delta\sigma$	ΔP_{ve}	ΔR	ΔC
Δ HP	−0.02	−0.15	1.20	0.51
Δ SV	3.05	0.52	−1.04	1.19

5   Conclusion and Discussion

In this manuscript we performed a consistent longitudinal estimation of cardiac model parameters for 21 patient-specific hearts at two different instants within a 1 h time span, from clinical data. This was done through two successive parameter estimation problems: we first estimated 6 parameters to fit the simulated Stroke Volume, the Septal Shortening and the Mean and Diastolic Pressures to their values at the first instant. Then we reused the estimated values of the fixed parameters at this step and performed a second estimation of 4 parameters to fit values of Stroke Volume and Pressures at the second instant. This was done in parallel for the 21 hearts in around two days and a maximum of 150 simulations of the 3D model per patient.

From those personalised hearts, we identified relationships between the estimated parameters and the simulated pressure and volume outputs, and linked their evolution between these two instants to classical physiological phenomena. Then we extracted a law which computes changes of electromechanical parameters from changes of stroke volume and heart rate with constant pressure. This law allows in particular to easily simulate the changes observed between the two instants without having to perform the parameter estimation step at the second instant. This was evaluated in a leave-one-out test and showed that it can predict accurately changes in the model parameters.

A first direct continuation of this work would be to quantify (from further data) to what extent this law holds for changes of cardiac outputs which are more important (digestion can be seen as a 'mild' exercise and it is known for example that blood pressure rises during more intense exercises). Finally, for future patients, it could also be interesting to evaluate to what extent the changes in both the Stroke Volume and the Heart Rate can be predicted, and use our law to simulate the predicted heartbeats.

Ackowledgements. This work has been partially funded by the EU FP7-funded project MD-Paedigree (Grant Agreement 600932) and contributes to the objectives of the ERC advanced grant MedYMA (2011-291080).

References

1. Laughlin, M.H.: Cardiovascular response to exercise. Am. J. Physiol. **277**(6 Pt 2), S244–S259 (1999)
2. Chabiniok, R., et al.: Multiphysics and multiscale modelling, data-model fusion and integration of organ physiology in the clinic: ventricular cardiac mechanics. Interface Focus **6**(2), 20150083 (2016)
3. Hauser, J.A., et al.: Comprehensive assessment of the global and regional vascular responses to food ingestion in humans using novel rapid MRI. Am. J. Physiol. Regul. Integr. Comp. Physiol. **310**(6), R541–R545 (2016)
4. Otsuki, T., et al.: Contribution of systemic arterial compliance and systemic vascular resistance to effective arterial elastance changes during exercise in humans. Acta physiologica **188**(1), 15–20 (2006)

5. Albert, R.E., et al.: The response of the peripheral venous pressure to exercise in congestive heart failure. Am. Heart J. **43**(3), 395–400 (1952)
6. Molléro, R., et al.: Propagation of myocardial fibre architecture uncertainty on electromechanical model parameter estimation: a case study. In: van Assen, H., Bovendeerd, P., Delhaas, T. (eds.) FIMH 2015. LNCS, vol. 9126, pp. 448–456. Springer, Cham (2015). doi:10.1007/978-3-319-20309-6_51
7. Westerhof, N., et al.: The arterial windkessel. Med. Biol. Eng. Comput. **47**(2), 131–141 (2009)
8. Sermesant, M., Konukoğlu, E., Delingette, H., Coudière, Y., Chinchapatnam, P., Rhode, K.S., Razavi, R., Ayache, N.: An anisotropic multi-front fast marching method for real-time simulation of cardiac electrophysiology. In: Sachse, F.B., Seemann, G. (eds.) FIMH 2007. LNCS, vol. 4466, pp. 160–169. Springer, Heidelberg (2007). doi:10.1007/978-3-540-72907-5_17
9. Pernod, E., et al.: A multi-front eikonal model of cardiac electrophysiology for interactive simulation of radio-frequency ablation. Comput. Graph. **35**(2), 431–440 (2011)
10. Chapelle, D., et al.: Energy-preserving muscle tissue model: formulation and compatible discretizations. Int. J. Multiscale Comput. Eng. **10**(2), 189–211 (2012)
11. Marchesseau, S.: Simulation of patient-specific cardiac models for therapy planning. Thesis, Ecole Nationale Supérieure des Mines de Paris (2013)
12. Mollero, R., Pennec, X., Delingette, H., Ayache, N., Sermesant, M.: A multiscale cardiac model for fast personalisation and exploitation. In: Ourselin, S., Joskowicz, L., Sabuncu, M.R., Unal, G., Wells, W. (eds.) MICCAI 2016. LNCS, vol. 9902, pp. 174–182. Springer, Cham (2016). doi:10.1007/978-3-319-46726-9_21
13. Peherstorfer, B., et al.: Survey of multifidelity methods in uncertainty propagation, inference, and optimization (2016)

One Mesh to Rule Them All: Registration-Based Personalized Cardiac Flow Simulations

Alexandre This[1,2,3(✉)], Ludovic Boilevin-Kayl[2,3], Hernán G. Morales[1],
Odile Bonnefous[1], Pascal Allain[1], Miguel A. Fernández[2,3],
and Jean-Frédéric Gerbeau[2,3]

[1] Medisys, Philips Research, Suresnes, France
`alexandre.this@philips.com`
[2] Inria Paris, 75012 Paris, France
[3] Sorbonne Universités, UPMC Univ Paris 6,
UMR 7598 LJLL, 75005 Paris, France

Abstract. The simulation of cardiac blood flow using patient-specific geometries can help for the diagnosis and treatment of cardiac diseases. Current patient-specific cardiac flow simulations requires a significant amount of human expertise and time to pre-process image data and obtain a case ready for simulations. A new procedure is proposed to alleviate this pre-processing by registering a unique generic mesh on patient-specific cardiac segmentations and transferring appropriately the spatiotemporal dynamics of the ventricle. The method is applied on real patient data acquired from 3D ultrasound imaging. Both a healthy and a pathological conditions are simulated. The resulting simulations exhibited physiological flow behavior in cardiac cavities. The experiments confirm a significant reduction in pre-processing work.

Keywords: Patient-specific · Numerical simulation · Registration · Mitral regurgitation

1 Introduction

In order to reduce the deadly impact of cardiovascular diseases, efforts are made to improve diagnosis and treatments of patients. Numerical simulations of blood flow within the cardiac cavities provide a better understanding of the intraventricular hemodynamics and hence could be used to improve the diagnosis or predict the outcome of treatments [1].

The current trend for patient-specific computational fluid dynamics (CFD) is to rely on image data to personalize the models [2,3]. Several sequential steps are usually required. The segmentation of images is nowadays almost automated thanks to robust image processing algorithms. The resulting segmented surfaces, however, often requires specific manual edition as well as a volumic meshing step before being ready for numerical simulations. Those manual edition might involve cleaning the mesh, refining regions of interest, including valves, labeling

M. Pop and G.A. Wright (Eds.): FIMH 2017, LNCS 10263, pp. 441–449, 2017.
DOI: 10.1007/978-3-319-59448-4_42

the mesh surfaces for boundary condition and extruding inlets and outlets to reduce the effect of boundary conditions. Additionally, in the context of cardiac flow simulations, those pre-processing steps also involve the extraction and manipulation of the dynamics of the segmented surface in order to provide appropriate displacement boundary conditions. It has been reported in a recent review of patient-specific cardiac flow simulations that around "20–50 hours of human effort are needed for these pre-processing steps" [1] and similar claims have been made by others [13].

A new procedure is presented in order to reduce pre-processing time. First, real-time 3D echocardiography (RT3DE) data acquired on a patient were used to register a pre-processed generic mesh. Mesh morphing method have been used previously in bone mechanics [11,12]. Here we propose to use a similar methodology to obtain a patient-specific geometry from US image segmentations. The spatiotemporal data from the patient describing the ventricle dynamics are then transferred on the registered mesh to perform a patient-specific CFD simulation. Modifying a simplified valve model, the proposed framework allows to deliver physiological pressures during the isovolumic phases and this model has further been modified to allow the modeling of mitral regurgitation (MR), a pathology referring to a defect of the mitral complex that causes a backflow of blood into the atrium during ventricular systole, without any additional modifications of the generic mesh.

2   Methods

2.1   Mathematical Modeling

The generic computational domain (see Fig. 1) includes the left ventricle (LV), the left atrium (LA) and a part of the ascending aorta. Both the aortic and the mitral valves were also included in the domain. Blood flow was described using

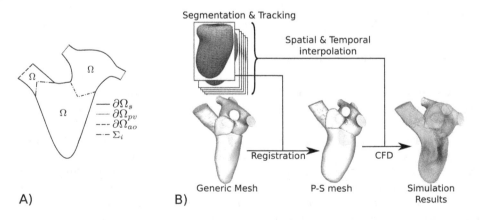

Fig. 1. (A) Problem domain and (B) pipeline schematic

the Navier-Stokes equations for a Newtonian incompressible fluid. In order to account for the motion of the computational domain, the equations were considered in their Arbitrary Lagrangian Eulerian (ALE) formulation. The aortic and mitral valves were modeled using a simplified resistive immersed surfaces (RIS) model [4], for which a dissipative term is added in the system. This dissipative term can be interpreted as a penalization of the fluid velocity for which the penalty parameter depends on the open or closed state of the valve. In this work, two extensions of the RIS model are considered. First, to model valve regurgitation, a space and time varying resistive term $R(\boldsymbol{x}, t)$ is considered allowing a finer control on which immersed elements are affected. The second extension concerns the isovolumic phases of the heart cycle. Generally speaking, when a fluid is enclosed in a cavity under Dirichlet boundary conditions, the pressure is only determined up to a constant. Numerically, the RIS model overcomes this issue by imposing a Dirichlet boundary condition via penalization. Nevertheless, this pressure is not correctly determined. The RIS model has been enhanced so that the intraventricular pressure is corrected, fitting a given pressure function $\overline{P_v}(t)$ inside the ventricle. This additional pressure data can be obtained by several means (e.g.: measurements or numerical simulations). Defining $+$ as the distal part (the outside of the ventricle), $\boldsymbol{n}_\Sigma$ as the outer-pointing normal vector of the surface Σ and $[[\cdot]]_\Sigma$ as the jump operator across the surface Σ, the interface condition on the valves are:

$$[[\boldsymbol{\sigma}(\boldsymbol{u}, p) \cdot \boldsymbol{n}_\Sigma]]_\Sigma = -R_\Sigma(\boldsymbol{x}, t)(\boldsymbol{u} - \boldsymbol{w}) + (P_\Sigma^+ - \overline{P_v}(t)) \cdot \boldsymbol{n}_\Sigma \qquad (1)$$

The full problem is therefore written as:

$$
\begin{cases}
\rho\left(\dfrac{\partial \boldsymbol{u}}{\partial t}\bigg|_{\mathcal{A}} + (\boldsymbol{u} - \boldsymbol{w}) \cdot \nabla \boldsymbol{u}\right) - \nabla \cdot \boldsymbol{\sigma}(\boldsymbol{u}, p) \\
\qquad = \displaystyle\sum_{i=\{o,c\}} \left[-R_{\Sigma_i}(\boldsymbol{x}, t)(\boldsymbol{u} - \boldsymbol{w}) + (P_{\Sigma_i}^+ - \overline{P_v}(t)) \cdot \boldsymbol{n}_{\Sigma_i}\right]\delta_{\Sigma_i} \quad , \ \Omega(t) \\
\hfill \nabla \cdot \boldsymbol{u} = 0 \quad , \ \Omega(t) \\
\hfill \boldsymbol{u} = \boldsymbol{u}_s \quad , \ \partial\Omega_s \\
\hfill \boldsymbol{\sigma}(\boldsymbol{u}, p) \cdot \boldsymbol{n} = -P_{pv}\boldsymbol{n} \quad , \ \partial\Omega_{pv} \\
\hfill \boldsymbol{\sigma}(\boldsymbol{u}, p) \cdot \boldsymbol{n} = -P_{ao}\boldsymbol{n} \quad , \ \partial\Omega_{ao}
\end{cases}
$$

Here, ρ is the fluid density, $\boldsymbol{u}$ is the fluid velocity, $\boldsymbol{w}$ is the domain velocity, $\boldsymbol{\sigma}(\boldsymbol{u}, p) = -p\boldsymbol{I} + 2\mu\boldsymbol{\epsilon}(\boldsymbol{u})$ is the fluid Cauchy stress tensor where p is the fluid pressure, μ is the fluid dynamic viscosity and $\boldsymbol{\epsilon}(\boldsymbol{u}) = \frac{1}{2}\left(\nabla \boldsymbol{u} + \nabla \boldsymbol{u}^T\right)$ is the strain rate tensor. $\boldsymbol{u}_s$ is the velocity satisfying a no-slip boundary condition on the solid surfaces, P_{pv} is the pulmonary vein pressure and P_{ao} is the aortic pressure calculated from a coupled 0D Windkessel RCR model. This system was discretized using P1/P1 stabilized finite elements. The surface displacements are extended within the left heart cavities using an appropriate incremental elastic operator whose Lamé parameters are updated element-wise to stiffen the smallest elements ([14,15]). This approach allows to efficiently handle large deformations without compromising the validity of the mesh.

2.2   Generic Mesh Registration

The presented work aims to bypass the tedious patient-specific mesh and data pre-processing. The main idea, described in this section, is to design a unique generic mesh and to combine it with image data to automatically generate patient-specific geometries and boundary conditions.

The computational domain was generated from the *Zygote 3D Human Heart Model*[1] obtained from highly resolved MRI and CT images. The computational domain is restricted to the LV, LA and a portion of the ascending aorta (Fig. 1). While the original model contained segmentation of the aortic and mitral valve in their closed state, the open mitral valve configuration was not available. Therefore, it was modeled based on anatomical data [5,6] using the CAD software *Salome*[2]. Particular attention has been paid to proper design of valvular leaflets substructure as recommended in [6]. The model was then post-processed using the software *3-Matic* and a tetrahedral mesh was generated using *GHS3D*. This computational domain is referred as the *generic* mesh, which can be used in combination with any patient surface segmentation.

Using the software *Qlab - 3DQAdvanced Plugin (Philips, Andover, MA)*, segmentation and tracking of the left ventricle is automatically performed on RT3DE images. A sequence of surfaces depicting the left ventricle is obtained with a time discretization of 25 Hz. By interpolating the position of the vertices using cubic splines, a time discretization fitting the required simulation time steps is recovered. $D^s_{t_n \rightarrow t_{n+1}}$ denotes the displacement field of the vertices of the segmented mesh s at time t_n to the vertices at time t_{n+1}.

In order to adapt the generic mesh to the patient ventricular surface segmentations, a combination of rigid and non-rigid registration technique was used. A time reference t_0 was chosen as the starting point of the simulation. A coarse rigid registration step was first performed to align the generic mesh ventricle surface onto the ventricle segmentation. This alignment was achieved by matching the apexes, longitudinal and radial axis of both geometries. A non-rigid registration step was then performed. In this second step, it is convenient to preserve the generic mesh properties in order to remain suitable for CFD simulations. In that regard, the large deformation diffeomorphic metric mapping (LDDMM) framework implemented in the open-source software *Deformetrica* [7] was used to find a diffeomorphism that maps the generic LV surface to the segmented LV surface. This diffeomorphism is described as a displacement field $D^g_{gen \rightarrow t_0}$ where the superscript g denotes the definition on all the vertices of the generic ventricle mesh. The fields D^s were then spatially interpolated onto the generic mesh using the normalized radial basis function (RBF) framework (2) with a gaussian RBF ϕ and a smoothing parameter λ.

$$D^g_{t_n \rightarrow t_{n+1}}(\boldsymbol{x}) = \frac{\sum_i D^s_{t_n \rightarrow t_{n+1}}(\boldsymbol{x_i}) \phi(d_i, \lambda)}{\sum_i \phi(d_i, \lambda)} \text{ with } d_i = \|\boldsymbol{x} - \boldsymbol{x_i}\| \qquad (2)$$

---

[1] http://www.3dscience.com.
[2] http://www.salome-platform.org.

Exploiting the ALE formulation, the displacement field $D^g_{gen \to t_0}$ was used as a boundary condition of an initial mesh motion step allowing the full computational mesh to deform and fit the segmentation. The full problem described in Sect. 2.1 was then solved for all time steps using the displacement fields $D^g_{t_n \to t_{n+1}}$ as the ventricle surface boundary condition.

3 Experiment

We evaluated the proposed pipeline on one patient's dataset. Echocardiographic images were acquired using an IE33 ultrasound system (Philips, Andover, MA) equipped with a 1–5 MHz transthoracic matrix array transducer (xMATRIX x5-1). RT3DE images were reconstructed as volumes over one cardiac cycle from acquisitions of sub-volumes over four cycles. The volumes were automatically segmented and tracked as previously described and used as sequence of 25 surface meshes. The cardiac cycles were replicated to allow the computation of several heart cycles. The generic mesh was composed of 178,348 elements for the proposed experiment. The registration process of Sect. 2.1 was applied to register the LV surface of the generic mesh onto the LV segmented surface. Time intervals of 1 ms were used for the time interpolation, and the spatial interpolation free parameter λ was taken as the average Euclidean distance between vertices of the segmented mesh.

Once the appropriate displacement fields D^g were computed, the fluid simulation problem described in Sect. 2.2 was solved using the finite element library *FELiScE*[3]. The blood parameters were $\mu = 4$ Pa s and $\rho = 1060$ kg/m^3. The simulation was run with a timestep of 1 ms. P_{pv} was set to 10 mmHg while the 0D RCR model has the following parameter: $R_{proximal} = 150$, $R_{distal} = 2300$, $C = 0.9$, $P_{venous} = 0$ mmHg and $P_{distal,t=0} = 70$ mmHg. Two heart cycles (0.8 s each) were simulated and the reported results were taken from the second one. Another simulation was performed with a circular hole of 0.5 cm^2 in the mitral valve representing a severe MR [8].

4 Results and Discussions

The results of the registration steps are shown in Fig. 2. The rigid registration aligns the two ventricles and the non-rigid registration process succeeds to deform the generic mesh onto the segmented mesh. It should be noted that the open-source software used allows for the tuning of an accuracy parameter. While this experiment uses a coarse kernel size, as recommended in the software documentation, the algorithm still provides good matching accuracy.

Time and space interpolation are then performed to produce the displacement fields D^g at all times. The overlap between the image segmentation and the deformed generic mesh under the influence of the displacement fields D^g is shown on Fig. 3. This exhibits the ability of the proposed method to fit the

---

[3] http://felisce.gforge.inria.fr.

Fig. 2. Rigid and non-rigid step of the registration of the ventricle surfaces. The white surface is the generic mesh and the red surface is the output of the image segmentation

Fig. 3. Overlap of the segmented mesh and the deformed generic mesh over one cardiac cycle

generic mesh onto the segmented surfaces. The complete pipeline was performed without human intervention in approximately 30 min. Compared to state of the art methods, it brings several advantages: it decreases the amount of work needed between the acquisition of the images and the launch of the CFD simulation and removes the need of a meshing expert.

The CFD simulations took 8 hours on a laptop using a *Intel Core i7-4810MQ* cadenced at 2.8 GHz. The evolution of blood during the cardiac cycles is depicted in Fig. 4. When the ventricular volume starts increasing, the mitral valve opens and blood enters the LV. The maximum blood velocity reaches approximately 100 cm/s in the mitral valve area. One can notice the presence of a vortex that detaches from the tip of the mitral valve. Blood then slows down as the E-wave finishes. A second vortex appears inside the LV in the late diastole corresponding to the A-wave. Those complex flow structures are coherent with the ones reported in the literature [2,9]. In particular, the presence of the A-wave vortex is described in [2]. While the contraction of the atrium has not been simulated, the increase in LV volume recovered by the segmentation allows this behavior to appear in the presented simulation. As the LV contracts, the blood is then ejected at a speed of approximately 125 cm/s through the aorta. The blood finally slows down before the next cardiac cycle.

Figure 5 depicts the simulation including regurgitation. During the systole, a strong jet of blood comes back to the atrium through the mitral valve orifice. As reported in the litterature, the maximum velocity is obtained just downstream of the mitral valve regurgitation (*vena contracta*). The jet then hits the walls of the atrium. In the ventricular region, the blood velocity is forming flattened

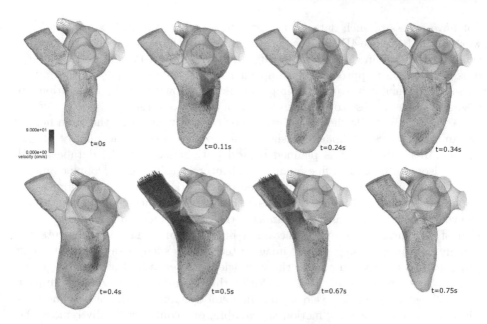

Fig. 4. Blood velocity vector field and magnitude on a cut plane

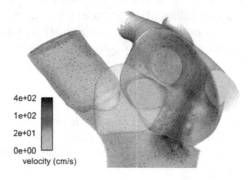

Fig. 5. Regurgitation case: flow convergence region and jet in the atrium

hemispherical isovelocity surfaces. This is coherent with the litterature on mitral quantification for such a hole geometry [8,10].

One of the limitations of the proposed method is that it includes two free parameters (λ and the kernel size of the non-rigid algorithm). Fortunately, the impact of those parameter is intuitive (e.g. λ for the RBF interpolation has a smoothing effect on the interpolation). Nevertheless, no quantitative analysis was performed in the presented work and further investigation should be performed. In this study, the 4D ventricular surface has been used to register a generic mesh that also includes a portion of the ascending aorta and the atrium. Therefore, one could argue that the rigid template of the atrium is arbitrary. While we did not have access to such data for the present article, a natural extension of the work would be to use 4D atrial segmentation within the same pipeline to improve the model. While in the current setting, the atrial geometry is indeed arbitrary, it is still a richer approach than imposing lumped atrial model at the mitral valve. In particular it avoids the imposition of unknown boundary conditions in a zone extremely close to the mitral valve, a region of interest.

Additionally, while the CFD simulations correctly captured the main hemodynamics features, comparisons should be done to assess the quality of the results. In particular, it is planned to add patient cases in the database and comparisons of hemodynamic quantities of interest using color Doppler images will allow to further validate the proposed framework. A more general limitation of patient-specific methods is concerning the loss of displacement information of the ventricle in the imaging stage. Indeed, the displacement in the normal direction of the surface is properly recovered but the tangential displacements are usually not (unless using specific imaging techniques such as tagged-MRI). The physiological twisting motion of the ventricle may hence be lost in the segmentation stage. Further work need to address this issue by quantifying the impact of the tangential ventricle velocity on the hemodynamics. Using a baseline simulation where the twisting motion is available, one could artificially remove the twisting motion and compare the main hemodynamic features of the two flow solutions. This problem could be solved by using anatomical landmarks visible in the images and using that information to constrain the segmentation process. Finally, to properly model patient-specific regurgitation, image data could be used to recover the patient-specific mitral valve regurgitant hole geometries.

5   Conclusion

A procedure was proposed to speed up the mesh pre-processing prior to CFD simulations of patient-specific cardiac blood flow. This procedure combines a rigid and non-rigid registration to morph a pre-computed unique generic mesh using patient-specific data. The ventricle dynamics is then transferred onto the registered mesh to prescribe appropriate boundary conditions. This pipeline was tested on RT3DE data and performed in around 30 min (improvement of around two order of magnitude compared to what is reported in the literature). To evaluate the pipeline, two CFD simulations were performed. The first simulation was done on a healthy patient, and in the second one, a hole in the mitral valve was added to obtain a severe MR. In both simulations, results showed flow patterns and hemodynamics quantities within the known cardiac physiology and blood velocities were coherent with the literature.

Acknowledgements. The authors gratefully acknowledge Mathieu De Craene, Èric Lluch and Hélène Langet from Philips Reasearch - Medisys for their support in acquiring patient data and their use, as well as for reviewing the presented manuscript. We would like to also acknowledge the help of INRIA - M3DISIM research team for sharing with us pressure curves generated from complex electro-mechanical simulations.

References

1. Mittal, R., Seo, J.H., Vedula, V., Choi, Y.J., Liu, H., Huang, H.H., George, R.T.: Computational modeling of cardiac hemodynamics: current status and future outlook. J. Comput. Phys. **305**, 1065–1082 (2016)
2. Chnafa, C., Mendez, S., Nicoud, F.: Image-based large-eddy simulation in a realistic left heart. Comput. Fluids **94**, 173–187 (2014)
3. Bavo, A.M., Pouch, A.M., Degroote, J., Vierendeels, J., Gorman, J.H., Gorman, R.C., Segers, P.: Patient-specific CFD simulation of intraventricular haemodynamics based on 3D ultrasound imaging. Biomed. Eng. OnLine **15**(1), 107–122 (2016)
4. Astorino, M., Hamers, J., Shadden, S.C., Gerbeau, J.-F.: A robust and efficient valve model based on resistive immersed surfaces. Int. J. Numer. Meth. Biomed. Eng. **28**(9), 937–959 (2012)
5. Ranganathan, N., Lam, J.H., Wigle, E.D., Silver, M.D.: Morphology of the human mitral valve. II. The valve leaflets. Circulation **41**(3), 459–467 (1970)
6. Votta, E., Caiani, E., Veronesi, F., Soncini, M., Montevecchi, F.M., Redaelli, A.: Mitral valve finite-element modelling from ultrasound data: a pilot study for a new approach to understand mitral function and clinical scenarios. Philos. Trans. R. Soc. Ser. A Math. Phys. Eng. Sci. **366**(1879), 3411–3434 (2008)
7. Durrleman, S., Prastawa, M., Charon, N., Korenberg, J.R., Joshi, S., Gerig, G., Trouv, A.: Morphometry of anatomical shape complexes with dense deformations and sparse parameters. NeuroImage **101**(8), 35–49 (2014)
8. Lambert, A.S.: Proximal isovelocity surface area should be routinely measured in evaluating mitral regurgitation: a core review. Anesth. Analg. **105**(4), 940–943 (2007)
9. Hong, G.R., Pedrizzetti, G., Tonti, G., Li, P., Wei, Z., Kim, J.K., Vannan, M.A.: Characterization and quantification of vortex flow in the human left ventricle by contrast echocardiography using vector particle image velocimetry. JACC Cardiovasc. Imaging **1**(6), 705–717 (2008)
10. This, A., Morales, H.G., Bonnefous, O.: Proximal isovelocity surface for different mitral valve hole geometries. In: ECCOMAS Congress 2016 - Proceedings of the 7th European Congress on Computational Methods in Applied Sciences and Engineering 1, pp. 155–163 (2016)
11. Couteau, B., Payan, Y., Lavallée, S.: The mesh-matching algorithm: an automatic 3D mesh generator for finite element structures. J. Biomech. **33**(8), 1005–1009 (2000)
12. Bucki, M., Lobos, C., Payan, Y.: A fast and robust patient specific finite element mesh registration technique: application to 60 clinical cases. Med. Image Anal. **14**(3), 303–317 (2010)
13. Doost, S., Ghista, D., Su, B., Zhong, S., Morsi, Y.: Heart blood flow simulation a perspective review. Biomed. Eng. Online **15**(1), 101 (2016)
14. Stein, K., Tezduyar, T., Benney, R.: Mesh moving techniques for fluid-structure interactions with large displacements. J. Appl. Mech. **70**(1), 58 (2003)
15. Landajuela, M., Vidrascu, M., Chapelle, D., Fernández, M.A.: Coupling schemes for the FSI forward prediction challenge: comparative study and validation. Int. J. Numer. Meth. Biomed. Eng. (2016)

Estimating 3D Ventricular Shape From 2D Echocardiography: Feasibility and Effect of Noise

Gabriel Bernardino[1]([✉]), Constantine Butakoff[1], Marta Nuñez-Garcia[1],
Sebastian Imre Sarvari[2], Merida Rodriguez-Lopez[3,4], Fatima Crispi[3,4],
Miguel Ángel González Ballester[1,5], Mathieu De Craene[6], and Bart Bijnens[1,5]

[1] DTIC, UPF, Barcelona, Spain
gabriel.bernardino@upf.edu
[2] Department of Cardiology, Oslo University Hospital, Rikshospitalet, Oslo, Norway
[3] Fetal i+D Fetal Medicine Research Center, IDIBAPS, Barcelona, Spain
[4] BCNatal, Hospital Clínic and Hospital Sant Joan de Déu, Barcelona, Spain
[5] ICREA, Barcelona, Spain
[6] Philips Research, Paris, France

Abstract. Many cardiac diseases are associated with changes in ventricular shape. However, in daily practice, the heart is mostly assessed by 2D echocardiography only. While 3D techniques are available, they are rarely used. In this paper we analyze to which extent it is possible to obtain the 3D shape of a left ventricle (LV) using measurements from 2D echocardiography. First, we investigate this using synthetic datasets, and afterwards, we illustrate it in clinical 2D echocardiography measurements with corresponding 3D meshes obtained using 3D echocardiography. We demonstrate that standard measurements taken in 2D allow quantifying only the ellipsoidal shape of the ventricle, and that capturing other shape features require either additional geometrical measurements or clinical information related to shape remodelling. We show that noise in the measurements is the primary cause for poor association between the measurements and the LV shape features and that an estimated 10% level of noise on the 2D measurements limits the recoverability of shape. Finally we show that clinical variables relating to the clinical history can substitute the lack of geometric measurements, thus providing alternatives for shape assessment in daily practice.

Keywords: LV shape · LV measurements · Echocardiography · Shape prediction · LV measurement accuracy

1 Introduction

Many cardiac pathologies are associated with changes in the cardiovascular system and in particular the shape of cardiac ventricles [1]. Some pathologies are associated to very particular changes in the shape, such as infarctions producing a concavity/aneurysm in one of the walls of the left ventricle (LV) [2]. Additionally, some conditions determined in fetal life (e.g. tetralogy of Fallot [3] or

© Springer International Publishing AG 2017
M. Pop and G.A. Wright (Eds.): FIMH 2017, LNCS 10263, pp. 450–460, 2017.
DOI: 10.1007/978-3-319-59448-4_43

intra-uterine growth restriction [4]) shown to result in (sometimes subtle) shape changes that are related to the severity of remodeling and thus might have clinical predictive value above more commonly used parameters like volumes and ejection fraction.

In daily practice, cardiac imaging is primarily carried out by 2D echocardiography. 2D echo has a good temporal resolution and a reasonable spatial resolution. It is also cheap and can be operated at bedside. Its problem is that it only allows to image a slice of the heart in each acquisition. This makes it difficult to assess the shape, as the whole heart is not observable from a single acquisition, but several views must be combined.

Even if 3D imaging techniques are available that give a complete view of the heart and might thus be preferable, they have important drawbacks in daily practice. 3D echo is nowadays not used in daily clinical practice due to its limited temporal and spatial resolution and need for extra transducers and high-end scanners. It is also more prone to having poor acoustic window and poor image quality. Other 3D technologies, MRI and CT also have drawbacks. They have better volume reproducibility and less imaging artifacts as compared to 2D echo, but their temporal resolution is much worse. Moreover, they are not portable, expensive and, in case of CT, use ionizing radiation.

In this paper, we investigate whether it is possible to infer the 3D shape of the human left ventricle from the 2D echo measurements with a linear regression approach. We will use not only the 2D measurements related to the geometry, but also functional and clinical parameters. This approach reveals which parts of the shape can be recovered in 2D and what confidence we can expect on those predictions. We aim to know in which situations 3D imaging is required, and in which 2D gives enough information.

To that end, we first create a synthetic dataset for studying the influence of noise on the measurements and, afterwards, we illustrate the use of our regression approach on a population of pre-adolescents born after intrauterine growth restriction (IUGR) where both 3D shapes and 2D measurements are available. Specifically, we want to study how the following shape patterns reported in the literature for these patients [5] can be best predicted from 2D measurements: the overall size of the Left Ventricle (LV), its sphericity, and a lateral shift of the apex with respect to the LV base. We also study in this paper the influence of increasing noise levels on the 2D measurements.

Although the problem of inferring the complete 3D shape from 2D images or an incomplete part of the 3D shape has already been addressed by the community (e.g. for the liver [6], or the femur [7]), this is, to our knowledge, the first attempt to infer the LV shape from 2D clinical measurements obtained through echocardiography, including the analysis of the impact of measurement noise on the prediction.

2   Data Description

2.1   IUGR Patient Data

We used the dataset of [5] of 152, 7–13 years old, individuals (58 with Intrauterine Growth Restriction (IUGR) and 94 controls). For each individual we have a 3D mesh of the end-diastolic LV acquired using 3D echo (Echopac version 108.1.6) and an independent complete 2D echo study with the standard clinical measurements. After removing cases with incomplete measurements, 116 patients remained. The summary of the population can be found in Table 2. The meshes in this dataset have point correspondence, containing 720 nearly regular faces with an average face area of $13.9\,\mathrm{mm}^2$. Each patient has 90 measurements related to the function and geometry of the LV, other chambers, as well as clinical history. Some examples of measurements included in the previous classification are showed in Table 1. A more complete overview of the variables included can be found in [8].

2.2   Generation of the Synthetic Database

In order to study the effect of noise on the measurements, we have generated a synthetic database matching the main sources of variability found in the IUGR

Table 1. Classification of the 2D measurements included in the study

Class	#measurements	Examples
Geometry of the LV	18	ES volume, ED volume, internal dimension, long axis length, basal diameter
Function of the LV	14	Cardiac output, cardiac index, mitral A speed
Geometry of the other parts	17	LA area, RV volume
Function of the other parts	18	TAPSE, Tricuspid E speed, AR duration
Clinical history	13	Age, sex, IUGR label, height, weight

Table 2. Summary of the IUGR population.

Variable	Mean	Std	Min	Max
Age (years)	10.51	1.70	7.24	13.23
Weight (kg)	38.32	9.96	19.10	64.00
Height (m)	1.42	0.12	1.16	1.74
LV volume (ml)	80.16	18.17	44.00	126.00
LV axial diameter (mm)	66.90	6.68	48.30	87.00

population: size, sphericity and septal bulging (an asymmetrical *half moon* deformation). We first created a *template LV* as a sphere cropped at 3/4 of its height. We estimated the distribution of the scaling coefficients that the IUGR meshes follow by affinely registering the *template LV* to each LV mesh, and storing the scaling coefficients of the affine transform. With those we estimated a Multivariate Normal distribution C. We randomly assigned every individual to either the pathological (P) or control (H) group. Individuals belonging to the pathological group were given a stronger *half moon* shape pattern.

The *template LV* was first positioned so that the cropped section is perpendicular to the y axis, the *half moon* deformation was further introduced along the x axis using the following deformation for very point $\mathbf{p}(x, y, z)$ of the mesh:

$$\phi(x, y, z) = (|x - x_{mean}|(y - y_{min})(y - y_{max}), 0, 0) \tag{1}$$

where x_{mean} refers to the mean value of the x coordinates, y_{min} and y_{max} refer to the minimal and maximal y coordinate over the points of the template. The deformation is then normalized to have unit L_2 norm. The generation of a synthetic mesh was performed as follows:

1. Sample scaling coefficients c from C, scale the *template LV* using c and obtain the cropped ellipsoid e;
2. Generate the label H/P, with 50% probability for each class;
3. Sample μ from $D_H = \mathcal{N}(0\,\mathrm{cm},\ 0.3\,\mathrm{cm}^2)$ if the label is H, or from $D_P = \mathcal{N}(1.5\,\mathrm{cm},\ 0.4\,\mathrm{cm}^2)$ if the label is P;
4. The synthetic mesh is $m := e + \mu\phi$, where ϕ is the *half moon* deformation vector. We modeled the noise in the 3D mesh by a random displacement at every vertex with mean norm of 1 mm.

2.3   Measurements

The 2D measurements in the synthetic dataset are taken from the 3D shape, and they contain similar information as real measurements. They are not directly comparable to the real 2D measurements, because the latter are obtained from the echocardiographic image. We used 3 different sets of synthetic measurements:

1. *Basic measurements:* they are equivalent to the volume, long axis length, basal diameter and biggest internal diameter as described in the European Association of Cardiovascular Imaging (EACVI) clinical recommendations ([8]). In Fig. 1a, there is a graphical description of how the linear measurements were taken in the mesh.
2. *Extra measurements:* basic measurements, with the internal diameter divided in two segments from the furthest points to the long axis (Fig. 1b). Those distances are specific to the *half moon* pattern.
3. *Random measurements:* basic measurements and 5 distances between random pairs of points and 5 diameters at different heights along the principal axis.

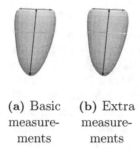

(a) Basic
measure-
ments

(b) Extra
measure-
ments

Fig. 1. Linear measurements taken for the synthetic meshes. The basic measurements correspond roughly to the long axis (black), internal dimension (orange) and basal diameter (purple). In the extra measurements, the internal dimension is split in the distances from the extremes to the long axis (orange and cyan). (Color figure online)

Optionally the H/P label used in the generation was added to the geometrical measurements.

For the IUGR dataset, 3 types of measurements, obtained from the 2D echo study, were used:

1. Left ventricular end-diastolic volume, obtained using the biplane method, and 3 linear measurements: long axis length, basal diameter and biggest internal diameter.
2. Measurements from the point 1 and the label of the IUGR condition.
3. All the measurements available in the 2D study, including function and geometry of all parts of the heart.

3 3D Shape Regression from 2D Measurements

This section describes how we learn a regression from a set of 2D measurements to the 3D shape. It also defines the metrics that will be used later to quantitatively compare regression results.

3.1 Shape Predictor Training

In order to carry out 3D shape prediction from the 2D echo measurements, we trained a linear regressor using the standardized 2D measurements as predictor variables. As output, we used the coordinates of the mesh we want to reconstruct after projection on a standard PCA basis. PCA has indeed the advantage of giving both projection and reconstruction linear operators. To make the PCA representation more compact, we kept PCA modes that explained 99% of the variance. As input, the PCA takes the concatenation of all mesh vertex coordinates. We used the Elastic Nets algorithm to learn the matrix of regression coefficients $\hat{\mathbf{W}}^*$ and bias vector $\hat{\mathbf{b}}^*$:

$$\hat{\mathbf{W}}^*, \hat{\mathbf{b}}^* = \arg\min_{\mathbf{W},\mathbf{b}} \left(\frac{1}{n} \|\hat{\mathbf{Y}} - \mathbf{X}\mathbf{W} - \hat{\mathbf{b}}\|_{\text{fro}}^2 + \lambda_1 \|\mathbf{W}\|_{\mathbf{21}} + \lambda \|\mathbf{W}\|_{\text{fro}}^2 \right) \quad (2)$$

where $\|\mathbf{W}\|_{\mathrm{fro}}$ is the Frobinus norm, $\|\mathbf{W}\|_{21} = \sum_i \|\mathbf{W_i}\|_2$, n is the number of samples used for training, X are the 2D measurements arranged in rows and $\hat{Y}$ are the PCA coordinates after projection. We kept only the modes explaining 99% of the variance. The λ_1 and λ_2 were determined using 3-Fold cross-validation with the training data. From Eqn. 2, we learn a regressor predicting PCA coordinates from the input 2D measurement. The resulting shape is then simply obtained by applying the PCA reconstruction to these regressed coordinates.

3.2 Evaluation

We used 9 different noise levels (between 0 and 0.5) for 2D measurements in the synthetic dataset. For every level, we have added to every measurement a white Gaussian noise with a mean μ_i and a variance $\sigma^2 = (\alpha\mu_i)^2$, where α is the noise level. We generated 10 different synthetic datasets and report in Sect. 4 the mean and 95% CI of theirs results. The quality of the regression in the IUGR dataset was evaluated using Leave-One-Out cross-validation and an independently generated test-set of size 50 for the synthetic dataset. We use the R^2 determination coefficient as a quality metric. It is defined as $R^2 = 1 - var(\mathbf{Y}_{pred} - \mathbf{Y})/var(\mathbf{Y})$, where $\mathbf{Y}$ is the real response and $\mathbf{Y}_{pred} = \mathbf{XW}^*$. The R^2 coefficient is the percentage of the response space variance that can be predicted via the regression. Given a linear deformation column mode $\mathbf{m}$, the R^2 of the variability associated with that mode is defined similarly: $R_m^2 = 1 - var(\mathbf{Y}_{pred}\mathbf{m})/var(\mathbf{Y}\ \mathbf{m})$

4 Results

4.1 PCA Modes

As PCA was used as encoding of the 3D shape for learning the regression (Sect. 3.1), we show here PCA modes on both the IUGR and the synthetic data. For IUGR we have previously aligned the meshes using the Procrustes algorithm (preserving the scale). The 4 largest PCA modes and the percentage of variability they represent are shown in the top row of Fig. 2. The two modes with the largest variability are elongation and sphericity. The third mode appears to be associated with a 3D segmentation artifact due to the partial coverage of the outflow tract in many cases. The fourth mode is a non-symmetric deformation where the apex is shifted with respect to the base with a small inclination of the base, creating an impression of a *half moon* ventricular shape due to a bulging septum. The bottom row of Fig. 2 shows the main PCA modes learned on the synthetic data.

4.2 Synthetic Data

We first evaluated how the choice of 2D measurements influences the regression quality. Figure 3 plots the R^2 coefficient (Sect. 3.2) for the first 3 PCA modes

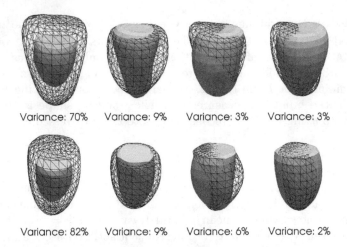

Fig. 2. Top row: main PCA modes learned on the IUGR dataset. Bottom row: the biggest 4 modes of the PCA of the synthetic dataset. All modes are displayed at ±3 STD

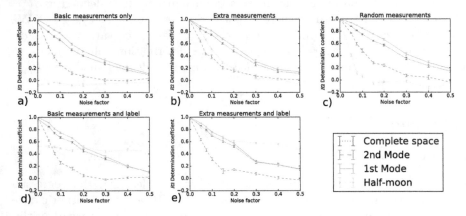

Fig. 3. Influence of the noise and choice of measurements on the regression. For each set of measurements, we plot the R^2 coefficient for specific PCA modes as a function of the noise. All these results were obtained with the synthetic dataset. See Sect. 2.3 for a definition of all types of measurements combined here. (Color figure online)

as a function of noise for all possible combinations of measurements defined in Sect. 2.3. Then, we investigated how training affects the quality of the regression. First we tested whether adding more training data or selecting another regression algorithm would affect the performance. Figure 4.a shows the effect of changing the training data size for the basic set of measurements and Fig. 4.d for the random set. We also tested how different linear regression algorithms [9] compared in terms of R^2 values: PLS, Ridge, Elastic Nets, classic Linear Regression, Gaussian Kernel Ridge regression for both the basic measurements (Fig. 4.c)

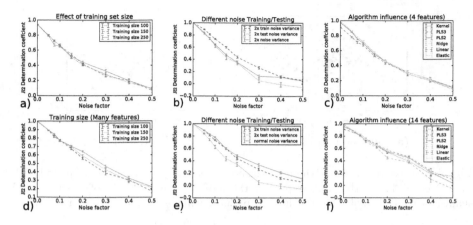

Fig. 4. Effect of: error in training set (a, d), train data size (b, e) and regression algorithm (c, f). The upper row corresponds to the basic measurements and the lower to the random measurements. All these results were obtained with the synthetic dataset.

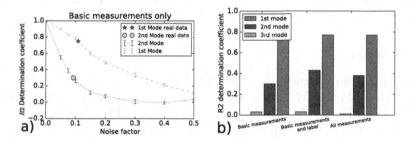

Fig. 5. (a) R^2 coefficients for the 1st and 2nd PCA mode obtained in the real data compared with the synthetic results for their equivalent PCA modes. (b) R^2 for the 3 first PCA modes for different sets of measurements.

and the random ones (Fig. 4.f). Finally we studied the influence of noise on the training and testing steps. For the basic set of parameters, we have added to the training set noise with a variance equal to twice the testing set noise and viceversa (Fig. 4.b and e).

4.3   IUGR Data

We used 4 measurements (volume, long axis length, basal diameter and internal dimension) in the IUGR dataset that are equivalent to the basic measurements of the synthetic dataset. We have performed regression using those measurements to compare real and synthetic results. In Fig. 5.a we plot the observed R^2 values of the real data for the first two PCA modes along with the synthetic R^2 curves for the same PCA modes. We can see that the noise factor is around 10%, which coincides with the results obtained by D'hooge et al. in [10]. We have also compared the regression results when different types of measurements are used

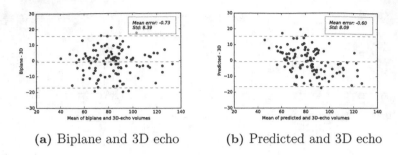

(a) Biplane and 3D echo   (b) Predicted and 3D echo

Fig. 6. Bland-Altman plot of the biplane and predicted shape volumes compared with the 3D echo. The dotted lines represent the 95% CI. The volumes of the predicted shapes have a RMSE of 8.11 ml, while the biplanes volume measurements have 8.32 ml.

as predictors. Results show that, with 116 data samples, the best combination of parameters is the basic geometric measurements and the clinical label. Using all the measurements available introduces error, as the algorithm is not able to choose which measurements are relevant for the regression.

4.4   Volumes

We computed the volumes of the predicted meshes and compared them with the 3D echo volumes. We evaluated whether they were closer to the volumes acquired from 3D echo than the ones obtained using the biplane method. Results showed that there was no significant precision difference between both volumes estimations. Figure 6 shows the Bland-Altman plots of the volumes obtained with the biplane method and the volumes of the predicted shapes compared with the 3D echo.

5   Discussion

The most common shape representation for the LV is a cropped ellipsoid, as commonly assumed in many measurements done in clinical practice. Our results confirm the validity of this assumption since the PCA analysis of the IUGR dataset (Fig. 2, top row) give modes coinciding with the longitudinal and radial scaling of a mean shape similar to a cropped ellipsoid. Although these modes explain 82% of the total variance, they are not always the most useful parameters for shape assessment to evaluate cardiac pathological remodeling. More complex shape features are encoded by the next modes. As those have much less variability than the main ellipsoidal modes, they are more sensitive to noise and harder to recover. Results in Fig. 3.a show that with the geometrical measurements inspired from EACVI recommendations, only the main ellipsoidal modes can be estimated. A consequence of this might be that the predicted shape does not provide a better volume estimation than the biplane method (Fig. 6). Biplane method is also accurate when the shapes are symmetric, but it is less precise for asymmetric shapes. Findings in the IUGR data were similar (Fig. 3.b).

When using only the basic geometrical measurements, neither increasing the training dataset size (Fig. 4.a), nor removing the training noise (Fig. 4.e), nor changing the algorithm (Fig. 4.c) altered the regression quality. This suggests that most of the regression error is induced by the error on 2D input measurements. Adding new measurements allows to improve shape accuracy. An example of this is the *half moon* pattern, that can not be estimated from basic measurements (dotted cyan graph in Fig. 3.a), but can be retrieved when specific measurements sensitive to this pattern or many random measurements are added. Another positive effect of adding measurements is that we can use redundant information to make the regression more resilient to noise. In Fig. 3 we can see that for the 1st mode, R^2 decreases slower when random measurements are used (c) than when only the basic ones are available (a). We do not have to restrict to geometrical measurements as geometry predictors, we can use measurements related to the clinical history: Fig. 3.d and e show how adding a label related to the clinical diagnosis improves the regression of the *half moon* feature. A similar behavior was observed for the IUGR dataset in Fig. 4.b.

6   Conclusion

We have shown that it is possible to predict 3D shape from 2D measurements, and how the quality of the prediction is intrinsically limited by the amount of noise in the input measurements. The uncertainty present in the measurements used in daily clinical practice is of such magnitude that it hampers recovering subtle cardiac remodeling features. This can be improved if more measurements specific to the deformation we aim to recover are taken. Alternatively, additional geometric measurements, possibly random, or some clinical history parameters can be used to improve the accuracy of the 3D reconstructed shape. In any case, a sensitivity analysis as done in this paper is important to determine the minimal size of the training population and the optimal choice of features for a given population.

Acknowledgements. This study was partially supported by the Spanish Ministry of Economy and Competitiveness (grant TIN2014-52923-R; Maria de Maeztu Units of Excellence Programme - MDM-2015-0502), FEDER and the European Union Horizon 2020 Programme for Research and Innovation, under grant agreement No. 642676 (CardioFunXion).

References

1. Konstam, M.A., Kramer, D.G., Patel, A.R., Maron, M.S., Udelson, J.E.: Left ventricular remodeling in heart failure: current concepts in clinical significance and assessment. JACC Cardiovasc. Imaging **4**(1), 98–108 (2011)
2. Di Donato, M., Castelvecchio, S., Kukulski, T., Bussadori, C., Giacomazzi, F., Frigiola, A., Menicanti, L.: Surgical ventricular restoration: left ventricular shape influence on cardiac function, clinical status, and survival. Ann. Thorac. Surg. **87**, 455–461 (2009)
3. Ye, D.H., Desjardins, B., Hamm, J., Litt, H., Pohl, K.M.: Regional manifold learning for disease classification. IEEE Trans. Med. Imaging **33**(6), 1236–1247 (2014)
4. Crispi, F., Bijnens, B., Figueras, F., Bartrons, J., Eixarch, E., Le Noble, F., Ahmed, A., Gratacós, E.: Fetal growth restriction results in remodeled and less efficient hearts in children. Circulation **121**(22), 2427–2436 (2010)
5. Sarvari, S.I., Rodriguez-Lopez, M., Nuñez-Garcia, M., Sitges, M., Sepulveda-Martinez, A., Camara, O., Butakoff, C., Gratacos, E., Bijnens, B., Crispi, F.: Persistence of cardiac remodeling in Preadolescents with fetal growth restriction. Circ. Cardiovasc. Imaging 10(1) (2017)
6. Kistler, M., Bonaretti, S., Pfahrer, M., Niklaus, R., Büchler, P.: The virtual skeleton database: an open access repository for biomedical research and collaboration. J. Med. Internet Res. **15**(11) (2013). e245
7. Blanc, R., Seiler, C., Székely, G., Nolte, L.P., Reyes, M.: Statistical model based shape prediction from a combination of direct observations and various surrogates: application to orthopaedic research. Med. Image Anal. **16**(6), 1156–1166 (2012)
8. Lancellotti, P., Cosyns, B.: The EACVI Echo Handbook. The European Society of Cardiology Textbooks Series. Oxford University Press, Oxford (2015)
9. Murphy, K.: Machine Learning: A Probabilistic Perspective. MIT Press, Cambridge (2012)
10. D'hooge, J., Barbosa, D., Gao, H., Claus, P., Prater, D., Hamilton, J., Lysyansky, P., Abe, Y., Ito, Y., Houle, H., Pedri, S., Baumann, R., Thomas, J., Badano, L.P.: Two-dimensional speckle tracking echocardiography standardization efforts based on synthetic ultrasound data. Eur. Heart J. Cardiovasc. Imaging **17**, 693–701 (2016)

Combining Deformation Modeling and Machine Learning for Personalized Prosthesis Size Prediction in Valve-Sparing Aortic Root Reconstruction

Jannis Hagenah[1,2]([⊠]), Michael Scharfschwerdt[3], Achim Schweikard[1],
and Christoph Metzner[4]

[1] Institute for Robotics and Cognitive Systems, University of Lübeck,
Lübeck, Germany
hagenah@rob.uni-luebeck.de
[2] Graduate School for Computing in Medicine and Life Sciences,
University of Lübeck, Lübeck, Germany
[3] Department of Cardiac Surgery, University Hospital Schleswig-Holstein,
Lübeck, Germany
[4] Science and Technology Research Institute, University of Hertfordshire,
Hatfield, UK

Abstract. Finding the individually optimal prosthesis size is an intricate task during valve-sparing aortic root reconstruction. Previous work has shown that machine learning based prosthesis size prediction is possible. However, the very high demands on the underlying training data set prevent the application in a clinical setting. In this work, the authors present an alternative approach combining simplified deformation modeling with machine learning to mimic the surgeon's decision making process. Compared to the previously published approach, the new method provides a similar prediction accuracy whith a dramatic decrease of demand on the training data. This is an important step towards the clinical application of machine learning based planning of personalized valve-sparing aortic root reconstruction.

Keywords: Aortic valve · Valve-sparing aortic root reconstruction · Machine learning · Computer assisted surgery · Personalized medicine

1 Introduction

Patients with the diagnosis of a pathological dilation of the aortic root often suffer from aortic valve insufficiency [1]. This insufficiency arises from the changing valve geometry due to the dilation of the surrounding tissue while the valve leaflets stay constant in their size and shape. For these patients, valve-sparing aortic root reconstruction presents an alternative to valve replacement. The goal of this surgery is to remodel the original size of the aortic root using a graft prosthesis. The patients leaflets are spared and attached to the prosthesis. Hence,

© Springer International Publishing AG 2017
M. Pop and G.A. Wright (Eds.): FIMH 2017, LNCS 10263, pp. 461–470, 2017.
DOI: 10.1007/978-3-319-59448-4_44

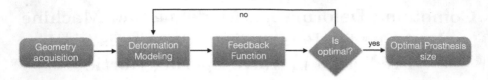

Fig. 1. Flowchart of the proposed method. The individual dilated geometry is manipulated according to a deformation model and rated by a feedback function. The geometry is deformed iteratively until the score is optimal.

the superior hemodynamic characteristics of the original valve can be preserved, improving patient's outcome compared to valve replacement [2].

However, choosing the optimal prosthesis size is a challenging task for the surgeon. The original shape of the aortic root before the dilation is unknown and only the dilated state can be assessed. The estimation of the healthy state is further complicated by the high interpatient variability of the aortic root shape and valve geometry. Therefore, a pre-operative planning tool, which predicts the optimal prosthesis size for the current patient, could greatly improve the outcome of this procedure.

A common approach to develop such a tool is utilizing biomechanical simulations to evaluate the individual valves movement for different prosthesis sizes virtually [3]. One problem of this approach are the complex biomechanical properties of the aortic valve leaflets which are not yet completely understood. Hence, the choice of realistic simulation parameters is very uncertain. Additionally, the approach suffers from high computational complexity and cannot be fully automated since the user needs to score the valve movement for each specific prosthesis.

An alternative approach is to optimize the aortic root deformation using machine learning. Due to the implicit, data-driven optimization of the dilation behaviour, the biomechancial parameters of the valve do not have to be set explicitly. In a first study, porcine aortic roots were examined using transesophageal ultrasound (TEE) in a healthy and an artificially dilated state [4]. On this dataset, a regression model was trained to predict the healthy root diameter based on the dilated valve geometry. The study showed that learning the root deformation and predicting the optimal prosthesis size using machine learning is possible [4]. To the author's best knowledge, this direct estimation method is the first implementation of a patient-specific prediction of the optimal prosthesis size for valve-sparing aortic root reconstruction.

However, the drawback of this method is its demand on the training dataset. To train the regression model, the aortic root shape of a number of individuals must be available in both states, the dilated and the healthy one. Such a dataset is hard to obtain for human patients. Hence, the clinical application of this is a distant prospect.

The purpose of this paper is the introduction of a hybrid approach combining simplified aortic root deformation modeling and machine learning. It is inspired by the surgeon's decision making process, avoids biomechanical uncertainties and minimizes the demand on the underlying data base.

2   Materials and Methods

The proposed method is inspired by the surgeon's decision making process. This process can be separated into four steps:

(1) Obtaining the patient's individual dilated aortic root geometry
(2) Estimation of the valve geometry if it would be attached to a prosthesis
(3) Rating the correspondence between the resulting geometry and a healthy behaviour
(4) If the rating was good enough, the optimal prosthesis size is found. Otherwise, repeat with decreasing prosthesis size until geometry is rated as healthy

The proposed approach implements the single steps of this decision making pipeline (cf. Fig. 1). In the following, these steps are explained in more detail.

2.1   Data Acqusition

Transesophageal ultrasound (TEE) is the gold standard for aortic valve examination [5]. Accordingly, the underlying data for this study was acquired in a setup comparable to [6], performing TEE examinations on ex-vivo porcine aortic roots. After 3D image acquisition, the aortic root was manually dilated following [4] to approximate pathological dilation. Afterwards, 3D images of the dilated aortic root were obtained. The full dataset consists of 3D ultrasound images of 24 porcine aortic roots in the healthy and the dilated state, respectively.

2.2   Deformation Modeling

The aim of this step is to model the deformation of the aortic valve geometry in correspondance to a change in the root diameter. The proposed method utilizes strong simplifications of the aortic valve geometry and behaviour. Thus, it is assumed that the valve geometry can be described meaningful enough based on four landmarks: The three commissure points P_1, P_2, P_3 and the coaptation point P_{coap}. The commissure points are the points where the leaflets are attached to the root wall while the coaptation point is defined as the point were all three leaflets are touching each other (cf. Fig. 2 a, c). Accordingly, the individual valve geometry is given by three vectors $L_1, L_2, L_3 \in \mathbb{R}^3$ pointing from P_{coap} to one of the coaptation points, respectively. After transformation to spherical coordinates with P_{coap} as the origin, the valve geometry $\mathbf{V}$ is given by

$$\mathbf{V} = \begin{pmatrix} L_{1,r} & L_{2,r} & L_{3,r} \\ L_{1,\theta} & L_{2,\theta} & L_{3,\theta} \\ L_{1,\varphi} & L_{2,\varphi} & L_{3,\varphi} \end{pmatrix}, \tag{1}$$

where r, θ and φ are the spherical coordinates. While the aortic root wall dilates, the leaflet size is not affected by the deformation [2]. Hence, it can be assumed that the length of the vectors L_1, L_2 and L_3 stays constant over all relevant

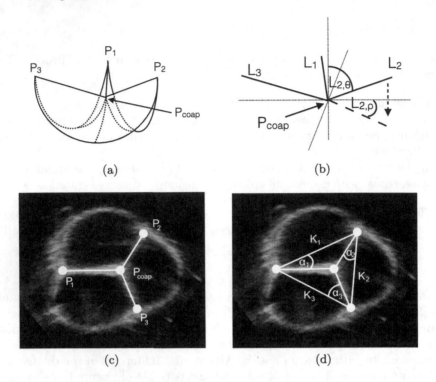

Fig. 2. (a) Sketched aortic valve in the closed state with the three commissure points P_1, P_2, P_3 and the coaptation point P_{coap}. (b) Geometry description based on the three vectors L_1, L_2 and L_3. (c) Slice of ultrasound volume image of an aortic root with the three commissure points P_1, P_2, P_3 and the coaptation point P_{coap}, projected to the x-y-plane. (d) Slice of ultrasound volume image of an aortic root with the features K_i and $\alpha_i, i = 1, 2, 3$, projected to the x-y-plane.

deformation steps. Following this assumption, the influence of the dilation on the valve geometry can be described as a change of the orientation of L_1, L_2 and L_3. This can be reached by varying two parameters of the spherical geometry description of every vector: θ, i.e. the angle to the vertical axis, and φ, i.e. the angle of the rotation around the vertical axis (cf. Fig. 2 b). Accordingly, the deformation of the valve geometry $\mathbf{V}$ can be modeled as a transformation T given as

$$\mathbf{V}' = T(\mathbf{V}, \Delta\theta_1, \Delta\theta_2, \Delta\theta_3, \Delta\varphi_1, \Delta\varphi_2, \Delta\varphi_3)$$
$$= \begin{pmatrix} L_{1,r} & L_{2,r} & L_{3,r} \\ L_{1,\theta} - \Delta\theta_1 & L_{2,\theta} - \Delta\theta_2 & L_{3,\theta} - \Delta\theta_3 \\ L_{1,\varphi} - \Delta\varphi_1 & L_{2,\varphi} - \Delta\varphi_2 & L_{3,\varphi} - \Delta\varphi_3 \end{pmatrix}. \qquad (2)$$

The definition of $\Delta\theta_i$ and $\Delta\varphi_i, i = 1, 2, 3$ induces different deformation models. In the following, three models of different complexity are presented.

Homogeneous Dilation (HD). In this model it is assumed that the orientation of the plane formed by the commissure points P_1, P_2, P_3 stays constant. The change of these points cartesian z coordinate is the same for all three points without allowing any rotation around the vertical axis. This corresponds to a homogenous dilatation of all three sinuses, i.e. the areas of the root wall enclosing one leaflet, and can be derived by setting

$$\Delta\theta_i = \arccos\left(\frac{L_{1,r}}{L_{i,r}}(\cos(L_{1,\theta} + \Delta\theta_1) - \cos L_{1,\theta} + \cos L_{i,\theta})\right) - L_{i,\theta}, \; i = 2, 3. \quad (3)$$

The two values $\Delta\theta_2$ and $\Delta\theta_3$ are dependant of $\Delta\theta_1$, which is the only model parameter. $\Delta\varphi_j, j = 1, 2, 3$ are set to 0. Hence, the HD model has only one degree of freedom (DOF).

Angle-Compensated Homogenous Dilation (AHD). As mentioned above, the HD model does not take different dilations of the sinuses into account. This can be compensated by allowing small rotations around the vertical axis. This results in three independent parameters $\Delta\varphi_i, i = 1, 2, 3$ within a range of $\pm10°$. The parameters $\Delta\theta_i, i = 1, 2, 3$ are set as in the HD model. This leads to a model with four DOF.

Free Deformation (FD). In this model, the parameters $\Delta\theta_i, i = 1, 2, 3$ are decoupled. Hence, all orientation parameters $\Delta\theta_i$ and $\Delta\varphi_i, i = 1, 2, 3$ can change independently within given ranges. With 6 DOF, the FD model offers the highest complexity of the presented deformation models.

With this simplified valve geometry description and the defined transformation models, parameterized, realistic deformations of arbitrary aortic valve geometries can be calculated efficiently. Through simplifying assumptions, biomechanical uncertainties are avoided. To evaluate the models, the dilated geometries from the data base were deformed to match the corresponding healthy geometries as close as possible, followed by a comparison of the resulting root diameter with the healthy reference.

2.3   Feedback Function

While the developed deformation models allow the prediction of the valve's geometry after a prosthesis induced deformation, there is a need for a feedback function that rates the healthiness of an arbitrary aortic valve geometry. Due to a lack of basic knowledge about the complex valve behaviour, the use of machine learning is proposed to achieve a data-driven estimation of this relationship.

For that purpose, a suitable feature space has to be defined. Due to the description of the valve geometry based on four anatomical landmarks, all features should be extractable from these points. Instead of using the raw point coordinates, meta-descriptors were defined and used as features. This allows for a feature space dimensionality reduction, position- and orientation-invariant features and the integration of prior knowledge, i.e. geometrical descriptors known to have a strong relationship to the valve's functionality [4].

In this study, two features were defined: $\frac{K}{L_f}$ and $\frac{\alpha}{L_f}$. K is the mean of the three commissure point distances, while α is the mean over the three angles $\alpha_i, i = 1, 2, 3$ that describe the angle between the vector pointing from the ith commissure point to its clockwise neighbor and the vector from the ith commissure point to the coaptation point (cf. Fig. 2 d). Lf is the leaflets free edge length, calculated as the sum over the lengths of the three vectors L_1, L_2 and L_3. Due to the assumption that L_f stays constant, the defined features measure K and α relative to the valves specific leaflets free edge length.

Two different feedback function concepts were examined: a binary feedback and a continous feedback. While the binary feedback function gives back a distinct binary flag whether the geometry is healthy or not (classification problem), the continous feedback functions output is a continous score to rate the healthiness of the valve (regression problem).

The binary feedback function was derived from the data using a Support Vector Machine (SVM) [7] utilizing the open-source library *LIBSVM* [8]. Non-linear relationships in the data were adressed by choosing a radial basis function kernel. The SVM-parameter C and the kernel-parameter γ were optimized utilizing a leave-one-out-method on the training data set with an iterative gridsearch on multiple grids, ranging steadily from coarse to fine (initial grid: 5 values within [1, 1000] and [0.001, 1] for C and γ, respectively, finer grids centered around optimal parameters of coarser grid from last iteration).

This SVM model learns a seperation of valve geometries in healthy and dilated ones. To provide reasonable classification results for this purpose, the training set needs to include healthy and dilated valve geometry samples.

Since a binary separation of valve geometries is not very realistic, a continuous feedback function was used additionally. This leads to a continous healthiness score. To this end, the feedback function was learned using a Gaussian Mixture Model (GMM) [9]:

(1) For every training sample point $j \in [1, n]$, define a Gaussian distribution $\mathcal{N}(\mu_{\mathbf{j}}, \Sigma_{\mathbf{j}})$
(2) Set μ_j to the samples feature space coordinates and Σ_j to a diagonal matrix with the estimated measurement noise of each feature at the corresponding diagonal entry (10% of the feature's mean value)
(3) Derive feedback function F by superimposing these distributions with equal weighting:

$$F(\mathbf{x}) = \frac{1}{n} \sum_{j=1}^{n} \mathcal{N}(\mu_j, \Sigma_j). \tag{4}$$

This completely data-driven feedback function combines low computation times with a simpler and much more practical data base. The score predicted from the GMM model corresponds to the probability of an arbitrary valve geometry to be healthy. The probability to be dilated is included implicitly in the low score values. Hence, a training data set only consisting of healthy geometry samples is sufficient and no dilated samples are required.

2.4   Surgery Planning

The surgery planning can be interpreted as an optimization problem, where the objective function is given by the feedback function. The solution space is constrained by the geometries that can be reached by deforming the individual dilated valve geometry according to a deformation model. The deformation model defines a trajectory in the feature space with its origin in the observed dilated valve geometry. The optimal prosthesis size corresponds to the point on this trajectory with maximal feedback output (cf. Fig. 3). The relationship between a valve geometry and the corresponding prosthesis size is approximated by the diameter of the circumcircle of the triangle formed by the three commissure points. In the case of the binary feedback function, the optimal solution is defined as the valve geometry classified as healthy with the smallest deformation, i.e. the intersection between the trajectory and the class border.

In this initial study, the optimization problem was solved using a gridsearch method (discretization: $0.006°$ for $\Delta\varphi_i$ and $\frac{L_{i,\theta}}{100}°$ for $\Delta\theta_i, i = 1, 2, 3$). This corresponds to a valve diameter change (i.e. change of prosthesis size) of about $0.073\,\mathrm{mm}$ for each step of θ and about $5.2 \cdot 10^{-5}\,\mathrm{mm}$ for each step of φ. This fine discretization provides an acceptable avoidance of local optima.

The evaluation of the surgery planning method was performed on all 24 aortic roots using a 10-fold cross validation. During this method, optimal prosthesis diameters were predicted for the dilated valve geometries and compared to the natural root diameter received from the corresponding healthy geometry, serving as a reference.

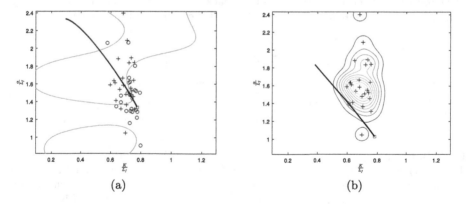

(a)   (b)

Fig. 3. (a) Example of surgery planning with binary feedback function. The HD model defines a trajectory (black) in the feature space starting at the dilated geometry (red square). The binary feedback function (blue contour) is trained on the healthy (+) and dilated (o) training samples and evaluated along the trajectory to find the optimal geometry (red diamond). (b) Example of surgery planning with continous feedback function. The HD model defines a trajectory (black) in the feature space starting at the dilated geometry (red square). The continous feedback function (contour lines) is trained on the healthy (+) training samples only and evaluated along the trajectory to find the optimal geometry (red diamond). (Color figure online)

3   Results

A fundamental assumption of the method is that the length of the vectors $L_i, i = 1, 2, 3$ stays constant during the dilation. To confirm this statement, a two-sample t-test (95% level of significance) on the L_i derived from the healthy and the dilated data was performed. We found no significant differences (L_1 : $t = -0.4978$, $p > 0.05$; L_2 : $t = -1.1455$, $p > 0.05$; L_3 : $t = -0.0405$, $p > 0.05$).

To evaluate the consistency of the presented deformation models, they were applied to reproduce the deformation observed in the experiments. The mean errors of the predicted diameters for all valves were 0.37 mm for HD, 0.09 mm for AHD and 0.25 mm for the FD model.

The presented deformation modeling method was evaluated on the data set aiming on the prediction accuracy of the optimal root diameter for each dilated valve geometry with the corresponding healthy geometry as a reference. Table 1 shows the mean prediction error and the number of matches, i.e. the relative number of predictions with an error of less than 1 mm, for all combinations of deformation models and feedback functions. The continous feedback combined with the HD model reaches the highest prediction accuracy.

A comparison of the previously proposed direct estimation method and the deformation modeling method with the binary as well as the continous feedback function is shown in Table 2. Using the continous feedback function,

Table 1. Evaluation of the deformation modeling approach for all combinations of the two feedback functions (binary and continous) and the three deformation models (HD, AHD and FD). For each combination of feedback function and deformation model the mean prosthesis size prediction error in mm and the number of matches in %.

	HD	AHD	FD
Binary	21 %	25 %	21 %
	3.36 mm	2.81 mm	4.38 mm
Continous	63 %	50 %	13 %
	1.64 mm	1.64 mm	7.60 mm

Table 2. Comparison of the direct estimation (DE) method [4], the deformation modeling (DM) with binary feedback function (AHD model) and the deformation modeling with continous feedback function (HD model) subject to the prediction accuracy and the needed training data. The deformation models were chosen to provide optimal results.

Method	Matches	Training data set
DE	63 %	Dilated and healthy valve geometry of each subject in training set
DM, binary	25 %	Dilated and healthy valve geomtries, not necessarily from the same subject
DM, continous	63 %	Healthy valve geometries

the deformation modeling approach provides the same prediction accuracy while only healthy samples are requested in the training phase.

4 Discussion

The evaluation of the three deformation models shows that all models reach a mean diameter prediction error of less than 0.5 mm. This indicates that all three models are capable of mimicking realistic aortic root deformations.

Table 1 shows that for the HD and the AHD deformation model, the continous feedback function clearly outperforms the binary feedback. The probable reason for that is that the optimal solution does not have to lie near the class border. Combining the HD deformation model with a continous feedback function provides the best prediction results. This might be related to the fact that the information about the individual dilated valve geometry disappears with growing size of the search space, which is the case with increasing DOF of the model and once more emphasizes the advantage of simplification.

The comparison with the direct estimation approach shows that using a continous feedback function, the deformation modeling approach reaches comparable results based on a much simpler training data set. The construction of a clinical data base for the direct estimation approach is nearly impossible or at least extremely time consuming. In constrast, the deformation modeling method utilizing a continous feedback function only relies on healthy valve geometries, which can be easily extracted from clinical TEE examinations. Hence, the application of simplified deformation modeling and GMMs could push the pre-operative planning of valve-sparing aortic root reconstruction towards clinical application.

However, the prediction accuracy still has to be improved. One reason for the relatively low maximal accuracy could be a biased training data set. Due to the manual identification of the landmarks in the ultrasound images, significant observer-specific errors are possible. Hence, an automatic valve geometry extraction could improve the method. A more detailed study of the influence of observer errors on the predictor performance would be interesting and relevant for future work, but was out of the scope of this paper. Additionally, the approach could be enhanced by a larger training data set. This would provide a more realistic representation of the underlying distribution of valve geometries and a higher dimensional feature space could be explored.

5 Conclusion

In this work, a new approach for pre-operative planning of valve-sparing aortic root reconstruction was presented. The approach combines simplified deformation modeling with machine learning to mimic the surgeon's decision making process and represents a first alternative to the direct estimation approach. Three deformation models and two feedback functions were defined, evaluated and compared to the previously described direct estimation approach. The results show that the presented method provides comparable results while the demand on the training data is dramatically decreased, taking a step towards clinical application for the prediction of personalized aortic root prosthesis sizes.

Acknowledgement. The authors would like to thank Ingvild Detjens and Erik Werrmann for their help conducting the experiments. This publication is a result of the ongoing research within the LUMEN research group, which is funded by the German Bundesministerium für Bildung und Forschung (BMBF) (FKZ 13EZ1140A/B). LUMEN is a joint research project of Lübeck University of Applied Sciences and University of Lübeck and represents an own branch of the Graduate School for Computing in Medicine and Life Sciences of University of Luübeck.

References

1. David, T.E.: Aortic valve sparing operations. Semin. Thorac. Cardiovasc. Surg. **23**(2), 146–148 (2011). doi:10.1053/j.semtcvs.2011.08.002. WB Saunders
2. Scharfschwerdt, M., Sievers, H.H., Hussein, A., Kraatz, E.G., Misfeld, M.: Impact of progressive sinotubular junction dilatation on valve competence of the 3F Aortic and Sorin Solo stentless bioprosthetic heart valves. Eur. J. Cardio Thorac. Surg. **37**(3), 631–634 (2010). doi:10.1016/j.ejcts.2009.09.010
3. Labrosse, M.R., Beller, C.J., Boodhwani, M., Hudson, C., Sohmer, B.: Subject-specific finite-element modeling of normal aortic valve biomechanics from 3D+t TEE images. Med. Image Anal. **20**(1), 162–172 (2015). doi:10.1016/j.media.2014.11.003
4. Hagenah, J., Werrmann, E., Scharfschwerdt, M., Ernst, F., Metzner, C.: Prediction of individual prosthesis size for valve-sparing aortic root reconstruction based on geometric features. In: IEEE 38th Annual International Conference of the Engineering in Medicine and Biology Society (EMBC), pp. 3273–3276 (2016). doi:10.1109/EMBC.2016.7591427
5. Hall, T., Pallav, S., Sudhir, W.: The role of transesophageal echocardiography in aortic valve preserving procedures. Indian Heart J. **66**(3), 327–333 (2014). doi:10.1016/j.ihj.2014.05.001
6. Hagenah, J., Scharfschwerdt, M., Stender, B., Ott, S., Friedl, R., Sievers, H.H., Schlaefer, A.: A setup for ultrasound based assessment of the aortic root geometry. Biomed. Eng./Biomedizinische Technik (2013). doi:10.1515/bmt-2013-4379
7. Burges, C.J.: A tutorial on support vector machines for pattern recognition. Data Min. Knowl. Discov. **2**(2), 121–167 (1998). doi:10.1023/A:1009715923555
8. Chang, C.C., Lin, C.J.: LIBSVM: a library for support vector machines. ACM Trans. Intell. Syst. Technol. (TIST) **2**(3), 27 (2011). doi:10.1145/1961189.1961199
9. Marin, J.M., Mengersen, K., Robert, C.P.: Bayesian modelling and inference on mixtures of distributions. Handb. Stat. **25**, 459–507 (2005). doi:10.1016/S0169-7161(05)25016-2

Assessment of Haemodynamic Remodeling in Fetal Aortic Coarctation Using a Lumped Model of the Circulation

Paula Giménez Mínguez[1]([✉]), Bart Bijnens[1], Gabriel Bernardino[1], Èric Lluch[2], Iris Soveral[3], Olga Gómez[3], and Patricia Garcia-Canadilla[1]

[1] Univesitat Pompeu Fabra, Barcelona, Spain
paula.gimenez01@estudiant.upf.edu
[2] Philips Research Medisys, Suresnes, France
[3] Fetal i+D Fetal Medical Research Center, Barcelona, Spain

Abstract. Introduction: Aortic coarctation is one of the most difficult cardiac defects to diagnose before birth, and it accounts for 8% of congenital heart diseases. Antenatal diagnosis is crucial for early treatment of the neonate and to decrease the risk of morbidity and mortality; however the fetal hemodynamic changes are not fully understood and current imaging methods are limited to accurately diagnosis this congenital defect. Objective: We propose to use a lumped model of the fetal circulation to provide insights into the hemodynamic changes in fetuses with aortic coarctation, and thus helping to improve its diagnosis. Methods: To achieve this goal a patient-specific lumped model of the fetal circulation was implemented in OpenCOR, including the modeling of different types and degrees of aortic coarctation. A parametric study of degree and type of coarctation was performed, where blood flow distribution, cerebroplacental ratio, pressure drop over the coarctation and left ventricular pressure were quantified. Results: Obvious changes in the fetal hemodynamics were observed only from 80% of coarctation, corresponding to the clinically used cut-off for pressure drop of 20 mmHg. Furthermore, the observed hemodynamic changes were different depending on the location and degree of the coarctation.

Keywords: Fetal circulation · Modeling · Aortic coarctation

1 Introduction

Aortic Coarctation accounts for around 8% of heart defects. This congenital disease consists on the narrowing of the distal aortic arch in mild cases; thus reducing the blood flow in the fetal aortic arch [1]. According to different studies [2, 3], hypoplasia of the whole aortic arch is also observed in most severe cases.

Coarctation of the aorta remains one of the cardiac defects more difficult to diagnose before birth. Prenatal diagnosis is of critical importance to improve survival and reduce morbidity [4]. In most severe cases, suspicion rises when there is ventricular disproportion, however; a slight degree of physiological disproportion appears during the third trimester of gestation [5]. Consequently, current echography methods lead to both, high false-positive and false-negative rate during diagnosis [1, 4]. Additionally, the

© Springer International Publishing AG 2017
M. Pop and G.A. Wright (Eds.): FIMH 2017, LNCS 10263, pp. 471–480, 2017.
DOI: 10.1007/978-3-319-59448-4_45

hemodynamic remodeling induced by the disease is yet not fully understood, partly due to the wide variety of fetal aortic coarctation types and degrees of severity.

Accordingly, there is a need to develop new technologies that allow accurate prenatal diagnosis of aortic coarctation and that help clinicians to better understand the disease. Currently, in clinical practice ultrasonographic evaluation, including measurement of right and left ventricular size, isthmus and ductal diameters, is used to diagnose aortic coarctation. Still, these measurements have low sensitivity and specificity [1, 6]. On the other hand, patient-specific lumped models of the fetal circulation have shown to be a great approach to understand the underlying mechanisms of fetal hemodynamics, both under healthy and pathological conditions, such as shown in Garcia-Canadilla et al. [7, 8].

Thereby, our purpose is to re-implement the fetal circulation model proposed in [8] in CellML using OpenCOR, an open-source cross-platform modelling environment [9], which favors model reuse amongst researchers. Besides, the modular structure of CellML allows users to easily add/remove new components to the model for future applications. The model will be used to simulate fetal circulation subjected to aortic coarctation.

2   Materials and Methods

2.1   Patient-Specific Model

Lumped Model of the Fetal Circulation
The fetal circulation model [8] was implemented in OpenCOR (version 0.5) using the CellML language. The ascending aorta of the original model was split into 3 different arterial segments: ascending aorta, ascending and descending aortic arches as shown in Fig. 1. Moreover, a resistor distal from the aortic-ductal junction was included, representing distal aortic coarctation. The length of the segment was fixed at a 10% of the total ascending aorta length, and the radius was set initially to be equal to the radius of the ascending aorta. These changes in the model made possible the simulation of a variety of aortic coarctation scenarios.

The equivalent lumped model circuit was built by interconnecting arterial segments and vascular beds. Each arterial segment contained a capacitor (C) representing arterial compliance, a resistor (R) representing resistance to blood flow through the arterial segment, and an inductor (L) describing blood inertia. The vascular bed components consisted of a three-element Windkessel model, which included one resistor (R_c), representing resistance of the characteristic impedance of the feeding artery, leading to a parallel circuit with a resistor (R_p) and a capacitor (C_p) describing peripheral resistance and compliance respectively.

Furthermore, CellML language was used to implement the improved version of the fetal circulation model, thanks to its modular structure that allow easily defining interconnected individual components (such as arterial segments and vascular beds), establishing connections between them, and importing the individual components into the model. Regarding the OpenCOR solver we used CVODE with a time step of 0.001 s.

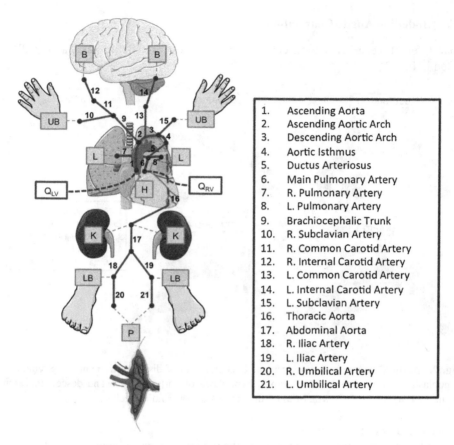

1.	Ascending Aorta
2.	Ascending Aortic Arch
3.	Descending Aortic Arch
4.	Aortic Isthmus
5.	Ductus Arteriosus
6.	Main Pulmonary Artery
7.	R. Pulmonary Artery
8.	L. Pulmonary Artery
9.	Brachiocephalic Trunk
10.	R. Subclavian Artery
11.	R. Common Carotid Artery
12.	R. Internal Carotid Artery
13.	L. Common Carotid Artery
14.	L. Internal Carotid Artery
15.	L. Subclavian Artery
16.	Thoracic Aorta
17.	Abdominal Aorta
18.	R. Iliac Artery
19.	L. Iliac Artery
20.	R. Umbilical Artery
21.	L. Umbilical Artery

Fig. 1. Lumped model scheme of the fetal circulation consisting on 21 artery segments (lines), and 12 vascular beds (boxes). Two blood flow inputs were considered: Left and right ventricular outflows. UB: upper body, B: brain, L: lungs, H: heart, K: kidneys, LB: lower body, P: Placenta.

Patient-Specificity

Patient-specific data were used to build the model, including: output flow from left (Q_{LV}) and right (Q_{RV}) ventricles, fetal gestational age (GA) and the estimated fetal weight (EFW). All the data were obtained from the same control fetus as used to validate the model described in [7].

All the electrical components of the equivalent circuit were fitted to a control healthy fetus of 33.28 weeks of GA and an EFW of 2250 gr. The dimensions of all arterial segments (radius, length and thickness), Young's Moduli, blood viscosity, and electrical components of all vascular beds (R_p, C_p), were calculated according to the GA and EFW of the fetus, as described in [8]. Particularly, ascending aorta, and aortic arch new segments' dimensions were calculated based on [10].

2.2   Modelling Aortic Coarctation

Four different scenarios of aortic coarctation were considered in this study andare illustrated in Fig. 2.

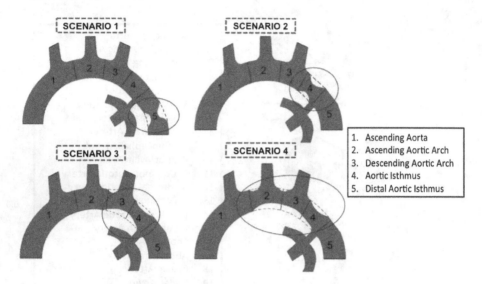

Fig. 2. Aortic Coarctation types. Scenario 1: coarctation of distal aortic isthmus; Scenario 2: hypoplasia of aortic isthmus; Scenario 3: hypoplasia of aortic isthmus and descending arch; Scenario 4: hypoplasia of aortic isthmus, descending and ascending arches.

Modeling different scenarios

To model the first scenario, a reduction in the radius of the aortic segment located distal from the aortic-ductal junction was introduced, simulating different degrees of coarctation ranging from 0% (normal conditions) to 90%. Scenarios 2, 3 and 4 where modeled by reducing the radius of the corresponding arterial segment/s denoted in Fig. 2.

Modeling ventricular disproportion in aortic arch hypoplasia

In addition, ventricular disproportion was modeled in aortic arch hypoplasia(Scenario 4), by simulating different proportions of right (RV) and left ventricular (LV) outflows: 50–50% (no disproportion), 40–60%, 30–70% and 20–80%, respectively.

Parametric Study of Coarctation Degree

The effect of aortic coarctation in the fetal circulation was analyzed by decreasing the radius of the different regions a given percentage. The following percentages of narrowing were considered: 0% (normal conditions), 50%, 70%, 80%, and 90%. Cerebro-placental ratio, pressure drop over the narrowed area and left ventricular pressure were measured for each scenario. Cerebro-placental ratio was computed as the maximum systolic brain flow over the maximum systolic placental flow. Pressure drop was calculated as the difference between the maximum input and the maximum output pressures of the narrowed area. Moreover, blood flow in the aortic isthmus (AoI), middle cerebral artery (MCA) and umbilical artery (UA) were also obtained. In the present model, MCA blood flow was defined as the 75% of the internal common carotid artery outflow.

3   Results

3.1   Modeling Different Scenarios

The AoI blood flow showed a progressive reversal as the percentage of coarctation increased in the distal aortic isthmus (Scenario 1, Fig. 3A). In the rest of scenarios, blood flow in the AoI was reduced until it became 0 ml/s for a 100% of coarctation, as expected. It can be observed that no differences were notice until a 70% of stenosis was reached. Regarding MCA and UA blood flows (Fig. 3B, C), they increased and decreased respectively when increasing the amount of coarctation. The increment in MCA blood flow was larger in the most severe scenario corresponding to a tubular hypoplasia (Scenario 4), as well as in the case of coarctation distal to the aortic-ductal junction (Scenario 1); while reduction in UA blood flow was much more significant in Scenario 1.

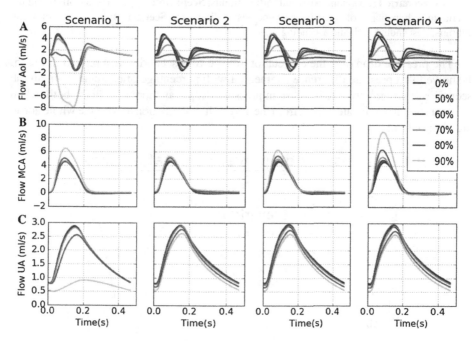

Fig. 3. Blood flow through: aortic isthmus (AoI), middle cerebral artery (MCA) and umbilical artery (UA); for different percentages of stenosis. Scenario 1: coarctation of distal aortic isthmus; Scenario 2: hypoplasia of aortic isthmus; Scenario 3: hypoplasia of aortic isthmus and descending arch; Scenario 4: hypoplasia of aortic isthmus, descending arch and ascending arch.

The cerebro-placental ratio (Fig. 4A) augmented with stenosis for all coarctation types; however, under distal isthmus coarctation the most severe increment was attained for 90% stenosis. The left ventricular pressure (Fig. 4B) also increased with coarctation;

it was observed that in aortic isthmus-arch hypoplasia (Scenario 4), when LV and RV output were similar the highest pressure was reached, at a non-physiological 90 mmHg for 90% stenosis.

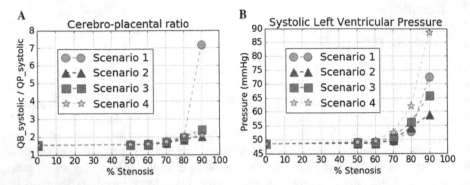

Fig. 4. A. Cerebro-placental ratio. **B.** Left ventricular pressure; for different percentages of stenosis. **Scenario 1:** coarctation of distal aortic isthmus; **Scenario 2:** hypoplasia of aortic isthmus; **Scenario 3:** hypoplasia of aortic isthmus and descending arch; **Scenario 4:** hypoplasia of aortic isthmus, descending arch and ascending arch.

The pressure drop over the coarctation for the four scenarios was plotted in Fig. 5. It showed that pressure difference augmented with percentage of coarctation, in all the scenarios. Furthermore, the increase in pressure drop was more significant both, when the coarctation was located distal from the aortic-ductal junction and for the whole aortic arch hypoplasia.

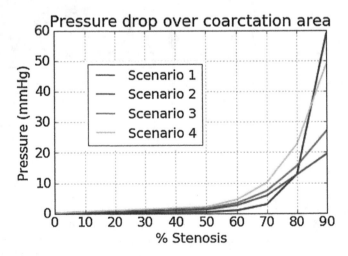

Fig. 5. Maximum pressure drop over area of coarctation, against percentage of stenosis. Dark blue: coarctation of distal aortic isthmus; Green: hypoplasia of aortic isthmus, Red: hypoplasia of aortic isthmus and descending arch; Light blue: hypoplasia of aortic isthmus, descending arch and ascending arch. (Color figure online)

3.2   Modeling Ventricular Disproportion in Aortic Arch Hypoplasia

In clinical scenarios, the LV stroke volume (and size) decreases when the coarctation becomes more drastic. The driving force (as can be seen from Fig. 4B) is likely the increase in LV pressure. Therefore we also simulated this RV/LV imbalance. One can see that reducing the amount of flow leaving the left ventricle favored the retrograde flow in the AoI, in aortic arch hypoplasia (Fig. 6A); while increased stenosis resulted in a decrement of the AoI retrograde flow. Regarding the MCA blood flow (Fig. 6B), it was increased by stenosis; nevertheless, under severe ventricular disproportion (20–80%), MCA flow decreased with stenosis. Moreover, it was noticed that UA flow augmented with stenosis, under ventricular disproportion (Fig. 6C). The more severe the disproportion, the higher the increment with stenosis.

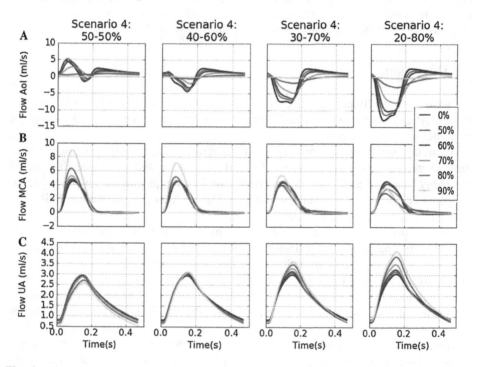

Fig. 6. Blood flow through aortic isthmus (AoI), middle cerebral artery (MCA) and umbilical artery (UA), for different percentages of stenosis. Hypoplasia of aortic isthmus, descending arch and ascending arch, simulated for 50–50%, 40–60%, 30–70% and 20–80% proportion of left-right ventricle flow respectively.

Regarding cerebro-placental ratio (Fig. 7A) it was observed to decrease with stenosis as ventricular disproportion became more severe. On the other hand, left ventricular pressure (Fig. 7B) started to normalize. Ventricular disproportion favored the decrement of left ventricular pressure, also enhanced by coarctation.

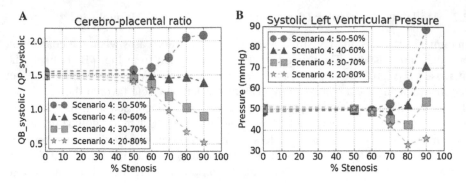

Fig. 7. A. Cerebro-placental ratio. **B.** Left ventricular pressure; for different percentages of stenosis. Hypoplasia of aortic isthmus, descending arch and ascending arch, simulated for 50–50%, 40–60%, 30–70% and 20–80% proportion of left-right ventricle flow respectively.

4   Discussion

We have successfully implemented a lumped model of the fetal circulation in CellML/ OpenCOR and simulated the hemodynamic effects of different degrees and geometries of in-utero aortic coarctations.

Coarctation of the aorta is one of the most difficult cardiac defects to diagnose before birth. Antenatal diagnosis is crucial for early treatment and to reduce the risk of morbidity and mortality. For this purpose, we have implemented apatient-specific lumped model of the fetal circulation in OpenCOR, including the modeling of different types and degrees of aortic coarctation.

The results show that the current lumped model of the fetal circulation provides insight in the underlying mechanism of aortic coarctation. The reversal of the systolic AoI flow observed when coarctation is located distal from the aortic-ductal junction agrees with previous studies [1]. As a consequence, the blood to the brain increases while the blood flow to the lower body and placenta decreases. Since the coarctation is located just after the aortic-ductal junction, most of the blood coming through the ductus is redirected towards the upper body due to the narrowing of the distal aortic arch. The increase in MCA blood flow is also observed in the other scenarios, but more strongly in aortic isthmus and arch hypoplasia. Thus, as coarctation increases, blood from the ascending aorta can only reach the brachiocephalic trunk and just a small amount of blood passes through the narrowed arch towards the descending aorta. However, keeping RV/LV output balanced led to non-physiological LV pressures.

We also noticed that reducing the amount of outflow of the left ventricle, normalized the LV pressures and led to retrograde flow in the AoI, as supported by previous studies [11]. Nevertheless, when the whole aortic arch is hypoplastic (Scenario 4), less AoI retrograde flow is observed when increasing stenosis. Though it seems contradictory, MCA blood flow increases with stenosis, since the flow through the aortic arch is reduced and it favors blood flow towards the brachiocephalic trunk. This is again partly normalized when the relative amount of left ventricular outflow decreases up to 20% and the blood flow in the MCA is also reduced. On the other hand, blood flow in the UA increases with the amount of stenosis as well as when the difference between right and left relative amount of outflow rises, showing that more blood goes from the ductus arteriosus to the lower body and placenta.

Concerning the pressure drop over the narrowed area, we perceived that at around 80% stenosis the pressure gradient was around 20 mmHg. In current clinical practice, a coarctation (as measured postnatally) is considered significant from this value onwards, indicating that hemodynamic changes observed in our model agree with clinically established knowledge [12]. Therefore, the simulations provide more insights into the hemodynamics of aortic coarctation. Moreover this model provides patient-specific pressure gradient estimations, which are challenging to obtain in the clinic.

On the other hand, implementation of the model in CellML/OpenCOR reduces the computation time [9], making possible real time simulations, and it facilitates its reutilization for future applications.

The main limitation of this model is that fetal circulation is still quite simplified. However, future lines of work aim to overcome this limitation by modeling the heart, and thus, increasing resemblance to reality. Moreover, we haven't analyzed the effect of oxygen and vasoconstriction, which can also have an effect on the results.

5 Conclusion

In this paper, we proposed the use of a lumped model of the fetal circulation to evaluate the hemodynamic changes in fetal aortic coarctation. Given the insight obtained, this tool might help to improve prenatal diagnosis of aortic coarctation.

Acknowledgments. This study was partially supported by the Spanish Ministry of Economy and Competitiveness (grant TIN2014-52923-R; Maria de Maeztu Units of Excellence Programme - MDM-2015-0502), FEDER and the European Union Horizon 2020 Programme for Research and Innovation, under grant agreement No. 642676 (CardioFunXion).

References

1. Buyens, A., Gyselaers, W., Coumans, A., Al Nasiry, S., Willekes, C., Boshoff, D., Witters, I., et al.: Difficult prenatal diagnosis: fetal coarctation. FVV ObGyn. **4**, 230–236 (2012)
2. Espinoza, J., Romero, R., Kusanovic, J.P., Gotsch, F., et al.: Prenatal diagnosis of coarctation of the aorta with the multiplanar display and B-flow imaging using 4-dimensional sonography. J. Ultrasound Med. **28**, 1375–1378 (2009). https://doi.org/10.7863/jum.2009.28.10.1375
3. Achiron, R., Zimand, S., Rotstein, Z., et al.: Fetal aortic arch measurements between 14 and 38 weeks' gestation: in-utero ultrasonographic study. Ultrasound Obstet. Gynecol. **15**, 226–230 (2000). https://doi.org/10.7863/jum.2009.28.10.1375
4. Matsui, H., Mellander, M., Roughton, M., et al.: Morphological and physiological predictors of fetal aortic coarctation. Circulation **118**, 1793–1801 (2008). https://doi.org/10.1161/CIRCULATIONAHA.108.787598

5. Kenny, J.F., Plappert, T., Sutton, M.S.J., et al.: Changes in intracardiac blood flow velocities and right and left ventricular stroke volumes with gestational age in the normal human fetus: a prospective doppler echocardiographic study. Circulation **74**, 1208–1216 (1986). https://doi.org/10.1161/01.CIR.74.6.1208

6. Mukai, Y., Samura, O., Teraoka, Y., Sasaki, M.: Prenatal diagnosis of coarctation of the aorta using the three-vessel view by calculating the ratio of the diameter of the aortic root to that of the pulmonary artery. Ultrasound Obs. Gynecol. **44**, 1793–1801 (2014). https://doi.org/10.1002/uog.14118

7. Garcia-Canadilla, P., Rudenick, P.A., Crispi, F., Cruz-Lemini, M., Palau, G., Camara, O., et al.: A computational model of the fetal circulation to quantify blood redistribution in intrauterine growth restriction. PLoS Comput. Biol. **10**, e1003667 (2014). doi:10.1371/journal.pcbi.1003667

8. Garcia-Canadilla, P., Rudenick, P.A., Crispi, F., Cruz-Lemini, M., Triunfo, S., Nadal, A., et al.: Patient-specific estimates of vascular and placental properties in growth-restricted fetuses based on a model of the fetal circulation. Placenta **36**, 981–989 (2015). doi:10.1016/j.placenta.2015.07.130

9. Hunter, P.: OpenCOR Tutorial (2016)

10. Szpinda, M.: Length growth of the various aortic segments in human foetuses. Folia Morphol. **67**, 245–250 (2008)

11. Yamamoto, Y., Khoo, N.S., Brooks, P.A., Savard, W., et al.: Severe left heart obstruction with retrograde arch flow influences fetal cerebral and placental blood flow. Ultrasound Obstet. Gynecol. **42**, 294–299 (2013). doi:10.1002/uog.12448

12. Beauchesne, L.M., Connolly, H.M., et al.: Coarctation of the aorta: outcome of pregnancy. J. Am. Coll. Cardiol. **38**, 1728–1733 (2001). https://doi.org/10.1016/S0735-1097(01)01617-5

3D Motion Modeling and Reconstruction of Left Ventricle Wall in Cardiac MRI

Dong Yang[1]([✉]), Pengxiang Wu[1], Chaowei Tan[1], Kilian M. Pohl[3],
Leon Axel[2], and Dimitris Metaxas[1]

[1] Department of Computer Science, Rutgers University,
Piscataway, NJ 08854, USA
don.yang.mech@gmail.com
[2] Department of Radiology, New York University, New York, NY 10016, USA
[3] Center for Health Sciences, SRI International, Menlo Park, CA 94025, USA

Abstract. The analysis of left ventricle (LV) wall motion is a critical step for understanding cardiac functioning mechanisms and clinical diagnosis of ventricular diseases. We present a novel approach for 3D motion modeling and analysis of LV wall in cardiac magnetic resonance imaging (MRI). First, a fully convolutional network (FCN) is deployed to initialize myocardium contours in 2D MR slices. Then, we propose an image registration algorithm to align MR slices in space and minimize the undesirable motion artifacts from inconsistent respiration. Finally, a 3D deformable model is applied to recover the shape and motion of myocardium wall. Utilizing the proposed approach, we can visually analyze 3D LV wall motion, evaluate cardiac global function, and diagnose ventricular diseases.

Keywords: Quantitative shape modeling · 3D motion reconstruction · Left ventricle (LV) · Cardiac MRI

1 Introduction

Cardiovascular diseases, such as ventricular dyssynchrony, heart attack, and congestive heart failure, are one of the major causes for human death all over the world. A comprehensive analysis of 3D heart wall motion is fundamental for understanding the ventricular functioning mechanism, and essential for early prevention and accurate treatment of the related diseases. However, classical diagnosic tests, including electrocardiogram (ECG), echocardiography (echo), chest X-ray, and cardiac catheterization, are not able to provide sufficient spatial information with adequate resolution for motion modeling. The 3D echocardiography can be currently used to study cardiac motion and strains, but its visual appearance may not be clear due to limited imaging quality. In the present study, we adapt images from cardiac cine magnetic resonance imaging (MRI), which is an a non-invasive imaging technique to visualize the heart conditions both in time and space. The sequence of 2D MRI acquired along the long-axis (LAX)

© Springer International Publishing AG 2017
M. Pop and G.A. Wright (Eds.): FIMH 2017, LNCS 10263, pp. 481–492, 2017.
DOI: 10.1007/978-3-319-59448-4_46

and the full cardiac cycle provides a complementary view of the shape and function of the left ventricle (LV) compared to the sequence of 2D MRI acquired across the short-axis (SAX) and time. Thus, analyzing a set of 2D cine MRI sequence can provide a feasible way to fully recover 3D LV wall motion.

In order to analyze the global function and the regional heart wall motion, the contours of the epicardium and endocardium of the LV must be annotated or delineated, either by human experts or machines. However, the annotation procedure is time-consuming and tedious for doctors and physicians, which then becomes a bottleneck for extraction of functional cardiac data in the clinical practice. An automatic method for LV segmentation (or contour extraction), which would reduce both manual labor and annotation time dramatically, has been sought for decades to increase the clinical efficiency of cardiac MRI. Recently, some scholars have proposed several methods to address them. For example, Paragios proposed a level-set method for cardiac MR segmentation [4] with the gradient vector flow and geodesic active contour model. Jolly also introduced an automatic segmentation method for both CT and MR images, using multi-stage graph cut optimization in the image plane [5]. In addition, Zhu et al. developed a statistical model, named subject-specific dynamic model (SSDM), to handle the cardiac dynamics and shape variation [6]. Although the ring-shaped structure formed by the paired epicardium and endocardium contours is fairly simple, the cardiac MR imaging quality can be inconsistent, because of factors such as different acquisition settings or potential artifacts introduced by respiration during the slow acquisition process. Furthermore, the endocardial contour is intrinsically somewhat ill-defined, due to the presence of the papillary muscles and trabeculations, which tend to be considered as part of the ventricular cavity. Thus, the contours of the LV wall segmentation may need to be estimated even when the local image contrast is partially corrupted; conventional intensity-based segmentation methods may fail in such cases. Moreover, the prevailing approaches [11–13] are mostly concerned with the SAX MR slices. Without further study of the LAX slices and slice alignment, the calculation of the global functions for the LV may not be accurate.

Removal of motion artifacts caused by varying respiration is another important issue to accommodate for analyzing the function of the heart. Although the cine MR sequences are captured at fixed spatial locations during breath-holding, it is unlikely that the respiration phase would remain the same at different slices of the cine MRI. The MR slices at different locations are inevitably misaligned with spatial offsets and in-plane deformation. Such misalignment issues can seriously affect the precision and representativeness of a 3D heart model that is built up on the unaligned MR sequences. Therefore, we need to solve this image registration problem between different MR slices. In [2], Lotjonen et al. proposed an alignment method maximizing normalized mutual information of image appearances between SAX and LAX slices. However, the optimization procedure is highly non-convex and easily falls into a local minimum. Garlapati et al. [10] proposed an effective method to solve the misalignment problems in brain imaging, based on the local boundary detection. It may not be applicable in our case because the boundary of LV wall is not always clear in cardiac MRI.

Assuming the in-plane segmentation and slice alignment are accommodated, 3D shape modeling and motion reconstruction of LV wall are the next steps in analysis. The 3D shape and motion provide quantitative and visual characteristics to study the normal and abnormal heart functioning mechanisms in a more comprehensive way, compared with echo. Park et al. studied the shape and motion of the LV using a volumetric deformable model based on tagging MRI [1]. The dynamic deformation of the ventricular wall is computed with Lagrangian dynamics and finite element method.

In this paper, we present a novel approach to reconstruct 3D shape and motion of the LV wall for understanding ventricular functioning mechanisms. First, we adopt a fully convolutional network (FCN) to extract epicardium and endocardium contours from the MR slices. Second, we develop a new algorithm to align MR slices in space, compensating the respiration effect. Finally, a deformable model is utilized to recover the 3D shape and motion of the LV wall with Lagrangian dynamics.

2   Myocardium Contour Extraction

In our approach, LV segmentation is defined as a pixel-wise semantic classification problem, that is, segmentation with class labels. The pixels of myocardium muscle within a semi-ring shape (formed by the epicardium and endocardium) are labelled as one class; pixels of blood pool and other contents are labelled as another class. We adopt the fully convolutional network (FCN), U-net [3], as the learning model following the end-to-end convention during the training and testing. The initial segmentation results are shown in Fig. 1, which don't all resemble the golden standard (Fig. 2).

We enforce strong shape constraints for the segmented contours resulting from the previous step, since the ring-shaped structure of the LV contours is an important prerequisite and the smoothness of contours needs further refinement.

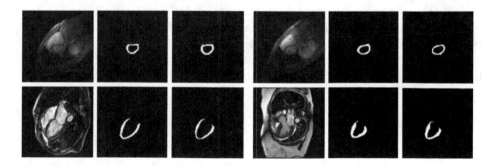

Fig. 1. For example segmentation: the raw images (left) are segmented by FCN (middle), which are close to the gold standard (right).

Fig. 2. Left: clusters in the shape pool; right: mean shape.

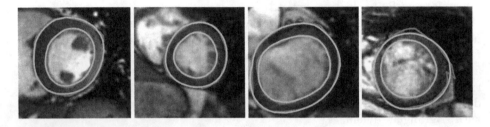

Fig. 3. Four sample results before and after applying the group sparsity constraints: red contours are the results from the proposed fully convolutional network (FCN), green ones are the refined results after applying group sparsity constraints. (Color figure online)

However, the raw prediction from the FCN sometimes forms ring-shapes with unreasonable patterns, e.g., zig-zag curves or the intersection of two contours, as shown in the fourth example of Fig. 3. The initial shape is generated from the shape pool and can be reliably placed in the image plane even when the appearance cue is misleading. The shapes of the LV wall vary from the phase of end-diastole (ED) to that of end-systole (ES), and from the slices near the aorta to those near LV apex. For instance, the contours close to the aorta may be partially merged together, particularly in the membranous portion of the interventricular septum, and the myocardium muscle close to the LV apex is thinner compared to the muscle at other locations (although the typically oblique intersection of the image plane with the apical LV wall in SAX images can result in apparent increased wall thickness, due to volume averaging). We cluster training shapes into different groups by the geometry and muscle thickness, and compute the refined contours of testing data by optimizing the dictionary learning formulation with the group sparsity constraints shown in Eq. 1:

$$\underset{x,e,\beta}{\text{minimize}} \left\{ \|T(y,\beta) - Dx - e\|_2^2 + \lambda_1 \sum_{s \subseteq S} \|x_s\|_2 + \lambda_2 \|e\|_1 \right\} \tag{1}$$

where $T(y, \beta)$ is the similarity transformation with parameter β for aligning the initial shape y, generated by FCN, to the mean shape of the shape pool. Matrix $D = [d_1, d_2, \cdots, d_k]$ represents the training shape pool, column vector $d_i \in \mathbb{R}^{3n}$

contains the coordinates of n vertices on the contours. $S = \{1, 2, \cdots, k\}$ is the set of indices of x. The clustering process divides S into several non-overlap subsets, $S = \bigcup_i S_i$, $S_i \bigcap S_j = \emptyset, \forall i \neq j$. Vector $x \in \mathbb{R}^k$ contains the weights for the linear combination of shapes in the pool. x_{S_i} is the sub-vector for the group $S_i \in S$, and the term $\sum_{S_i \subseteq S} \|x_{S_i}\|_2$ is a standard group-sparsity regularization ($l_{2,1}$ norm). Vector e models the non-Gaussian error in the case that partial contour information is missing. λ_1 and λ_2 control the weights of two sparsity terms. After solving the optimization, the myocardium contours are refined with the most correlated shapes from a small group of shapes from the training pool. The sample results are shown in Fig. 3. The similar process is conducted for the LAX slices as well.

3 Rigid Image Registration for Spatial Alignment

The heart motion under respiration is mainly a rigid-body translation in the craniocaudal (CC) direction, with minimum deformation [7]. Therefore, we assume that in-plane rigid translation is sufficient to compensate the respiration effect for SAX. We also assume the offset of one cardiac phase in a slice can be applied for all cardiac phases at the location, because the respiration phase is almost identical in one-slice acquisition with breath-holding. For simplicity, the registration is carried out only at the state of end-diastolic (ED) for all slices simultaneously.

We propose a novel slice alignment algorithm, described in Algorithm 1, to adjust both SAX and LAX slices, using the contours from the previous step and slice intersection relations. Since SAX slices are almost parallel to each other, we take intersections between SAX and LAX slices, or different LAX slices, into consideration. At SAX slice s, the corresponding image plane is T_s and 2D contours (epicardium and endocardium) are v_s. Contours v_l in LAX slice l have intersection points p_l with slice s. Then, the closest points $p_s \in v_s$ are computed corresponding to all points in p_l. $\|p_s - p_l\|_2 = 0$ ideally if no respiration effect exists during the acquisition. However, as shown in Fig. 4, p_s and p_l may not intersect with each other. The difference $p_s - p_l$ provides the direction to shift the image plane (or shift the contours equivalently). Computing all the intersection points from LAX contours, the final translation displacement can be determined by taking the average on $p_s - p_l$. The procedure is analogous for LAX slices. The whole procedure is repeated if the marginal update of alignment is greater than a fixed threshold. The complete algorithm, shown in Algorithm 1, is guaranteed to converge to a stable condition where most intersection points are on the in-plane contours among all slices and frames.

4 3D Shape Modeling and Motion Reconstruction

Deriving 3D shape and motion of LV wall from the well-aligned contours of different slices is essential for understanding heart functioning mechanism. Analyzing motion of a sequence of 2D contours along an axis and time is able to show some

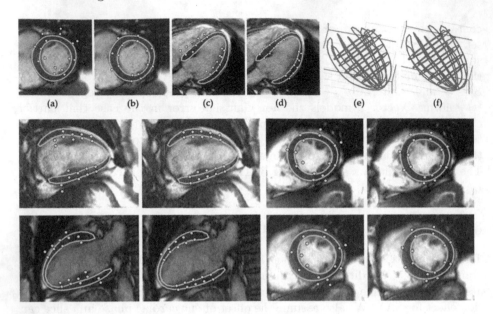

Fig. 4. Results (before and after) MR slice alignment. (a,b): SAX myocardium contours and intersection points with LAX contours; (c,d): LAX myocardium contours and intersection points with SAX contours; (e,f): all contour points in 3D space; bottom: four sample slices with intersection points before and after alignment.

Algorithm 1. Joint alignment of 2D MR short- and long-axis slices

---

    Data: all 2D contours $\mathbf{v}$ on different image planes $\mathbf{T}$
    Result: in-plane translation $(\delta x, \delta y)_s$ for each MR slice s
1   initial step coefficient $\gamma = 0.5$, initial gap threshold $\theta = 0.1$;
2   initial $(\delta x, \delta y)_s = (0, 0)$;
3   compute intersection points $\mathbf{p}$ of $\mathbf{v}$ and $\mathbf{T}$;
4   compute the closest in-plane points $\mathbf{p}' \in \mathbf{v}$ to $\mathbf{p}$;
5   iteration index $i = 1$, maximum iteration number $i_{max} = 100$;
6   **while** $i \leq i_{max}$ *and* $\|\mathbf{p} - \mathbf{p}'\| \leq \theta$ **do**
7       $i \leftarrow i + 1$;
8   **for** *each slice s* **do**
9           $\delta x_s \leftarrow \delta x_s + \gamma \cdot \sum (\mathbf{p}'_s - \mathbf{p}_s)_x$;
10         $\delta y_s \leftarrow \delta y_s + \gamma \cdot \sum (\mathbf{p}'_s - \mathbf{p}_s)_y$;
11         $\mathbf{v}_s \leftarrow \mathbf{v}_s + (\delta x, \delta y)_s$;
12   **end**
13   compute intersection points $\mathbf{p}$ of $\mathbf{v}$ and $\mathbf{T}$;
14   compute the closest in-plane points $\mathbf{p}' \in \mathbf{v}$ to $\mathbf{p}$;
15 **end**

---

characteristics of heart motion. However, 2D image slices, at the same location but at different phases of the cardiac cycle, actually may present different parts of heart, due to the 3D ventricular motion. Thus, the sequence of 2D MRI slices

---

Algorithm 2. LV wall motion computation over the whole cardiac cycle.

Data: all 2D contours $\mathbf{v}$ in space and time, a 3D reference shell shape S_0
Result: shapes of LV wall at all cardiac phases

1 compute initial ED shape S from S_0 to $\mathbf{v}_{ED}$ using non-rigid CPD;
2 initial threshold $\theta > 0$, initial overall displacement update $\Delta > \theta$;
3 **while** $\Delta > \theta$ **do**
4 $\quad$ compute the closest point sets $\mathbf{u} \in S$ corresponding to points in $\mathbf{v}_{ED}$;
5 $\quad$ calculate forces based on difference $\mathbf{v}_{ED} - \mathbf{u}$;
6 $\quad$ interpolate forces f for vertices $V \in S$;
7 $\quad$ calculate $\dot{q}$ and update q;
8 $\quad$ $S' \leftarrow$ update surface mesh S using q;
9 $\quad$ $\Delta \leftarrow \|S' - S\|$, and $S \leftarrow S'$;
10 **end**
11 $S_{initial} \leftarrow S$;
12 **for** *cardiac phase t from ED to ES* **do**
13 $\quad$ initial overall displacement update $\Delta > \theta$;
14 $\quad$ **while** $\Delta > \theta$ **do**
15 $\quad\quad$ compute the closest point set $\mathbf{u} \in S_{initial}$ of points in $\mathbf{v}_t$;
16 $\quad\quad$ calculate forces based on difference $\mathbf{v}_t - \mathbf{u}$;
17 $\quad\quad$ interpolate force f_t for vertices in $S_{initial}$, calculate $\dot{q}$ and update q;
18 $\quad\quad$ $S' \leftarrow$ update surface mesh $S_{initial}$ using q;
19 $\quad\quad$ $\Delta \leftarrow \|S' - S_{initial}\|$;
20 $\quad\quad$ $S_t \leftarrow S'$;
21 $\quad\quad$ $S_{initial} \leftarrow S'$;
22 $\quad$ **end**
23 **end**

---

does not show the true pattern of heart dynamics (shape, strain, etc.). In order to achieve better analysis, we recover the 3D LV wall shapes over the whole cardiac cycle from the sparse in-plane contours. We propose a new method, shown in Algorithm 2, to reconstruct 3D LV shapes and motion, adopting the deformable model. We use the rigid point-wise registration method, coherent point drifting (CPD) [8], to initialize the 3D shape for the cardiac phase of end-diastolic (ED) from a reference shape towards the aligned contours in space. The shapes for the whole cardiac cycle are computed along the direction from ED to end-systolic (ES). Next we construct the deformable model directly on the triangular mesh from results of CPD registration. The point locations of the deformable model are a function of time t and vector q:

$$x(q, t) = c + R(s + d) \tag{2}$$

where c is the origin of local coordinates, R is a rotation matrix, s and d are global and local deformation, respectively. q is defined as a vector of parameters in kinematics and dynamics and $\dot{x} = L\dot{q}$, where matrix L is derived from Eq. 2. According to Lagrangian dynamics, we have the following equation:

$$D\dot{q} + Kq = f_t \tag{3}$$

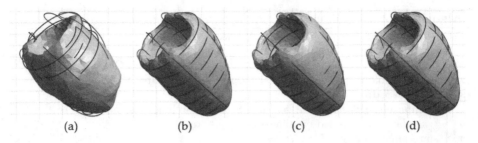

Fig. 5. 3D yellow models are the LV models, red curves are the 2D aligned contours from SAX and LAX slices in space. (a) Initial model from the referenced LV model at the phase of ED using CPD; (b) fitted model for the phase of ED using deformable model based on the contours; (c) LV model at the phase $k-1$ and contours at the phase k; (d) final fitted model at phase k. (Color figure online)

where D is the damping matrix, and K is the stiffness matrix. The external force f_t at phase t is proportional to the Euclidean distance between contour points and initial shape S within a local neighborhood. Once we have the initial shape, we can update the deformable model and the corresponding mesh by solving Eq. 3. Therefore, the shape at each phase can be computed using the computed shape of the previous phase as initialization for deformation (Fig. 5). Then, we can recover the whole motion of the LV wall phase-by-phase with proper smoothness (guaranteed by the deformable model).

5   Experiments

We used a cardiac MRI dataset containing MR image sets of 22 normal volunteers and 3 patients for the initial study. The patients all had heart failure with dyssynchrony, and were scheduled for cardiac resynchronization therapy (CRT). We manually annotated LV contours for all LAX and SAX images at each location over different cardiac phases, except the slice planes that did not cut through the LV. Image size varied between 224×204 pixels and 240×198 pixels, and its resolution varied from 1.17 mm to 1.43 mm. In total, 25 subjects (approximately 5625 images, both SAX and LAX) were used from our dataset, randomly divided into training set (20 subjects) and testing set (5 subjects). The SAX and LAX network models were trained separately. A 5-fold cross-validation was used in the training set. We compared the results for FCN and the proposed methods, using Dice's coefficient as the evaluation metric for segmentation. The result in Table 1 shows that the proposed method has better performance than FCN, because the shapes of output contours are regularized. For the motion reconstruction, some manual adjustment of segmented contours is necessary in terms of accuracy, which takes a few minutes for each case, on average. Once the adjustment of contours is finished, we conduct the processing steps without any further update for the contours.

Table 1. Evaluation results

Dice's coefficient	Our dataset	Challenge dataset
U-Net (mean)	0.70	0.53
U-Net (std)	0.07	0.15
Proposed method (mean)	0.86	0.70
Proposed method (std)	0.04	0.12

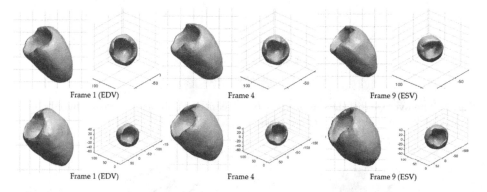

Fig. 6. Two views of LV model at three frames: first row for a volunteer, second row for a patient.

We also evaluated our methods with the public dataset from the cardiac MRI segmentation challenge of MICCAI 2009 [9], as well. The dataset, from the Sunnybrook Health Sciences Center, contains 45 cine SAX slices, covering both normal and abnormal cardiac conditions. Image size is 256 × 256 pixels, and its resolution varies from 1.2500 mm to 1.3672 mm. Expert annotations of endocardium and epicardium contours are provided for some slices at EDV and ESV phases. We only evaluated the cases where both endocardium and epicardium annotations are given for the same image. The dataset is divided into three subsets: training, validation, and online, following the standard nested cross-validation. We trained our model with the training set (135 images), evaluate the model with the evaluation set (138 images) and tested with the online set (147 images). Accuracy was measured with the Dice's coefficient, as well shown in Table 1. The accuracy is slightly less than previous experiments since the training set is fairly small.

The average distance between the contour points and the reconstructed model is utilized as the metric to evaluate the performance of the rigid alignment. The result for the whole dataset along the cardiac phase is shown in the Fig. 7. We find that the distance at each time point is much smaller when applying alignment than that without any alignment. This means our alignment strategy well improves the consistency of contours in 3D space well. The model from the aligned contours is also improved, as shown in Fig. 8.

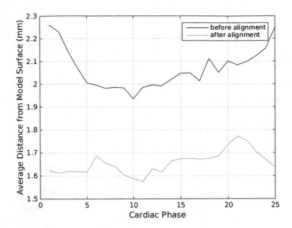

Fig. 7. The average distance (in *mm*) between contour points and reconstructed model along the full cardiac cycle for the whole dataset.

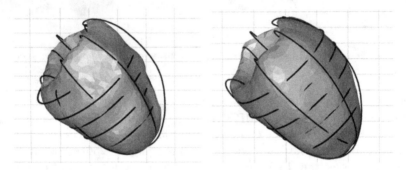

Fig. 8. Left: the model reconstructed from the contours without aligned; right: the model from the contours with alignment. The model shape with alignment becomes more proper and smooth comparing result without alignment.

Figure 6 shows the reconstructed shapes at different frames of the cardiac cycle. There is a clear difference between LV motions of normal volunteers and those of patients with heart dyssynchrony. In the ES phase, the LV contracts well to pump the blood out for normal people; whereas, it does not deform as much for patients, which means the patients' hearts are unable to function properly. Based on the reconstructed model, we can study the LV volumes along time for normal volunteers and patients shown in Fig. 9. Comparing with normal people, the patient's LV contains more blood and it does not contract much during the cardiac motion (which also can be proved by the ejection fraction rate: 55% for a normal volunteer and 28% for a patient). 2D myocardium contours in tagged MR slices, which are useful for further studying the interior dynamics of the LV

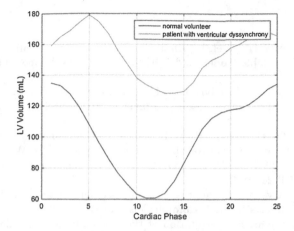

Fig. 9. LV volume change along time within a full cardiac cycle for a normal volunteer and a patient with heart dyssynchrony.

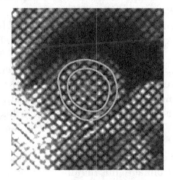

Fig. 10. Intersected contours of fitted model on a tagged MRI slice.

wall, can also be located and mutually registered, based on the reconstructed 3D LV model and its intersection with the MR planes (as shown in Fig. 10).

6 Conclusion

In the paper, we proposed a novel approach to reconstruct 3D shape and motion of LV wall from cardiac cine MRI. The approach is effective and efficient with few user interactions. We may extend the proposed approach to the tagged MRI, whose alignment is still challenging due to the imaging artifacts. We call for the future extension to explore our approach to other potential applications.

Acknowledgments. This research has been supported by NIH grant (R01 HL 127661). We thank our colleagues from CBIM at Rutgers University who provided insight and expertise that greatly assisted the research, and Ms. Yan Chen for comments that greatly improved the manuscript.

References

1. Park, J., Metaxas, D., Axel, L.: Analysis of left ventricular wall motion based on volumetric deformable models and MRI-SPAMM. Med. Image Anal. **1**(1), 53–71 (1996)
2. Lotjonen, J., Pollari, M., Kivisto, S., Lauerma, K.: Correction of motion artifacts from cardiac cine magnetic resonance images. Acad. Radiol. **12**(10), 1273–1284 (2005)
3. Ronneberger, O., Fischer, P., Brox, T.: U-Net: convolutional networks for biomedical image segmentation. In: Navab, N., Hornegger, J., Wells, W.M., Frangi, A.F. (eds.) MICCAI 2015. LNCS, vol. 9351, pp. 234–241. Springer, Cham (2015). doi:10.1007/978-3-319-24574-4_28
4. Paragios, N.: A variational approach for the segmentation of the left ventricle in cardiac image analysis. Int. J. Comput. Vis. **50**(3), 345–362 (2002)
5. Jolly, M.P.: Automatic segmentation of the left ventricle in cardiac MR and CT images. Int. J. Comput. Vis. **70**(2), 151–163 (2006)
6. Zhu, Y., Papademetris, X., Sinusas, A.J., Duncan, J.S.: Segmentation of the left ventricle from cardiac MR images using a subject-specific dynamical model. IEEE Trans. Med. Imaging **29**(3), 669–687 (2010)
7. McLeish, K., Hill, D.L., Atkinson, D., Blackall, J.M., Razavi, R.: A study of the motion and deformation of the heart due to respiration. IEEE Trans. Med. Imaging **21**(9), 1142–1150 (2002)
8. Myronenko, A., Song, X.: Point set registration: coherent point drift. IEEE Trans. Pattern Anal. Mach. Intell. **32**(12), 2262–2275 (2010)
9. Radau, P., Lu, Y., Connelly, K., Paul, G., Dick, A.J., Wright, G.A.: Evaluation framework for algorithms segmenting short axis cardiac MRI. The MIDAS Journal - Cardiac MR Left Ventricle Segmentation Challenge. http://hdl.handle.net/10380/3070
10. Garlapati, R.R., Mostayed, A., Joldes, G.R., Wittek, A., Doyle, B., Miller, K.: Towards measuring neuroimage misalignment. Comput. Biol. Med. **64**, 12–23 (2015)
11. Queiros, S., Barbosa, D., Engvall, J., Ebbers, T., Nagel, E., Sarvari, S.I., Claus, P., Fonseca, J.C., Vilaa, J.L., D'hooge, J.: Multi-centre validation of an automatic algorithm for fast 4D myocardial segmentation in cine CMR datasets. Eur. Heart J. Cardiovasc. Imaging **17**(10), 1118–1127 (2016)
12. Suinesiaputra, A., Cowan, B.R., Al-Agamy, A.O., Elattar, M.A., Ayache, N., Fahmy, A.S., Khalifa, A.M., Medrano-Gracia, P., Jolly, M.P., Kadish, A.H., Lee, D.C.: A collaborative resource to build consensus for automated left ventricular segmentation of cardiac MR images. Med. Image Anal. **18**(1), 50–62 (2014)
13. Petitjean, C., Dacher, J.N.: A review of segmentation methods in short axis cardiac MR images. Med. Image Anal. **15**(2), 169–184 (2011)

Modeling of Myocardium Compressibility and its Impact in Computational Simulations of the Healthy and Infarcted Heart

Joao S. Soares[1], David S. Li[1], Eric Lai[2], Joseph H. Gorman III[2], Robert C. Gorman[2], and Michael S. Sacks[1(✉)]

[1] Center for Computational Simulation, Institute for Computational Engineering and Sciences, University of Texas at Austin, Austin, TX, USA
msacks@ices.utexas.edu
[2] Gorman Cardiovascular Research Group, Perelman School of Medicine, University of Pennsylvania, Philadelphia, PA, USA

Abstract. Simulation of heart function requires many components, including accurate descriptions of regional mechanical behavior of the normal and infarcted myocardium. Myocardial compressibility has been known for at least two decades, however its experimental measurement and incorporation into computational simulations has not yet been widely utilized in contemporary cardiac models. In the present work, based on novel in-vivo ovine experimental data, we developed a specialized compressible model that reproduces the peculiar unimodal compressible behavior of myocardium. Such simulations will be extremely valuable to understand etiology and pathophysiology of myocardium remodeling and its impact on tissue-level properties and organ-level cardiac function.

Keywords: Myocardium · Cardiac simulation · Compressibility

1 Introduction

Myocardial infarction (MI) induces maladaptive remodeling of the left ventricle (LV), causing dilation, wall thinning, change in mechanical properties, and loss of contractile function [1]. Simulation technologies can potentially lead the way for in-silico based models of the heart for many therapeutic applications. One of the truly unique advantages of computational modeling of the heart is its ability to estimate parameters that cannot be measured directly. One of these parameters, ventricular wall stress, may be the single most important indicator of ventricular myocardial function [2]. Specifically, a computational platform to accurately evaluate the effect of MI on cardiac function impairment, acutely and chronically, would be extremely valuable to understand etiology and pathophysiology of myocardium remodeling, its impact on tissue-level properties and organ-level cardiac function, and ultimately, to improve virtual surgery technologies, medical device development, as well as to provide quantitative risk stratification tools for these interventions.

© Springer International Publishing AG 2017
M. Pop and G.A. Wright (Eds.): FIMH 2017, LNCS 10263, pp. 493–501, 2017.
DOI: 10.1007/978-3-319-59448-4_47

494   J.S. Soares et al.

However, material modeling of myocardium, and its critical numerical implementation, remains an area where much progress is required. One particular challenging feature of functioning myocardium is the interaction of coronary flow with myocardial contractility and compressibility. Increases in perfusion pressure have been shown to increase myocardial tissue volume strain and decrease its compliance [3]. During in vivo function, active muscle contraction influences flow in the coronary vessels, microcirculation, and venous collection systems embedded in the myocardium. Systole has been shown to inhibit coronary flow perfusion [4], and diastole introduces an "intramyocardial pump" that adds sucking action on arterial blood as a result of the previous systole and compensates for the lower or even retrograde systolic flow [5]. While volumetric changes in the myocardium during the cardiac cycle have been known to occur for at least two decades [6], their incorporation into cardiac simulations and realization of this important effect in cardiac function has yet to be fully accomplished. In the following study we present novel experimental studies, material modeling, and numerical simulations as first step in developing more realistic cardiac models.

2   Methods

We have developed an in-silico model of MI using a comprehensive model pipeline (Fig. 1) and based on extensive datasets from a single ovine heart collected in vivo and ex vivo. All data collection of this in silico heart model was performed at the Visible Heart Laboratory (VHL, University of Minnesota), in three steps: (1) in vivo, (2) ex situ, and (3) ex vivo. Heart cavity pressures and volumes were acquired in vivo with catheterization and somicrometry. Epicardial electrical activity was measured with monophasic action potentials and diverse 2D echocardiography imaging was conducted to

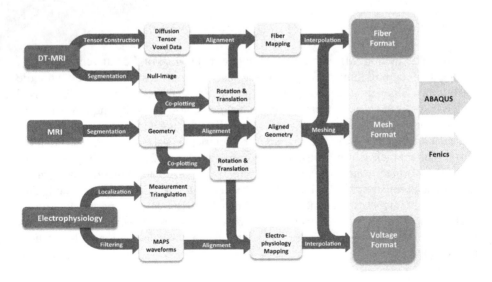

Fig. 1. Complete modeling pipeline from the single heart data source.

obtain validating datasets of in vivo heart function. The ex situ step was conducted in an isolated heart flow loop replicating in vivo conditions but where much better access was possible for extensive data collection. Subsequently, the heart was fixed under end-diastolic pressures with valves coapted and imaged. Magnetic resonance images (MRI) at end-diastole (Fig. 2a) were segmented (Fig. 2b) to create a finite element (FE) mesh (Fig. 2c). Diffusion tensor MRI (DTMRI) data (Fig. 2d) was aligned with the FE mesh (Fig. 2e) and employed to prescribe principal fiber direction of the FE model for the specification of the anisotropic material law (Fig. 2f).

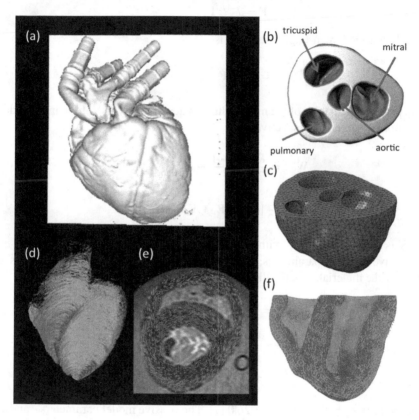

Fig. 2. (a) magnetic resonance imaging at the end diastolic configuration; (b) segmentation of the biventricular heart up to the mitral valve plane; (c) FE mesh of the biventricular heart model; (d) diffusion-tensor magnetic resonance imaging; (e) alignment and overlay of the MRI-DTMRI datasets; and (f) determination of principal material directions of the FE model for proper specification of transversely isotropic material properties and active fiber contraction.

As a first step, we utilized a conventional incompressible transversely isotropic Fung-based hyperelastic model for the passive mechanical properties of myocardium of the form

$$\mathbf{T} = -\frac{\partial \mathrm{W}^{\mathrm{vol}}}{\partial J} + \frac{1}{J}\tilde{\mathbf{F}}\frac{\partial \mathrm{W}^{\mathrm{dev}}}{\partial \tilde{\mathbf{E}}}\tilde{\mathbf{F}}^{\mathrm{T}} + \frac{1}{J}\mathbf{F}\mathbf{S}^{\mathrm{act}}\mathbf{F}^{\mathrm{T}}, \tag{1}$$

where the Jacobian of the motion is

$$J = \det \mathbf{F}, \tag{2}$$

$\mathbf{F}$ is the deformation gradient, the deviatoric deformation gradient is defined as

$$\tilde{\mathbf{F}} = J^{-1/3}\mathbf{F}, \tag{3}$$

and the deviatoric Green-Lagrange strain is

$$\tilde{\mathbf{E}} = \frac{1}{2}\left(\tilde{\mathbf{F}}^{\mathrm{T}}\tilde{\mathbf{F}} - 1\right). \tag{4}$$

The passive mechanics of the myocardium is described by its volumetric and deviatoric responses, given by respectively

$$\mathrm{W}^{\mathrm{vol}} = \frac{K}{2}\left(\frac{J^2 - 1}{2} - \ln J\right), \tag{5}$$

$$\mathrm{W}^{\mathrm{dev}} = \frac{c}{2}\left[\exp\left(\alpha\tilde{\mathbf{E}} \cdot \mathbb{C}\tilde{\mathbf{E}}\right) - 1\right], \tag{6}$$

where K is the bulk modulus (with $K \gg c$ enforcing material incompressibility), c and α are passive material parameters, and 4th order tensor $\mathbb{C}$ characterizes the transverse isotropy of the material, specifically resulting in

$$\mathrm{W}^{\mathrm{dev}} = \frac{c}{2}\left[\exp\left(\alpha Q\right) - 1\right], \tag{7}$$

where

$$Q = A_1 E_{11}^2 + A_2\left(E_{22}^2 + E_{33}^2 + 2E_{23}^2\right) + A_3\left(E_{12}^2 + E_{13}^2\right), \tag{8}$$

with $A_1 = 12.0, A_2 = 8.0$, and $A_3 = 26.0$ [7]. The passive model parameters c and α are calibrated to match the general end-diastolic pressure volume relationship obtained by Klotz et al. [8].

The active stress responsible for myocardium contraction follows Hunter-McCulloch-Ter Keurs model [9] and is given by

$$\mathbf{S}^{\mathrm{act}} = T_{Ca^{2+}}(\mathbf{x}, t)[1 + \beta(\lambda - 1)]\mathbf{m} \otimes \mathbf{m}, \tag{9}$$

where m is the fiber direction and local fiber stretch is

$$\lambda = \mathbf{m} \cdot \mathbf{F}^{\mathrm{T}}\mathbf{F}\mathbf{m}. \tag{10}$$

Time and space-dependent active contraction $T_{Ca^{2+}}(\mathbf{x}, t)$ is driven by epicardial electrical activity measured with monophasic action potentials (MAP). A total of 71 measurement (over cardiac cycle and synchronized with the QRS complex of ECG) were taken at scattered locations of the LV and RV epicardium. The measurement locations were employed to define an interpolating triangulation, and subsequently, the MAP waveform at each individual FE was obtained with barycentric coordinates in the triangulation. The pressure-volume (PV) loop, measured through catheterization and sonomicrometry, was used to best-fit the modulation of active contraction. We did not

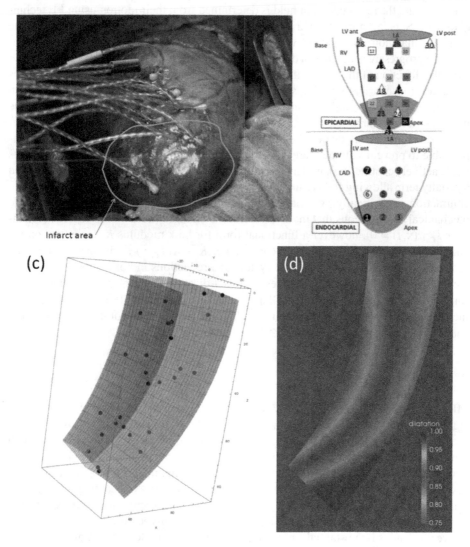

Fig. 3. (a) sonocrystal study in the ovine model; (b) sonocrystal layout; (c) prolate spheroid wedge used for strain analysis; and (d) regional dilatation for normal myocardium showing regional variations of volume change.

perform systematic coupling of the electrical and mechanical aspects of functioning myocardium, but we have used measured electrical data to modulate spatially and temporally the heterogeneity of contraction in active myocardium to simulate the in vivo PV loop.

We conducted novel regional experiments using an ovine model with 27 sonocrystals implanted in a 3D array in the LV free wall (Fig. 3a, b) [10]. The positions of the sonocrystals during the cardiac cycle in vivo can yield a direct measurement of the changes of volume of myocardial tissue. However, a richer analysis can be achieved with a prolate spheroidal coordinate wedge defined in the region delimited by the sonocrystals (Fig. 3c), and the regional strain field is determined from their motion using FE techniques with Sobolev norm regularization [11]. From the strain field we determined the regional dilatation (normalized volume change, Fig. 3d). Experimental data demonstrated a progressive decrease in dilatation from the epi- to the endocardial surfaces, and from the direction of the base to the apex. These results indicated that myocardium does not deform isochorically during systole and myocardial tissue volume decreases (with volumetric changes ranging from 1.0 to ~ 0.75), so that conventional compressible material models cannot be used as $J \leq 1$ still needs to be enforced.

To account for this distinctive type of compressible material behavior (i.e. incompressible to prevent volume increases but compressible to allow for volume reductions while actively contracting), we have developed a specialized material model that utilized a penalty term allowing for a reduction in volume only and was modulated by the active contraction. Specifically, we have enforced material incompressibility to passive mechanical deformations that involve volumetric changes (i.e. $J = 1$ is enforced whenever $T_{Ca^{2+}}(\mathbf{x}, t) = 0$), however a functional form for bulk modulus K dependent on the amount of active contraction was chosen, i.e. $K \equiv K(T_{Ca^{2+}})$ such that the material becomes compressible during active systole and reductions in volume occur. Because we are interested in modeling changes from end-diastole to peak-systole, a linear form for K (decreasing from $K^{inc} = 10$ GPa for nearly-incompressible myocardium) with increasing T_{Ca} was sufficient to obtain the observed reduction in volume observed from end-diastole to peak-systole, when $K^{systole} = 0.64$ GPa.

3   Results

MRI, DTMRI, MAP, and in vivo PV loops allow the construction of a physiologically significant in silico heart model. Passive mechanical properties compared favorably with previous literature, time-modulation of active contraction shows good agreement with typical calcium-activated contraction, and qualitative comparison with in vivo 2D echo validated the baseline healthy model [12]. Testing of the material model was done in several steps, starting with a basic 3D cube to investigate how axial contraction produces transverse expansion (Fig. 4). Note that as the level of compressibility increases, the level of transversal expansion decreases. The impact of myocardial compressibility in active mechanics is substantial – not only incompressible models are unable to replicate the volumetric changes during systole, but also the required activation T_{Ca} to replicate the experimentally measured PV loop was substantially overestimated (Fig. 5).

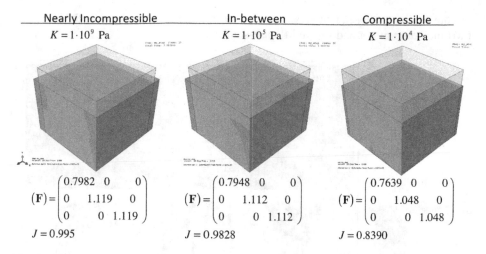

Fig. 4. The effects of the same axial contraction on the corresponding transverse stretches (F_{22} and F_{33}) and volumetric changes (J). Material compressibility impacts the relationship between active contraction and wall thickening.

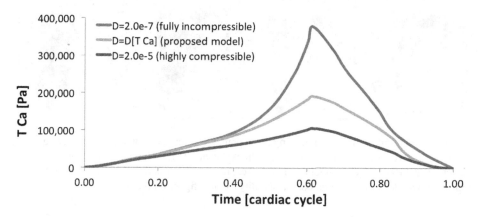

Fig. 5. Active contraction T_{Ca} inversely determined with the observed PV loop ($D = 2/K$). Active contraction increases up to a maximum at peak-systole, and relaxes over diastole. Incompressible myocardium requires substantially higher active contractions to achieve similar cavity volumes.

We simulated stiffening and loss of contractility in infarcted myocardium by prescribing a region of infarct observed experimentally in the ovine model of MI (shown as medium MI in Fig. 6) and determined the magnitudes of each effect by inversely matching the infarcted PV loop. In vivo function in the baseline and infarct models was quantitatively validated with the transmural strain smooth fields determined with sonocrystal arrays. Changes in properties (passive mechanics and active contraction) in infarcted and border-zone regions around the LV apex to best-fit the in vivo MI PV loop resulted in good agreement with transmural strain field observed in vivo and with previous findings. Subsequently, we have conducted a parametric study on the intensity of MI including minor and severe MI case-studies (Fig. 6).

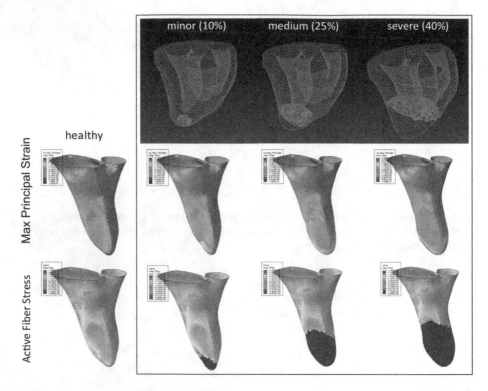

Fig. 6. Simulations of myocardium using the new material model showing the effects of gradual increases in myocardial infarction and the increase of apical LV chamber dilatation due to the reduction in local active contraction (blue). Accuracy of such simulations are enhanced with appropriate modeling of material compressibility. (Color figure online)

4 Discussion

The employment of a complete dataset from a single heart to develop an in-silico model avoids unnecessary image registration between MRI and DTMRI images. The coupling between electrical activity and active contraction is significant to heart biomechanics and is needed to obtain physiological realistic heart function. Myocardium is extensively perfused with distensible vessels (approximately 10–20% of the total volume). Increases in stiffness/tension of the surrounding tissue, as occurs during muscle contraction, causes the fluid to be extruded from these vessels [6]. Most importantly, the finding that perfused myocardium is compressible implies that results from analyses that assume incompressibility are not realistic. Not only the ventricular walls modeled with incompressible myocardium will be thicker at peak systole (as the relationship between fiber shortening and wall thickening is highly sensitive to volumetric changes) but also ventricular and myofiber stress and active tension will be substantially overestimated – the former is crucial for accurate representation of ventricular kinematics and the latter is a key factor in myofiber stress and ventricular remodeling. To simulate MI, we have found that both impairment of contraction and stiffening of myocardium were necessary to replicate appropriately MI biomechanics. Subsequent remodeling mechanisms of collagen fiber re-orientation that change the material anisotropy locally may be important and are subject of future studies.

Acknowledgments. National Institutes of Health grants R01 HL068816, HL089750, HL070969, HL108330, and HL063954.

References

1. Gorman, R.C., Jackson, B.M., Burdick, J.A., Gorman, J.H.: Infarct restraint to limit adverse ventricular remodeling. J. Cardiovasc. Transl. Res. **4**, 73–81 (2011)
2. Pasque, M.K.: Invited commentary. Ann. Thorac. Surg. **76**, 1180 (2003)
3. May-Newman, K., Omens, J.H., Pavelec, R.S., McCulloch, A.D.: Three-dimensional transmural mechanical interaction between the coronary vasculature and passive myocardium in the dog. Circ. Res. **74**, 1166–1178 (1994)
4. Downey, J.M., Kirk, E.S.: Inhibition of coronary blood flow by a vascular waterfall mechanism. Circ. Res. **36**, 753–760 (1975)
5. Spaan, J.A.: Coronary diastolic pressure-flow relation and zero flow pressure explained on the basis of intramyocardial compliance. Circ. Res. **56**, 293–309 (1985)
6. Yin, F.C., Chan, C.C., Judd, R.M.: Compressibility of perfused passive myocardium. Am. J. Physiol. **271**, H1864–H1870 (1996)
7. Wang, V.Y., Lam, H.I., Ennis, D.B., Cowan, B.R., Young, A.A., Nash, M.P.: Modelling passive diastolic mechanics with quantitative MRI of cardiac structure and function. Med. Image Anal. **13**, 773–784 (2009)
8. Klotz, S., Dickstein, M.L., Burkhoff, D.: A computational method of prediction of the end-diastolic pressure-volume relationship by single beat. Nat. Protoc. **2**, 2152–2158 (2007)
9. Hunter, P.J., McCulloch, A.D., ter Keurs, H.E.: Modelling the mechanical properties of cardiac muscle. Prog. Biophys. Mol. Biol. **69**, 289–331 (1998)
10. Jackson, B.M., Gorman, J.H., Moainie, S.L., Guy, T.S., Narula, N., Narula, J., John-Sutton, M.G., Edmunds Jr., L.H., Gorman, R.C.: Extension of borderzone myocardium in postinfarction dilated cardiomyopathy. J. Am. Coll. Cardiol. **40**, 1160–1171 (2002)
11. Smith, D.B., Sacks, M.S., Vorp, D.A., Thornton, M.: Surface geometric analysis of anatomic structures using biquintic finite element interpolation. Ann. Biomed. Eng. **28**, 598–611 (2000)
12. McGarvey, J.R., Mojsejenko, D., Dorsey, S.M., Nikou, A., Burdick, J.A., Gorman 3rd, J.H., Jackson, B.M., Pilla, J.J., Gorman, R.C., Wenk, J.F.: Temporal changes in infarct material properties: An in vivo assessment using magnetic resonance imaging and finite element simulations. Ann. Thorac. Surg. **100**, 582–589 (2015)

Characterizing Patterns of Response During Mild Stress-Testing in Continuous Echocardiography Recordings Using a Multiview Dimensionality Reduction Technique

Mariana Nogueira[1,3](✉), Gemma Piella[2], Sergio Sanchez-Martinez[2], Hélène Langet[1], Eric Saloux[4], Bart Bijnens[3], and Mathieu De Craene[1]

[1] Medisys, Philips Research, Paris, France
mariana.nogueira@philips.com
[2] Simbiosys, ETIC, Universitat Pompeu Fabra, Barcelona, Spain
[3] Physense, ETIC, Universitat Pompeu Fabra, Barcelona, Spain
[4] Centre Universitaire Hospitalier de Caen, Caen, France

Abstract. In this paper, we capture patterns of response to cardiac stress-testing using a multiview dimensionality reduction technique that allows the compact representation of patient response to stress, regarding multiple features over consecutive cycles, as a low-dimensional trajectory. In this low-dimensional space, patients can be compared and clustered in distinct healthy and pathological responses, and the patterns that characterize each of them can be reconstructed. Experiments were performed on (a) synthetic data simulating different types of response and (b) a real acquisition during a cold pressor test. Results show that the proposed approach allows the clustering of healthy and pathological responses, as well as the reconstruction of characteristic patterns of such responses, in terms of multiple features of interest.

Keywords: Stress echo · Strain · Multiview · Dimensionality reduction · MKL

1 Introduction

Characterizing cardiac response to increased activity or adverse situations is key in the analysis of Heart Failure (HF) etiologies. For this matter, the clinical value of stress-testing is well established: in clinical practice, standardized stress-invoking protocols (e.g. dobutamine challenge [15], cycling/running with controlled heart rate, exercise time or generated power [3]) are adopted. The strictly protocolized nature of these tests makes the collection of measurements of cardiac function at only a few well-defined time-points/stress-levels of the test (e.g. 4 in the case of dobutamine challenges) sufficient to analyze how the heart copes with the induced stress. These tests are, however, difficult to implement at a large scale (i.e. in all patients), due to the required time, equipment and

© Springer International Publishing AG 2017
M. Pop and G.A. Wright (Eds.): FIMH 2017, LNCS 10263, pp. 502–513, 2017.
DOI: 10.1007/978-3-319-59448-4_48

staff (and thus cost). Handgrip [7] or cold pressure testing [14] are, on the other hand, cheap, fast, and easy ways to invoke stress. However, they are difficult to standardize, and thus not reproducible with regard to timings/intensity levels of the stress challenge. This makes the classical 'single time-point' measurements at different stress levels infeasible, implying a continuous data acquisition throughout the test (e.g. 40–60 cycles). In this context, the assessment of cardiac response to stress needs to be performed based on the quantification of trends rather than on amplitude differences. Naturally, this kind of acquisition raises other challenges, such as the considerably increased amount of data involved, and well-known issues related to their processing, such as image artifacts caused by breathing motion in echocardiographic sequences. In this paper, we propose an approach based on multiview dimensionality reduction for the analysis of response to stress in such contexts. We start by defining features of interest to be collected at each consecutive cycle, for each patient, such as heart rate and deformation features. Then, the approach allows the projection of several patients onto a space where the response to stress of each patient is compactly represented as a low-dimensional trajectory, that encodes change patterns in the defined features. The main objective, once this space is obtained, is to discriminate healthy and pathological responses, and to reconstruct the patterns in the features of interest that characterize them.

As a first step, a synthetic dataset was generated, so as to obtain a sufficient number of patients to evaluate the performance of the approach in terms of the proposed objectives, i.e. its capability of clustering different types of response and reconstructing the main patterns that characterize them. Then, we used a real echocardiographic sequence, acquired on a healthy volunteer during a cold pressor test, to illustrate the applicability of the proposed approach in a real context.

2 Synthetic Data

The importance of heart rate (HR) and left-ventricular (LV) deformation patterns for the analysis of response to stress has been demonstrated in several clinical studies [6,15]. In particular, the longitudinal deformation function of the LV is presumed to be one of the earliest to be reduced in several cardiovascular pathologies [8]. For these reasons, we selected HR and the global longitudinal strain (GLS) curve as features of interest to monitor over each patient's stress test. GLS is here defined as the change in longitudinal size of the LV during the cardiac cycle, relative to the end-diastolic size. The features were collected for each consecutive cycle, namely one HR value and one vector holding the evolution of GLS throughout the cycle.

2.1 Model

The model for generating synthetic GLS curves is based on 4 control points adjusted in time and amplitude, and piecewise cubic spline interpolation. Let us first consider 6 keypoints $p_i = (t_i, A_i), i = 0, .., 5$, of the GLS curve, as illustrated in Fig. 1.

Here, $\boldsymbol{p_0} = (0,0)$ and $\boldsymbol{p_5} = (t_5, 0)$ correspond, respectively, to the start and end points of the cardiac cycle, $\boldsymbol{p_0}$ coinciding with mitral valve closure. Furthermore, we define t_1 as the instant when GLS slope effectively becomes negative, t_2 as the instant of aortic valve closure (AVC), t_3 as the instant when GLS slope turns positive (start of relaxation), and t_4 as the

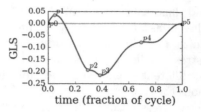

Fig. 1. Synthetic GLS curve model.

start of atrial contraction (AC). Since we capture the variation of the HR as a distinct feature, all GLS curves were normalized in time and all t_i ranged between 0 and 1 for all cycles. Timings and amplitudes of $\boldsymbol{p_1}$ to $\boldsymbol{p_4}$ can be adjusted to reflect stress-induced changes as described in the literature [15]. A few additional points were defined to maintain key features of the GLS curves, whose coordinates were defined proportionally to $\boldsymbol{p_1}$, $\boldsymbol{p_2}$, $\boldsymbol{p_3}$ and $\boldsymbol{p_4}$, and thus passively followed these active control points. Each synthetic GLS curve results then from the piecewise cubic spline interpolation of the set of points defined by the extremes of the cycle $\boldsymbol{p_0}$ and $\boldsymbol{p_5}$, the active points $\boldsymbol{p_1}$ to $\boldsymbol{p_4}$, and the remaining passive points. Since the extremes are fixed and the passive points change passively with the active points, the GLS curve needs only 8 parameters to be defined, which are the timings and amplitudes of $\boldsymbol{p_1}$ to $\boldsymbol{p_4}$.

2.2   Dataset Characteristics

To generate physiologically consistent GLS curves, we took as reference previous clinical studies where physiological and pathological responses to stress are characterized [15]. Three types of response to stress were recreated: one normal (Fig. 2a), and two pathological with the following signatures: post-systolic shortening (PSS, i.e. continuation of shortening after AVC – Fig. 2b) and the combination of PSS and prolonged early relaxation/delayed AC (Fig. 2c).

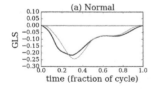 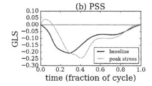 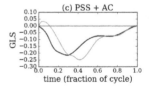

Fig. 2. Three types of responses to stress were considered in the generated synthetic GLS curves: (a) normal, (b) pathological with PSS, (c) pathological with PSS and prolonged early relaxation/delayed AC.

For each type of response, 5 synthetic patients were generated. To include inter-patient and inter-acquisition variability, the parameters of the model, the total number of cycles, the point where stress was introduced, and the time it took to reach peak stress, were slightly varied among patients. In terms of HR, the response was considered normal for all patients (≈ 60 beats per minute (bpm) baseline; ≈ 120bpm peak stress). The responses were considered to be approximately linear in time from baseline to peak stress. An example of a synthetic sequence of GLS curves and corresponding HR values is illustrated in Fig. 3.

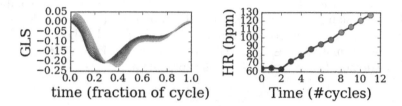

Fig. 3. The 2-feature sample representation of a synthetic patient: sequence of GLS curves and respective HR values, from rest to peak stress (blue to green). (Color figure online)

In summary, we generated a synthetic dataset consisting of 15 patients with a 2-feature sample representing each of their (≈ 15) consecutive cycles. The dimensionalities associated with those 2 features are 1 (HR) and 75 (GLS curve). For analyzing the main trends and modes of variation in the data, and how they are clustered, it is convenient to obtain a more compact representation of the data. For that reason, we performed dimensionality reduction.

3   Dimensionality Reduction Methodology

Within dimensionality reduction approaches, unsupervised methods are particularly suited for analyzing the main trends and modes of variation in the data, and discover how they are clustered. Furthermore, a non-linear method was preferred, for the sake of robustness to possible data distribution geometries where linear methods such as Principal Component Analysis [5] might deliver limited performances. Within non-linear methods, those categorized as graph embedding algorithms (e.g. Isomap [13], Laplacian Eigenmaps [1] (LEM), Locally Linear Embedding [10]) are particularly popular. However, all of the above-mentioned methods are prepared for a single multivariate input. Given that the GLS is a multivariate feature with a functional structure, concatenating our two features into a single multivariate input does not seem to be the most appropriate way to deal with our data. Instead, a multiview approach was considered more suited.[1] We thus selected the unsupervised formulation of the Multiple Kernel

---

[1] In the context of multiview learning, the term view refers to each such independently considered feature. The terms view and feature are hereafter used interchangeably.

Learning (MKL) algorithm for dimensionality reduction [9], which can be seen as a multiview generalization of the non-linear method LEM. In addition, MKL has been shown to perform well in several multiview dimensionality reduction problems [9], including cardiovascular applications [11].

3.1 Formalism of Unsupervised MKL

Let us consider N input samples, each one consisting of F uni/multivariate features. For each feature $f = 1, ..., F$, an affinity matrix $W^f \in \mathbb{R}^{N \times N}$ is computed using the Gaussian kernel, which encodes the similarity among samples. Let us now express W^f as the set of its columns, $W^f = [W_1^f, ..., W_N^f]$, $W_i^f \in \mathbb{R}^N$, $i = 1, ..., N$, and let matrix $\mathbb{K}^i \in \mathbb{R}^{N \times F}$ be defined as $\mathbb{K}^i = [W_i^1, ..., W_i^F]$. The mapping of a sample i to the output space is expressed as

$$y_i = A^T \mathbb{K}^i \beta \,, \tag{1}$$

where A is the projection matrix to the output space and $\beta \in \mathbb{R}^F$ contains the normalized weights of each feature f in the mapping. Let $\mathbb{W}$ be a linear or non-linear combination of all the feature-wise similarity matrices W^f (we used $\mathbb{W} = \frac{1}{F} \sum_f W^f$). The entry $\mathbb{W}_{ij}$ corresponds then to a similarity coefficient between samples i and j in the input space based on their F features. The goal is to map the data onto a lower-dimensional space where samples that are close in the input space remain close in the output space. Extending the idea of Laplacian Eigenmaps [1], the optimal embedding can be obtained by finding A and β which minimize

$$\sum_{ij} \|A^T \mathbb{K}^i \beta - A^T \mathbb{K}^j \beta\|^2 \mathbb{W}_{ij} \,. \tag{2}$$

Thus, close samples in the input space (high $\mathbb{W}_{ij}$) will be enforced to remain close in the output space, so as to minimize the product $\|y_i - y_j\|^2 \mathbb{W}_{ij}$. Matrix $\mathbb{W}$ is often made sparse, so that pairs of samples that are very distant do not contribute to the final projection.

Lin et al. [9] proposed an iterative two-step approach that alternately solves the minimization for β and for A. To better control and understand the effects of weighting the features in the obtained projections, we withdrew β as minimization argument, tuned its value, and solved the minimization of (2) for A through a generalized eigenvalue problem (first step of the minimization strategy proposed by Lin et al. [9]). The first dimensions of the obtained space correspond to the eigenvectors with lowest associated eigenvalues, and encode the main modes of variation of the data.

3.2 Multiscale Kernel Regression

After the optimal mapping is obtained, multiscale kernel regression (MKR) [2,4] can be used to associate an output-space sample with its corresponding form in the input feature space. This is done based on the similarity of such output-space sample to all others in the output space, and their known representation in the input space. By studying the effects of moving a sample along a dimension of the output space in its input space representation, we can analyze the modes of variation of each input feature encoded in such output-space dimension, and relate output-space trajectories with specific patterns in the input features.

3.3 Generalizing for Multiple Views and Multiple Samples per Subject

MKL has been applied before in echocardiography by Sanchez-Martinez et al. [11] for the analysis of the main modes of variation in the myocardial velocity traces among healthy patients and patients suffering from HF with preserved ejection fraction, under rest and stress conditions. In [11], each patient was represented by one sample with 6 different views, which included 4 cyclewise velocity curves (basal/septal regions of the LV at rest/submaximal exercise) and 2 vectors providing information on the timing of cardiac phases.

In this paper, in addition to multiple patients, we consider multiple temporal samples per patient, which draw a trajectory from rest to stress. In this context, we have interest in analyzing the modes of variation in the input features both among patients and over time, or, in other words, in analyzing and comparing patient trajectories. For that, we need to map all the patients onto the same space. Assuming that all patients lie in a common manifold, one possible approach is applying MKL having as inputs affinity matrices W^f that compare all samples (i.e. cardiac cycles) of all patients. More specifically, each input sample consists of two views (GLS and HR) from a cycle c of a patient p, and it is indexed according to $i(p, c) = \sum_{q=1}^{p-1} N_q + c$, where N_q represents the total number of cycles of patient q. Conveniently, this approach does not require an equal number of samples from the different patients, nor identical sampling grids.

In this context, the proposed approach comprises the following steps:

1. Collecting the 2-view (HR value and GLS curve) sample corresponding to each of the consecutive cycles of each of the patients;
2. Building feature-wise affinity matrices W^f, $f = \{GLS, HR\}$ comparing all samples of all patients;
3. Tuning Gaussian kernel bandwidths (σ^f) and sparsity of W to adjust the sensitivity of the algorithm to the order of amplitudes of the sought modes of variation;
4. Normalizing the affinity matrices by variance before being fed to the MKL algorithm;
5. Tuning β and finding the projection matrix A of the data which minimizes the objective function in (2);
6. Applying data projection.
7. Performing MKR to obtain the modes of variation encoded in each dimension of the new space.

Once all samples are projected to the output space, the temporal trajectory of each patient in the new space can be obtained by connecting his samples over time. These trajectories can be analyzed dimension-wise. Then, within the first dimensions (which encode the main variations in the data), we can perform a combined analysis of the trajectories and the corresponding modes to search for those more relevant for the characterization/discrimination of responses to stress.

4   Experiments with Synthetic Data

In the experiments with synthetic data, HR and GLS curves for each of the ($\approx$15) consecutive cycles of each of the 15 synthetic patients were collected as described in Sect. 2.2. We then applied our MKL extension described in Sect. 3.3, each HR value and GLS curve per cycle/patient being considered as an input sample. MKL projects these input samples to a low dimensional space, where trajectories over time can be reconstructed. Figure 4 shows, in the right column, these trajectories for some dimensions of the MKL output space (#1, #4, #5). In these plots, each curve represents one patient, and the color corresponds to the patient class (healthy/PSS/PSS+AC label, see Sect. 2.2). For interpreting what each dimension of the output space relates to in the input samples, we applied MKR as described in Sect. 3.2. Through MKR, we were able to reconstruct the modes of variation in GLS curves and HR associated to each dimension of the output space. In the left and middle columns of Fig. 4 we show the results regarding three GLS and HR modes that we considered important in the characterization/discrimination of different types of response to stress. Trajectories over time (right column) for each mode, combined with physiological interpretation of the mode can reveal important trends in the data. For example, in the first dimension, the trajectory plot shows that the 3 groups of 5 patients experiment a similar upwards trajectory over consecutive cycles. The HR and GLS modes associated to this dimension indicate that it encodes a mix of the different pathological responses included in the database (PSS, AC amplitude and timing). A mapping between the trajectory and the corresponding GLS and HR is plotted by the colorbar on the right. By looking at the colorbar, we can relate the shift in the output coordinate with the color shift of the GLS curves and HR values: it reveals an increase in GLS peak amplitude and in HR over time as the main factor of response to stress. As this mode is common to both healthy and pathological, further modes are needed to differentiate the populations. In the 4th and 5th dimensions of the output space, trajectories diverge over time between healthy and pathological populations. The 4th dimension represents increasing levels of PSS, whereas the 5th dimension corresponds to increasing AC delays in the pathological trajectories. These results show that, while considerably reducing the dimensionality of the data – we moved from a space where each sample consisted of 2 views, represented by a scalar and a 75-sized vector, to a space where each sample is represented by a single 3-coordinate vector – we can reconstruct important patterns of response to stress.

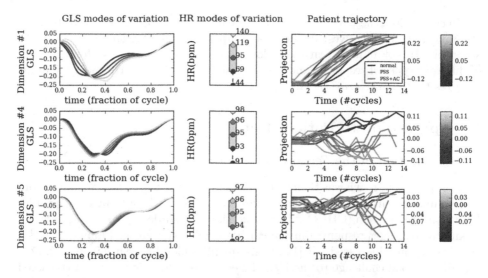

Fig. 4. Synthetic data: modes of variation of GLS curves and HR and patient trajectories over 3 dimensions of the output space. MKL parameters: $\beta_{GLS} = 0.98$; $\beta_{HR} = 0.02$; σ^f: average f-wise 4-NN distance; sparsity of $\mathbb{W}$: for each sample i, the 10% highest $\mathbb{W}_{ij}$ entries were preserved. MKR parameters: The GLS and HR modes were obtained at increments of the standard deviation σ_d (from $-2\sigma_d$ to $2\sigma_d$) for each dimension d. The colorbars link output coordinates to GLS curves and HR values. (Color figure online)

Furthermore, to investigate how patients were clustered in the output space, we computed the distance of each patient to each group in a leave-one out experiment. The trajectories of each patient were first averaged in the output space to obtain a single output point per patient. Then, for each patient, the averages of the distances to the k nearest neighbors (k-NN) within each group were used as patient-group distance estimates. A scatter plot of these distances is shown in Fig. 5, suggesting that the output space is able to discriminate the 3 groups defined in Sect. 2.2.

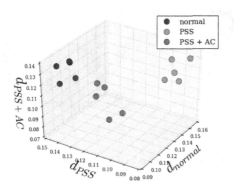

Fig. 5. Mapping of synthetic patients based on their output-space k-NN distances to each patient group ($k = 3$).

In conclusion, these results show that the proposed approach succeeds to meet the initially set objectives: it allows a compact representation of responses to stress in terms of multiple features as low-dimensional trajectories, the clustering of different types of response, and the reconstruction of the patterns in the input features that characterize them.

5   Application to Patient Data

5.1   Collection of the Features of Interest

The proposed approach was then tested on echocardiographic data acquired from one volunteer during a cold pressor stress test, provided by Centre Hospitalier Universitaire de Caen (CHUC). The immersion of a subjects's arm in iced water is known to trigger responses in the cardiovascular system, including arteriolar constriction and increased HR [14]. Consequently, blood pressure increases, posing an afterload challenge to the LV. The echocardiographic recording consisted of over 4000 apical 4-chamber view frames corresponding to about 60 consecutive cycles, and respective ECG traces (Fig. 6a). The LV myocardium was segmented on the first frame and its deformation was tracked over consecutive frames using the Sparse Demons registration algorithm [12]. GLS was computed as the relative change in longitudinal size of the LV during the cardiac cycle. The start and end points of cardiac cycles were defined by the timings of the R-peaks of the ECG. An inter-cycle registration was first performed (i.e. among the initial frames of all cycles), followed by the intra-cycle registration (i.e. among consecutive frames within each cycle), so as to prevent high error accumulation. Motion artifacts were addressed through drift correction. It is worth referring that, given the considerable size of the frame sequence and the breathing motion artifacts that are strongly amplified with stress, performing a quality tracking over the whole sequence represents a big challenge. HR information was extracted from the ECG. We assume that stress was introduced around the 30th cycle, when HR shows a sudden sharp increase (Fig. 6b).

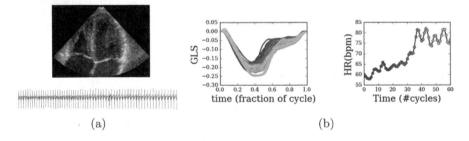

(a)   (b)

Fig. 6. Patient data. (a) Echo frame and ECG from a cold pressor test acquisition (an animated version is available at http://goo.gl/WGCJpt). (b) Extracted GLS curves and corresponding HR values, from rest to peak stress (blue to green). (Color figure online)

5.2   Experiments

After the collection of the 2-view samples for each consecutive cycle of the acquisition, the methodological steps in Sect. 3.3 were applied. Given that we had data
from one single patient, we sought modes, and trajectories over time in such
modes, that correlated with the timing of stress and/or known physiological
GLS patterns. In Fig. 7, we observe that the trajectory in the first dimension of
the output space is very correlated with the timing of stress, as a clear upwards
motion starts around the 30th cycle. Interestingly, the corresponding GLS mode
of variation reveals two pathological signatures of stress-response that had been
introduced in the synthetic dataset: PSS and late AC. Looking at the colorbar,
an upwards trajectory corresponds to increasing HR and reinforcing these GLS
signatures. Indeed, to cope with the acute afterload challenge, the patient's heart
developed inotropic mechanisms similar to some typically observed in hypertensive patients (chronic afterload challenge). Given that this is a quite demanding
challenge for the heart, it is not uncommon to find traces of these mechanisms
even in normal patients. In this context, it is rather how accentuated the pathological signatures are, or the combination with abnormal changes in other features, that distinguish physiological adaptations from pathological responses.
With the second dimension, we illustrate how data artifacts/tracking errors can
affect some of the modes: while the trajectory is clearly affected at the time
of stress, it oscillates in end-systolic and AC peak amplitudes, preventing ten-
dency analysis. Thus, although an acquisition from a single healthy patient was
insufficient to recreate the type of analysis led with the synthetic data, with this
experiment we *(i)* confirmed that we are able to extract from a true ultrasound
acquisition the same features we used in the synthetic case, i.e. the simulated

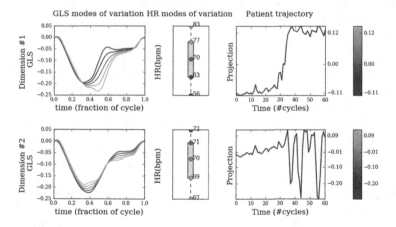

Fig. 7. Patient data: modes of variation of GLS curves and HR and patient trajectories
over the first 2 dimensions of the output space. MKL parameters: $\beta_{GLS} = 0.9$; $\beta_{HR} =
0.1$; σ^f: average f-wise 6-NN distance; sparsity of $\mathbb{W}$: for each sample i, the 26% highest
$\mathbb{W}_{ij}$ entries were preserved. MKR parameters: The GLS and HR modes were obtained
at increments of the standard deviation σ_d (from $-2\sigma_d$ to $2\sigma_d$) for each dimension
d. The colorbars link output coordinates to GLS curves and HR values. (Color figure
online)

features can be realistically extracted; *(ii)* were able to recover patterns of response in the GLS curve that have a clear physiological interpretation.

6   Conclusions

Results suggest that multiview dimensionality reduction may be interesting for representing patient response to stress over time as a low-dimensional trajectory encoding fundamental modes of variation in features that we have interest in monitoring, such as global left-ventricular deformation and heart rate. Moreover, it can be used to characterize and discriminate different types of response, as illustrated with a synthetic population. Results of experiments with real data were consistent with typical patterns of response, although some modes of variation and trajectories are naturally disturbed by artifacts in the input data (e.g. breathing). Further work will target reducing their impact on the analysis.

Acknowledgements. This work is supported by the European Union Horizon 2020 Programme for Research and Innovation, under grant agreement No. 642676 (CardioFunXion).

References

1. Belkin, M., Niyogi, P.: Laplacian eigenmaps for dimensionality reduction and data representation. Neural Comput. **15**(6), 1373–1396 (2003)
2. Bermanis, A., et al.: Multiscale data sampling and function extension. Appl. Comput. Harmonic Anal. **34**(1), 15–29 (2013)
3. Davidavicius, G., et al.: Can regional strain and strain rate measurement be performed during both dobutamine and exercise echocardiography, and do regional deformation responses differ with different forms of stress testing? J. Am. Soc. Echocardiogr. **16**(4), 299–308 (2003)
4. Duchateau, N., Craene, M., Sitges, M., Caselles, V.: Adaptation of multiscale function extension to inexact matching: application to the mapping of individuals to a learnt manifold. In: Nielsen, F., Barbaresco, F. (eds.) GSI 2013. LNCS, vol. 8085, pp. 578–586. Springer, Heidelberg (2013). doi:10.1007/978-3-642-40020-9_64
5. Fisher, R.A.: The use of multiple measurements in taxonomic problems. Ann. Eugenics **7**(2), 179–188 (1936)
6. Haemers, P., et al.: Further insights into blood pressure induced premature beats: Transient depolarizations are associated with fast myocardial deformation upon pressure decline. Heart Rhythm **12**(11), 2305–2315 (2015)
7. Helfant, R.H., et al.: Effect of sustained isometric handgrip exercise on left ventricular performance. Circulation **44**(6), 982–993 (1971)
8. Langeland, S.: Experimental validation of a new ultrasound method for the simultaneous assessment of radial and longitudinal myocardial deformation independent of insonation angle. Circulation **112**(14), 2157–2162 (2005)
9. Fuh, C., Lin, Y.Y., Liu, T.L.: Multiple kernel learning for dimensionality reduction. IEEE Trans. Pattern Anal. Mach. Intell. **33**(6), 1147–1160 (2011)
10. Roweis, S.T.: Nonlinear dimensionality reduction by locally linear embedding. Science **290**(5500), 2323–2326 (2000)

11. Sanchez-Martinez, S., et al.: Characterization of myocardial motion patterns by unsupervised multiple kernel learning. Med. Image Anal. **35**, 70–82 (2017)
12. Somphone, O., et al.: Fast myocardial motion and strain estimation in 3d cardiac ultrasound with sparse demons, April 2013
13. Tenenbaum, J.B.: A global geometric framework for nonlinear dimensionality reduction. Science **290**(5500), 2319–2323 (2000)
14. Velasco, M., et al.: The cold pressor test. Am. J. Ther. **4**(1), 34–38 (1997)
15. Voigt, J.-U.: Strain-rate imaging during dobutamine stress echocardiography provides objective evidence of inducible ischemia. Circulation **107**(16), 2120–2126 (2003)

Author Index